AF443501

ENVIRONMENTAL BIOTECHNOLOGY

ENVIRONMENTAL BIOTECHNOLOGY

REZA MARANDI
ALI SHAERI

2009

SBS Publishers & Distributors Pvt. Ltd.
New Delhi

ISBN 13 - 9788189741877

First Published in India in 2007

Published by:
SBS PUBLISHERS & DISTRIBUTORS PVT. LTD.
2/9, Ground Floor, Ansari Road, Darya Ganj,
New Delhi – 110002, INDIA
Tel: 23289119, 41563911
Email: mail@sbspublishers.com

Printed in India by Syndicate Binders, Noida.

Preface

The term 'Biotechnology' is used to cover the use of living things in industry, technology, medicine or agriculture. It is used in the production of foods and medicines, the removal of wastes and the creation of renewable energy sources. Biotechnology is being applied in many areas which includes for human uses, such as, the development of new diagnostic and therapeutic tools, DNA profiling and cloning; in food, for example, enhancing specific characteristics of crops; and in the environment, to create bacteria that break down waste and in many other areas. It is important to think about the benefits and risks of biotechnology and genetic modification. Biotechnology has potential risks as well as potential benefits. It therefore needs close scrutiny to ensure that it is applied safely. Biotechnology has the potential to have both positive and negative impacts on the environment. It can be used to support work on recovering threatened species and controlling or even eradicating introduced predators and pests. Organisms can even be engineered to remove wastes and pollution from the environment. It is important to consider the ideas that scientific decisions are never without risk and that they can be coloured by the particular world-view of the scientists who are trying to solve the problem. For example, a palaeontologist who studies the history of the earth over millions of year, brings a very different understanding of species extinction to an environmental debate than an environmental biologist. It is also important to think about actions taken based on conflicting advice and how they are weighed up, as these decisions may have the potential to deprive future generations of their right to determine some aspects of their lives. Biotechnology as it relates to the environment usually means introducing a new organism into an existing situation. The food and agriculture section of biotechnology also investigates case studies on genetically modified organisms (GMOs) and the potential environmental impact of releasing these organisms. It is important to make a proper assessment of risks and benefits before the release of any bioengineered organism to prevent environmental damage and to preserve our biodiversity.

Environmental Biotechnology is the use of living organisms or biological processes in a wide variety of applications across many industries. Technologies can be applied to industrial processes to improve them and to sustain and manage the environment as well as solve environmental problems. So, environmental biotechnology is when biotechnology is applied to and used to study the natural environment. Environmental biotechnology could also imply that one try to harness biological process for commercial uses and exploitation. The International Society for Environmental Biotechnology defines environmental biotechnology as "the development, use and regulation of biological systems for remediation of contaminated environments (land, air, water), and for environment-friendly processes (green manufacturing technologies and sustainable development)". The environmental biotechnologists and engineers provide expertise in microbial ecology and environmental engineering. They are generally supported by the environmental remediation technology. They focus on research in real-time direct environmental assessment and biological treatment, bioremediation, natural attenuation and other areas, such as, hydroecological engineering; molecular microbial ecology; environment and ecosystem processes; environmental resources; environmental perturbation; global environmental problems; structural and functional dynamics of microbial life, environmental biotechnology and engineering; air pollution and its control; water pollution and its control; soil waste and soil pollution management, radiation pollution control; biopollution or bioinvasion; hazardous materials/hazardous wastes; environmental pollution prevention using bioindicators and biological markers; biosorption of metals; green technology, biofertilizers and biopesticides; composting; vermicompost and phoshocompost; green technology and phytoremediation, green technology and xenobiotics; green technology, biosensors, biochips and biosurfactents, bio-polymers and bio-plastics; bioleaching and biomining; biomethanation; biofuel and biodiesel; sewage treatment; role of information technology in environment and human health; environmental bio-nanotechnology; and a case study of status environmental biotechnology in Australia.

This publication manages to address most of the prominent issues related to environmental biotechnology in contemporary world. It provides readers with a complete overview of the said subject with a completely new perspective. In sum, this book serves the purpose of an up-to-date reference book on the said subject of environmental biotechnology. This book will be helpful to students, scholars, doctoral candidates, policy analysts, NGO practitioners, etc. interested in the biotechnology studies. This publication provides unique resources for understanding the very many areas of environmental biotechnology

Reza Marandi
Ali Shaeri

Contents

Contents

1

Introduction

1.1 Biotechnology

Biotechnology is technology based on biology, especially when used in agriculture, food science, and medicine. The United Nations Convention on Biological Diversity defines biotechnology as:

> Any technological application that uses biological systems, living organisms, or derivatives thereof, to make or modify products or processes for specific use.

Biotechnology is often used to refer to genetic engineering technology of the 21st century, however the term encompasses a wider range and history of procedures for modifying biological organisms according to the needs of humanity, going back to the initial modifications of native plants into improved food crops through artificial selection and hybridization. Bioengineering is the science upon which all Biotechnological applications are based. With the development of new approaches and modern techniques, traditional biotechnology industries are also acquiring new horizons enabling them to improve the quality of their products and increase the productivity of their systems. Before 1971, the term, *biotechnology*, was primarily used in the food processing and agriculture industries. Since the 1970s, it began to be used by the Western scientific establishment to refer to laboratory-based techniques being developed in biological research, such as recombinant DNA or tissue culture-based processes, or horizontal gene transfer in living plants, using vectors such as the *Agrobacterium* bacteria to transfer DNA into a host organism. In fact, the term should be used in a much broader sense to describe the whole range of methods, both ancient and modern, used to manipulate organic materials to reach the demands of food production. So the term could be defined as, "The application of indigenous and/or scientific knowledge to the management of (parts of) microorganisms, or of cells and tissues of higher organisms, so that these supply goods and services of use to the food industry and its consumers.Biotechnology combines disciplines like genetics, molecular biology, biochemistry, embryology and cell biology, which are in turn linked to practical disciplines like chemical engineering, information technology, and robotics. Patho-biotechnology describes the exploitation of pathogens or pathogen derived compounds for beneficial effect.

1.2 What is Biotechnology?

Biotechnology can be summarized as the manipulation of living organisms to produce goods and services. Although the technology has received widespread media coverage in the past few years, it is a technology with a long history, dating as far back as 6000 B.C. Advances in science and technology have transformed traditional biotechnology techniques, such as selective breeding, hybridization and mutagenesis, into modern ones, such as recombinant DNA techniques and tissue culture. This transformation has opened the door to more varied applications in areas such as health care, the environment, forestry, industrial processes and others. Some developments to watch for include research into nutritionally enhanced GM foods and transgenic animals, bio chips and protein drugs.

1.3 Traditional vs. Modern Biotechnology

When Hungarian engineer Karl Ereky coined the word "biotechnology" in 1919 to describe the production of goods from raw materials with the aid of living organisms, he was describing techniques that had been employed by human beings for centuries. This includes techniques such as selective breeding, fermentation and hybridization. However, the term biotechnology has evolved to include modern techniques with new applications, to what is now known as modern biotechnology. Modern biotechnology encompasses techniques such as recombinant DNA technology (genetic engineering), tissue culture and others. These modern techniques also take considerably less time to achieve the desirable changes in living organisms and carry greater precision than traditional techniques.

1.4 Traditional vs. Modern Biotechnology

Traditional Biotechnology

Traditional biotechnology refers to a number of ancient ways of using living organisms to make new products or modify existing ones. In its broadest definition, traditional biotechnology can be traced back to human's transition from hunter-gatherer to farmer. As farmers, humans collected wild plants and cultivated them and the best yielding strains were selected for growing the following seasons. As humans discovered more plant varieties and traits or characteristics, they gradually became adept at breeding specific plant varieties over several years and sometimes generations, to obtain desired traits such as disease resistance, better taste and higher yield. With the domestication of animals, ancient farmers applied the same breeding techniques to obtain desired traits among animals over generations. Centuries ago, people accidentally discovered how to make use of natural processes that occur all the time within living cells. Although they had no scientific explanation for the processes, they applied the results they saw to their domestic lives. They discovered, for example, that food matures in a way that changes its taste and content, and makes it less perishable. Hence, through a process later called fermentation, flour dough becomes leavened in the making of bread, grape juice becomes wine, and milk stored in bags made from camels' stomachs turns into cheese.

Through trial and error and later through advances in technology, people learned to control these processes and make large quantities of biotechnology products. Advances in science enabled the transfer of these mostly domestic techniques into industrial applications and the discovery of new techniques. Examples of traditional biotechnology techniques include selective breeding, hybridization and fermentation.

Modern Biotechnology

Modern biotechnology refers to a number of techniques that involve the intentional manipulation of genes, cells and living tissue in a predictable and controlled manner to generate changes in the genetic make-up of an organism or produce new tissue. Examples of these techniques include: recombinant DNA techniques (rDNA or genetic engineering), tissue culture and mutagenesis. Modern biotechnology began with the 1953 discovery of the structure of deoxyribonucleic acid (DNA) and the way genetic information is passed from generation to generation. This discovery was made possible by the earlier discovery of genes (discrete, independent units that transmit traits from parents to offspring) by Gregor Mendel. These discoveries laid the groundwork for the transition from traditional to modern biotechnology. They made it possible to produce desired changes in an organism through the direct manipulation of its genes in a controlled and less time-consuming fashion in comparison to traditional biotechnology techniques. These discoveries, coupled with advances in technology and science (such as biochemistry and physiology), opened up the possibilities for new applications of biotechnology which were unknown with traditional forms.

Difference between Modern Biotechnology and Traditional Biotechnology?

Modern Knowledge

Before the discovery of genes and DNA, genetic changes in organisms (including plants) were carried out at the organism level. For example, a plant with the desirable trait was cross-bred with other plants in the hope that through cross-pollination, the desirable traits would be transferred to the offspring of the parent plants. In modern biotechnology, achieving desired traits in an organism is done mostly at the gene level. Hence, the gene responsible for the desired trait is identified, transferred and inserted into the organism at the cell level, to produce genetic changes. Also, in other modern techniques of biotechnology such as mutagenesis, past knowledge of causes of mutations, known as mutagens, (such as exposure to radiation or temperature extremes) has been harnessed to generate intentional changes in the genetic make-up of a cell or plant tissue. For example, mutation breeding is a biotechnology technique commonly used to develop plants with novel traits. In mutation breeding, plant tissues are exposed to powerful mutagens in hopes of causing beneficial changes in the genetic make-up of the plant cells and then exposed to the conditions under which the plants would have to grow (such as pesticides, limited amounts of water and so forth). Those plants which experienced beneficial mutations survive the exposure to the conditions and are bred and developed into plant lines.

Applications of Biotechnology

Modern biotechnology has many more applications than traditional biotechnology. Traditional biotechnology focused mainly on food and agriculture through techniques such as selective breeding and fermentation. Science and technology advances have broadened the scope in which biotechnology can be applied to include the environment (bio-remediation and bio-filtration), human and animal health (development of vaccines and drugs, gene therapy, diagnostics), energy (production of bio-fuels), forestry (production of bio-pesticides and genetically modified tree seedlings) and other areas.

Techniques of Biotechnology

Advances in science and technology have led to the development of new types of biotechnology techniques that did not exist in traditional biotechnology. For instance, modern biotechnology includes such techniques as recombinant DNA techniques (also known as genetic engineering), mutagenesis breeding, hybridoma technology and tissue culture. Most of the techniques used in modern biotechnology have been developed as a result of the increased knowledge of genetics and microorganisms. For example, in the traditional biotechnology technique of selective breeding, only plants that were similar (of the same species) were cross-bred. However, through recombinant DNA technology in modern biotechnology, genes can be transferred between unrelated species, for example between bacteria and humans. Therefore, modern techniques of biotechnology have widened the gene pool from which genes with desired traits can be obtained for transfer into organisms lacking those traits. Also, modern techniques such as recombinant DNA techniques and mutagenesis have made it possible to develop novel products in larger amounts than was possible in traditional biotechnology. For instance, in rDNA technology, genes with useful (desirable) traits from plants and animals are transferred into microorganisms, such as yeasts and bacteria, that are easy to grow in large quantities. This technique is used in producing genetically modified bacteria to produce humulin, human insulin used to treat diabetes. Since the discovery of insulin in 1921, humans have relied on the use of animal insulin to treat human diabetes. However, some diabetic patients have had adverse reactions to it. There was also concern about the possible decline in the production of animal-derived insulin. Hence, researchers formulated a way of synthesizing humulin by inserting the insulin gene into bacterial cells to produce insulin that is chemically identical to its natural counterpart. The right form of insulin can now be obtained in much larger quantities than before.

Time and Precision

The discovery of the structure of the DNA molecule and the way genetic information is passed from generation to generation has made the transfer of genes and therefore desired traits between organisms using modern biotechnology more precise and less time consuming than traditional biotechnology. Although it is not guaranteed that every cell used in the process incorporates the transferred gene, the probability of cells incorporating the desired gene is much higher using modern biotechnology techniques than with traditional biotechnology. In addition, using traditional techniques it took many generations and was a lot more time consuming to produce plants and animals with desired traits, since those techniques were mostly based on trial and error. Knowledge of the actual genes being transferred, coupled with modern technology, has significantly reduced the time it takes to obtain the same results in traditional biotechnology.

1.5 History

The most practical use of biotechnology, which is still present today, is the cultivation of plants to produce food suitable to humans. Agriculture has been theorized to have become the dominant way of producing food since the Neolithic Revolution. The processes and methods of agriculture have been refined by other mechanical and biological sciences since its inception. Through early biotechnology farmers were able to select the best suited and highest-yield crops to produce enough food to support a growing population. Other uses of biotechnology were required as crops and fields became increasingly large and difficult to maintain. Specific organisms and organism byproducts were used to fertilize, restore nitrogen, and control pests. Throughout the use of agriculture farmers have inadvertently altered the genetics of their crops through introducing them to new environments and breeding them with other plants—one of the first forms of biotechnology. Cultures such as those in Mesopotamia, Egypt, and India developed the process of brewing beer. It is still done by the same basic method of using malted grains (containing enzymes) to convert starch from grains into sugar and then adding specific yeasts to produce beer. In this process the carbohydrates in the grains were broken down into alcohols such as ethanol. Ancient Indians also used the juices of the plant Ephedra Vulgaris and used to call it Soma. Later other cultures produced the process of Lactic acid fermentation which allowed the fermentation and preservation of other forms of food. Fermentation was also used in this time period to produce leavened bread. Although the process of fermentation was not fully understood until Louis Pasteur's work in 1857, it is still the first use of biotechnology to convert a food source into another form. Combinations of plants and other organisms were used as medications in many early civilizations. Since as early as 200 BC, people began to use disabled or minute amounts of infectious agents to immunize themselves against infections. These and similar processes have been refined in modern medicine and have led to many developments such as antibiotics, vaccines, and other methods of fighting sickness. In the early twentieth century scientists gained a greater understanding of microbiology and explored ways of manufacturing specific products. In 1917, Chaim Weizmann first used a pure microbiological culture in an industrial process, that of manufacturing corn starch using *Clostridium acetobutylicum* to produce acetone, which the United Kingdom desperately needed to manufacture explosives during World War I. The field of modern biotechnology is thought to have largely begun on June 16, 1980, when the United States Supreme Court ruled that a genetically-modified microorganism could be patented in the case of *Diamond v. Chakrabarty.* Indian-born Ananda Chakrabarty, working for General Electric, had developed a bacterium (derived from the *Pseudomonas* genus) capable of breaking down crude oil, which he proposed to use in treating oil spills. Revenue in the industry is expected to grow by 12.9 per cent in 2008. Another factor influencing the biotechnology sector's success is improved intellectual property rights legislation—and enforcement—worldwide, as well as strengthened demand for medical and pharmaceutical products to cope with an ageing, and ailing, U.S. population. Rising demand for biofuels is expected to be good news for the biotechnology sector, with the Department of Energy estimating ethanol usage could reduce U.S. petroleum-derived fuel consumption by up to 30 per cent by 2030. The biotechnology sector has allowed the U.S. farming industry to rapidly increase its supply of corn and soybeans—the main inputs into biofuels—by developing genetically-modified seeds which are resistant to pests and drought. By boosting farm productivity, biotechnology plays a crucial role in ensuring that biofuel production targets are met .

1.6 What is Biotechnology

A Brief History of Biotechnology

The term biotechnology was coined in 1919 by a Hungarian engineer called Karl Ereky. However, its origins date back further than that. The history of biotechnology can be divided into two eras: traditional and modern biotechnology. Traditional biotechnology dates back thousands of years, to early farming societies in which people collected seeds of plants with the most desirable traits for planting the following year. This practice is now known as selective breeding. The same selective breeding practices were used by early Babylonians, Egyptians, and Romans to improve livestock. As far back as 6000 B.C., natural processes such as fermentation, in which microorganisms such as bacteria, yeasts and moulds play a critical role, were used to produce bread, beer and wine. Gregor Mendel's study of genetics, using seed and plant experiments at the end of the 19th century, gave the first indications of the cross from traditional to modern biotechnology. He discovered that traits are transmitted from parents to offspring by discrete, independent units, later called genes. His observations laid the groundwork for the field of genetics. In 1943, the first direct evidence that deoxyribonucleic acid, or DNA, carried genetic information was discovered. However, it wasn't until 1953 that the mystery of the structure of DNA and the way genetic information is passed from generation to generation was unlocked by the discovery of the structure of DNA by James Watson and Francis Crick. The era of 'modern biotechnology', which involves manipulation of genes from living organisms in more precise and controlled ways than traditional biotechnology, began with their discovery. In 1985, genetically engineered plants resistant to insects, viruses, and bacteria were tested for the first time. Since then, many genetically engineered plants have been developed, successfully field tested and received food, livestock feed and environmental safety approval in many countries, including Canada. It was also in 1985 that a plan for mapping and sequencing the human genome was made. The goal of this Human Genome Initiative, which was launched in 1990, was to map all of the 80,000 to 100,000 human genes by the year 2003.

Biotechnology Techniques

Biotechnology can be best understood as an umbrella term referring to a variety of techniques, both traditional and modern. Note that some traditional techniques, such as selective breeding, hybridization and mutagenesis, are used in current applications of biotechnology, only with increased scientific knowledge and more advanced technology. These include:

❖ *Fermentation*: The breakdown of complex organic substances by micro-organisms, in the absence of oxygen. The process is energy-yielding.

❖ *Selective breeding*: The breeding of selected plants and animals to produce offspring with desired traits. The offspring with the desired traits are then used as breeding stock for the next generation and so on, until offspring that express the desired traits are obtained.

❖ *Hybridization*: The production of offspring, known as hybrids, from genetically dissimilar parents. The object of hybridization is to combine desirable genes found in two or more different varieties to produce pure-breeding offspring superior in many respects to the parental types.

- ❖ *Mutagenesis*: The use of mutagens (such as exposure to radiation, temperature extremes and certain chemicals), to cause changes in the genetic make-up of cells, possibly resulting in new desirable, inheritable traits.

- ❖ *Recombinant DNA (rDNA) techniques*: The application of genetic techniques to produce desirable traits in living organisms using other living organisms such as bacteria. Examples of these techniques include the use of: restriction endonucleases (enzymes, produced by bacteria, that break foreign DNA molecules with the gene of interest into fragments which recombine with complementary molecules from a different source to form a recombinant DNA molecule); vectors (plasmids, often bacterial, or viruses that carry a piece of DNA into a bacterium for cloning purposes); and gene guns (called DNA particle guns and used to fire genes into cells by coating the genes onto tiny gold or tungsten particles and firing them from the barrel of the gun).

- ❖ *Tissue Culture*: A biological technique in which fragments of plant or animal tissue or cells are transferred to an artificial environment, free of other organisms, in which they continue to survive and function for reproduction, chemical production and medical research.

Applications of Biotechnology

Modern biotechnology techniques are currently being used in many areas such as food, agriculture, health care, forestry, the environment, minerals, oil and gas recovery and industrial processes to develop new products and processes, and to modify existing ones. Summarized below are some of the ways that modern biotechnology techniques have been applied in these areas.

Food and Agriculture

One of the most extensive applications of biotechnology has been in agriculture. Biotechnology techniques such as recombinant DNA techniques and mutagenesis have been used to develop plants with novel traits. These traits include herbicide tolerance and pest, insect and virus resistance. Biotechnology techniques have been used to produce biopesticides which are toxic to targeted plant pests. Also, experiments into genetic modification of aquatic organisms such as salmon, for such novel traits as enhanced growth, have been carried out using recombinant DNA techniques.

Health Care

To date, applications of biotechnology in health care have focused on fighting disease using the human body's own 'weapons'. Biotechnology medicines and therapies synthesize proteins, enzymes, antibodies and other substances which occur naturally in the human body, to fight infections and diseases. However, biotechnology also uses other living organisms, that is, plant and animal cells, viruses and yeasts to help produce human medicines. There are four main areas in health care in which biotechnology is currently being used: medicines, vaccines, diagnostics and gene therapy. Several biotechnology medicines have been developed and approved to treat such diseases as anaemia, growth deficiency in children, hemophilia, diabetes and others. Biotechnology companies have also developed therapeutic products for infectious agents such as HIV, hepatitis B and influenza. Biotechnology is responsible for new vaccines such as the hepatitis B vaccine which was approved in the U.S. to fight the hepatitis B virus. Biotechnology diagnostics have been used to detect a wide variety of diseases and genetic conditions. For example, gene therapy is used diagnostically to screen donated blood to protect the blood supply from HIV and hepatitis. Biotechnology diagnostics are also used in home pregnancy tests. In a home pregnancy test kit, a protein called a monoclonal antibody (MAb), binds to Human Chorionic Gonadotrophin (HCG), causing a colour change. HCG is present in a woman's urine only during pregnancy. In gene therapy, genes are used as 'drugs' to treat hereditary genetic disorders. Faulty or missing genes can be replaced to prevent the occurrence of a genetic disease. Current gene therapy is primarily experiment-based, with a few human clinical trials, such as for the treatment of cystic fibrosis, in the early stages.

Environment

Biotechnology applications in the environment focus on using living organisms primarily to treat waste and prevent pollution. Examples of these applications include bio-filtration and bio-remediation. Bio-filtration refers to the use of microorganisms to remove complex pollutants from the air emissions and waste water discharges of many manufacturing processes. Bio-remediation refers to a number of processes that use living organisms to degrade toxic waste into harmless by-products such as water, carbon dioxide and other materials. Examples of bio-remediation processes include:

- ❖ *Bio-stimulation*: A technique which involves introducing nutrients to stimulate the growth of waste-eating microorganisms already present in the environment at a waste site.

- ❖ *Bio-augmentation:* A technique whereby microorganisms, not normally present in the ecosystem of a contaminated site, are added to the site to clean up pollutants. The microorganisms added may be natural or genetically modified. The organisms die when their food is used up.

- ❖ *Phytoremediation:* A technique which uses certain plants and fungi, planted at waste sites, to naturally reclaim the sites without using chemical treatment, incineration or land filling. These plants and fungi are used because they flourish by accumulating metals and other waste materials present in the soil.

Forestry

To date, only a few biotechnology-derived products and processes used in forestry have been commercialized. Most biotechnology applications in forestry are still in the research and development stage. However, some commercial applications include the development of bacterial bio-pesticides such as *Bacillus thuringiensis* (Bt), as an alternative to chemical pesticides. Tissue culture technology is also being used to produce genetically modified seedlings for forest regeneration. For example, BCRI Inc., a biotechnology company in British Columbia, is using tissue culture to produce weevil-resistant Sitka spruce. Companies are also developing bio-fertilizers such as plant growth-promoting rhizobia bacterial fertilizers. In addition, genetic engineering has been used to develop traits of agronomic interest such as pest and disease resistance, within trees. The pulp and paper industry has used enzymes and microorganisms in their bio-bleaching and de-inking processes.

Industrial Processes

Biotechnology has been applied in a variety of industrial processes in different ways, particularly in the use of biocatalysts in manufacturing processes. Biocatalysts are substances that initiate or modify the rate of a biological process and are generally consumed in the process. Some examples of industrial processes where biotechnology has been applied include chemicals, starch/grain processing, cleaning and textile industrial processes. In the chemicals industry, biotechnology has been used to produce commodity and specialty chemicals. It is also used in the starch and grain processing industries through the use of enzymes to turn starch into glucose and fructose. Corn and other grains can be converted to sweeteners such as high-fructose corn syrup and maltose syrup, using enzymes. Biotechnology applications have been used to produce ethanol from grain. Furthermore, biotechnology is used in the textile industry for the finishing of fabrics and garments. In the pulp and paper industry, bio-pulping is used to manufacture some products.

Metals and Minerals Recovery

Biotechnology has been used in the mining industry in such processes as bio-leaching and metal bio-remediation and recovery. Bio-leaching is the use of bacteria to extract valuable metals. Metal bio-remediation and recovery refers to the use of biologically created enzymes in the degreasing process to recover metals of value.

Energy

To date, biotechnology applications in the energy industry include the production of cleaner coal and petroleum by removing sulfur and thus reducing the environmental contaminants released during combustion. Biotechnology has also been used widely to produce bio fuels such as bio-ethanol and bio-diesel. Bio fuels are alcohols, ethers, esters and other organic chemicals made from biomass such as herbaceous and woody plants, agricultural and forestry residues and industrial waste. These fuels can be used for electricity and as fuels for transportation.

1.7 Applications

Biotechnology has applications in four major industrial areas, including health care (medical), crop production and agriculture, non food (industrial) uses of crops and other products (e.g. biodegradable plastics, vegetable oil, biofuels), and environmental uses.

For example, one application of biotechnology is the directed use of organisms for the manufacture of organic products (examples include beer and milk products). Another example is using naturally present bacteria by the mining industry in bioleaching. Biotechnology is also used to recycle, treat waste, clean up sites contaminated by industrial activities (bioremediation), and also to produce biological weapons.

A series of derived terms have been coined to identify several branches of biotechnology, for example:

- *Red biotechnology* is applied to medical processes. Some examples are the designing of organisms to produce antibiotics, and the engineering of genetic cures through genomic manipulation.

- *Green biotechnology* is biotechnology applied to agricultural processes. An example would be the selection and domestication of plants via micropropagation. Another example is the designing of transgenic plants to grow under specific environmental conditions or in the presence (or absence) of certain agricultural chemicals. One hope is that green biotechnology might produce more environmentally friendly solutions than traditional industrial agriculture. An example of this is the engineering of a plant to express a pesticide, thereby eliminating the need for external application of pesticides. An example of this would be Bt corn. Whether or not green biotechnology products such as this are ultimately more environmentally friendly is a topic of considerable debate.

- *White biotechnology*, also known as industrial biotechnology, is biotechnology applied to industrial processes. An example is the designing of an organism to produce a useful chemical. Another example is the using of enzymes as industrial catalysts to either produce valuable chemicals or destroy hazardous/polluting chemicals. White biotechnology tends to consume less in resources than traditional processes used to produce industrial goods.

- *Blue biotechnology* is a term that has been used to describe the marine and aquatic applications of biotechnology, but its use is relatively rare.

- The investments and economic output of all of these types of applied biotechnologies form what has been described as the *bioeconomy*.

- *Bioinformatics* is an interdisciplinary field which addresses biological problems using computational techniques, and makes the rapid organization and analysis of biological data possible. The field may also be referred to as *computational biology*, and can be defined as, "conceptualizing biology in terms of molecules and then applying informatics techniques to understand and organize the information associated with these molecules, on a large scale." Bioinformatics plays a key role in various areas, such as functional genomics, structural genomics, and proteomics, and forms a key component in the biotechnology and pharmaceutical sector.

Medicine

In medicine, modern biotechnology finds promising applications in such areas as

- ❖ pharmacogenomics;
- ❖ drug production;
- ❖ genetic testing; and
- ❖ gene therapy.

Pharmacogenomics

Pharmacogenomics is the study of how the genetic inheritance of an individual affects his/her body's response to drugs. It is a coined word derived from the words "pharmacology" and "genomics". It is hence the study of the relationship between pharmaceuticals and genetics. The vision of pharmacogenomics is to be able to design and produce drugs that are adapted to each person's genetic makeup.

Pharmacogenomics results in the following benefits:

1. *Development of tailor-made medicines.* Using pharmacogenomics, pharmaceutical companies can create drugs based on the proteins, enzymes and RNA molecules that are associated with specific genes and diseases. These tailor-made drugs promise not only to maximize therapeutic effects but also to decrease damage to nearby healthy cells.

2. *More accurate methods of determining appropriate drug dosages.* Knowing a patient's genetics will enable doctors to determine how well his/ her body can process and metabolize a medicine. This will maximize the value of the medicine and decrease the likelihood of overdose.

3. *Improvements in the drug discovery and approval process.* The discovery of potential therapies will be made easier using genome targets. Genes have been associated with numerous diseases and disorders. With modern biotechnology, these genes can be used as targets for the development of effective new therapies, which could significantly shorten the drug discovery process.

4. *Better vaccines.* Safer vaccines can be designed and produced by organisms transformed by means of genetic engineering. These vaccines will elicit the immune response without the attendant risks

of infection. They will be inexpensive, stable, easy to store, and capable of being engineered to carry several strains of pathogen at once.

Pharmaceutical Products

Most traditional pharmaceutical drugs are relatively simple molecules that have been found primarily through trial and error to treat the symptoms of a disease or illness. Biopharmaceuticals are large biological molecules known as proteins and these usually target the underlying mechanisms and pathways of a malady (but not always, as is the case with using insulin to treat type 1 diabetes mellitus, as that treatment merely addresses the symptoms of the disease, not the underlying cause which is autoimmunity); it is a relatively young industry. They can deal with targets in humans that may not be accessible with traditional medicines. A patient typically is dosed with a small molecule *via* a tablet while a large molecule is typically injected. Small molecules are manufactured by chemistry but larger molecules are created by living cells such as those found in the human body: for example, bacteria cells, yeast cells, animal or plant cells. Modern biotechnology is often associated with the use of genetically altered microorganisms such as *E. coli* or yeast for the production of substances like synthetic insulin or antibiotics. It can also refer to transgenic animals or transgenic plants, such as Bt corn. Genetically altered mammalian cells, such as Chinese Hamster Ovary (CHO) cells, are also used to manufacture certain pharmaceuticals. Another promising new biotechnology application is the development of plant-made pharmaceuticals. Biotechnology is also commonly associated with landmark breakthroughs in new medical therapies to treat hepatitis B, hepatitis C, cancers, arthritis, haemophilia, bone fractures, multiple sclerosis, and cardiovascular disorders. The biotechnology industry has also been instrumental in developing molecular diagnostic devices than can be used to define the target patient population for a given biopharmaceutical. Herceptin, for example, was the first drug approved for use with a matching diagnostic test and is used to treat breast cancer in women whose cancer cells express the protein HER2. Modern biotechnology can be used to manufacture existing medicines relatively easily and cheaply. The first genetically engineered products were medicines designed to treat human diseases. To cite one example, in 1978 Genentech developed synthetic humanized insulin by joining its gene with a plasmid vector inserted into the bacterium *Escherichia coli*. Insulin, widely used for the treatment of diabetes, was previously extracted from the pancreas of abattoir animals (cattle and/or pigs). The resulting genetically engineered bacterium enabled the production of vast quantities of synthetic human insulin at relatively low cost, although the cost savings was used to increase profits for manufacturers, not passed on to consumers or their healthcare providers. According to a 2003 study undertaken by the International Diabetes Federation (IDF) on the access to and availability of insulin in its member countries, synthetic 'human' insulin is considerably more expensive in most countries where both synthetic 'human' and animal insulin are commercially available: e.g. within European countries the average price of synthetic 'human' insulin was twice as high as the price of pork insulin. Yet in its position statement, the IDF writes that "there is no overwhelming evidence to prefer one species of insulin over another" and "[modern, highly-purified] animal insulins remain a perfectly acceptable alternative. Modern biotechnology has evolved, making it possible to produce more easily and relatively cheaply human growth hormone, clotting factors for hemophiliacs, fertility drugs, erythropoietin and other drugs. Most drugs today are based on about 500 molecular targets. Genomic knowledge of the genes involved in diseases, disease pathways, and drug-response sites are expected to lead to the discovery of thousands more new targets.

Genetic Testing

Genetic testing involves the direct examination of the DNA molecule itself. A scientist scans a patient's DNA sample for mutated sequences. There are two major types of gene tests. In the first type, a researcher may design short pieces of DNA ("probes") whose sequences are complementary to the mutated sequences. These probes will seek their complement among the base pairs of an individual's genome. If the mutated sequence is present in the patient's genome, the probe will bind to it and flag the mutation. In the second type, a researcher may conduct the gene test by comparing the sequence of DNA bases in a patient's gene to disease in healthy individuals or their progeny.

Genetic testing is now used for:

❖ Determining sex
❖ Carrier screening, or the identification of unaffected individuals who carry one copy of a gene for a disease that requires two copies for the disease to manifest
❖ Prenatal diagnostic screening
❖ Newborn screening
❖ Presymptomatic testing for predicting adult-onset disorders
❖ Presymptomatic testing for estimating the risk of developing adult-onset cancers

* ❖ Confirmational diagnosis of symptomatic individuals
* ❖ Forensic/identity testing

Some genetic tests are already available, although most of them are used in developed countries. The tests currently available can detect mutations associated with rare genetic disorders like cystic fibrosis, sickle cell anemia, and Huntington's disease. Recently, tests have been developed to detect mutation for a handful of more complex conditions such as breast, ovarian, and colon cancers. However, gene tests may not detect every mutation associated with a particular condition because many are as yet undiscovered, and the ones they do detect may present different risks to different people and populations.

Controversial Questions

Several issues have been raised regarding the use of genetic testing:

1. *Absence of cure.* There is still a lack of effective treatment or preventive measures for many diseases and conditions now being diagnosed or predicted using gene tests. Thus, revealing information about risk of a future disease that has no existing cure presents an ethical dilemma for medical practitioners.
2. *Ownership and control of genetic information.* Who will own and control genetic information, or information about genes, gene products, or inherited characteristics derived from an individual or a group of people like indigenous communities? At the macro level, there is a possibility of a genetic divide, with developing countries that do not have access to medical applications of biotechnology being deprived of benefits accruing from products derived from genes obtained from their own people. Moreover, genetic information can pose a risk for minority population groups as it can lead to group stigmatization. At the individual level, the absence of privacy and anti-discrimination legal protections in most countries can lead to discrimination in employment or insurance or other misuse of personal genetic information. This raises questions such as whether genetic privacy is different from medical privacy.
3. *Reproductive issues.* These include the use of genetic information in reproductive decision-making and the possibility of genetically altering reproductive cells that may be passed on to future generations. For example, germline therapy forever changes the genetic make-up of

an individual's descendants. Thus, any error in technology or judgment may have far-reaching consequences. Ethical issues like designer babies and human cloning have also given rise to controversies between and among scientists and bioethicists, especially in the light of past abuses with eugenics.

4. *Clinical issues.* These center on the capabilities and limitations of doctors and other health-service providers, people identified with genetic conditions, and the general public in dealing with genetic information.
5. *Effects on social institutions.* Genetic tests reveal information about individuals and their families. Thus, test results can affect the dynamics within social institutions, particularly the family.
6. Conceptual and philosophical implications regarding human responsibility, free will vis-à-vis genetic determinism, and the concepts of health and disease.

Gene Therapy

Gene therapy may be used for treating, or even curing, genetic and acquired diseases like cancer and AIDS by using normal genes to supplement or replace defective genes or to bolster a normal function such as immunity. It can be used to target somatic (i.e., body) or germ (i.e., egg and sperm) cells. In somatic gene therapy, the genome of the recipient is changed, but this change is not passed along to the next generation. In contrast, in germline gene therapy, the egg and sperm cells of the parents are changed for the purpose of passing on the changes to their offspring.

There are basically two ways of implementing a gene therapy treatment:

1. *Ex vivo*, which means "outside the body"—Cells from the patient's blood or bone marrow are removed and grown in the laboratory. They are then exposed to a virus carrying the desired gene. The virus enters the cells, and the desired gene becomes part of the DNA of the cells. The cells are allowed to grow in the laboratory before being returned to the patient by injection into a vein.
2. *In vivo*, which means "inside the body"—No cells are removed from the patient's body. Instead, vectors are used to deliver the desired gene to cells in the patient's body. Currently, the use of gene therapy is limited. Somatic gene therapy is primarily at the experimental stage. Germline therapy is the subject of much discussion but it is not being actively investigated in larger animals and human beings. As of June

2001, more than 500 clinical gene-therapy trials involving about 3,500 patients have been identified worldwide. Around 78 per cent of these are in the United States, with Europe having 18 per cent. These trials focus on various types of cancer, although other multigenic diseases are being studied as well. Recently, two children born with severe combined immunodeficiency disorder ("SCID") were reported to have been cured after being given genetically engineered cells. Gene therapy faces many obstacles before it can become a practical approach for treating disease. At least four of these obstacles are as follows:

1. *Gene delivery tools.* Genes are inserted into the body using gene carriers called vectors. The most common vectors now are viruses, which have evolved a way of encapsulating and delivering their genes to human cells in a pathogenic manner. Scientists manipulate the genome of the virus by removing the disease-causing genes and inserting the therapeutic genes. However, while viruses are effective, they can introduce problems like toxicity, immune and inflammatory responses, and gene control and targeting issues.

2. *Limited knowledge of the functions of genes.* Scientists currently know the functions of only a few genes. Hence, gene therapy can address only some genes that cause a particular disease. Worse, it is not known exactly whether genes have more than one function, which creates uncertainty as to whether replacing such genes is indeed desirable.

3. *Multigene disorders and effect of environment.* Most genetic disorders involve more than one gene. Moreover, most diseases involve the interaction of several genes and the environment. For example, many people with cancer not only inherit the disease gene for the disorder, but may have also failed to inherit specific tumor suppressor genes. Diet, exercise, smoking and other environmental factors may have also contributed to their disease.

4. *High costs.* Since gene therapy is relatively new and at an experimental stage, it is an expensive treatment to undertake. This explains why current studies are focused on illnesses commonly found in developed countries, where more people can afford to pay for treatment. It may take decades before developing countries can take advantage of this technology.

Human Genome Project

The Human Genome Project is an initiative of the U.S. Department of Energy ("DOE") that aims to generate a high-quality reference sequence for the entire human genome and identify all the human genes. The DOE and its predecessor agencies were assigned by the U.S. Congress to develop new energy resources and technologies and to pursue a deeper understanding of potential health and environmental risks posed by their production and use. In 1986, the DOE announced its Human Genome Initiative. Shortly thereafter, the DOE and National Institutes of Health developed a plan for a joint Human Genome Project ("HGP"), which officially began in 1990. The HGP was originally planned to last 15 years. However, rapid technological advances and worldwide participation accelerated the completion date to 2003 (making it a 13 year project). Already it has enabled gene hunters to pinpoint genes associated with more than 30 disorders.

Cloning

Cloning involves the removal of the nucleus from one cell and its placement in an unfertilized egg cell whose nucleus has either been deactivated or removed. There are two types of cloning:

1. *Reproductive cloning.* After a few divisions, the egg cell is placed into a uterus where it is allowed to develop into a fetus that is genetically identical to the donor of the original nucleus.
2. *Therapeutic cloning.* The egg is placed into a Petri dish where it develops into embryonic stem cells, which have shown potentials for treating several ailments.

In February 1997, cloning became the focus of media attention when Ian Wilmut and his colleagues at the Roslin Institute announced the successful cloning of a sheep, named Dolly, from the mammary glands of an adult female. The cloning of Dolly made it apparent to many that the techniques used to produce her could someday be used to clone human beings. This stirred a lot of controversy because of its ethical implications.

Agriculture

Improve Yield from Crops

Using the techniques of modern biotechnology, one or two genes may be transferred to a highly developed crop variety to impart a new character that would increase its yield (30). However, while increases in crop yield are the most obvious applications of modern biotechnology in agriculture, it is also the most difficult one. Current genetic engineering techniques work best for effects that are controlled by a single gene. Many of the genetic characteristics associated with yield (e.g., enhanced growth) are controlled by a large number of genes, each of which has a minimal effect on the overall yield (31). There is, therefore, much scientific work to be done in this area.

Reduced Vulnerability of Crops to Environmental Stresses

Crops containing genes that will enable them to withstand biotic and abiotic stresses may be developed. For example, drought and excessively salty soil are two important limiting factors in crop productivity. Biotechnologists are studying plants that can cope with these extreme conditions in the hope of finding the genes that enable them to do so and eventually transferring these genes to the more desirable crops. One of the latest developments is the identification of a plant gene, At-DBF2, from thale cress, a tiny weed that is often used for plant research because it is very easy to grow and its genetic code is well mapped out. When this gene was inserted into tomato and tobacco see RNA interference cells, the cells were able to withstand environmental stresses like salt, drought, cold and heat, far more than ordinary cells. If these preliminary results prove successful in larger trials, then At-DBF2 genes can help in engineering crops that can better withstand harsh environments (32). Researchers have also created transgenic rice plants that are resistant to rice yellow mottle virus (RYMV). In Africa, this virus destroys majority of the rice crops and makes the surviving plants more susceptible to fungal infections (33).

Increased Nutritional Qualities of Food Crops

Proteins in foods may be modified to increase their nutritional qualities. Proteins in legumes and cereals may be transformed to provide the amino acids needed by human beings for a balanced diet (34). A good example is the work of Professors Ingo Potrykus and Peter Beyer on the so-called Goldenrice(discussed below).

Improved Taste, Texture or Appearance of Food

Modern biotechnology can be used to slow down the process of spoilage so that fruit can ripen longer on the plant and then be transported to the consumer with a still reasonable shelf life. This improves the taste, texture and appearance of the fruit. More importantly, it could expand the market for farmers in developing countries due to the reduction in spoilage. The first genetically modified food product was a tomato which was transformed to delay its ripening (35). Researchers in Indonesia, Malaysia, Thailand, Philippines and Vietnam are currently working on delayed-ripening papaya in collaboration with the University of Nottingham and Zeneca (36). Biotechnology in cheese production : enzymes produced by micro-organisms provide an alternative to animal rennet—a cheese coagulant—and an alternative supply for cheese makers. This also eliminates possible public concerns with animal-derived material, although there is currently no plans to develop synthetic milk, thus making this argument less compelling. Enzymes offer an animal-friendly alternative to animal rennet. While providing comparable quality, they are theoretically also less expensive. About 85 million tons of wheat flour is used every year to bake bread. By adding an enzyme called maltogenic amylase to the flour, bread stays fresher longer. Assuming that 10-15 per cent of bread is thrown away, if it could just stay fresh another 5-7 days then 2 million tons of flour per year would be saved. That corresponds to 40 per cent of the bread consumed in a country such as the USA. This means more bread becomes available with no increase in input. In combination with other enzymes, bread can also be made bigger, more appetizing and better in a range of ways.

Reduced Dependence on Fertilizers, Pesticides and Other Agrochemicals

Most of the current commercial applications of modern biotechnology in agriculture are on reducing the dependence of farmers on agrochemicals. For example, *Bacillus thuringiensis* (Bt) is a soil bacterium that produces a protein with insecticidal qualities. Traditionally, a fermentation process has been used to produce an insecticidal spray from these bacteria. In this form, the Bt toxin occurs as an inactive protoxin, which requires digestion by an insect to be effective. There are several Bt toxins and each one is specific to certain target insects. Crop plants have now been engineered to contain and express the genes for Bt toxin, which they produce in its active form. When a susceptible insect ingests the transgenic crop cultivar expressing the Bt protein, it stops feeding and soon thereafter dies as a result of the Bt toxin binding to its gut wall. Bt corn is now commercially available in a number of countries to control corn borer (a lepidopteran insect), which is otherwise controlled by spraying (a more difficult process). Crops have also been genetically engineered to acquire tolerance to broad-spectrum herbicide. The lack of cost-effective herbicides with broad-spectrum activity and no crop injury was a consistent limitation in crop weed management. Multiple applications of numerous herbicides were routinely used to control a wide range of weed species detrimental to agronomic crops. Weed management tended to rely on preemergence—that is, herbicide applications were sprayed in response to expected weed infestations rather than in response to actual weeds present. Mechanical cultivation and hand weeding were often necessary to control weeds not controlled by herbicide applications. The introduction of herbicide tolerant crops has the potential of reducing the number of herbicide active ingredients used for weed management, reducing the number of herbicide applications made during a season, and increasing yield due to improved weed management and less crop injury. Transgenic crops that express tolerance to glyphosate, glufosinate and bromoxynil have been developed. These herbicides can now be sprayed on transgenic crops without inflicting damage on the crops while killing nearby weeds (37). From 1996 to 2001, herbicide tolerance was the most dominant trait introduced to commercially available transgenic crops, followed by insect resistance. In 2001, herbicide tolerance deployed in soybean, corn and cotton accounted for 77 per cent of the 626,000 square kilometres planted to transgenic crops; Bt crops accounted for 15 per cent; and "stacked genes" for herbicide tolerance and insect resistance used in both cotton and corn accounted for 8 per cent (38).

Production of Novel Substances in Crop Plants

Biotechnology is being applied for novel uses other than food. For example, oilseed can be modified to produce fatty acids for detergents, substitute fuels and petrochemicals. Potatos, tomatos, rice, tobacco, lettuce, safflowers, and other plants have been genetically-engineered to produce insulin and certain vaccines. If future clinical trials prove successful, the advantages of edible vaccines would be enormous, especially for developing countries. The transgenic plants may be grown locally and cheaply. Homegrown vaccines would also avoid logistical and economic problems posed by having to transport traditional preparations over long distances and keeping them cold while in transit. And since they are edible, they will not need syringes, which are not only an additional expense in the traditional vaccine preparations but also a source of infections if contaminated. In the case of insulin grown in transgenic plants, it is well-established that the gastrointestinal system breaks the protein down therefore this could not currently be administered as an edible protein. However, it might be produced at significantly lower cost than insulin produced in costly, bioreactors. For example, Calgary, Canada-based SemBioSys Genetics, Inc. reports that its safflower-produced insulin will reduce unit costs by over 25 per cent or more and reduce the capital costs associated with building a commercial-scale insulin manufacturing facility by approximately over $100 million compared to traditional biomanufacturing facilities.

Criticism

There is another side to the agricultural biotechnology issue however. It includes increased herbicide usage and resultant herbicide resistance, "super weeds," residues on and in food crops, genetic contamination of non-GM crops which hurt organic and conventional farmers, damage to wildlife from glyphosate, etc.

Biological Engineering

Biotechnological engineering or biological engineering is a branch of engineering that focuses on biotechnologies and biological science. It includes different disciplines such as biochemical engineering, biomedical engineering, bio-process engineering, biosystem engineering and so on. Because of the novelty of the field, the definition of a bioengineer is still undefined. However, in general it is an integrated approach of fundamental biological sciences and traditional engineering principles. Bioengineers are often employed to scale up bio processes from the laboratory scale to the manufacturing scale. Moreover, as with most engineers, they often deal with management, economic and legal issues. Since patents and regulation (e.g. FDA regulation in the U.S.) are very important issues for biotech enterprises, bioengineers are often required to have knowledge related to these issues. The increasing number of biotech enterprises is likely to create a need for bioengineers in the years to come. Many universities throughout the world are now providing programmes in bioengineering and biotechnology (as independent programmes or specialty programmes within more established engineering fields).

Bioremediation and Biodegradation

Biotechnology is being used to engineer and adapt organisms especially microorganisms in an effort to find sustainable ways to clean up contaminated environments. The elimination of a wide range of pollutants and wastes from the environment is an absolute requirement to promote a sustainable development of our society with low environmental impact. Biological processes play a major role in the removal of contaminants and biotechnology is taking advantage of the astonishing catabolic versatility of microorganisms to degrade/convert such compounds. New methodological breakthroughs in sequencing, genomics, proteomics, bioinformatics and imaging are producing vast amounts of information. In the field of Environmental Microbiology, genome-based global studies open a new era providing unprecedented *in silico* views of metabolic and regulatory networks, as well as clues to the evolution of degradation pathways and to the molecular adaptation strategies to changing environmental conditions. Functional genomic and metagenomic approaches are increasing our understanding of the relative importance of different pathways and regulatory networks to carbon flux in particular environments and for particular compounds and they will certainly accelerate the development of bioremediation technologies and biotransformation processes. Marine environments are especially vulnerable since oil spills of coastal regions and the open sea are poorly containable and mitigation is difficult. In addition to pollution through human activities, millions of tons of petroleum enter the marine environment every year from natural seepages. Despite its toxicity, a considerable fraction of petroleum oil entering marine systems is eliminated by the hydrocarbon-degrading activities of microbial communities, in particular by a remarkable recently discovered group of specialists, the so-called hydrocarbonoclastic bacteria (HCB).

Notable Researchers and Individuals

- ❖ *Canada*: Frederick Banting, Lap-Chee Tsui, Tak Wah Mak, Lorne Babiuk
- ❖ *Europe*: Paul Nurse, Jacques Monod, Francis Crick
- ❖ *Finland*: Leena Palotie
- ❖ *Iceland*: Kari Stefansson
- ❖ *India*: Kiran Mazumdar-Shaw (Biocon)
- ❖ *Ireland*: Timothy O'Brien, Dermot P Kelleher
- ❖ *Mexico*: Francisco Bolívar Zapata, Luis Herrera-Estrella
- ❖ *U.S.*: David Botstein, Craig Venter, Sydney Brenner, Eric Lander, Leroy Hood, Robert Langer, James J. Collins, Roger Beachy, Herbert Boyer, Michael West, Thomas Okarma, James D. Watson

1.8 Environmental biotechnology

Environmental biotechnology is when biotechnology is applied to and used to study the natural environment. Environmental biotechnology could also imply that one try to harness biological process for commercial uses and exploitation. The International Society for Environmental Biotechnology defines environmental biotechnology as "the development, use and regulation of biological systems for remediation of contaminated environments (land, air, water), and for environment-friendly processes (green manufacturing technologies and sustainable development)".

Significance Towards Industrial Biotechnology

Consider an environment in which pollution of a particular type is maximum. Let us consider the effluents of a starch industry (aka *Sago industry*) which has mixed up with a local water body like a lake or pond. We find huge deposits of starch which are not so easily taken up for degradation by micro-organisms except for a few exemptions. We isolate a few micro-organisms from the polluted site and scan for any significant changes in their genome like mutations or evolutions. The modified genes are then identified. This is done because, the isolate would have adapted itself to degrade/utilize the starch better than other microbes of the same genus. Thus, the resultant genes are cloned onto industrially significant micro-organisms and are used for more economically significant processess like in pharmaceutical industry, fermentations.etc. Similar situations can be elucitated like in the case of oil spills in the oceans which require cleanup, microbes isolated from oil rich environments like oil wells, oil transfer pipelines.etc have been found having the potential to degrade oil or use it as an energy source. Thus they serve as a remedy to oil spills. Still another elucidation would be in the case of microbes isolated from pesticide rich soils These would be capable of utilizing the pesticides as energy source and hence when mixed along with bio-fertilizers, would serve as excellent insurance against increased pesticide-toxicity levels in agricultural platform. But the counter argument would be that whether these newly introduced microorganisms would create an imbalance in the environment concerned.The mutual harmony in which the organisms in that particular environment existed may have to face alteration and we should be extremely careful so as to not disturb the mutual relationships already existing in the environment to which we are introducing the newly discovered and cloned microorganisms. Analysis of both the benefits and the disadvantages would pave way for an improvised version of environmental biotechnology. After all it is the environment that we strive to protect.

Environment and Ecosystem Processes

2.1 Environment

Environment may refer to:

- ❖ Environment (biophysical), the physical and biological factors along with their chemical interactions that affect an organism.
- ❖ Natural environment, all living and non-living things that occur naturally on Earth.
- ❖ Built environment, constructed surroundings that provide the setting for human activity, ranging from the large-scale civic surroundings to the personal places.
- ❖ Social environment, the culture that an individual lives in, and the people and institutions with whom they interact.
- ❖ Environmental science, the study of the interactions among the physical, chemical and biological components of the environment.
- ❖ Environmental psychology.
- ❖ Environmental determinism.
- ❖ Environmental policy.
- ❖ Environmental quality.
- ❖ Environmental art.

In computing:

- ❖ Environment variable, the set of environments defined in a process.
- ❖ Runtime environment, a virtual machine state which provides software services for processes or programmes while a computer is running.
- ❖ Integrated development environment, a type of computer software that assists computer programmers in developing software.
- ❖ Desktop environment, in computing, is graphical user interface to the computer.

2.2 Natural Environment

The natural environment, commonly referred to simply as the environment, is a terminology that comprises all living and non-living things that occur naturally on Earth or some region thereof. This term includes a few key components:

1. Complete ecological units that function as natural systems without massive human intervention, including all vegetation, animals, microorganisms, rocks, atmosphere and natural phenomena that occur within their boundaries.
2. Universal natural resources and physical phenomena that lack clear-cut boundaries, such as air, water, and climate, as well as energy, radiation, electric charge, and magnetism, not originating from human activity.

The natural environment is contrasted with the built environment, which comprises the areas and components that are strongly influenced by man. A geographical area is regarded as a natural environment (with an indefinite article), if the human impact on it is kept under a certain limited level (similar to section 1 above). This level depends on the specific context, and changes in different areas and contexts. The term wilderness, on the other hand, refers to areas without human intervention.

Challenges

It is the common understanding of *natural environment* that underlies environmentalism—a broad political, social, and philosophical movement that advocates various actions and policies in the interest of protecting what nature remains in the natural environment, or restoring or expanding the role of nature in this environment. While true wilderness is increasingly rare, *wild* nature (e.g., unmanaged forests, uncultivated grasslands, wildlife, wildflowers) can be found in many locations previously inhabited by humans.

Goals commonly expressed by environmental scientists include:

*   reduction and clean up of pollution, with future goals of zero pollution;
*   cleanly converting nonrecyclable materials into energy through direct combustion or after conversion into secondary fuels;
*   reducing societal consumption of non-renewable fuels;
*   development of alternative, green, low-carbon or renewable energy sources;

- ❖ conservation and sustainable use of scarce resources such as water, land, and air;
- ❖ protection of representative or unique or pristine ecosystems;
- ❖ preservation of threatened and endangered species extinction;
- ❖ the establishment of nature and biosphere reserves under various types of protection; and, most generally, the protection of biodiversity and ecosystems upon which all human and other life on earth depends.

Very large development projects—also called megaprojects—pose special challenges and risks to the natural environment. Major dams and power plants are cases in point. The challenge to the environment from such projects is growing because more and bigger megaprojects are being built, in developed and developing nations alike. Recently, there has been a strong concern about climate change such as global warming caused by anthropogenic releases of greenhouse gases, most notably carbon dioxide, and their interactions with humans and the natural environment. Efforts here have focused on the mitigation of greenhouse gases that are causing climatic changes (e.g. through the Climate Change Convention and the Kyoto Protocol), and on developing adaptative strategies to assist species, ecosystems, humans, regions and nations in adjusting to the Effects of global warming. A more profound challenge, however, is to identify the natural environmental dynamics in contrast to environmental changes not within natural variances. A common solution is to adapt a static view neglecting natural variances to exist. Methodologically this view could be defended when looking at processes which change slowly and short time series, while the problem arrives when fast processes turns essential in the object of the study.

2.3 Ecosystem Ecology

Ecosystematic ecology is the integrated study of biotic and abiotic components of ecosystems and their interactions within an ecosystem framework. This science examines how ecosystems work and relates this to their components such as chemicals, bedrock, soil, plants, and animals. Ecosystem ecology examines physical and also biological structures and examines how these ecosystem characteristics interact with each other. Ultimately, this helps us understand how to maintain high quality water and economically viable commodity production in this and many other ecosystems. A major focus of ecosystem ecology is on functional processes, ecological mechanisms that maintain the structure and services produced by ecosystems. These include primary productivity (production of biomass), decomposition, and trophic interactions. Studies of ecosystem function have greatly improved human understanding of sustainable production of forage, fiber, fuel, and provision of water. Functional processes are mediated by regional-to-local level climate, disturbance, and management thus ecosystem ecology provides a powerful framework for identifying ecological mechanisms that interact with global environmental problems, especially global warming and degradation of surface water. This article will describe the context of ecosystem ecology and provide an overview of the mechanisms that maintain ecosystem structure and function.

Ecosystems and Scale

Ecosystems are difficult entities to define theoretically or to delineate in space For example, consider the forest in Figure 2.1. When standing on the stream bank, one can easily see two ecosystems, an aquatic one where fish, insects, and algae interact, and the other a terrestrial one with trees, another community of insects, and perhaps herbivores and predators such as deer and coyote. Although these communities appear distinct they interact intimately. Insects may be aquatic for certain parts of their life-cycle and emerge to become herbivores of the vegetation and prey for many predators. Riparian trees utilize stream water for growth and ztheir leaf litter is an important flux of energy and nutrients to a rich community of benthic invertebrates. The distinction becomes even less clear when streams flood and deposit nutrient rich sediment on flood planes and scour other areas clean of biota and soil. This example demonstrates several important aspects of ecosystems:

Figure 2.1: A riparian forest in the White Mountains, New Hampshire (USA).

1. Ecosystem boundaries are often nebulous and may fluctuate in time

2. Organism within ecosystems are dependent on ecosystem level biological and physical processes and
3. adjacent ecosystems closely interact and often are interdependent for maintenance of community structure and functional processes that maintain productivity and biodiversity.

These characteristics also introduce practical problems into natural resource management. Who will manage which ecosystem? Will timber cutting in the forest degrade recreational fishing in the stream? These questions are difficult for land managers to address while the boundary between ecosystems remains unclear even though decisions in one ecosystem will affect the other. We need better understanding of the interactions and interdependencies of these ecosystems and the processes that maintain them before we can begin to address these questions. Ecosystem ecology is an inherently interdisciplinary field of study. An individual ecosystem is composed of populations of organisms, interacting within communities, and contributing to the cycling of nutrients and the flow of energy. The ecosystem is the principle unit of study in ecosystem ecology. Population, community, and physiological ecology provide many of the underlying biological mechanisms influencing ecosystems and the processes they maintain. Cycling of energy and matter at the ecosystem level are often examined in ecosystem ecology but, as a whole this science is defined more by subject matter than by scale. Ecosystem ecology approaches organisms and abiotic pools of energy and nutrients as an integrated system which distinguishes it from associated sciences such as biogeochemistry. Biogeochemistry and hydrology focus on several fundamental ecosystem processes such as biologically mediated chemical cycling of nutrients and physical-biological cycling of water. Ecosystem ecology forms the mechanistic basis for regional or global processes encompassed by landscape-to-regional hydrology, global biogeochemistry, and earth system science.

History

Ecosystem ecology is philosophically and historically rooted in terrestrial ecology. The ecosystem concept has evolved rapidly during the last 100 years with important ideas developed by Fredrick Clements, a botanist who argued for specific definitions of ecosystems and that physiological processes were responsible for their development and persistence. Although most of Clements ecosystem definitions have been greatly revised by contemporary ecologists, the idea that physiological processes are fundamental to ecosystem structure and function remains central to ecology. Later work by Eugene Odum and

Howard T. Odum quantified flows of energy and matter at the ecosystem level, thus documenting the general ideas proposed by Clements and his contemporary Charles Elton, the intellectual father of the "food web" concept. In this model, energy flows through the whole system were dependent on biotic and abiotic interactions of each individual component, (inorganic pools of nutrients, etc). Later work demonstrated that these interactions and flows applied to nutrient cycles, changed over the course of succession, and held powerful controls over ecosystem productivity. Transfers of energy and nutrients are innate to ecological systems regardless of whether they are aquatic or terrestrial. Thus, ecosystem ecology has emerged from important biological studies of plants, animals, terrestrial, aquatic, and marine ecosystems.

Ecosystem Services

Ecosystem services are ecologically mediated functional processes essential to sustaining healthy societies. Water provision and filtration, production of biomass in forestry, agriculture, and fisheries, and removal of greenhouse gases such as carbon dioxide (CO_2) from the atmosphere are examples of ecosystem services essential to public health and economic opportunity. Nutrient cycling is a process fundamental to agricultural and forest production. However, like most ecosystem processes, nutrient cycling is not an ecosystem characteristic which can be "dialed" to the most desirable level. Maximizing production in degraded systems is an overly simplistic solution to the complex problems of hunger and economic security. For instance, intensive fertilizer use in the midwestern United States has resulted in degraded fisheries in the Gulf of Mexico. Regrettably, a "green revolution" of intensive chemical fertilization has been recommended for agriculture in developed and developing countries. These short-sighted strategies risk alteration of ecosystem processes that may be difficult to restore, especially when applied at broad scales without adequate assessment of impacts. Ecosystem processes may take many years to recover from significant disturbance. For instance, large-scale forest clearance in the northeastern United States during the 18th and 19th centuries has altered soil texture, dominant vegetation, and nutrient cycling in ways that impact forest productivity in the present day. An appreciation of the importance of ecosystem function in maintenance of productivity, whether in agriculture or forestry, is needed in conjunction with plans for restoration of essential processes. Improved knowledge of ecosystem function will help to achieve long-term sustainability and stability in the poorest parts of the world.

How do Ecosystems Work?

Biomass productivity is one of the most apparent and economically important ecosystem functions. Biomass accumulation begins at the cellular level via photosynthesis. Photosynthesis requires water and consequently, global patters of annual biomass production are correlated with annual precipitation. Amounts of productivity are also dependent on the overall capacity of plants to capture sunlight which is directly correlated with plant leaf area and leaf N content. Net primary productivity (NPP) is the primary measure of biomass accumulation within an ecosystem. Net primary productivity can be calculated by a simple formula where total amount of productivity is adjusted for total productivity losses through maintenance of biological processes:

$$NPP = GPP - Rplant$$

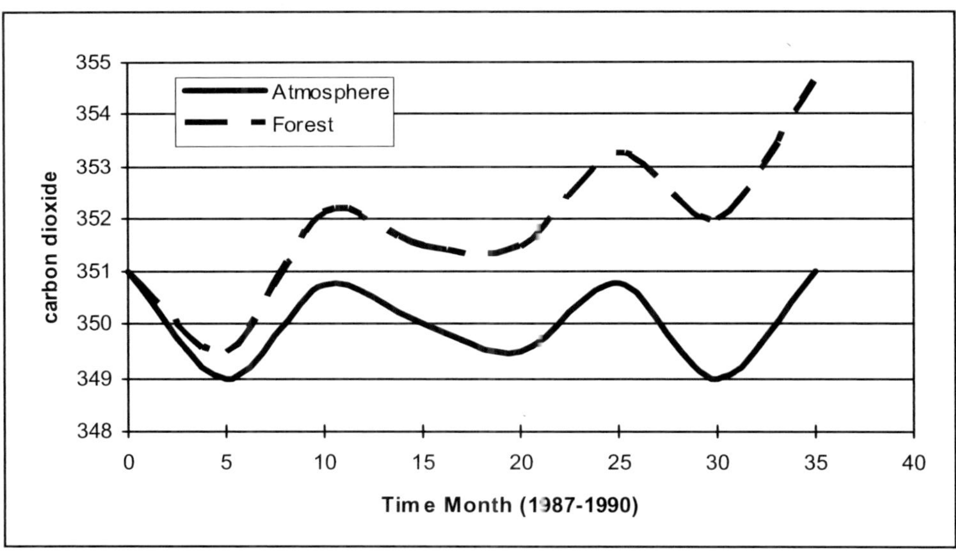

Figure 2.2: Seasonal and annual changes in ambient carbon dioxide (CO_2) concentration at Mauna Loa Hawaii (Atmosphere) and above the canopy of a deciduous forest in Massachusetts (Forest). Data show clear seasonal trends associated with periods of high and low NPP and an overall annual increase of atmospheric CO_2. Data approximates of those reported by Keeling and Whorf and Barford.

Where GPP is gross primary productivity and Rplant is photosynthate (Carbon) lost via cellular respiration. NPP is difficult to measure but a new technique known as eddy co-variance has shed light on how natural ecosystems influence the atmosphere. Figure 2.2 shows seasonal and annual changes in CO_2 concentration measured at Mauna Loa, Hawaii from approximately 1987 to 1990. CO_2 concentration steadily increased but within-year variation has been greater than the annual increase since measurements began in 1957. These variations were thought to be due to seasonal uptake of CO_2 during summer months. A newly developed technique for assessing ecosystem NPP has confirmed seasonal variation are driven by seasonal changes in CO_2 uptake by vegetation. This has led many scientists and policy makers to speculate that ecosystems can be managed to ameliorate problems with global warming. This type of management may include reforesting or altering forest harvest schedules many parts of the world.

Decomposition and Nutrient Cycling

Decomposition and nutrient cycling are fundamental to ecosystem biomass production. Most natural ecosystems are nitrogen (N) limited and biomass production is closely correlated with N turnover. Typically external input of nutrients is very low and efficient recycling of nutrients maintains productivity. Decomposition of plant litter accounts for the majority of nutrients recycled through ecosystems (Figure 2.3). Rates of plant litter decomposition are highly dependent on litter quality; high concentration of phenolic compounds, especially lignin, in plant litter has a retarding effect on litter decomposition. More complex C compounds are decomposed more slowly and may take many years to completely breakdown. Decomposition is typically described with exponential

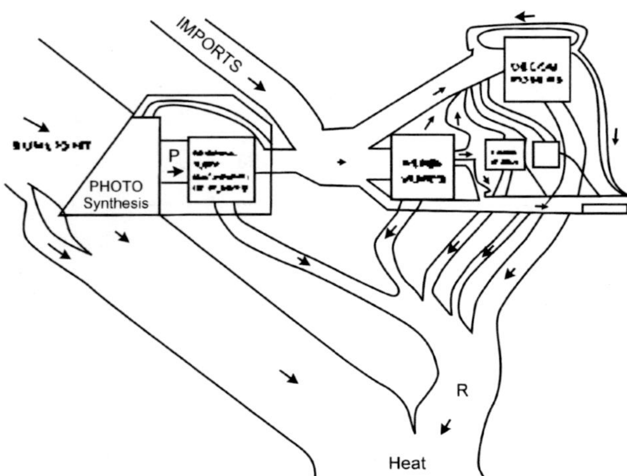

Figure 2.3: Energy and matter flows through an ecosystem, adapted from the Silver Springs model. H are herbivores, C are carnivores, TC are top carnivores, and D are decomposers. Squares represent biotic pools and ovals are fluxes or energy or nutrients from the system.

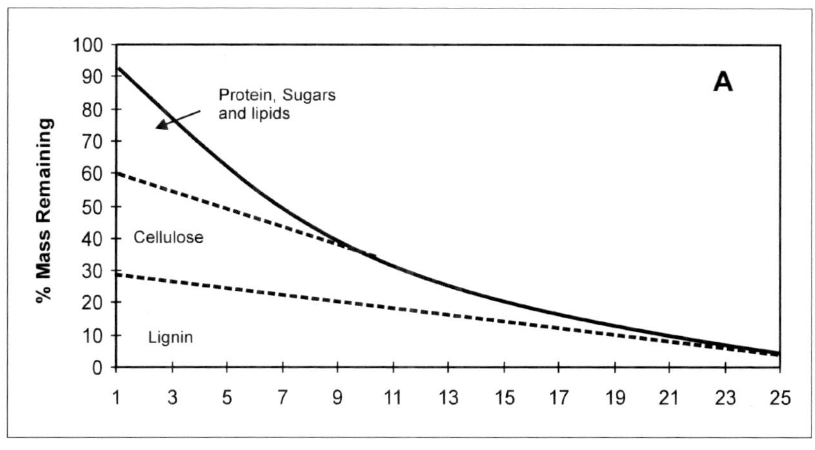

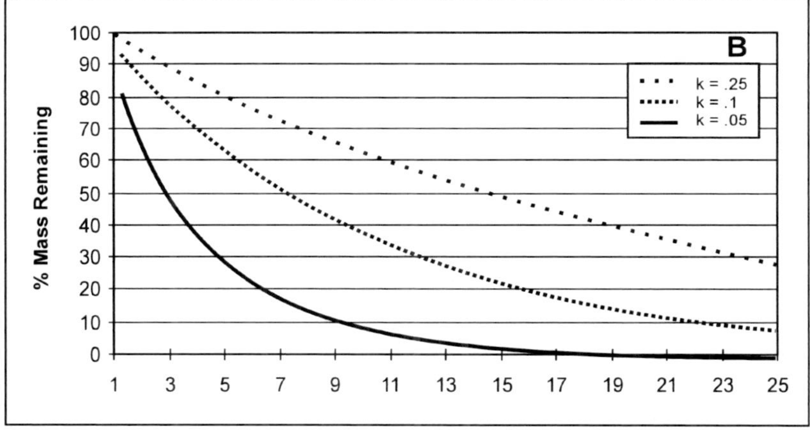

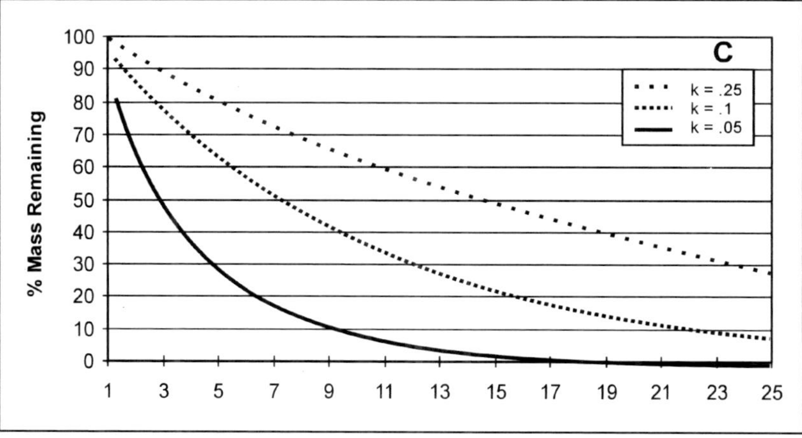

Figure 2.4: Dynamics of decomposing plant litter (A) described with an exponential model (B) and a combined exponential-linear model (C).

Globally, rates of decomposition are mediated by litter quality and climate. Ecosystems dominated by plants with low-lignin concentration often have rapid rates of decomposition and nutrient cycling (Chapin *et al.* 1982). Simple carbon (C) containing compounds are preferentially metabolized by decomposer microorganisms which results in rapid initial rates of decomposition. However, these models do not reflect simultaneous linear and non-linear decay processes which likely occur during decomposition. For instance, proteins, sugars and lipids decompose exponentially, but lignin decays at a more linear rate Thus, litter decay is probably inaccurately predicted by the most simplistic models. A simple alternative model presented in Figure 2.4C shows significantly more rapid decomposition that the standard model of . Better understanding of decomposition models is an important research area of ecosystem ecology because this process is closely tied to nutrient supply and the overall capacity of ecosystems to sequester CO_2 from the atmosphere. decomposed more slowly and may take many years to completely breakdown. Decomposition is typically described with exponential

Trophic Dynamics

Trophic dynamics refers to process of energy and nutrient transfer between organisms. Trophic dynamics is an important part of the structure and function of ecosystems. Figure 2.3 shows energy transferred for an ecosystem at Silver Springs, Florida. Energy gained by primary producers (plants, P) is consumed by herbivores (H), which are consumed by carnivores (C), which are themselves consumed by "top-carnivores"(TC). One of the most obvious patterns in Figure 2.3 is that as one moves up to higher trophic levels (i.e. from plants to top-carnivores) the total amount of energy decreases. Plants exert a "bottom-up" control on the energy structure of ecosystems by determining the total amount of energy that enters the system. However, predators can also influence the structure of lower trophic levels from the top-down. So called top-down effects can dramatically shift dominant species in terrestrial and marine systems The interplay and relative strength of top-down vs. bottom-up controls on ecosystem structure and function is an important area of research in the greater field of ecology. Trophic dynamics can strongly influence rates of decomposition and nutrient cycling in time and in space. For example, herbivory can increase litter decomposition and nutrient cycling via direct changes in litter quality and altered dominant vegetation. Insect herbivory has been shown to increase rates of decomposition and nutrient turnover due to changes in litter quality and increased frass inputs However, insect outbreak does not always increase nutrient cycling.

Stadler showed that C rich honeydew produced during aphid outbreak can result in increased N immobilization by soil microbes thus slowing down nutrient cycling and potentially limiting biomass production. North atlantic marine ecosystems have been greatly altered by overfishing of cod. Cod stocks crashed in the 1990's which resulted in increases in their prey such as shrimp and snow crab Human intervention in ecosystems has resulted in dramatic changes to ecosystem structure and function. These changes are occurring rapidly and have unknown consequences for economic security and human well-being.

Applications: Why does this Science Matter?

Lessons from two Central American Cities

The biosphere has been greatly altered by the demands of human societies. Ecosystem ecology plays an important role in understanding and adapting to the most pressing current environmental problems. Restoration ecology and ecosystem management are closely associated with ecosystem ecology. Restoring highly degraded resources depends on integration of functional mechanisms of ecosystems.

Without these functions intact, economic value of ecosystems is greatly reduced and potentially dangerous conditions may develop in the field. For example, areas within the mountainous western highlands of Guatemala are more susceptible to catastrophic landslides and crippling seasonal water shortages due to loss of forest resources. In contrast, cities such as Totonicapán that have preserved forests through strong social institutions have greater local economic stability and overall greater human well-being.

This situation is striking considering that these areas are close to each other, the majority of inhabitants are of Mayan descent, and the topography and overall resources are similar. This is a case of two groups of people managing resources in fundamentally different ways. Ecosystem ecology provides the basic science needed to avoid degradation and to restore ecosystem processes that provide for basic human needs.

2.4 Ecosystem Valuation

Valuation can be a useful tool that aids in evaluating different options that a natural resource manager might face. Because ecological resources and services are so varied in their composition, it is often difficult to examine them on the same level. However, after they are assigned a value, an environmental resource or service can then be compared to any other item with a respective value. Ecosystem valuation is the process by which policymakers assign a value—monetary or otherwise—to environmental resources or to the outputs and/or services provided by those resources. For example, a mountain forest may provide environmental services by preventing downstream flooding. Environmental resources and/or services are particularly hard to quantify due to their intangible benefits and multiple value options. It is almost impossible to attach a specific value to some of the experiences we have in nature, such as viewing a beautiful sunset. Problems also exist when a resource can be used for multiple purposes, such as a tree—the wood is valued differently if it is used for flood control versus if it is used for building a house. The quantity of a resource must also be taken into consideration because value can change depending how much of a resource is available. An example of this might be in preventing the first "unit" of pollution if we have a pristine air environment. Preventing the first unit of pollution is not valued very highly because the environment can easily recover. However, if the pollution continues until the air is becoming toxic to its surroundings; the value of preserving clean air by preventing additional pollution is going to be increasingly valued. Within economics, value is generally defined as the amount of alternate goods a person is willing to give up in order to get one "additional unit" of the good in question. An individual's preference for certain goods may either be stated or revealed. In the case of stated preferences, the amount of money a person is willing to pay for a good determines the value because that money could otherwise be used to purchase other goods. However, value may also be determined by simply ranking the alternatives according to the amount of benefit each will produce. Revealed preferences can be measured by examining a person's behaviour when it is not possible to use market pricing. There are typically two ways to assign value to environmental resources and services—use and non-use—and there are approaches to measuring environmental benefits based on these defined values. When environmental resources or services are being used, it is easier to observe the price consumers are willing to pay for the conservation or preservation of those resources. Market or opportunity cost pricing can be used when there are tangible products to measure, such as the amount of fish caught in a lake. Replacement cost can also be used, calculated based on any expenses incurred to reverse environmental damage. Hedonic pricing will measure the effect that negative environmental qualities have on the price of related market goods. When evaluating non-use value, contingent valuation is employed through the use of surveys that attempt to assess an individual's willingness to pay for a resource that they do not consume. The process of environmental resource or service valuation provides a way to compare alternative proposals, but it is not without problems. All valuation techniques encompass a great deal of uncertainty: flaws can exist in the methods of assigning value accurately due to a wide number of variables and it is difficult to compartmentalize and measure environmental and natural resources and/or services withing an ecosystem that functions as an interconnected web. Yet challenging, it allows policymakers to make decisions based on specific comparisons (typically monetary) rather than some other arbitrary basis. The government has placed increasing emphasis on cost-effective laws and projects; therefore, establishing a common measure by which to evaluate alternatives is essential.

2.5 Ecosystem Services

Ecosystem services are processes by which the natural environment produces resources useful to people, as to economic services. They include:

- ❖ Provision of clean water and air
- ❖ Flood control
- ❖ Pollination of crops
- ❖ Mitigation of environmental hazards
- ❖ Pest and disease control
- ❖ Carbon sequestration
- ❖ Aesthetic, cultural and ethical values associated with biodiversity.

The concept of ecosystem services is similar to that of natural capital. The Millennium Ecosystem Assessment released in 2005 showed that 60 per cent of ecosystem services are used in a way that destroys them.

3

Environmental Resources: Air, Water and Soil

3.1 Natural resource

Natural resources are naturally occurring substances that are considered valuable in their relatively unmodified (natural) form. A natural resource's value rests in the amount of the material available and the demand for it. There are 2 types of natural resources—renewable and non-renewable. Natural resource are natural capital converted to commodity inputs to infrastructural capital processes. They include soil, timber, oil, minerals, and other goods taken more or less from the Earth. Both extraction of the basic resource and refining it into a purer, directly usable form, (e.g., metals, refined oils) are generally considered natural-resource activities, even though the latter may not necessarily occur near the former. A nation's natural resources often determine its wealth in the world economic system, by determining its political influence. Developed nations are those which are less dependent on natural resources for wealth, due to their greater reliance on infrastructural capital for production. However, some see a resource curse whereby easily obtainable natural resources could actually hurt the prospects of a national economy by fostering political corruption. Political corruption can negatively impact the national economy because time is spent giving bribes or other economically unproductive acts instead of the generation of generative economic activity. There also tends to be concentrations of ownership over specific plots of land that have proven to yield natural resources. In recent years, the depletion of natural capital and attempts to move to sustainable development have been a major focus of development agencies. This is of particular concern in rainforest regions, which hold most of the Earth's natural biodiversity—irreplaceable genetic natural capital. Conservation of natural resources is the major focus of natural capitalism, environmentalism, the ecology movement, and Green Parties. Some view this depletion as a major source of social unrest and conflicts in developing nations.

3.1 Earth's atmosphere

The Earth's atmosphere is a layer of gases surrounding the planet Earth and retained by the Earth's gravity. It contains roughly (by molar content/volume) 78.08 per cent nitrogen, 20.95 per cent oxygen, 0.93 per cent argon, 0.038 per cent carbon dioxide, trace amounts of other gases, and a variable amount (average around 1%) of water vapor. This mixture of gases is commonly known as air. The atmosphere protects life on Earth by absorbing ultraviolet solar radiation and reducing temperature extremes between day and night. There is no definite boundary between the atmosphere and outer space. It slowly becomes thinner and fades into space. Three quarters of the atmosphere's mass is within 11 km of the planetary surface. In the United States, people who travel above an altitude of 80.5 km (50 statute miles) are designated astronauts. An altitude of 120 km (~75 miles or 400,000 ft) marks the boundary where atmospheric effects become noticeable during re-entry. The Kármán line, at 100 km (62 miles or 328,000 ft), is also frequently regarded as the boundary between atmosphere and outer space.

Temperature and Layers

The temperature of the Earth's atmosphere varies with altitude; the mathematical relationship between temperature and altitude varies among five different atmospheric layers (ordered highest to lowest, the ionosphere is part of the thermosphere):

❖ *Exosphere*: from 500-1000 km (300-600 mi) up to 10,000 km (6,000 mi), free-moving particles that may migrate into and out of the magnetosphere or the solar wind.

Exobase Boundary

❖ *Ionosphere:* is the part of the atmosphere that is ionized by solar radiation. It plays an important part in atmospheric electricity and forms the inner edge of the magnetosphere. It has practical importance because, among other functions, it influences radio propagation to distant places on the Earth. It is located in the thermosphere and is responsible for auroras.

Thermopause Boundary

❖ *Thermosphere:* from 80-85 km (265,000-285,000 ft) to 640+ km (400+ mi), temperature increasing with height.

Mesopause Boundary

❖ **Mesosphere:** From the Greek word "ìÝóïò" meaning middle. The mesosphere extends from about 50 km (160,000 ft) to the range of 80 to 85 km (265,000-285,000 ft), temperature decreasing with height. This is also where most meteors burn up when entering the atmosphere.

Stratopause Boundary

❖ *Stratosphere:* From the Latin word "stratus" meaning a spreading out. The stratosphere extends from the troposphere's 7 to 17 km (23,000-60,000 ft) range to about 50 km (160,000 ft). Temperature increases with height. The stratosphere contains the ozone layer, the part of the Earth's atmosphere which contains relatively high concentrations of ozone. "Relatively high" means a few parts per million—much higher than the concentrations in the lower atmosphere but still small compared to the main components of the atmosphere. It is mainly located in the lower portion of the stratosphere from approximately 15 to 35 km (50,000-115,000 ft) above Earth's surface, though the thickness varies seasonally and geographically.

Tropopause Boundary

❖ *Troposphere:* From the Greek word "ôñÝðù" meaning to turn or change. The troposphere is the lowest layer of the atmosphere; it begins at the surface and extends to between 7 km (23,000 ft) at the poles and 17 km (60,000 ft) at the equator, with some variation due to weather factors. The troposphere has a great deal of vertical mixing due to solar heating at the surface. This heating warms air masses, which makes them less dense so they rise. When an air mass rises the pressure upon it decreases so it expands, doing work against the opposing pressure of the surrounding air. To do work is to expend energy, so the temperature of the air mass decreases. As the temperature decreases, water vapor in the air mass may condense or solidify, releasing latent heat that further uplifts the air mass. This process determines the maximum rate of decline of temperature with height, called the adiabatic lapse rate. It contains roughly 80 per cent of the total mass of the atmosphere. 50 per cent of the total mass of the atmosphere is located in the lower 5 km of the troposphere.

The average temperature of the atmosphere at the surface of Earth is 15 °C (59 °F).

Pressure and Thickness

The average atmospheric pressure, at sea level, is about 101.3 kilopascals (about 14.7 psi); total atmospheric mass is 5.1480 × 10 kg . Atmospheric pressure is a direct result of the total weight of the air above the point at which the pressure is measured. This means that air pressure varies with location and time, because the amount (and weight) of air above the earth varies with location and time. However the *average* mass of the air above a square meter of the earth's surface is known to the same high accuracy as the total air mass of 5148.0 teratonnes and area of the earth of 51007.2 megahectares, namely 5148.0/510.072 = 10.093 metric tonnes or 14.356 lbs (mass) per square inch. This is about 2.5 per cent below the officially standardized unit atmosphere (1 atm) of 101.325 kPa or 14.696 psi, and corresponds to the mean pressure not at sea level but at the mean base of the atmosphere as contoured by the earth's terrain.

Atmospheric pressure decreases with height, dropping by 50 per cent at an altitude of about 5.6 km (18,000 ft). For comparison: the highest mountain, Mount Everest, is higher, at 8.8 km, which is why it is so difficult to climb without supplemental oxygen. Equivalently, about 50 per cent of the total atmospheric mass is within the lowest 5.6 km. This pressure drop is approximately exponential, so that pressure decreases by approximately half every 5.6 km. However, because of changes in temperature throughout the atmospheric column, as well as the fact that the force of gravity begins to decrease at great altitudes, a single equation does not model atmospheric pressure through all altitudes (it is modeled in 7 exponentially decreasing layers, in the equations given above). Even in the exosphere, the atmosphere is still present (as can be seen for example by the effects of atmospheric drag on satellites). The equations of pressure by altitude in the above references can be used directly to estimate atmospheric thickness. However, the following published data are given for reference:

❖ 50 per cent of the atmosphere by mass is below an altitude of 5.6 km.

- ❖ 90 per cent of the atmosphere by mass is below an altitude of 16 km. The common altitude of commercial airliners is about 10 km.
- ❖ 99.99997 per cent of the atmosphere by mass is below 100 km. The highest X-15 plane flight in 1963 reached an altitude of 354,300 ft (108,000 m).

Therefore, most of the atmosphere (99.9997%) is below 100 km, although in the rarefied region above this there are auroras and other atmospheric effects.

Composition

Composition of Dry Atmosphere, by Volume
ppmv: parts per million by volume

Gas	Volume
Nitrogen (N_2)	780,840 ppmv (78.084%)
Oxygen (O_2)	209,460 ppmv (20.946%)
Argon (Ar)	9,340 ppmv (0.9340%)
Carbon dioxide (CO_2)	383 ppmv (0.0383%)
Neon (Ne)	18.18 ppmv (0.001818%)
Helium (He)	5.24 ppmv (0.000524%)
Methane (CH_4)	1.745 ppmv (0.0001745%)
Krypton (Kr)	1.14 ppmv (0.000114%)
Hydrogen (H_2)	0.55 ppmv (0.000055%)
Not included in above dry atmosphere:	
Water vapor (H_2O)	-0.25% over full atmosphere, typically 1% to 4% near surface

Minor components of air not listed above include

Gas	Volume
nitrous oxide	0.3 ppmv (0.00005%)
xenon	0.09 ppmv (9x10 %)
ozone	0.0 to 0.07 ppmv (0%-7x10 %)
nitrogen dioxide	0.02 ppmv (2x10 %)
iodine	0.01 ppmv (1x10 %)
carbon monoxide	trace
ammonia	trace

The mean molar mass of air is 28.97 g/mol. Note that the composition figures above are by volume-fraction (V%), which for ideal gases is equal to mole-fraction (that is, fraction of total molecules). By contrast, *mass-fraction* abundances of gases, particularly for gases with significantly different molecular (molar) mass from that of air will differ from those by volume. For example, in air, helium is 5.2 ppm by *volume-fraction* and *mole-fraction*, but only about (4/29) × 5.2 ppm = 0.72 ppm by *mass-fraction*.

Heterosphere

Below the turbopause at an altitude of about 100 km (not far from the mesopause), the Earth's atmosphere has a more-or-less uniform composition (apart from water vapor) as described above; this constitutes the homosphere. However, above about 100 km, the Earth's atmosphere begins to have a composition which varies with altitude. This is essentially because, in the absence of mixing, the density of a gas falls off exponentially with increasing altitude, but at a rate which depends on the molar mass. Thus higher mass constituents, such as oxygen and nitrogen, fall off more quickly than lighter constituents such as helium, molecular hydrogen, and atomic hydrogen. Thus there is a layer, called the heterosphere, in which the earth's atmosphere has varying composition. As the altitude increases, the atmosphere is dominated successively by helium, molecular hydrogen, and atomic hydrogen. The precise altitude of the heterosphere and the layers it contains varies significantly with temperature. After loss of the hydrogen, helium and other hydrogen-containing gases from early Earth due to the Sun's radiation, primitive Earth was devoid of an atmosphere. The first atmosphere was formed by outgassing of gases trapped in the interior of the early Earth, which still goes on today in volcanoes.

Density and Mass

The density of air at sea level is about 1.2 kg/m^3(1.2 g/L). Natural variations of the barometric pressure occur at any one altitude as a consequence of weather. This variation is relatively small for inhabited altitudes but much more pronounced in the outer atmosphere and space due to variable solar radiation. The atmospheric density decreases as the altitude increases. This variation can be approximately modeled using the barometric formula. More sophisticated models are used by meteorologists and space agencies to predict weather and orbital decay of satellites. The average mass of the atmosphere is about 5 quadrillion metric tons or 1/1,200,000 the mass of Earth. According to the National Center for Atmospheric Research, "The total mean mass of the atmosphere is 5.1480×10^{18} kg with an annual range due to water vapor of 1.2 or 1.5×10^{15} kg depending on whether surface pressure or water vapor data are used; somewhat smaller than the previous estimate. The mean mass of water vapor is estimated as 1.27×10^{16} kg and the dry air mass as $5.1352 \pm 0.0003 \times 10^{18}$ kg."

Evolution on Earth

The history of the Earth's atmosphere prior to one billion years ago is poorly understood and an active area of scientific research. The following discussion presents a plausible scenario. The modern atmosphere is sometimes referred to as Earth's "third atmosphere", in order to distinguish the current chemical composition from two notably different previous compositions. The original atmosphere was primarily helium and hydrogen. Heat from the still-molten crust, and the sun, plus a probably enhanced solar wind, dissipated this atmosphere.

About 4.4 billion years ago, the surface had cooled enough to form a crust, still heavily populated with volcanoes which released steam, carbon dioxide, and ammonia. This led to the early "second atmosphere", which was primarily carbon dioxide and water vapor, with some nitrogen but virtually no oxygen. This second atmosphere had approximately 100 times as much gas as the current atmosphere, but as it cooled much of the carbon dioxide was dissolved in the seas and precipitated out as carbonates. The later "second atmosphere" contained largely nitrogen and carbon dioxide. However, simulations run at the University of Waterloo and University of Colorado in 2005 suggest that it may have had up to 40 per cent hydrogen. It is generally believed that the greenhouse effect, caused by high levels of carbon dioxide and methane, kept the Earth from freezing.

One of the earliest types of bacteria was the cyanobacteria. Fossil evidence indicates that bacteria shaped like these existed approximately 3.3 billion years ago and were the first oxygen-producing evolving phototropic organisms. They were responsible for the initial conversion of the earth's atmosphere from an anoxic state to an oxic state (that is, from a state without oxygen to a state with oxygen) during the period 2.7 to 2.2 billion years ago. Being the first to carry out oxygenic photosynthesis, they were able to produce oxygen while sequestering carbon dioxide in organic molecules, playing a major role in oxygenating the atmosphere.

Photosynthesising plants would later evolve and continue releasing oxygen and sequestering carbon dioxide. Over time, excess carbon became locked in fossil fuels, sedimentary rocks (notably limestone), and animal shells. As oxygen was released, it reacted with ammonia to release nitrogen; in addition, bacteria would also convert ammonia into nitrogen. But most of the nitrogen currently present in the atmosphere results from sunlight-powered photolysis of ammonia released steadily over the aeons from volcanoes.

As more plants appeared, the levels of oxygen increased significantly, while carbon dioxide levels dropped. At first the oxygen combined with various elements (such as iron), but eventually oxygen accumulated in the atmosphere, resulting in mass extinctions and further evolution. With the appearance of an ozone layer (ozone is an allotrope of oxygen) lifeforms were better protected from ultraviolet radiation. This oxygen-nitrogen atmosphere is the "third atmosphere". 200-250 million years ago, up to 35 per cent of the atmosphere was oxygen (as found in bubbles of ancient atmosphere were found in an amber).

This modern atmosphere has a composition which is enforced by oceanic blue-green algae as well as geological processes. O_2 does not remain naturally free in an atmosphere, but tends to be consumed (by inorganic chemical reactions, and by animals, bacteria, and even land plants at night), and CO_2 tends to be produced by respiration and decomposition and oxidation of organic matter. Oxygen would vanish within a few million years due to chemical reactions and CO_2 dissolves easily in water and would be gone in millennia if not replaced. Both are maintained by biological productivity and geological forces seemingly working hand-in-hand to maintain reasonably steady levels over millions of years (see *Gaia theory*).

Air Pollution

Air pollution is a chemical, physical (e.g. particulate matter) or biological agent that modifies the natural characteristics of the atmosphere in an unwanted way. Stratospheric ozone depletion due to air pollution (chiefly from chlorofluorocarbons) has long been recognized as a threat to human health as well as to the earth's ecosystems. Worldwide air pollution is responsible for large numbers of deaths and cases of respiratory disease. Enforced air quality standards, like the Clean Air Act in the United States, have reduced the presence of some pollutants. While major stationary sources are often identified with air pollution, the greatest source of emissions is actually mobile sources, principally the automobile. Gases such as carbon dioxide, methane, and fluorocarbons contribute to global warming, and these gases, or excess amounts of some emitted from fossil fuel burning, have recently been identified by the United States and many other countries (see Kyoto accord), as pollutants.

3.3 Air

Air means Earth's atmosphere. It is the clear gas we live in and breathe in. It has no colour or smell. It has weight. Air creates atmosphere pressure. There is no air in the vacuum and cosmos. Air is a mixture of 78.03 per cent Nitrogen, 20.99 per cent Oxygen, 0.94 per cent Argon, 0.03 per cent Carbon Dioxide, 0.01 per cent Hydrogen, 0.00123 per cent Neon, 0.0004 per cent Helium, 0.00005 per cent Krypton, 0.000006 per cent Xenon. Humans need the oxygen in the air to live. In our bodies, the lungs give oxygen to the blood, and give back carbon dioxide to the air. Plants and animals also need air to live. Air is very important to everyone. The wind is moving air. Air can be polluted from gas, smoke, and ash. This pollution may be one of the causes of global warming.

Air is also something that planes fly through. There are three things in air, Nitrogen (79%), oxygen (20%), and other gasses (1%) warm air goes up while cold air goes down.

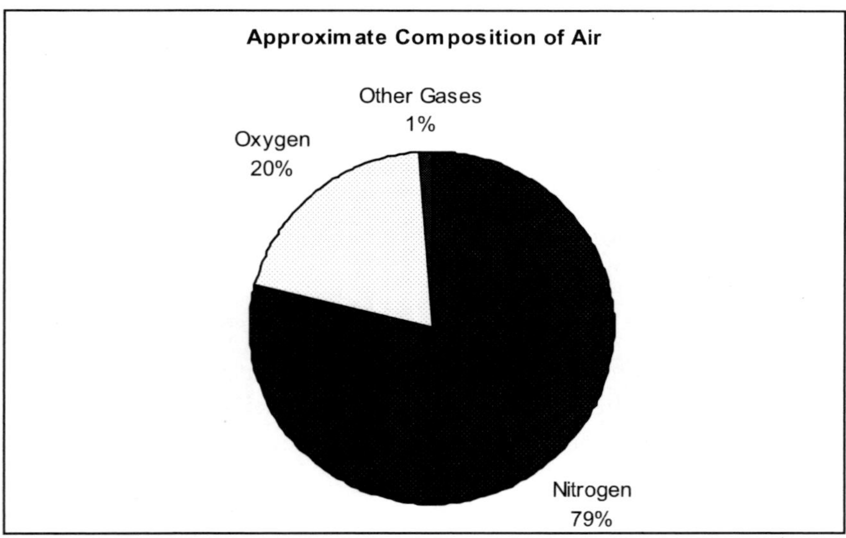

NB: "Other gases" includes carbon dioxide (0.3% and small proportion of other gases inclue argon and water vapour.

3.4 Water Resources

Definition and Terminology

Water resources are sources of water that are useful or potentially useful to humans. Uses of water includes agricultural, industrial, household, recreational and environmental activities. Virtually all of these human uses require fresh water. 97.5 per cent of water on the Earth is salt water, leaving only 2.5 per cent as fresh water of which over two thirds is frozen in glaciers and polar ice caps. The remaining unfrozen freshwater is mainly found as groundwater, with only a small fraction present above ground or in the air. Fresh water is a renewable resource, yet the world's supply of clean, fresh water is steadily decreasing. Water demand already exceeds supply in many parts of the world, and as world population continues to rise at an unprecedented rate, many more areas are expected to experience this imbalance in the near future. The framework for allocating water resources to water users (where such a framework exists) is known as water rights..

Water and Conflict

The only known example of an actual inter-state conflict over water took place between 2500 and 2350 BC between the Sumerian states of Lagash and Umma. Yet, despite the lack of evidence of international wars being fought over water alone, water has been the source of various conflicts throughout history. When water scarcity causes political tensions to arise, this is referred to as water stress. Water stress has led most often to conflicts at local and regional levels. Using a purely quantitative methodology, Thomas Homer-Dixon successfully correlated water scarcity and scarcity of available arable lands to an increased chance of violent conflict. Water stress can also exacerbate conflicts and political tensions which are not directly caused by water. Gradual reductions over time in the quality and/or quantity of fresh water can add to the instability of a region by depleting the health of a population, obstructing economic development, and exacerbating larger conflicts.

Conflicts and tensions over water are most likely to arise within national borders, in the downstream areas of distressed river basins. Areas such as the lower regions of China's Yellow River or the Chao Phraya River in Thailand, for example, have already been experiencing water stress for several years. Additionally, certain arid countries which rely heavily on water for irrigation, such as China, India, Iran, and Pakistan, are particularly at risk of water-related conflicts.. Political tensions, civil protest, and violence may also occur in reaction to water privatization. The Bolivian Water Wars of 2000 are a case in point.

Sources of Fresh Water

Surface Water

Surface water is water in a river, lake or fresh water wetland. Surface water is naturally replenished by precipitation and naturally lost through discharge to the oceans, evaporation, and sub-surface seepage. Although the only natural input to any surface water system is precipitation within its watershed, the total quantity of water in that system at any given time is also dependent on many other factors. These factors include storage capacity in lakes, wetlands and artificial reservoirs, the permeability of the soil beneath these storage bodies, the runoff characteristics of the land in the watershed, the timing of the precipitation and local evaporation rates. All of these factors also affect the proportions of water lost. Human activities can have a large impact on these factors. Humans often increase storage capacity by constructing reservoirs and decrease it by draining wetlands. Humans often increase runoff quantities and velocities by paving areas and channelizing stream flow. The total quantity of water available at any given time is an important consideration. Some human water users have an intermittent need for water. For example, many farms require large quantities of water in the spring, and no water at all in the winter. To supply such a farm with water, a surface water system may require a large storage capacity to collect water throughout the year and release it in a short period of time. Other users have a continuous need for water, such as a power plant that requires water for cooling. To supply such a power plant with water, a surface water system only needs enough storage capacity to fill in when average stream flow is below the power plant's need. Nevertheless, over the long term the average rate of precipitation within a watershed is the upper bound for average consumption of natural surface water from that watershed. Natural surface water can be augmented by importing surface water from another watershed through a canal or pipeline. It can also be artificially augmented from any of the other sources listed here, however in practice the quantities are negligible. Humans can also cause surface water to be "lost" (i.e. become unusable) through pollution. Canada is the country estimated to have the largest supply of fresh water in the world, followed by Brazil and Russia.

Sub-surface Water

Sub-Surface water, or groundwater, is fresh water located in the pore space of soil and rocks. It is also water that is flowing within aquifers below the water table. Sometimes it is useful to make a distinction between sub-surface water that is closely associated with surface water and deep sub-surface water in an aquifer (sometimes called "fossil water").

Sub-surface water can be thought of in the same terms as surface water: inputs, outputs and storage. The critical difference is that due to its slow rate of turnover, sub-surface water storage is generally much larger compared to inputs than it is for surface water. This difference makes it easy for humans to use sub-surface water unsustainably for a long time without severe consequences. Nevertheless, over the long term the average rate of seepage above a sub-surface water source is the upper bound for average consumption of water from that source.

The natural input to sub-surface water is seepage from surface water. The natural outputs from sub-surface water are springs and seepage to the oceans.

If the surface water source is also subject to substantial evaporation, a sub-surface water source may become saline. This situation can occur naturally under endorheic bodies of water, or artificially under irrigated farmland. In coastal areas, human use of a sub-surface water source may cause the direction of seepage to ocean to reverse which can also cause soil salinization. Humans can also cause sub-surface water to be "lost" (*i.e.* become unusable) through pollution. Humans can increase the input to a sub-surface water source by building reservoirs or detention ponds.

Water in the ground are in sections called aquifers. Rain rolls down and comes into these. Normally an aquifer is near to the equilibrium in its water content. The water content of an aquifier normally depends on the grain sizes. This means that the rate of extraction may be limited by poor permiability.

Desalination

Desalination is an artificial process by which saline water (generally sea water) is converted to fresh water. The most common desalination processes are distillation and reverse osmosis. Desalination is currently expensive compared to most alternative sources of water, and only a very small fraction of total human use is satisfied by desalination. It is only economically practical for high-valued uses (such as household and industrial uses) in arid areas. The most extensive use is in the Persian Gulf.

Frozen Water

Several schemes have been proposed to make use of icebergs as a water source, however to date this has only been done for novelty purposes. Glacier runoff is considered to be surface water.

Water Stress

The concept of water stress is relatively simple: According to the World Business Council for Sustainable Development, it applies to situations where there is not enough water for all uses, whether agricultural, industrial or domestic. Defining thresholds for stress in terms of available water per capita is more complex, however, entailing assumptions about water use and its efficiency. Nevertheless, it has been proposed that when annual per capita renewable freshwater availability is less than 1,700 cubic meters, countries begin to experience periodic or regular water stress. Below 1,000 cubic meters, water scarcity begins to hamper economic development and human health and well-being.

Population Growth

In 2000, the world population was 6.2 billion. The UN estimates that by 2050 there will be an additional 3 billion people with most of the growth in developing countries that already suffer water stress. Thus, water demand will increase unless there are corresponding increases in water conservation and recycling of this vital resource.

Increased Affluence

The rate of poverty alleviation is increasing especially within the two population giants of China and India. However, increasing affluence inevitably means more water consumption: from needing clean fresh water 24 hours a day, 7 days a week and basic sanitation service, to demanding water for gardens and car washing, to wanting jacuzzis or private swimming pools.

Expansion of Business Activity

Business activity ranging from industrialization to services such as tourism and entertainment continues to expand rapidly. This expansion requires increased water services including both supply and sanitation, which can lead to more pressure on water resources and natural ecosystems.

Rapid Urbanization

The trend towards urbanization is accelerating. Small private and septic tanks that work well in low-density communities are not feasible within high-density urban areas. Urbanization requires significant investment in water infrastructure in order to deliver water to individuals and to process the concentrations of wastewater—both from individuals and from business. These polluted and contaminated waters must be treated or they pose unacceptable public health risks. In 60 per cent of European cities with more than 100,000 people, groundwater is being used at a faster rate than it can be replenished. Even if some water remains available, it costs more and more to capture it.

Climate Change

Climate change could have significant impacts on water resources around the world because of the close connections between the climate and hydrologic cycle. Rising temperatures will increase evaporation and lead to increases in precipitation, though there will be regional variations in rainfall. Overall, the global supply of freshwater will increase. Both droughts and floods may become more frequent in different regions at different times, and dramatic changes in snowfall and snowmelt are expected in mountainous areas. Higher temperatures will also affect water quality in ways that are not well understood. Possible impacts include increased eutrophication. Climate change could also mean an increase in demand for farm irrigation, garden sprinklers, and perhaps even swimming pools.

Depletion of Aquifers

Due to the expanding human population, competition for water is growing such that many of the worlds major aquifers are becoming depleted. This is due both for direct human consumption as well as agricultural irrigation by groundwater. Millions of small pumps of all sizes are currently extracting groundwater throughout the world. Irrigation in dry areas such as northern China and India is supplied by groundwater, and is being extracted at an unsustainable rate. Cities that have experienced aquifer drops between 10 to 50 meters include Mexico City, Bangkok, Manila, Beijing, Madras and Shanghai.

Pollution and Water Protection

Water pollution is one of the main concerns of the world today. The governments of many countries have striven to find solutions to reduce this problem. Many pollutants threaten water supplies, but the most widespread, especially in underdeveloped countries, is the discharge of raw sewage into natural waters; this method of sewage disposal is the most common method in underdeveloped countries, but also is prevalent in quasi-developed countries such as China, India and Iran.

Sewage, sludge, garbage, and even toxic pollutants are all dumped into the water. Even if sewage is treated, problems still arise. Treated sewage forms sludge, which is sent out into the sea and dumped. In addition to sewage, chemicals dumped by industries and governments are another major source of water pollution.

Uses of Fresh Water

Uses of fresh water can be categorized as consumptive and non-consumptive (sometimes called "renewable"). A use of water is consumptive if that water is not immediately available for another use. Losses to sub-surface seepage and evaporation are considered consumptive, as is water incorporated into a product (such as farm produce). Water that can be treated and returned as surface water, such as sewage, is generally considered non-consumptive if that water can be put to additional use.

Agricultural

It is estimated that 69 per cent of world-wide water use is for irrigation, with 15-35 per cent of irrigation withdrawals being unsustainable . In some areas of the world irrigation is necessary to grow any crop at all, in other areas it permits more profitable crops to be grown or enhances crop yield. Various irrigation methods involve different trade-offs between crop yield, water consumption and capital cost of equipment and structures. Irrigation methods such as most furrow and overhead sprinkler irrigation are usually less expensive but also less efficient, because much of the water evaporates or runs off. More efficient irrigation methods include drip or trickle irrigation, surge irrigation, and some types of sprinkler systems where the sprinklers are operated near ground level. These types of systems, while more expensive, can minimize runoff and evaporation. Any system that is improperly managed can be wasteful. Another trade-off that is often insufficiently considered is salinization of sub-surface water.

Aquaculture is a small but growing agricultural use of water. Freshwater commercial fisheries may also be considered as agricultural uses of water, but have generally been assigned a lower priority than irrigation (see Aral Sea and Pyramid Lake). As global populations grow, and as demand for food increases in a world with a fixed water supply, there are efforts underway to learn how to produce more food with less water, through improvements in irrigation methods and technologies, agricultural water management, crop types, and water monitoring.

Industrial

It is estimated that 15 per cent of world-wide water use is industrial. Major industrial users include power plants, which use water for cooling or as a power source (i.e. hydroelectric plants), ore and oil refineries, which use water in chemical processes, and manufacturing plants, which use water as a solvent. The portion of industrial water usage that is consumptive varies widely, but as a whole is lower than agricultural use.

Household

It is estimated that 15 per cent of world-wide water use is for household purposes. These include drinking water, bathing, cooking, sanitation, and gardening. Basic household water requirements have been estimated by Peter Gleick at around 50 liters per person per day, excluding water for gardens.

Recreation

Recreational water use is usually a very small but growing percentage of total water use. Recreational water use is mostly tied to reservoirs. If a reservoir is kept fuller than it would otherwise be for recreation, then the water retained could be categorized as recreational usage. Release of water from a few reservoirs is also timed to enhance whitewater boating, which also could be considered a recreational usage. Other examples are anglers, water skiers, nature enthusiasts and swimmers. Recreational usage is usually non-consumptive. Golf courses are often targeted as using excessive amounts of water, especially in drier regions. It is, however, unclear whether recreational irrigation (which would include private gardens) has a noticeable effect on water resources. This is largely due to the unavailability of reliable data. Some governments, including the Californian Government, have labelled golf course usage as agricultural in order to deflect environmentalists' charges of wasting water. However, using the above figures as a basis, the actual statistical effect of this reassignment is close to zero.

Additionally, recreational usage may reduce the availability of water for other users at specific times and places. For example, water retained in a reservoir to allow boating in the late summer is not available to farmers during the spring planting season. Water released for whitewater rafting may not be available for hydroelectric generation during the time of peak electrical demand.

Environmental

Explicit environmental water use is also a very small but growing percentage of total water use. Environmental water usage includes artificial wetlands, artificial lakes intended to create wildlife habitat, fish ladders around dams, and water releases from reservoirs timed to help fish spawn. Like recreational usage, environmental usage is non-consumptive but may reduce the availability of water for other users at specific times and places. For example, water release from a reservoir to help fish spawn may not be available to farms upstream.

World Water Supply and Distribution

Food and water are two basic human needs. However, global coverage figures from 2002 indicate that, of every 10 people:

- ❖ Roughly 5 have a connection to a piped water supply at home (in their dwelling, plot or yard);
- ❖ 3 make use of some other sort of improved water supply, such as a protected well or public standpipe;
- ❖ 2 are unserved; and
- ❖ In addition, 4 out of every 10 people live without improved sanitation.

At Earth Summit 2002 governments approved a Plan of Action to:

- ❖ Halve by 2015 the proportion of people unable to reach or afford safe drinking water. The Global Water Supply and Sanitation Assessment 2000 Report (GWSSAR) defines "Reasonable access" to water as at least 20 liters per person per day from a source within one kilometer of the user's home.
- ❖ Halve the proportion of people without access to basic sanitation. The GWSSR defines "Basic sanitation" as private or shared but not public disposal systems that separate waste from human contact.

As the picture shows, in 2025, water shortages will be more prevalent among poorer countries where resources are limited and population growth is rapid, such as the Middle East, Africa, and parts of Asia. By 2025, large urban and peri-urban areas will require new infrastructure to provide safe water and adequate sanitation. This suggests growing conflicts with agricultural water users, who currently consume the majority of the water used by humans.

Generally speaking the more developed countries of North America, Europe and Russia will not see a serious threat to water supply by the year 2025, not only because of their relative wealth, but more importantly their populations will be better aligned with available water resources. North Africa, the Middle East, South Africa and northern China will face very severe water shortages due to physical scarcity and a condition of overpopulation relative to their carrying capacity with respect to water supply. Most of South America, Sub-Saharan Africa, Southern China and India will face water supply shortages by 2025; for these latter regions the causes of scarcity will be economic constraints to developing safe drinking water, as well as excessive population growth.

Economic Considerations

Water supply and sanitation require a huge amount of capital investment in infrastructure such as pipe networks, pumping stations and water treatment works. It is estimated that OECD nations need to invest at least USD 200 billion per year to replace aging water infrastructure to guarantee supply, reduce leakage rates and protect water quality.

International attention has focused upon the needs of the developing countries. To meet the Millennium Development Goals targets of halving the proportion of the population lacking access to safe drinking water and basic sanitation by 2015, current annual investment on the order of USD 10 to USD 15 billion would need to be roughly doubled. This does not include investments required for the maintenance of existing infrastructure.

Once infrastructure is in place, operating water supply and sanitation systems entails significant ongoing costs to cover personnel, energy, chemicals, maintenance and other expenses. The sources of money to meet these capital and operational costs are essentially either user fees, public funds or some combination of the two.

But this is where the economics of water management start to become extremely complex as they intersect with social and broader economic policy. Such policy questions are beyond the scope of this article, which has concentrated on basic information about water availability and water use. They are, nevertheless, highly relevant to understanding how critical water issues will affect business and industry in terms of both risks and opportunities.

Business Response

The World Business Council for Sustainable Development in its H2OScenarios engaged in a scenario building process to:

- ❖ Clarify and enhance understanding by business of the key issues and drivers of change related to water.
- ❖ Promote mutual understanding between the business community and non-business stakeholders on water management issues.
- ❖ Support effective business action as part of the solution to sustainable water management.

It concluded that:

- ❖ Business cannot survive in a society that thirsts.
- ❖ You don't have to be in the water business to have a water crisis.
- ❖ Business is part of the solution, and its potential is driven by its engagement.
- ❖ Growing water issues and complexity will drive up costs.

3.5 Soil

Soil is the naturally occurring, unconsolidated or loose covering of broken rock particles and decaying organic matter (humus) on the surface of the Earth, capable of supporting life. In simple terms, soil has three components: solid, liquid, and gas. The solid phase is a mixture of mineral and organic matter. Soil particles pack loosely, forming a soil structure filled with voids. The solid phase occupies about half of the soil volume. The remaining void space contains water (liquid) and air (gas). Soil is also known as earth: it is the substance from which our planet takes its name.

Characteristics

Soil colour is the first impression one has when viewing soil. Striking colours and contrasting patterns are especially memorable. The Red River in Louisiana carries sediment eroded from extensive reddish soils like Port Silt Loam in Oklahoma. Soil colour results from chemical and biological weathering. As the primary minerals in parent material weather, the elements combine into new and colourful compounds. Iron forms secondary minerals with a yellow or red colour; organic matter decomposes into brown compounds; and manganese, sulfur and nitrogen can form black mineral deposits. Soil structure is the arrangement of soil particles into aggregates. These may have various shapes, sizes and degrees of development or expression. Soil texture refers to sand, silt and clay composition. Sand and silt are the product of physical weathering while clay is the product of chemical weathering. Clay content is particularly influential on soil behaviour due to a high retention capacity for nutrients and water.

Formation

Soil formation, or pedogenesis, is the combined effect of physical, chemical, biological, and anthropogenic processes on soil parent material resulting in the formation of soil horizons. Soil is always changing. The long periods over which change occurs and the multiple influences of change mean that simple soils are rare. While soil can achieve relative stability in properties for extended periods of time, the soil life cycle ultimately ends in soil conditions that leave it vulnerable to erosion. Little of the soil continuum of the earth is older than Tertiary and most no older than Pleistocene. Despite the inevitability of soils retrogression and degradation, most soil cycles are long and productive. How the soil "life" cycle proceeds is influenced by at least five classic soil forming factors: regional climate, biotic potential, topography, parent material, and the passage of time. An example of soil development from bare rock occurs on recent lava flows in warm regions under heavy and very frequent rainfall. In such climates plants become established very quickly on basaltic lava, even though there is very little organic material. The plants are supported by the porous rock becoming filled with nutrient bearing water, for example carrying dissolved bird droppings or guano. The developing plant roots themselves gradually breaks up the porous lava and organic matter soon accumulates but, even before it does, the predominantly porous broken lava in which the plant roots grow can be considered a soil.

In Nature

Biogeography is the study of spatial variations in biological communities. Soils are a restricting factor as to what plants can grow in which environments. Soil scientists survey soils in the hope of understanding controls as to what vegetation can and will grow in a particular location. Geologists also have a particular interest in the patterns of soil on the surface of the earth. Soil texture, colour and chemistry often reflect the underlying geologic parent material and soil types often change at geologic unit boundaries. Buried paleosols mark previous land surfaces and record climatic conditions from previous eras. Geologists use this paleopedological record to understand the ecological relationships in past ecosystems. According to the theory of biorhexistasy, prolonged conditions conducive to forming deep, weathered soils result in increasing ocean salinity and the formation of limestone. Geologists use soil profile features to establish the duration of surface stability in the context of geologic faults or slope stability. An offset subsoil horizon indicates rupture during soil formation and the degree of subsequent subsoil formation is relied upon to establish time since rupture. Soil examined in shovel test pits is used by archaeologists for relative dating based on stratigraphy (as opposed to absolute dating). What is considered most typical is to use soil profile features to determine the maximum reasonable pit depth than needs to be examined for archaeological evidence in the interest of cultural resources management. Soils altered or formed by man (anthropic and anthropogenic soils) are also of interest to archaeologists.

Uses

Soil material is a critical component in the mining and construction industries. Soil serves as a foundation for most construction projects. Massive volumes of soil can be involved in surface mining, road building, and dam construction. Earth sheltering is the architectural practice of using soil for external thermal mass against building walls. Soil resources are critical to the environment, as well as to food and fiber production. Soil provides minerals and water to plants. Soil absorbs rainwater and releases it later thus preventing floods and drought. Soil cleans the water as it percolates. Soil is the habitat for many organisms. Waste management often has a soil component. Septic drain fields treat septic tank effluent uses aerobic soil processes. Landfills use soil for daily cover. Organic soils, especially peat, serve as a significant fuel resource. Both humans in many cultures and animals occasionally eat soil.

Degradation

Land degradation is a human induced or natural process which impairs the capacity of land to function. Soils are the critical component in land degradation when it involves acidification, contamination, desertification, erosion, or salination. While soil acidification of alkaline soils is beneficial, it degrades land when soil acidity lowers crop productivity and increases soil vulnerability to contamination and erosion. Soils are often initially acid because their parent materials were acid and initially low in the basic cations (calcium, magnesium, potassium, and sodium). Acidification occurs when these elements are removed from the soil profile by normal rainfall or the harvesting of crops. Soil acidification is accelerated by the use of acid-forming nitrogenous fertilizers and by the effects of acid precipitation.

Soil contamination at low levels are often within soil capacity to treat and assimilate. Many waste treatment processes rely on this treatment capacity. Exceeding treatment capacity can damage soil biota and limit soil function. Derelict soils occur where industrial contamination or other development activity damages the soil to such a degree that the land cannot be used safely or productively. Remediation of derelict soil uses principles of geology, physics, chemistry, and biology to degrade, attenuate, isolate, or remove soil contaminants and to restore soil functions and values. Techniques include leaching, air sparging, chemical amendments, phytoremediation, bioremediation, and natural attenuation.

Desertification is an environmental process of ecosystem degradation in arid and semi-arid regions, or as a result of human activity. It is a common misconception that droughts cause desertification. Droughts are common in arid and semiarid lands. Well-managed lands can recover from drought when the rains return. Soil management tools include maintaining soil nutrient and organic matter levels, reduced tillage and increased cover. These help to control erosion and maintain productivity during periods when moisture is available. Continued land abuse during droughts, however, increases land degradation. Increased population and livestock pressure on marginal lands accelerates desertification.

Soil erosional loss is caused by wind, water, ice, movement in response to gravity. Although the processes may be simultaneous, erosion is distinguished from weathering. Erosion is an intrinsic natural process, but in many places it is increased by human land use. Poor land use practices include deforestation, overgrazing, and improper construction activity. Improved management can limit erosion using techniques like limiting disturbance during construction, avoiding construction during erosion prone periods, intercepting runoff, terrace-building, use of erosion suppressing cover materials and planting trees or other soil binding plants.

A serious and long-running water erosion problem is in China, on the middle reaches of the Yellow River and the upper reaches of the Yangtze River. From the Yellow River, over 1.6 billion tons of sediment flow each year into the ocean. The sediment originates primarily from water erosion in the Loess Plateau region of northwest China.

Soil piping is a particular form of soil erosion that occurs below the soil surface. It is associated with levee and dam failure as well as sink hole formation. Turbulent flow removes soil starting from the mouth of the seep flow and subsoil erosion advances upgradient. The term sand boil is used to describe the appearance of the discharging end of an active soil pipe.

Soil salination is the accumulation of free salts to such an extent that it leads to degradation of soils and vegetation. Consequences include corrosion damage, reduced plant growth, erosion due to loss of plant cover and soil structure, and water quality problems due to sedimentation. Salination occurs due to a combination of natural and human caused processes. Aridic conditions favour salt accumulation. This is especially apparent when soil parent material is saline. Irrigation of arid lands is especially problematic. All irrigation water has some level of salinity. Irrigation, especially when it involves leakage from canals, often raise the underlying water table. Rapid salination occurs when the land surface is within the capillary fringe of saline groundwater. Salinity control involves flushing with higher levels of applied water in combination with tile drainage.

4

Environmental Perturbation: Air, Atmosphere and Environmental Issues

4.1 Environmental Pollution

Environmental pollution is any discharge of material or energy into water, land, or air that causes or may cause acute (short-term) or chronic (long-term) detriment to the Earth's ecological balance or that lowers the quality of life. Pollutants may cause primary damage, with direct identifiable impact on the environment, or secondary damage in the form of minor perturbations in the delicate balance of the biological food web that are detectable only over long time periods.

Until relatively recently in humanity's history, where pollution has existed, it has been primarily a local problem. The industrialization of society, the introduction of motorized vehicles, and the explosion of the human population, however, have caused an exponential growth in the production of goods and services. Coupled with this growth has been a tremendous increase in waste by-products. The indiscriminate discharge of untreated industrial and domestic wastes into waterways, the spewing of thousands of tons of particulates and airborne gases into the atmosphere, the "throwaway" attitude toward solid wastes, and the use of newly developed chemicals without considering potential consequences have resulted in major environmental disasters, including the formation of smog in the Los Angeles area since the late 1940s and the pollution of large areas of the Mediterranean Sea. Technology has begun to solve some pollution problems (see pollution control), and public awareness of the extent of pollution will eventually force governments to undertake more effective environmental planning and adopt more effective antipollution measures.

Different Types of Pollution

Water Pollution

Water pollution is the introduction into fresh or ocean waters of chemical, physical, or biological material that degrades the quality of the water and affects the organisms living in it. This process ranges from simple addition of dissolved or suspended solids to discharge of the most insidious and persistent toxic pollutants (such as pesticides, heavy metals, and nondegradable, bioaccumulative, chemical compounds).

Conventional

Conventional or classical pollutants are generally associated with the direct input of (mainly human) waste products. Rapid urbanization and rapid population increase have produced sewage problems because treatment facilities have not kept pace with need. Untreated and partially treated sewage from municipal wastewater systems and septic tanks in unsewered areas contribute significant quantities of nutrients, suspended solids, dissolved solids, oil, metals (arsenic, mercury, chromium, lead, iron, and manganese), and biodegradable organic carbon to the water environment.

Conventional pollutants may cause a myriad of water pollution problems. Excess suspended solids block out energy from the Sun and thus affect the carbon dioxide-oxygen conversion process, which is vital to the maintenance of the biological food chain. Also, high concentrations of suspended solids silt up rivers and navigational channels, necessitating frequent dredging. Excess dissolved solids make the water undesirable for drinking and for crop irrigation. Although essential to the aquatic habitat, nutrients such as nitrogen and phosphorus may also cause overfertilization and accelerate the natural aging process (eutrophication) of lakes. This acceleration in turn produces an overgrowth of aquatic vegetation, massive algal blooms, and an overall shift in the biologic community—from low productivity with many diverse species to high productivity with large numbers of a few species of a less desirable nature. Bacterial action oxidizes biodegradable organic carbon and consumes dissolved oxygen in the water. In extreme cases where the organic-carbon loading is high, oxygen consumption may lead to an oxygen depression: (less than 2 mg/l compared with 5 to 7 mg/l for a healthy stream) is sufficient to cause a fish kill and seriously to disrupt the growth of associated organisms that require oxygen to survive.

Non-conventional

The nonconventional pollutants include dissolved and particulate forms of metals, both toxic and nontoxic, and degradable and persistent organic carbon compounds discharged into water as a by-product of industry or as an integral part of marketable products. More than 13,000 oil spills of varying magnitude occur in the United States each year. Thousands of environmentally untested chemicals are routinely discharged into waterways; an estimated 400 to 500 new compounds are marketed each year. In addition, coal strip mining releases acid wastes that despoil the surrounding waterways. Nonconventional pollutants vary from biologically inert materials such as clay and iron residues to the most toxic and insidious materials such as halogenated hydrocarbons (DDT, kepone, mirex, and polychlorinated biphenyls—PCB). The latter group may produce damage ranging from acute biological effects (complete sterilization of stretches of waterways) to chronic sublethal effects that may go undetected for years. The chronic low-level pollutants are proving to be the most difficult to correct and abate because of their ubiquitous nature and chemical stability.

Thermal Pollution

Thermal pollution is the discharge of waste heat via energy dissipation into cooling water and subsequently into nearby waterways. The major sources of thermal pollution are fossil-fuel and nuclear electric-power generating facilities and, to a lesser degree, cooling operations associated with industrial manufacturing, such as steel foundries, other primary-metal manufacturers, and chemical and petrochemical producers. The discharge temperatures from electric-power plants generally range from 5 to 11 C degrees (9 to 20 F degrees) above ambient water temperatures. An estimated 90 per cent of all water consumption, excluding agricultural uses, is for cooling or energy dissipation.

The discharge of heated water into a waterway often causes ecologic imbalance, sometimes resulting in major fish kills near the discharge source. The increased temperature accelerates chemical-biological processes and decreases the ability of the water to hold dissolved oxygen. Thermal changes affect the aquatic system by limiting or changing the type of fish and aquatic biota able to grow or reproduce in the waters. Thus rapid and dramatic changes in biologic communities often occur in the vicinity of heated discharges.

Land Pollution

Land pollution is the degradation of the Earth's land surface through misuse of the soil by poor agricultural practices, mineral exploitation, industrial waste dumping, and indiscriminate disposal of urban wastes.

Soil Misuse

Soil erosion—a result of poor agricultural practices—removes rich humus topsoil developed over many years through vegetative decay and microbial degradation and thus strips the land of valuable nutrients for crop growth. Strip mining for minerals and coal lays waste thousands of acres of land each year, denuding the Earth and subjecting the mined area to widespread erosion problems. The increases in urbanization due to population pressure presents additional soil-erosion problems; sediment loads in nearby streams may increase as much as 500 to 1,000 times over that recorded in nearby undeveloped stretches of stream. Soil erosion not only despoils the Earth for farming and other uses, but also increases the suspended-solids load of the waterway. This increase interferes with the ecological habitat and poses silting problems in navigation channels, inhibiting the commercial use of these waters.

Solid Waste

In the United States in 1988 municipal wastes alone—that is, the solid wastes sent by households, business, and municipalities to local landfills and other waste-disposal facilities—equaled 163 million metric tons (1980 million U.S. tons), or 18 k (40lb) per person, according to figures released by the Environmental Protection Agency. Additional solid wastes accumulate from mining, industrial production, and agriculture. Although municipal wastes are the most obvious, the accumulations of other types of wastes are the most obvious, the accumulations of other types of waste are far greater, in many instances are more difficult to dispose of, and present greater environmental hazards.

The most common and convenient method of disposing of municipal solid wastes is in the sanitary landfill. The open dump, once a common eyesore in towns across the United States, attracted populations of rodents and other pests and often emitted hideous odors; it is now illegal. Sanitary landfills provide better aesthetic control and should be odor-free. Often, however, industrial wastes of unknown content are commingled with domestic wastes. Groundwater infiltration and contamination of water supplies with toxic chemicals have recently led to more active control of landfills and industrial waste disposal. Careful management of sanitary landfills, such as providing for leachate and runoff treatment as well as daily coverage with topsoil, has alleviated most of the problems of open dumping. In many areas, however, space for landfills is running out and alternatives must be found.

Recycling of materials is practical to some extent for much municipal and some industrial wastes, and a small but growing proportion of solid wastes is being recycled. When wastes are commingled, however, recovery becomes difficult and expensive. New processes of sorting ferrous and nonferrous metals, paper, glass, and plastics have been developed, and many communities with recycling programmes now require refuse separation. Crucial issues in recycling are devising better processing methods, inventing new products for the recycled materials, and finding new markets for them.

Incineration is another method for disposing of solid wastes. Advanced incinerators use solid wastes as fuel, burning quantities of refuse and utilizing the resultant heat to make steam for electricity generation. Wastes must be burned at very high temperatures, and incinerator exhausts must be equipped with sophisticated scrubbers and other devices for removing dioxins and other toxic pollutants. Problems remain, however: incinerator ash contains high ratios of heavy metals, becoming a hazardous waste in itself, and high-efficiency incinerators may discourage the use of recycling and other waste-reduction methods.

Composting is increasingly used to treat some agricultural wastes, as well as such municipal wastes as leaves and brush. Composting systems can produce usable soil conditioners, or humus, within a few months (see compost).

Pesticide Pollution

Pesticides are organic and inorganic chemicals originally invented and first used effectively to better the human environment by controlling undesirable life forms such as bacteria, pests, and foraging insects. Their effectiveness, however, has caused considerable pollution. The persistent, or hard, pesticides, which are relatively inert and nondegradable by chemical or biologic activity, are also bioaccumulative; that is, they are retained within the body of the consuming organism and are concentrated with each ensuing level of the biologic food chain. For example, DDT provides an excellent example of cumulative pesticide effects. (Although DDT use has been banned in the United States since 1972, it is still a popular pesticide in much of the rest of the world.) DDT may be applied to an area so that the levels in the surrounding environment are less than one part per billion. As bacteria or other microscopic organisms ingest and retain the pesticide, the concentration may increase several hundred-to a thousandfold. Concentration continues as these organisms are ingested by higher forms of life—algae, fish, shellfish, birds, or humans. The resultant concentration in the higher life forms may reach levels of thousands to millions of parts per billion.

Many pesticides are nondiscriminatory; that is, they are not specific for a particular plant or organism. A dramatic example of this effect is DDE (a product of the breakdown of DDT), which effectively inhibits the ability of birds to provide sufficient calcium deposits for their eggs, producing fragile shells and a high percentage of nested eggs that break prematurely. Another reported side effect of pesticides is their effect on the nervous system of animals and fish; they can cause instability, disorientation, and, in some cases, death. These examples are generally a result of relatively high body residuals producing acute (short-term) readily recordable results.

The long-term (chronic) effects of persistent pesticides are virtually unknown, but many scientists believe they are as much an environmental hazard as are the acute effects. Nonpersistent (readily degradable) pesticides or substitutes, insect sterilization techniques, hormone homologues that check or interfere with maturation stages, and introduction of animals that prey on the pests present a potentially brighter picture for pest control with significantly reduced environmental consequences.

Radiation Pollution

Radiation pollution is any form of ionizing or nonionizing radiation that results from human activities. The most well-known radiation results from the detonation of nuclear devices and the controlled release of energy by nuclear-power generating plants (see nuclear energy). Other sources of radiation include spent-fuel reprocessing plants, by-products of mining operations, and experimental research laboratories. Increased exposure to medical X rays and to radiation emissions from microwave ovens and other household appliances, although of considerably less magnitude, all constitute sources of environmental radiation.

Public concern over the release of radiation into the environment greatly increased following the disclosure of possible harmful effects to the public from nuclear weapons testing, the accident (1979) at the Three Mile Island nuclear-power generating plant near Harrisburg, Pa., and the catastrophic 1986 explosion at Chernobyl, a Soviet nuclear power plant. In the late 1980s, revelations of major pollution problems at U.S. nuclear weapons reactors raised apprehensions even higher.

The environmental effects of exposure to high-level ionizing radiation have been extensively documented through postwar studies on individuals who were exposed to nuclear radiation in Japan. Some forms of cancer show up immediately, but latent maladies of radiation poisoning have been recorded from 10 to 30 years after exposure. The effects of exposure to low-level radiation are not yet known. A major concern about this type of exposure is the potential for genetic damage.

Radioactive nuclear wastes cannot be treated by conventional chemical methods and must be stored in heavily shielded containers in areas remote from biological habitats. The safest of storage sites currently used are impervious deep caves or abandoned salt mines. Most radioactive wastes, however, have half-lives of hundreds to thousands of years, and to date no storage method has been found that is absolutely infallible.

Noise Pollution

Noise pollution has a relatively recent origin. It is a composite of sounds generated by human activities ranging from blasting stereo systems to the roar of supersonic transport jets. Although the frequency (pitch) of noise may be of major importance, most noise sources are measured in terms of intensity, or strength of the sound field. The standard unit, one decibel (dB), is the amount of sound that is just audible to the average human. The decibel scale is somewhat misleading because it is logarithmic rather than linear; for example, a noise source measuring 70 dB is 10 times as loud as a source measuring 60 dB and 100 times as loud as a source reading 50 dB. Noise may be generally associated with industrial society, where heavy machinery, motor vehicles, and aircraft have become everyday items. Noise pollution is more intense in the work environment than in the general environment, although ambient noise increased an average of one dB per year during the 1980s. The average background noise in a typical home today is between 40 and 50 decibels. Some examples of high-level sources in the environment are heavy trucks (90 dB at 15 m/50 ft), freight trains (75 dB at 15 m/50 ft), and air conditioning (60 dB at 6 m/20 ft).

The most readily measurable physiological effect of noise pollution is damage to hearing, which may be either temporary or permanent and may cause disruption of normal activities or just general annoyance. The effect is variable, depending upon individual susceptibility, duration of exposure, nature of noise (loudness), and time distribution of exposure (such as steady or intermittent). On the average an individual will experience a threshold shift (a shift in an individual's upper limit of sound detectability) when exposed to noise levels of 75 to 80 dB for several hours. This shift will last only several hours once the source of noise pollution is removed. A second physiologically important level is the threshold of pain, at which even short-term exposure will cause physical pain (130 to 140 dB). Any noise sustained at this level will cause a permanent threshold shift or permanent partial hearing loss. At the uppermost level of noise (greater than 150 dB), even a single short-term blast may cause traumatic hearing loss and physical damage inside the ear.

Although little hard information is available on the psychological side effects of increased noise levels, many researchers attribute increased irritability, lower productivity, decreased tolerance levels, increased incidence of ulcers, migraine headaches, fatigue, and allergic responses to continued exposures to high-level noises in the workplace and the general environment.

Air Pollution

Air pollution is the accumulation in the atmosphere of substances that, in sufficient concentrations, endanger human health or produce other measured effects on living matter and other materials. Among the major sources of pollution are power and heat generation, the burning of solid wastes, industrial processes, and, especially, transportation. The six major types of pollutants are carbon monoxide, hydrocarbons, nitrogen oxides, particulates, sulfur dioxide, and photochemical oxidants.

Local and Regional

Smog has seriously affected more persons than any other type of air pollution. It can be loosely defined as a multisource, widespread air pollution that occurs in the air of cities. Smog, a contraction of the words smoke and fog, has been caused throughout recorded history by water condensing on smoke particles, usually from burning coal. The infamous London fogs—about 4,000 deaths were attributed to the severe fog of 1952—were smog of this type. Another type, ice fog, occurs only at high latitudes and extremely low temperatures and is a combination of smoke particles and ice crystals.

As a coal economy has gradually been replaced by a petroleum economy, photochemical smog has become predominant in many cities. Its unpleasant properties result from the irradiation by sunlight of hydrocarbons (primarily unburned gasoline emitted by automobiles and other combustion sources) and other pollutants in the air. Irradiation produces a long series of photochemical reactions (see photochemistry). The products of the reactions include organic particles, ozone, aldehydes, ketones, peroxyacetyl nitrate, and organic acids and other oxidants. Sulfur dioxide, which is always present to some extent, oxidizes and hydrates to form sulfuric acid and becomes part of the particulate matter. Furthermore, automobiles are polluters even in the absence of photochemical reactions. They are responsible for much of the particulate material in the air; they also emit carbon monoxide, one of the most toxic constituents of smog.

All types of smog decrease visibility and, with the possible exception of ice fog, are irritating to the respiratory system. Statistical studies indicate that smog is a contributor to malignancies of many types. Photochemical smog produces eye irritation and lacrimation and causes severe damage to many types of vegetation, including important crops. Acute effects include an increased mortality rate, especially among persons suffering from respiratory and coronary ailments. Air pollution also has a deleterious effect on works of art (see art conservation and restoration).

Air pollution on a regional scale is in part the result of local air pollution—including that produced by individual sources, such as automobiles—that has spread out to encompass areas of many thousands of square kilometers. Meteorological conditions and landforms can greatly influence air-pollution concentrations at any given place, especially locally and regionally. For example, cities located in bowls or valleys over which atmospheric inversions form and act as imperfect lids are especially likely to suffer from incidences of severe smog. Oxides of sulfur and nitrogen, carried long distances by the atmosphere and then precipitated in solution as acid rain, can cause serious damage to vegetation, waterways, and buildings.

Global

Humans also pollute the atmosphere on a global scale, although until the early 1970s little attention was paid to the possible deleterious effects of such pollution. Measurements in Hawaii suggest that the concentration of carbon dioxide in the atmosphere is increasing at a rate of about 0.2 per cent every year. The effect of this increase may be to alter the Earth's climate by increasing the average global temperature. Certain pollutants decrease the concentration of ozone occurring naturally in the stratosphere, which in turn increases the amount of ultraviolet radiation reaching the Earth's surface. Such radiation may damage vegetation and increase the incidence of skin cancer. Examples of stratospheric contaminants include nitrogen oxides emitted by supersonic aircraft and chlorofluorocarbons used as refrigerants and aerosol-can propellants. The chlorofluorocarbons reach the stratosphere by upward mixing from the lower parts of the atmosphere (see ozone layer). It is believed that these chemicals are responsible for the noticeable loss of ozone over the polar regions that has occurred in the 1980s.

4.2 Air pollution

Air pollution is the human introduction into the atmosphere of chemicals, particulate matter, or biological materials that cause harm or discomfort to humans or other living organisms, or damage the environment. Air pollution causes deaths and respiratory disease. Air pollution is often identified with major stationary sources, but the greatest source of emissions is actually mobile sources, mainly automobiles. Gases such as carbon dioxide, which contribute to global warming, have recently gained recognition as pollutants by climate scientists, while they also recognize that carbon dioxide is essential for plant life through photosynthesis.

The atmosphere is a complex, dynamic natural gaseous system that is essential to support life on planet Earth. Stratospheric ozone depletion due to air pollution has long been recognized as a threat to human health as well as to the Earth's ecosystems.

Pollutants

There are many substances in the air which may impair the health of plants and animals (including humans), or reduce visibility. These arise both from natural processes and human activity. Substances not naturally found in the air or at greater concentrations or in different locations from usual are referred to as *pollutants*. Pollutants can be classified as either primary or secondary. Primary pollutants are substances directly emitted from a process, such as ash from a volcanic eruption or the carbon monoxide gas from a motor vehicle exhaust. Secondary pollutants are not emitted directly. Rather, they form in the air when primary pollutants react or interact. An important example of a secondary pollutant is ground level ozone—one of the many secondary pollutants that make up photochemical smog. Note that some pollutants may be both primary and secondary: that is, they are both emitted directly and formed from other primary pollutants. Major primary pollutants produced by human activity include:

❖ Sulfur oxides (SO_x) especially sulfur dioxide are emitted from burning of coal and oil.

❖ Nitrogen oxides (NO_x) especially nitrogen dioxide are emitted from high temperature combustion. Can be seen as the brown haze dome above or plume downwind of cities.

❖ Carbon monoxide is colourless, odourless, non-irritating but very poisonous gas. It is a product by incomplete combustion of fuel such as natural gas, coal or wood. Vehicular exhaust is a major source of carbon monoxide.

❖ Carbon dioxide (CO_2), a greenhouse gas emitted from combustion.

❖ Volatile organic compounds (VOC), such as hydrocarbon fuel vapors and solvents.

❖ Particulate matter (PM), measured as smoke and dust. PM_{10} is the fraction of suspended particles 10 micrometers in diameter and smaller that will enter the nasal cavity. $PM_{2.5}$ has a maximum particle size of 2.5 μm and will enter the bronchies and lungs.

❖ Toxic metals, such as lead, cadmium and copper.

❖ Chlorofluorocarbons (CFCs), harmful to the ozone layer emitted from products currently banned from use.

❖ Ammonia (NH_3) emitted from agricultural processes.

❖ Odors, such as from garbage, sewage, and industrial processes.

❖ Radioactive pollutants produced by nuclear explosions and war explosives, and natural processes such as radon.

Secondary pollutants include:

❖ Particulate matter formed from gaseous primary pollutants and compounds in photochemical smog, such as nitrogen dioxide.

❖ Ground level ozone (O_3) formed from NOx and VOCs.

❖ Peroxyacetyl nitrate (PAN) similarly formed from NOx and VOCs.

Minor air pollutants include:

❖ A large number of minor hazardous air pollutants. Some of these are regulated in USA under the Clean Air Act and in Europe under the Air Framework Directive.

❖ A variety of persistent organic pollutants, which can attach to particulate matter.

Sources

Sources of air pollution refer to the various locations, activities or factors which are responsible for the releasing of pollutants in the atmosphere. These sources can be classified into two major categories which are:

Anthropogenic sources (human activity) mostly related to burning different kinds of fuel:

❖ "Stationary Sources" as smoke stacks of power plants, manufacturing facilities, municipal waste incinerators.

❖ "Mobile Sources" as motor vehicles, aircraft etc.

❖ Marine vessels, such as container ships or cruise ships, and related port air pollution.

❖ Burning wood, fireplaces, stoves, furnaces and incinerators.

❖ Oil refining, and industrial activity in general.

❖ Chemicals, dust and controlled burn practices in agriculture and forestry management, (see Dust Bowl).

❖ Fumes from paint, hair spray, varnish, aerosol sprays and other solvents.

❖ Waste deposition in landfills, which generate methane.

❖ Military, such as nuclear weapons, toxic gases, germ warfare and rocketry.

Natural Sources

❖ Dust from natural sources, usually large areas of land with little or no vegetation.

❖ Methane, emitted by the digestion of food by animals, for example cattle.

❖ Radon gas from radioactive decay within the Earth's crust.

❖ Smoke and carbon monoxide from wildfires.

❖ Volcanic activity, which produce sulfur, chlorine, and ash particulates.

Emission Factors

Air pollutant emission factors are representative values that attempt to relate the quantity of a pollutant released to the ambient air with an activity associated with the release of that pollutant. These factors are usually expressed as the weight of pollutant divided by a unit weight, volume, distance, or duration of the activity emitting the pollutant (e.g., kilograms of particulate emitted per megagram of coal burned). Such factors facilitate estimation of emissions from various sources of air pollution. In most cases, these factors are simply averages of all available data of acceptable quality, and are generally assumed to be representative of long-term averages.

The United States Environmental Protection Agency has published a compilation of air pollutant emission factors for a multitude of industrial sources. The United Kingdom, Australia, Canada and other countries have published similar compilations, as has the European Environment Agency.

Indoor Air Quality (IAQ)

A lack of ventilation indoors concentrates air pollution where people often spend the majority of their time. Radon (Rn) gas, a carcinogen, is exuded from the Earth in certain locations and trapped inside houses. Building materials including carpeting and plywood emit formaldehyde (H_2CO) gas. Paint and solvents give off volatile organic compounds (VOCs) as they dry. Lead paint can degenerate into dust and be inhaled. Intentional air pollution is introduced with the use of air fresheners, incense, and other scented items. Controlled wood fires in stoves and fireplaces can add significant amounts of smoke particulates into the air, inside and out. Indoor pollution fatalities may be caused by using pesticides and other chemical sprays indoors without proper ventilation.

Carbon monoxide (CO) poisoning and fatalities are often caused by faulty vents and chimneys, or by the burning of charcoal indoors. Chronic carbon monoxide poisoning can result even from poorly adjusted pilot lights. Traps are built into all domestic plumbing to keep sewer gas, hydrogen sulfide, out of interiors. Clothing emits tetrachloroethylene, or other dry cleaning fluids, for days after dry cleaning.

Though its use has now been banned in many countries, the extensive use of asbestos in industrial and domestic environments in the past has left a potentially very dangerous material in many localities. Asbestosis is a chronic inflammatory medical condition affecting the tissue of the lungs. It occurs after long-term, heavy exposure to asbestos from asbestos-containing materials in structures. Sufferers have severe dyspnea (shortness of breath) and are at an increased risk regarding several different types of lung cancer. As clear explanations are not always stressed in non-technical literature, care should be taken to distinguish between several forms of relevant diseases. According to the World Health Organisation (WHO), these may defined as; asbestosis, *lung cancer*, and *mesothelioma* (generally a very rare form of cancer, when more widespread it is almost always associated with prolonged exposure to asbestos).

Biological sources of air pollution are also found indoors, as gases and airborne particulates. Pets produce dander, people produce dust from minute skin flakes and decomposed hair, dust mites in bedding, carpeting and furniture produce enzymes and micrometre-sized fecal droppings, inhabitants emit methane, mold forms in walls and generates mycotoxins and spores, air conditioning systems can incubate Legionnaires' disease and mold, and houseplants, soil and surrounding gardens can produce pollen, dust, and mold. Indoors, the lack of air circulation allows these airborne pollutants to accumulate more than they would otherwise occur in nature.

Health Effects

The World Health Organization states that 2.4 million people die each year from causes directly attributable to air pollution; with 1.5 million of these deaths attributable to indoor air pollution. A study by the University of Birmingham has shown a strong correlation between pneumonia related deaths and air pollution from motor vehicles. Worldwide more deaths per year are linked to air pollution than to automobile accidents. Published in 2005 suggests that 310,000 Europeans die from air pollution annually. Direct causes of air pollution related deaths include aggravated asthma, bronchitis, emphysema, lung and heart diseases, and respiratory allergies. The US EPA estimates that a proposed set of changes in diesel engine technology (*Tier 2*) could result in 12,000 fewer *premature mortalities*, 15,000 fewer heart attacks, 6,000 fewer emergency room visits by children with asthma, and 8,900 fewer respiratory-related hospital admissions each year in the United States.

The worst short term civilian pollution crisis in India was the 1984 Bhopal Disaster. Leaked industrial vapors from the Union Carbide factory, belonging to Union Carbide, Inc., U.S.A., killed more than 2,000 people outright and injured anywhere from 150,000 to 600,000 others, some 6,000 of whom would later die from their injuries. The United Kingdom suffered its worst air pollution event when the December 4 Great Smog of 1952 formed over London. In six days more than 4,000 died, and 8,000 more died within the following months. An accidental leak of anthrax spores from a biological warfare laboratory in the former USSR in 1979 near Sverdlovsk is believed to have been the cause of hundreds of civilian deaths. The worst single incident of air pollution to occur in the United States of America occurred in Donora, Pennsylvania in late October, 1948, when 20 people died and over 7,000 were injured.

The health effects caused by air pollutants may range from subtle biochemical and physiological changes to difficulty in breathing, wheezing, coughing and aggravation of existing respiratory and cardiac conditions. These effects can result in increased medication use, increased doctor or emergency room visits, more hospital admissions and premature death. The human health effects of poor air quality are far reaching, but principally affect the body's respiratory system and the cardiovascular system. Individual reactions to air pollutants depend on the type of pollutant a person is exposed to, the degree of exposure, the individual's health status and genetics.

Effects on Cystic Fibrosis

A study from 1999 to 2000 by the University of Washington showed that patients near and around particulate matter air pollution had an increased risk of pulmonary exacerbations and decrease in lung function.   Patients were examined before the study for amounts of specific pollutants like P. aeruginosa or B. cepacia as well as their socioeconomic standing. Participants involved in the study were located in the United States in close proximity to an Environmental Protection Agency.During the time of the study 117 deaths were associated with air pollution. A trend was noticed that patients living closer or in large metropolitan areas to be close to medical help also had higher level of pollutants found in their system because of more emissions in larger cities. With cystic fibrosis patients already being born with decreased lung function everyday pollutants such as smoke emissions from automobiles, tobacco smoke and improper use of indoor heating devices could add to the dissemination of lung function.

Effects on COPD

Chronic Obstructive Pulmonary Disease (COPD) include diseases such as chronic bronchitis, emphysema, and some forms of asthma. Two researchers Holland and Reid conducted research on 293 male postal workers in London during the time of the 1952 London Fog incident and 477 male postal workers in the rural setting. The amount of the pollutant FEV1 was significantly lower in urban employees however lung function was decreased due to city pollutions such as car fumes and increased amount of cigarette exposure. It is believed that much like cystic fibrosis, by living in a more urban environment serious health hazards become more apparent. Studies have shown that in urban areas patients suffer mucus hypersecretion, lower levels of lung function, and more self diagnosis of chronic bronchitis and emphysema.

London Fog of 1952

In the matter of four days a combination of dense fog and sooty black coal smoke came over the London area. The fog was so dense residents of London could not see in front of them. The extreme reduction in visibility was accompanied by an increase in criminal activity as well as transportation delays and a virtual shut down of the city. During the 4 day period of the fog 12,000 are believed to have been killed.

Effects on Children

Cities around the world with high exposure to air pollutants has the possibility of children living within them to develop asthma, pneumonia and other lower respiratory infections as well as a low initial birth rate. Protective measures to ensure the youths health is being taken in countries such as New Delhi where buses now use compressed natural gas to help eliminate the "pea-soup" fog. Research by the World Health Organization shows there is the greatest concentration of particulate matter particles in countries with low economic world power and high poverty and population rates. Examples of these countries include Egypt, Sudan, Mongolia, and Indonesia. The Clean Air Act was passed in 1970, however in 2002 at least 146 million Americans were living in areas that did not meet at least one of the "criteria pollutants" laid out in the 1997 National Ambient Air Quality Standards.   Those pollutants included: ozone, particulate matter, sulfur dioxide, nitrogen dioxide, carbon monoxide, and lead. Because children are outdoors more and have higher minute ventilation they are more susceptible to the dangers of air pollution.

Reduction Efforts

There are various air pollution control technologies and urban planning strategies available to reduce air pollution. Efforts to reduce pollution from mobile sources includes primary regulation (many developing countries have permissive regulations)· expanding regulation to new sources (such as cruise and transport ships, farm equipment, and small gas-powered equipment such as lawn trimmers, chainsaws, and snowmobiles), increased fuel efficiency (such as through the use of hybrid vehicles), conversion to cleaner fuels (such as bioethanol, biodiesel, or conversion to electric vehicles).

Control Devices

The following items are commonly used as pollution control devices by industry or transportation devices. They can either destroy contaminants or remove them from an exhaust stream before it is emitted into the atmosphere.

- ❖ **Particulate control**
    - Mechanical collectors (dust cyclones, multicyclones)
    - Electrostatic precipitators
    - Baghouses
    - Particulate scrubbers
- ❖ **Scrubbers**
    - Baffle spray scrubber
    - Cyclonic spray scrubber
    - Ejector venturi scrubber

- ● Mechanically aided scrubber
- ● Spray tower
- ● Wet scrubber
- ❖ **NOx control**
  - ● Low NOx burners
  - ● Selective catalytic reduction (SCR)
  - ● Selective non-catalytic reduction (SNCR)
  - ● NOx scrubbers
  - ● Exhaust gas recirculation
  - ● Catalytic converter (also for VOC control)
- ❖ **VOC abatement**
  - ● Adsorption systems, such as activated carbon
  - ● Flares
  - ● Thermal oxidizers
  - ● Catalytic oxidizers
  - ● Biofilters
  - ● Absorption (scrubbing)
  - ● Cryogenic condensers
  - ● Vapor recovery systems
- ❖ **Acid Gas/SO$_2$ control**
  - ● Wet scrubbers
  - ● Dry scrubbers
  - ● Flue gas desulfurization
- ❖ **Mercury control**
  - ● Sorbent Injection Technology
  - ● Electro-Catalytic Oxidation (ECO)
  - ● K-Fuel
- ❖ **Dioxin and furan control**
- ❖ **Miscellaneous associated equipment**
  - ● Source capturing systems
  - ● Continuous emissions monitoring systems (CEMS)

Legal Regulation

In general, there are two types of air quality standards. The first class of standards (such as the U.S. National Ambient Air Quality Standards) set maximum atmospheric concentrations for specific pollutants. Environmental agencies enact regulations which are intended to result in attainment of these target levels. The second class (such as the North American Air Quality Index) take the form of a scale with various thresholds, which is used to communicate to the public the relative risk of outdoor activity. The scale may or may not distinguish between different pollutants.

Canada

In Canada, air quality is typically evaluated against standards set by the Canadian Council of Minister for the Environment (CCME), an inter-governmental body of federal, provincial and territorial Ministers responsible for the environment. The CCME has set Canada Wide Standards(CWS). These are:

- ❖ CWS for PM 2.5 = 30 µg/m3 (24 hour averaging time, by year 2010, based on 98th percentile ambient measurement annually, averaged over 3 consecutive years).
- ❖ CWS for *ozone* = 65 ppb (8-hour averaging time, by year 2010, achievement is based on the 4th highest measurement annually, averaged over 3 consecutive years.

Note that there is no consequence in Canada to not achieving these standards. In addition, these only apply to jurisdictions with populations greater than 100,000. Further, provinces and territories may set more stringent standards than those set by the CCME.

European Union

National Emission Ceilings (NEC) for certain atmospheric pollutants are regulated by Directive 2001/81/EC (NECD). As part of the preparatory work associated with the revision of the NECD, the European Commission is assisted by the NECPI working group (National Emission Ceilings—Policy Instruments).

United Kingdom

Air quality targets set by the UK's Department for Environment, Food and Rural Affairs (DEFRA) are mostly aimed at local government representatives responsible for the management of air quality in cities, where air quality management is the most urgent. The UK has established an air quality network where levels of the key air pollutants are published by monitoring centers. Air quality in Oxford, Bath and London is particularly poor. One controversial study performed by the Calor Gas company and published in the Guardian newspaper compared walking in Oxford on an average day to smoking over sixty light cigarettes.

More precise comparisons can be collected from the UK Air Quality Archive which allows the user to compare a cities management of pollutants against the national air quality objectives set by DEFRA in 2000.

Localized peak values are often cited, but average values are also important to human health. The UK National Air Quality Information Archive offers almost real-time monitoring of "current maximum" air pollution measurements for many UK towns and cities. This source offers a wide range of constantly updated data, including:

- Hourly Mean Ozone ($\mu g/m^3$)
- Hourly Mean Nitrogen dioxide ($\mu g/m^3$)
- Maximum 15-Minute Mean Sulphur dioxide ($\mu g/m^3$)
- 8-Hour Mean Carbon monoxide (mg/m^3)
- 24-Hour Mean PM_{10} ($\mu g/m^3$ Grav Equiv)

DEFRA acknowledges that air pollution has a significant effect on health and has produced a simple banding index system is used to create a daily warning system that is issued by the BBC Weather Service to indicate air pollution levels. DEFRA has published guidelines for people suffering from respiratory and heart diseases.

United States

In the 1960s, 70s, and 90s, the United States Congress enacted a series of Clean Air Acts which significantly strengthened regulation of air pollution. Individual U.S. states, some European nations and eventually the European Union followed these initiatives. The Clean Air Act sets numerical limits on the concentrations of a basic group of air pollutants and provide reporting and enforcement mechanisms.

In 1999, the United States EPA replaced the Pollution Standards Index (PSI) with the Air Quality Index (AQI) to incorporate new PM2.5 and Ozone standards.

The effects of these laws have been very positive. In the United States between 1970 and 2006, citizens enjoyed the following reductions in annual pollution emissions:

- carbon monoxide emissions fell from 197 million tons to 89 million tons,
- nitrogen oxide emissions fell from 27 million tons to 19 million tons,
- sulfur dioxide emissions fell from 31 million tons to 15 million tons,
- particulate emissions fell by 80 per cent, and
- lead emissions fell by more than 98 per cent.

In an October 2006 letter to EPA, the agency's independent scientific advisors warned that the ozone smog standard "needs to be substantially reduced" and that there is "no scientific justification" for retaining the current, weaker standard. The scientists unanimously recommended a smog threshold of 60 to 70 ppb after they conducted an extensive review of the evidence.

The EPA has proposed, in June 2007, a new threshold of 75 ppb. This falls short of the scientific recommendation, but is an improvement over the current standard.

Polluting industries are lobbying to keep the current (weaker) standards in place. Environmentalists and public health advocates are mobilizing to support compliance with the scientific recommendations.

The National Ambient Air Quality Standards are pollution thresholds which trigger mandatory remediation plans by state and local governments, subject to enforcement by the EPA.

An outpouring of dust layered with man-made sulfates, smog, industrial fumes, carbon grit, and nitrates is crossing the Pacific Ocean on prevailing winds from booming Asian economies in plumes so vast they alter the climate. Almost a third of the air over Los Angeles and San Francisco can be traced directly to Asia. With it comes up to three-quarters of the black carbon particulate pollution that reaches the West Coast.

Statistics

Most Polluted Cities

Most Polluted World Cities by PM

Particulate matter, $\mu g/m^3$ (2004)	City
169	Cairo, Egypt
150	Delhi, India
128	Kolkata, India (Calcutta)
127	Tehran
125	Tianjin, China
123	Chongqing, China
109	Kanpur, India
109	Lucknow, India
104	Jakarta, Indonesia
101	Shenyang, China

Air pollution is usually concentrated in densely populated metropolitan areas, especially in developing countries where environmental regulations are generally relatively lax. However, even populated areas in developed countries attain unhealthy levels of pollution.

Carbon Dioxide Emissions

Total CO_2 Emissions

10^6 Tons of CO_2 per year:

- United States: 2,790
- China: 2,680
- Russia: 661
- India: 583
- Japan: 400
- Germany: 356
- Australia: 226
- South Africa: 222
- United Kingdom: 212
- South Korea: 185

Per capita CO_2 emissions

Tons of CO_2 per year per capita:

- Australia: 10
- United States: 8.2
- United Kingdom: 3.2
- China: 1.8
- India: 0.5

Atmospheric Dispersion

The basic technology for analyzing air pollution is through the use of a variety of mathematical models for predicting the transport of air pollutants in the lower atmosphere. The principal methodologies are:

- Point source dispersion, used for industrial sources.
- Line source dispersion, used for airport and roadway air dispersion modeling.

❖ Area source dispersion, used for forest fires or duststorms.
❖ Photochemical models, used to analyze reactive pollutants that form smog.

The point source problem is the best understood, since it involves simpler mathematics and has been studied for a long period of time, dating back to about the year 1900. It uses a Gaussian dispersion model for buoyant pollution plumes to forecast the air pollution isopleths, with consideration given to wind velocity, stack height, emission rate and stability class (a measure of atmospheric turbulence). This model has been extensively validated and calibrated with experimental data for all sorts of atmospheric conditions.

The roadway air dispersion model was developed starting in the late 1950s and early 1960s in response to requirements of the National Environmental Policy Act and the U.S. Department of Transportation (then known as the Federal Highway Administration) to understand impacts of proposed new highways upon air quality, especially in urban areas. Several research groups were active in this model development, among which were: the Environmental Research and Technology (ERT) group in Lexington, Massachusetts, the ESL Inc. group in Sunnyvale, California and the California Air Resources Board group in Sacramento, California. The research of the ESL group received a boost with a contract award from the United States Environmental Protection Agency to validate a line source model using sulfur hexafluoride as a tracer gas. This programme was successful in validating the line source model developed by ESL inc. Some of the earliest uses of the model were in court cases involving highway air pollution, the Arlington, Virginia portion of Interstate 66 and the New Jersey Turnpike widening project through East Brunswick, New Jersey.

Area source models were developed in 1971 through 1974 by the ERT and ESL groups, but addressed a smaller fraction of total air pollution emissions, so that their use and need was not as widespread as the line source model, which enjoyed hundreds of different applications as early as the 1970s. Similarly photochemical models were developed primarily in the 1960s and 1970s, but their use was more specialized and for regional needs, such as understanding smog formation in Los Angeles, California.

Environmental Impacts

The greenhouse effect is a phenomenon whereby greenhouse gases, create a condition in the upper atmosphere causing a trapping of heat and leading to increased surface and lower tropospheric temperatures. It shares this property with many other gases, the largest overall forcing on Earth coming from water vapour. Other greenhouse gases include methane, hydrofluorocarbons, perfluorocarbons, chlorofluorocarbons, NOx, and ozone. Many greenhouse gases, contain carbon, and some of that from fossil fuels.

This effect has been understood by scientists for about a century, and technological advancements during this period have helped increase the breadth and depth of data relating to the phenomenon. Currently, scientists are studying the role of changes in composition of greenhouse gases from natural and anthropogenic sources for the effect on climate change.

A number of studies have also investigated the potential for long-term rising levels of atmospheric carbon dioxide to cause slight increases in the acidity of ocean waters and the possible effects of this on marine ecosystems. However, carbonic acid is a very weak acid, and is utilized by marine organisms during photosynthesis.

4.3 Water Pollution

Water pollution is the contamination of water bodies such as lakes, rivers, oceans, and groundwater caused by human activities, which can be harmful to organisms and plants which live in these water bodies. Although natural phenomena such as volcanoes, algae blooms, storms, and earthquakes also cause major changes in water quality and the ecological status of water, water is typically referred to as polluted when it impaired by anthropogenic contaminants and either does not support a human use (like serving as drinking water) or undergoes a marked shift in its ability to support its constituent biotic communities. Water pollution has many causes and characteristics. The primary sources of water pollution are generally grouped into two categories based on their point of origin. Point-source pollution refers to contaminants that enter a waterway through a discrete "point source". Examples of this category include discharges from a wastewater treatment plant, outfalls from a factory, leaking underground tanks, etc. The second primary category, non-point source pollution, refers to contamination that, as its name suggests, does not originate from a single discrete source. Non-point source pollution is often a cumulative effect of small amounts of contaminants gathered from a large area. Nutrient runoff in stormwater from sheet flow over an agricultural field, or metals and hydrocarbons from an area with high impervious surfaces and vehicular traffic are examples of non-point source pollution. The primary focus of legislation and efforts to curb water pollution for the past several decades was first aimed at point sources. As point sources have been effectively regulated, greater attention has come to be placed on non-point source contributions, especially in rapidly urbanizing/suburbanizing or developing areas.

The specific contaminants leading to pollution in water include a wide spectrum of chemicals, pathogens, and physical or sensory changes. While many of the chemicals and substances that are regulated may be naturally occurring (iron, manganese, etc) the concentration is often the key in determining what is a natural component of water, and what is a contaminant. Many of the chemical substances are toxic. Pathogens can produce waterborne diseases in either human or animal hosts. Alteration of water's physical chemistry include acidity, electrical conductivity, temperature, and eutrophication. Eutrophication is the fertilisation of surface water by nutrients that were previously scarce. Water pollution is a major problem in the global context. It has been suggested that it is the leading worldwide cause of deaths and diseases, and that it accounts for the deaths of more than 14,000 people daily.

Contaminants

Contaminants may include organic and inorganic substances.

Some organic water pollutants are:

- ❖ Insecticides and herbicides, a huge range of organohalide and other chemicals.
- ❖ Bacteria, often is from sewage or livestock operations.
- ❖ Food processing waste, including pathogens.
- ❖ Tree and brush debris from logging operations.
- ❖ VOCs (volatile organic compounds), such as industrial solvents, from improper storage.
- ❖ DNAPLs (dense non-aqueous phase liquids), such as chlorinated solvents, which may fall at the bottom of reservoirs, since they don't mix well with water and are more dense.
- ❖ Petroleum Hydrocarbons including fuels (gasoline, diesel, jet fuels, and fuel oils) and lubricants (motor oil) from oil field operations, refineries, pipelines, retail service station's underground storage tanks, and transfer operations. Note: VOCs include gasoline-range hydrocarbons.
- ❖ Detergents
- ❖ Various chemical compounds found in personal hygiene and cosmetic products.
- ❖ Disinfection by-products (DBPs) found in chemically disinfected drinking water.

Some inorganic water pollutants include:

- ❖ Heavy metals including acid mine drainage.
- ❖ Acidity caused by industrial discharges (especially sulfur dioxide from power plants).
- ❖ Pre-production industrial raw resin pellets, an industrial pollutant.
- ❖ Chemical waste as industrial by products.
- ❖ Fertilizers, in runoff from agriculture including nitrates and phosphates.
- ❖ Silt in surface runoff from construction sites, logging, slash and burn practices or land clearing sites.

Transport and Chemical Reactions of Water Pollutants

Most water pollutants are eventually carried by the rivers into the oceans. In some areas of the world the influence can be traced hundred miles from the mouth by studies using hydrology transport models. Advanced computer models such as SWMM or the DSSAM Model have been used in many locations worldwide to examine the fate of pollutants in aquatic systems. Indicator filter feeding species such as copepods have also been used to study pollutant fates in the New York Bight, for example. The highest toxin loads are not directly at the mouth of the Hudson River, but 100 kilometers south, since several days are required for incorporation into planktonic tissue. The Hudson discharge flows south along the coast due to coriolis force. Further south then are areas of oxygen depletion, caused by chemicals using up oxygen and by algae blooms, caused by excess nutrients from algal cell death and decomposition. Fish and shellfish kills have been reported, because toxins climb the foodchain after small fish consume copepods, then large fish eat smaller fish, etc. Each successive step up the food chain causes a stepwise concentration of pollutants such as heavy metals (e.g. mercury) and persistent organic pollutants such as DDT. This is known as biomagnification which is occasionally used interchangeably with bioaccumulation.

The big gyres in the oceans trap floating plastic debris. The North Pacific Gyre for example has collected the so-called "Great Pacific Garbage Patch" that is now estimated at two times the size of Texas. Many of these long-lasting pieces wind up in the stomachs of marine birds and animals. This results in obstruction of digestive pathways which leads to reduced appetite or even starvation.

Many chemicals undergo reactive decay or chemically change especially over long periods of time in groundwater reservoirs. A noteworthy class of such chemicals are the chlorinated hydrocarbons such as trichloroethylene (used in industrial metal degreasing and electronics manufacturing) and tetrachloroethylene used in the dry cleaning industry (note latest advances in liquid carbon dioxide in dry cleaning that avoids all use of chemicals). Both of these chemicals, which are carcinogens themselves, undergo partial decomposition reactions, leading to new hazardous chemicals (including dichloroethylene and vinyl chloride).

Groundwater pollution is much more difficult to abate than surface pollution because groundwater can move great distances through unseen aquifers. Non-porous aquifers such as clays partially purify water of bacteria by simple filtration (adsorption and absorption), dilution, and, in some cases, chemical reactions and biological activity: however, in some cases, the pollutants merely transform to soil contaminants. Groundwater that moves through cracks and caverns is not filtered and can be transported as easily as surface water. In fact, this can be aggravated by the human tendency to use natural sinkholes as dumps in areas of Karst topography.

There are a variety of secondary effects stemming not from the original pollutant, but a derivative condition. Some of these secondary impacts are:

- ❖ Silt bearing surface runoff from can inhibit the penetration of sunlight through the water column, hampering photosynthesis in aquatic plants.
- ❖ Thermal pollution can induce fish kills and invasion by new thermophilic species. This can cause further problems to existing wildlife.

Sampling and Monitoring

Sampling water can take several forms depending on the accuracy needed and the characteristics of the contaminant. Many contamination events are temporal and most commonly in association with rain events. For this reason 'grab' samples can be used as indicators, but are often inadequate for fully accessing contaminant concerns in a water body. Scientists gathering this type of data often employ auto-sampler devices that pump increments of water at either time or discharge intervals.

Regulatory Framework

In the UK there are common law rights (civil rights) to protect the passage of water across land unfettered in either quality of quantity. Criminal laws dating back to the 16th century exercised some control over water pollution but it was not until the *River (Prevention of pollution) Acts 1951-1961* were enacted that any systematic control over water pollution was established. These laws were strengthened and extended in the *Control of Pollution Act 1984* which has since been updated and modified by a series of further acts. It is a criminal offense to either pollute a lake, river, groundwater or the sea or to discharge any liquid into such water bodies without proper authority. In England and Wales such permission can only be issued by the Environment Agency and in Scotland by SEPA.

In the USA, concern over water pollution resulted in the enactment of state anti-pollution laws in the latter half of the 19th century, and federal legislation enacted in 1899. The Refuse Act of the federal Rivers and Harbors Act of 1899 prohibits the disposal of any refuse matter from into either the nation's navigable rivers, lakes, streams, and other navigable bodies of water, or any tributary to such waters, unless one has first obtained a permit. The Water Pollution Control Act, passed in 1948, gave authority to the Surgeon General to reduce water pollution.

Growing public awareness and concern for controlling water pollution led to enactment of the Federal Water Pollution Control Act Amendments of 1972. As amended in 1977, this law became commonly known as the Clean Water Act. The Act established the basic mechanisms for regulating contaminant discharge. It established the authority for the United States Environmental Protection Agency to implement wastewater standards for industry. The Clean Water Act also continued requirements to set water quality standards for all contaminants in surface waters. Further amplification of the Act continued including the enactment of the Great Lakes Legacy Act of 2002.

In 2004, the United States Environmental Protection Agency tested drinking water quality on commercial airline's aircraft and found that 15 percent of tested aircraft tested positive for total coliform bacteria, according to a press release issued on Friday March 28, 2008.

4.4 Soil Contamination

Soil contamination is caused by the presence of man-made chemicals or other alteration in the natural soil environment. This type of contamination typically arises from the rupture of underground storage tanks, application of pesticides, percolation of contaminated surface water to subsurface strata, oil and fuel dumping, leaching of wastes from landfills or direct discharge of industrial wastes to the soil. The most common chemicals involved are petroleum hydrocarbons, solvents, pesticides, lead and other heavy metals. This occurrence of this phenomenon is correlated with the degree of industrialization and intensity of chemical usage.

The concern over soil contamination stems primarily from health risks, both of direct contact and from secondary contamination of water supplies. Mapping of contaminated soil sites and the resulting cleanup are time consuming and expensive tasks, requiring extensive amounts of geology, hydrology, chemistry and computer modeling skills.

It is in North America and Western Europe that the extent of contaminated land is most well known, with many of countries in these areas having a legal framework to identify and deal with this environmental problem; this however may well be just the tip of the iceberg with developing countries very likely to be the next generation of new soil contamination cases.

The immense and sustained growth of the People's Republic of China since the 1970s has exacted a price from the land in increased soil pollution. The State Environmental Protection Administration believes it to be a threat to the environment, to food safety and to sustainable agriculture. According to a scientific sampling, 150 million mi (100,000 square kilometres) of China's cultivated land have been polluted, with contaminated water being used to irrigate a further 32.5 million mi (21,670 square kilometres) and another 2 million mi (1,300 square kilometres) covered or destroyed by solid waste. In total, the area accounts for one-tenth of China's cultivatable land, and is mostly in economically developed areas. An estimated 12 million tonnes of grain are contaminated by heavy metals every year, causing direct losses of 20 billion yuan (US$2.57 billion).

The United States, while having some of the most widespread soil contamination, has actually been a leader in defining and implementing standards for cleanup. Other industrialized countries have a large number of contaminated sites, but lag the U.S. in executing remediation. Developing countries may be leading in the next generation of new soil contamination cases.

Each year in the U.S., thousands of sites complete soil contamination cleanup, some by using microbes that "eat up" toxic chemicals in soil, many others by simple excavation and others by more expensive high-tech soil vapor extraction or air stripping. At the same time, efforts proceed worldwide in creating and identifying new sites of soil contamination, particularly in industrial countries other than the U.S., and in developing countries which lack the money and the technology to adequately protect soil resources.

Microanalysis of Soil Contamination

To understand the fundamental nature of soil contamination, it is necessary to envision the variety of mechanisms for pollutants to become entrained in soil. Soil particulates may be composed of a gamut of organic and inorganic chemicals with variations in cation exchange capacity, buffering capacity, and redox poise. For example, at the extremes, one has a sand component, a coarse grained, inert, and totally inorganic substance; whereas peat soils are dominated by a fine organic material, made of decomposing organic material and highly active. Most soils are mixtures of soil subtypes and thus have quite complex characteristics. There is also a great diversity of soil porosity, ranging from gravels to sands to silt to clay (in increasing order of porosity), pore size, and pore tortuosity (both in decreasing order). Finally there is a wide spectrum of chemical bonding or adhesion characteristics: each contaminant has a different interaction or bonding mechanism with a given soil type.

On balance, some contaminants may literally drain through soils such as sand and gravel and move to other soils or deeper aquifers, while polar or organic chemicals discharged into a clay soil will have a very high adsorption. Thus most soil contamination is the result of pollutants adhering to the soil particle surface, or lodging in interstices of a soil matrix. Clearly the equilibrium reached is a dynamic one, where new pollutants may lodge on new soil particles and the action of groundwater movement may over time transport some of the soil contaminants to other locations or depths.

Soil contamination results when hazardous substances are either spilled or buried directly in the soil or migrate to the soil from a spill that has occurred elsewhere. For example, soil can become contaminated when small particles containing hazardous substances are released from a smokestack and are deposited on the surrounding soil as they fall out of the air. Another source of soil contamination could be water that washes contamination from an area containing hazardous substances and deposits the contamination in the soil as it flows over or through it.

Health Effects

The major concern is that there are many sensitive land uses where people are in direct contact with soils such as residences, parks, schools and playgrounds. Other contact mechanisms include contamination of drinking water or inhalation of soil contaminants which have vaporized. There is a very large set of health consequences from exposure to soil contamination depending on pollutant type, pathway of attack and vulnerability of the exposed population. Chromium and obsolete pesticide formulations are carcinogenic to populations. Lead is especially hazardous to young children, in which group there is a high risk of developmental damage to the brainhile to all populations kidney damage is a risk.

Chronic exposure to at sufficient concentrations is known to be associated with higher incidence of leukemia. Obsolete pesticides such as mercury and cyclodienes are known to induce higher incidences of kidney damage, some irreversible; cyclodienes are linked to liver toxicity. Organophosphates and carbamates can induce a chain of responses leading to neuromuscular blockage. Many chlorinated solvents induce liver changes, kidney changes and depression of the central nervous system. There is an entire spectrum of further health effects such as headacheauseaatigue (physical)ye irritation and skin rash for the above cited and other chemicals.

Ecosystem Effects

Not unexpectedly, soil contaminants can have significant deleterious consequences for ecosystems. There are radical soil chemistry changes which can arise from the presence of many hazardous chemicals even at low concentration of the contaminant species. These changes can manifest in the alteration of metabolism of endemic microorganisms and arthropods resident in a given soil environment. The result can be virtual eradication of some of the primary food chain, which in turn have major consequences for predator or consumer species. Even if the chemical effect on lower life forms is small, the lower pyramid levels of the food chain may ingest alien chemicals, which normally become more concentrated for each consuming rung of the food chain. Many of these effects are now well known, such as the concentration of persistent DDT materials for avian consumers, leading to weakening of egg shells, increased chick mortality and potentially species extinction.

Effects occur to agricultural lands which have certain types of soil contamination. Contaminants typically alter plant metabolism, most commonly to reduce crop yields. This has a secondary effect upon soil conservation, since the languishing crops cannot shield the earth's soil mantle from erosion phenomena. Some of these chemical contaminants have long half-lives and in other cases derivative chemicals are formed from decay of primary soil contaminants.

RegulatoryFramework

United States of America

Until about 1970 there was little widespread awareness of the worldwide scope of soil contamination or its health risks. In fact, areas of concern were often viewed as unusual or isolated incidents. Since then, the U.S. has established guidelines for handling hazardous waste and the cleanup of soil pollution. In 1980 the U.S. Superfund/CERCLA established strict rules on legal liability for soil contamination. Not only did CERCLA stimulate identification and cleanup of thousands of sites, but it raised awareness of property buyers and sellers to make soil pollution a focal issue of land use and management practices.

While estimates of remaining soil cleanup in the U.S. may exceed 200,000 sites, in other industrialized countries there is a lag of identification and cleanup functions. Even though their use of chemicals is lower than industrialized countries, often their controls and regulatory framework is quite weak. For example, some persistent pesticides that have been banned in the U.S. are in widespread uncontrolled use in developing countries. It is worth noting that the cost of cleaning up a soil contaminated site can range from as little as about $10,000 for a small spill, which can be simply excavated, to millions of dollars for a widespread event, especially for a chemical that is very mobile such as perchloroethylene.

China

China, an economy that regularly records double digit annual economic growth, has little or no legislation to protect the environment. Currently, given the amount of land in question (up to one-tenth of China's cultivatable land may be polluted), the degree of the pollution in specific locations is unclear, making both prevention and remedy difficult. There are no laws or environmental standards regarding soil. Funding is limited, too, so there is little advanced scientific study of China's soil taking place. The severity of the pollution is not known by the public or business population, and the situation is most likely worsening as a result.

United Kingdom

Generic guidance commonly used in the UK are the Soil Guideline Values published by DEFRA and the Environment Agency. These are screening values that demonstrate the minimal acceptable level of a substance. Above this there can be no assurances in terms of significant risk of harm to human health. These have been derived using the Contaminated Land Exposure Asseeement Model (CLEA UK). Certain input parameters such as Health Criteria Values, age and land use are fed into CLEA UK to obtain a probablistic output.

Guidance by the Inter Departmental Committee for the Redevelopment of Contaminated Land (ICRCL) has been formally withdrawn by the Department for Environment, Food and Rural Affairs (DEFRA), for use as a prescriptive document to determine the potential need for remediation or further assessment. Therefore, no further reference is made to these former guideline values.

Other generic guidance that exists (to put the concentration of a particular contaminant in context), includes the United States EPA Region 9 Preliminary Remediation Goals (US PRGs), the US EPA Region 3 Risk Based Concentrations (US EPA RBCs) and National Environment Protection Council of Australia Guideline on Investigation Levels in Soil and Groundwater.

However international guidance should only be used in the UK with clear justification. This is because foreign standards are usually particular to that country due to drivers such as political policy, geology, flood regime and epidemiology. It is generally accepted by UK regulators that only robust scientific methods that relate to the UK should be used.

The CLEA model published by DEFRA and the Environment Agency (EA) in March 2002 sets a framework for the appropriate assessment of risks to human health from contaminated land, as required by Part IIA of the Environmental Protection Act 1990. As part of this framework, generic Soil Guideline Values (SGVs) have currently been derived for ten contaminants to be used as "intervention values". These values should not be considered as remedial targets but values above which further detailed assessment should be considered.

Three sets of CLEA SGVs have been produced for three different land uses, namely:

* residential (with and without plant uptake),
* allotments,
* commercial/industrial.

It is intended that the SGVs replace the former ICRCL values. It should be noted that the CLEA SGVs relate to assessing chronic (long term) risks to human health and do not apply to the protection of ground workers during construction, or other potential receptors such as groundwater, buildings, plants or other ecosystems. The CLEA SGVs are not directly applicable to a site completely covered in hardstanding, as there is no direct exposure route to contaminated soils.

To date, the first ten of fifty-five contaminant SGVs have been published, for the following: arsenic, cadmium, chromium, lead, inorganic mercury, nickel, selenium ethyl benzene, phenol and toluene. Draft SGVs for benzene, naphthalene and xylene have been produced but their publication is on hold. Toxicological data (Tox) has been published for each of these contaminants as well as for benzo[a]pyrene, benzene, dioxins, furans and dioxin-like PCBs, naphthalene, vinyl chloride, 1,1,2,2 tetrachloroethane and 1,1,1,2 tetrachloroethane, 1,1,1 trichloroethane, tetrachloroethene, carbon tetrachloride, 1,2-dichloroethane, trichloroethene and xylene. The SGVs for ethyl benzene, phenol and toluene are dependent on the soil organic matter (SOM) content (which can be calculated from the total organic carbon (TOC) content). As an initial screen the SGVs for 1 per cent SOM are considered to be appropriate.

Cleanup Options

Cleanup or remediation is analyzed by environmental scientists who utilize field measurement of soil chemicals and also apply computer models for analyzing transport and fate of soil chemicals. Thousands of soil contamination cases are currently in active cleanup across the U.S. as of 2006. There are several principal strategies for remediation:

- ❖ Excavate soil and remove it to a disposal site away from ready pathways for human or sensitive ecosystem contact. This technique also applies to dredging of bay muds containing toxins.
- ❖ Aeration of soils at the contaminated site (with attendant risk of creating air pollution).
- ❖ Bioremediation, involving microbial digestion of certain organic chemicals. Techniques used in bioremediation include landfarming, biostimulation and bioaugmentation soil biota with commercially available microflora.
- ❖ Extraction of groundwater or soil vapor with an active electromechanical system, with subsequent stripping of the contaminants from the extract.
- ❖ Containment of the soil contaminants (such as by capping or paving over in place).

5

Global Environmental Problems: Atmosphere and Environmental Issues

5.1 Natural Environment

The *natural environment*, commonly referred to simply as the environment, is all living and non-living things that occur naturally on Earth or some part of it (e.g. the natural environment in a country). This includes complete ecological units that function as natural systems without massive human intervention, including all vegetation, animals, microorganisms, rocks, atmosphere and natural phenomena that occur within their boundaries. And it includes universal natural resources and physical phenomena that lack clear-cut boundaries, such as air, water, and climate, as well as energy, radiation, electric charge, and magnetism, not originating from human activity.

5.2 Environmental Disaster

An environmental disaster is a disaster that is due to human activity and should not be confused with natural disasters. In this case, the impact of humans' alteration of the ecosystem has led to widespread and/or long-lasting consequences. It can include the deaths of animals (including humans) and plant systems, or severe disruption of human life, possibly requiring migration.

Difficulty in Categorization

Some might view, for example, the Three Gorges Dam as an environmental disaster, requiring the migration of 1 million people. Others might see it as beneficial to stop flooding.

Some might see the destruction of most of the North American forests as beneficial, as it cleared land for farming and other uses. In Ireland, the clearing of forest led to the formation of bogs, which some people like for their beauty, as well as their products such as peat moss. Others might see this deforestation as negative.

More cynical examples would be, for example, that Saddam Hussein felt it was beneficial to get rid of the Madan people by draining the Al-Hawizeh marsh, because they had joined the United States in the first Gulf War. Another example is the depopulation of the American Bison. It was thought by General William Sherman and others to be a good way to get rid of the American Indians living in the Great Plains, and would make way for the exploding population of the United States of America to take over the area. The pulverization of the Twin Towers with the Collapse of the World Trade Center and the release of Ground Zero dust propelled fine toxic dust into the vicinity; some have challenged whether this was an *industrial* disaster since this was a commercial site.

5.3 List of Environmental Issues

This is a list of environmental issues that are due to human activity. These articles relate to the anthropogenic effects on the natural environment.

- ❖ *Climate change*—Global warming, Fossil fuels, Sea level rise, Effects of the automobile on societies
- ❖ *Conservation*—Genetic erosion, Holocene extinction event, Invasive species, Species extinction, Habitat destruction, Habitat fragmentation, Pollinator decline, Coral bleaching, Whaling
- ❖ *Dams*—Environmental impacts of dams
- ❖ *Energy*—Energy conservation, Renewable energy, Efficient energy use, Renewable energy commercialization.
- ❖ *Genetic engineering*—Genetic pollution
- ❖ *Intensive farming*—Overgrazing, Irrigation, Monoculture
- ❖ *Land degradation*—Land pollution, Desertification
- ❖ *Soil*—Soil conservation, Soil erosion, Soil contamination, Soil salination
- ❖ *Nanotechnology*—Nanotoxicology, Nanopollution
- ❖ *Nuclear issues*—Nuclear fallout, Nuclear meltdown, Nuclear power
- ❖ *Overpopulation*—Burial
- ❖ *Ozone depletion*
- ❖ *Pollution*—Air pollution, Light pollution, Noise pollution, Thermal pollution
    - ● *Water pollution*—Acid rain, Eutrophication, Ocean dumping, Oil spills, Water crisis

- ❖ *Resource depletion*—Exploitation of natural resources
  - ● *Fishing*—Blast fishing, Bottom trawling, Cyanide fishing, Ghost nets, Illegal, unreported and unregulated fishing, Environmental effects of fishing, Overfishing
  - ● *Logging*—Clearcutting, Deforestation, Illegal logging
  - ● *Mining*—Acid mine drainage
- ❖ *Toxins*—Chlorofluorocarbons, DDT, Dioxin, Heavy metals, Herbicides, Pesticides, Toxics use reduction, Toxic waste
- ❖ *Urban sprawl*
- ❖ *Waste*—Waste disposal incidents

5.4 Ozone Depletion

Ozone depletion describes two distinct, but related observations: a slow, steady decline of about 4 percent per decade in the total amount of ozone in Earth's stratosphere since the late 1970s; and a much larger, but seasonal, decrease in stratospheric ozone over Earth's polar regions during the same period. The latter phenomenon is commonly referred to as the ozone hole. In addition to this well-known stratospheric ozone depletion, there are also tropospheric ozone depletion events, which occur near the surface in polar regions during spring.

The detailed mechanism by which the polar ozone holes form is different from that for the mid-latitude thinning, but the most important process in both trends is catalytic destruction of ozone by atomic chlorine and bromine. The main source of these halogen atoms in the stratosphere is photodissociation of chlorofluorocarbon (CFC) compounds, commonly called freons, and of bromofluorocarbon compounds known as halons. These compounds are transported into the stratosphere after being emitted at the surface. Both ozone depletion mechanisms strengthened as emissions of CFCs and halons increased.

CFCs and other contributory substances are commonly referred to as ozone-depleting substances (ODS). Since the ozone layer prevents most harmful UVB wavelengths (270-315 nm) of ultraviolet light (UV light) from passing through the Earth's atmosphere, observed and projected decreases in ozone have generated worldwide concern leading to adoption of the Montreal Protocol banning the production of CFCs and halons as well as related ozone depleting chemicals such as carbon tetrachloride and trichloroethane. It is suspected that a variety of biological consequences such as increases in skin cancer, damage to plants, and reduction of plankton populations in the ocean's photic zone may result from the increased UV exposure due to ozone depletion.

Ozone Cycle Overview

Three forms (or allotropes) of oxygen are involved in the ozone-oxygen cycle: Oxygen atoms (O or atomic oxygen), oxygen gas (O_2 or diatomic oxygen), and ozone gas (O_3 or triatomic oxygen). Ozone is formed in the stratosphere when oxygen molecules photodissociate after absorbing an ultraviolet photon whose wavelength is shorter than 240 nm. This produces two oxygen atoms. The atomic oxygen then combines with O_2 to create O_3. Ozone molecules absorb UV light between 310 and 200 nm, following which ozone splits into a molecule of O_2 and an oxygen atom. The oxygen atom then joins up with an oxygen molecule to regenerate ozone. This is a continuing process which terminates when an oxygen atom "recombines" with an ozone molecule to make two O_2 molecules:

$$O + O_3 \rightarrow 2\,O_2$$

The overall amount of ozone in the stratosphere is determined by a balance between photochemical production and recombination.

Ozone can be destroyed by a number of free radical catalysts, the most important of which are the hydroxyl radical (OH·), the nitric oxide radical (NO·) and atomic chlorine (Cl·) and bromine (Br·). All of these have both natural and anthropogenic (manmade) sources; at the present time, most of the OH· and NO· in the stratosphere is of natural origin, but human activity has dramatically increased the high in oxygen chlorine and bromine. These elements are found in certain stable organic compounds, especially chlorofluorocarbons (CFCs), which may find their way to the stratosphere without being destroyed in the troposphere due to their low reactivity. Once in the stratosphere, the Cl and Br atoms are liberated from the parent compounds by the action of ultraviolet light, e.g. ('h' is Planck's constant, 'i' is frequency of electromagnetic radiation)

$$CFCl_3 + h\upsilon \longrightarrow CFCl_2 + Cl$$

The Cl and Br atoms can then destroy ozone molecules through a variety of catalytic cycles. In the simplest example of such a cycle, a chlorine atom reacts with an ozone molecule, taking an oxygen atom with it (forming ClO) and leaving a normal oxygen molecule. A free oxygen atom then takes away the oxygen from the ClO, and the final result is an oxygen molecule and a chlorine atom, which then reinitiates the cycle. The chemical shorthand for these gas-phase reactions is:

$$Cl + O_3 \longrightarrow ClO + O_2$$
$$ClO + O \longrightarrow Cl + O_2$$

The net reaction is: $O_3 + O \rightarrow 2\,O_2$, the "recombination" reaction given above.

The overall effect is to increase the rate of recombination, leading to an overall decrease in the amount of ozone. For this particular mechanism to operate there must be a source of O atoms, which is primarily the photo dissociation of O_3; thus this mechanism is only important in the upper stratosphere where such atoms are abundant. More complicated mechanisms have been discovered that lead to ozone destruction in the lower stratosphere as well.

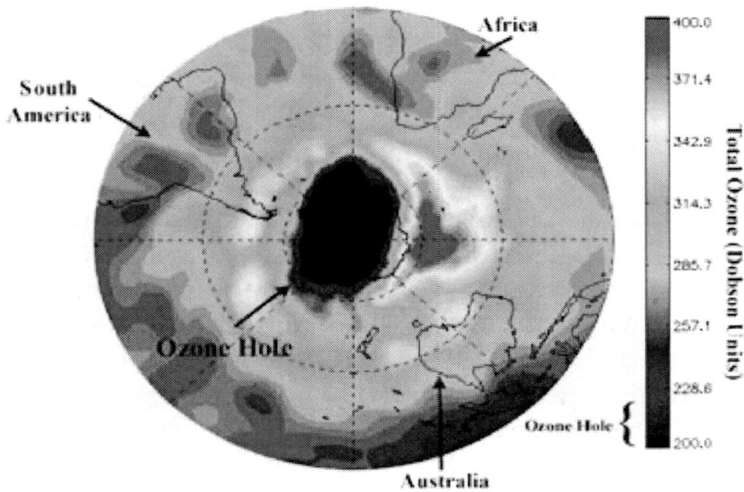

Figure 5.1: Total Ozone on September 29, 1997.

A single chlorine atom would keep on destroying ozone for up to two years (the time scale for transport back down to the troposphere) were it not for reactions that remove them from this cycle by forming reservoir species such as hydrogen chloride (HCl) and chlorine nitrate ($ClONO_2$). On a per atom basis, bromine is even more efficient than chlorine at destroying ozone, but there is much less bromine in the atmosphere at present. As a result, both chlorine and bromine contribute significantly to the overall ozone depletion. Laboratory studies have shown that fluorine and iodine atoms participate in analogous catalytic cycles. However, in the Earth's stratosphere, fluorine atoms react rapidly with water and methane to form strongly-bound HF, while organic molecules which contain iodine react so rapidly in the lower atmosphere that they do not reach the stratosphere in significant quantities. Furthermore, a single chlorine atom is able to react with 100,000 ozone molecules. This fact plus the amount of chlorine released into the atmosphere by chlorofluorocarbons (CFCs) yearly demonstrates how dangerous CFCs are to the environment.

Quantitative Understanding of the Chemical Ozone Loss Process

New research on the breakdown of a key molecule in these ozone-depleting chemicals, dichlorine peroxide (Cl2O2), calls into question the completeness of present atmospheric models of polar ozone depletion. Specifically, chemists at NASA's Jet Propulsion Laboratory in Pasadena, California, found in 2007 that the temperatures, and the spectrum and intensity of radiation present in the stratosphere created conditions insufficient to allow the rate of chemical-breakdown required to release chlorine radicals in the volume necessary to explain observed rates of ozone depletion. Instead, laboratory tests, designed to be the most accurate reflection of stratospheric conditions to date, showed the decay of the crucial molecule almost a magnitude lower than previously thought.

Observations on Ozone Layer Depletion

The most pronounced decrease in ozone has been in the lower stratosphere. However, the ozone hole is most usually measured not in terms of ozone concentrations at these levels (which are typically of a few parts per million) but by reduction in the total *column ozone*, above a point on the Earth's surface, which is normally expressed in Dobson units, abbreviated as "DU". Marked decreases in column ozone in the Antarctic spring and early summer compared to the early 1970s and before have been observed using instruments such as the Total Ozone Mapping Spectrometer (TOMS).

Reductions of up to 70 per cent in the ozone column observed in the austral (southern hemispheric) spring over Antarctica and first reported in 1985 (Farman *et al.* 1985) are continuing. Through the 1990s, total column ozone in September and October have continued to be 40-50 per cent lower than pre-ozone-hole values. In the Arctic the amount lost is more variable year-to-year than in the Antarctic. The greatest declines, up to 30 per cent, are in the winter and spring, when the stratosphere is colder.

Reactions that take place on polar stratospheric clouds (PSCs) play an important role in enhancing ozone depletion. PSCs form more readily in the extreme cold of Antarctic stratosphere. This is why ozone holes first formed, and are deeper, over Antarctica. Early models failed to take PSCs into account and predicted a gradual global depletion, which is why the sudden Antarctic ozone hole was such a surprise to many scientists.

In middle latitudes it is preferable to speak of ozone depletion rather than holes. Declines are about 3 per cent below pre-1980 values for 35-60°N and about 6 per cent for 35-60°S. In the tropics, there are no significant trends.

Ozone depletion also explains much of the observed reduction in stratospheric and upper tropospheric temperatures. The source of the warmth of the stratosphere is the absorption of UV radiation by ozone, hence reduced ozone leads to cooling. Some stratospheric cooling is also predicted from increases in greenhouse gases such as CO_2; however the ozone-induced cooling appears to be dominant.

Predictions of ozone levels remain difficult. The World Meteorological Organization Global Ozone Research and Monitoring Project—Report No. 44 comes out strongly in favour for the Montreal Protocol, but notes that a UNEP 1994 Assessment overestimated ozone loss for the 1994-1997 period.

Chemicals in the Atmosphere

CFCs in the Atmosphere

Chlorofluorocarbons (CFCs) were invented by Thomas Midgley in the 1920s. They were used in air conditioning/cooling units, as aerosol spray propellants prior to the 1980s, and in the cleaning processes of delicate electronic equipment. They also occur as by-products of some chemical processes. No significant natural sources have ever been identified for these compounds— their presence in the atmosphere is due almost entirely to human manufacture. As mentioned in the *ozone cycle overview* above, when such ozone-depleting chemicals reach the stratosphere, they are dissociated by ultraviolet light to release chlorine atoms. The chlorine atoms act as a catalyst, and each can break down tens of thousands of ozone molecules before being removed from the stratosphere. Given the longevity of CFC molecules, recovery times are measured in decades. It is calculated that a CFC molecule takes an average of 15 years to go from the ground level up to the upper atmosphere, and it can stay there for about a century, destroying up to one hundred thousand ozone molecules during that time.

Verification of Observations

Scientists have been increasingly able to attribute the observed ozone depletion to the increase of anthropogenic halogen compounds from CFCs by the use of complex chemistry transport models and their validation against observational data (e.g. SLIMCAT, CLaMS). These models work by combining satellite measurements of chemical concentrations and meteorological fields with chemical reaction rate constants obtained in lab experiments. They are able to identify not only the key chemical reactions but also the transport processes which bring CFC photolysis products into contact with ozone.

The Ozone Hole and its Causes

The Antarctic ozone hole is an area of the Antarctic stratosphere in which the recent ozone levels have dropped to as low as 33 per cent of their pre-1975 values. The ozone hole occurs during the Antarctic spring, from September to early December, as strong westerly winds start to circulate around the continent and create an atmospheric container. Within this polar vortex, over 50 per cent of the lower stratospheric ozone is destroyed during the Antarctic spring.

As explained above, the overall cause of ozone depletion is the presence of chlorine-containing source gases (primarily CFCs and related halocarbons). In the presence of UV light, these gases dissociate, releasing chlorine atoms, which then go on to catalyze ozone destruction. The Cl-catalyzed ozone depletion can take place in the gas phase, but it is dramatically enhanced in the presence of polar stratospheric clouds (PSCs).

These polar stratospheric clouds form during winter, in the extreme cold. Polar winters are dark, consisting of 3 months without solar radiation (sunlight). Not only lack of sunlight contributes to a decrease in temperature but also the polar vortex traps and chills air. Temperatures hover around or below $-80\,°C$. These low temperatures form cloud particles and are composed of either nitric acid (Type I PSC) or ice (Type II PSC). Both types provide surfaces for chemical reactions that lead to ozone destruction.

The photochemical processes involved are complex but well understood. The key observation is that, ordinarily, most of the chlorine in the stratosphere resides in stable "reservoir" compounds, primarily hydrogen chloride (HCl) and chlorine nitrate ($ClONO_2$). During the Antarctic winter and spring, however, reactions on the surface of the polar stratospheric cloud particles convert these "reservoir" compounds into reactive free radicals (Cl and ClO). The clouds can also remove NO_2 from the atmosphere by converting it to nitric acid, which prevents the newly formed ClO from being converted back into $ClONO_2$.

The role of sunlight in ozone depletion is the reason why the Antarctic ozone depletion is greatest during spring. During winter, even though PSCs are at their most abundant, there is no light over the pole to drive the chemical reactions. During the spring, however, the sun comes out, providing energy to drive photochemical reactions, and melt the polar stratospheric clouds, releasing the trapped compounds.

Most of the ozone that is destroyed is in the lower stratosphere, in contrast to the much smaller ozone depletion through homogeneous gas phase reactions, which occurs primarily in the upper stratosphere.

Warming temperatures near the end of spring break up the vortex around mid-December. As warm, ozone-rich air flows in from lower latitudes, the PSCs are destroyed, the ozone depletion process shuts down, and the ozone hole heals.

Interest in Ozone Layer Depletion

While the effect of the Antarctic ozone hole in decreasing the global ozone is relatively small, estimated at about 4 per cent per decade, the hole has generated a great deal of interest because:

- ❖ The decrease in the ozone layer was predicted in the early 1980s to be roughly 7 per cent over a sixty-year period.
- ❖ The sudden recognition in 1985 that there was a substantial "hole" was widely reported in the press. The especially rapid ozone depletion in Antarctica had previously been dismissed as a measurement error.
- ❖ Many were worried that ozone holes might start to appear over other areas of the globe but to date the only other large-scale depletion is a smaller ozone "dimple" observed during the Arctic spring over the North Pole. Ozone at middle latitudes has declined, but by a much smaller extent (about 4-5 per cent decrease).
- ❖ If the conditions became more severe (cooler stratospheric temperatures, more stratospheric clouds, more active chlorine), then global ozone may decrease at a much greater pace. Standard global warming theory predicts that the stratosphere will cool.
- ❖ When the Antarctic ozone hole breaks up, the ozone-depleted air drifts out into nearby areas. Decreases in the ozone level of up to 10 per cent have been reported in New Zealand in the month following the break-up of the Antarctic ozone hole.

Consequences of Ozone Layer Depletion

Since the ozone layer absorbs UVB ultraviolet light from the Sun, ozone layer depletion is expected to increase surface UVB levels, which could lead to damage, including increases in skin cancer. This was the reason for the Montreal Protocol. Although decreases in stratospheric ozone are well-tied to CFCs and there are good theoretical reasons to believe that decreases in ozone will lead to increases in surface UVB, there is no direct observational evidence linking ozone depletion to higher incidence of skin cancer in human beings. This is partly due to the fact that UVA, which has also been implicated in some forms of skin cancer, is not absorbed by ozone, and it is nearly impossible to control statistics for lifestyle changes in the populace.

Increased UV

Ozone, while a minority constituent in the earth's atmosphere, is responsible for most of the absorption of UVB radiation. The amount of UVB radiation that penetrates through the ozone layer decreases exponentially with the slant-path thickness/density of the layer. Correspondingly, a decrease in atmospheric ozone is expected to give rise to significantly increased levels of UVB near the surface.

Increases in surface UVB due to the ozone hole can be partially inferred by radiative transfer model calculations, but cannot be calculated from direct measurements because of the lack of reliable historical (pre-ozone-hole) surface UV data, although more recent surface UV observation measurement programmes exist (e.g. at Lauder, New Zealand).

Because it is this same UV radiation that creates ozone in the ozone layer from O_2 (regular oxygen) in the first place, a reduction in stratospheric ozone would actually tend to increase photochemical production of ozone at lower levels (in the troposphere), although the overall observed trends in total column ozone still show a decrease, largely because ozone produced lower down has a naturally shorter photochemical lifetime, so it is destroyed before the concentrations could reach a level which would compensate for the ozone reduction higher up.

Biological Effects of Increased UV and Microwave Radiation from a Depleted Ozone Layer

The main public concern regarding the ozone hole has been the effects of surface UV on human health. So far, ozone depletion in most locations has been typically a few percent and, as noted above, no direct evidence of health damage is available in most latitudes. Were the high levels of depletion seen in the ozone hole ever to be common across the globe, the effects could be substantially more dramatic. As the ozone hole over Antarctica has in some instances grown so large as to reach southern parts of Australia and New

Zealand, environmentalists have been concerned that the increase in surface UV could be significant.

Effects of Ozone Layer Depletion on Humans

UVB (the higher energy UV radiation absorbed by ozone) is generally accepted to be a contributory factor to skin cancer. In addition, increased surface UV leads to increased tropospheric ozone, which is a health risk to humans. The increased surface UV also represents an increase in the vitamin D synthetic capacity of the sunlight.

The cancer preventive effects of vitamin D represent a possible beneficial effect of ozone depletion. In terms of health costs, the possible benefits of increased UV irradiance may outweigh the burden.

1. *Basal and Squamous Cell Carcinomas:* The most common forms of skin cancer in humans, basal and squamous cell carcinomas, have been strongly linked to UVB exposure. The mechanism by which UVB induces these cancers is well understood—absorption of UVB radiation causes the pyrimidine bases in the DNA molecule to form dimers, resulting in transcription errors when the DNA replicates. These cancers are relatively mild and rarely fatal, although the treatment of squamous cell carcinoma sometimes requires extensive reconstructive surgery. By combining epidemiological data with results of animal studies, scientists have estimated that a one percent decrease in stratospheric ozone would increase the incidence of these cancers by 2 per cent.

2. *Malignant Melanoma:* Another form of skin cancer, malignant melanoma, is much less common but far more dangerous, being lethal in about 15 per cent-20 per cent of the cases diagnosed. The relationship between malignant melanoma and ultraviolet exposure is not yet well understood, but it appears that both UVB and UVA are involved. Experiments on fish suggest that 90 to 95 per cent of malignant melanomas may be due to UVA and visible radiation whereas experiments on opossums suggest a larger role for UVB. Because of this uncertainty, it is difficult to estimate the impact of ozone depletion on melanoma incidence. One study showed that a 10 per cent increase in UVB radiation was associated with a 19 per cent increase in melanomas for men and 16 per cent for women. A study of people in Punta Arenas, at the southern tip of Chile, showed a 56 per cent increase in melanoma and a 46 per cent increase in

nonmelanoma skin cancer over a period of seven years, along with decreased ozone and increased UVB levels.

3. *Cortical Cataracts:* Studies are suggestive of an association between ocular cortical cataracts and UV-B exposure, using crude approximations of exposure and various cataract assessment techniques. A detailed assessment of ocular exposure to UV-B was carried out in a study on Chesapeake Bay Watermen, where increases in average annual ocular exposure were associated with increasing risk of cortical opacity . In this highly exposed group of predominantly white males, the evidence linking cortical opacities to sunlight exposure was the strongest to date. However, subsequent data from a population-based study in Beaver Dam, WI suggested the risk may be confined to men. In the Beaver Dam study, the exposures among women were lower than exposures among men, and no association was seen. Moreover, there were no data linking sunlight exposure to risk of cataract in African Americans, although other eye diseases have different prevalences among the different racial groups, and cortical opacity appears to be higher in African Americans compared with whites.

4. *Increased Tropospheric Ozone:* Increased surface UV leads to increased tropospheric ozone. Ground-level ozone is generally recognized to be a health risk, as ozone is toxic due to its strong oxidant properties. At this time, ozone at ground level is produced mainly by the action of UV radiation on combustion gases from vehicle exhausts.

Effects on Crops

An increase of UV radiation would be expected to affect crops. A number of economically important species of plants, such as rice, depend on cyanobacteria residing on their roots for the retention of nitrogen. Cyanobacteria are sensitive to UV light and they would be affected by its increase.

Effects on Plankton

Research has shown a widespread extinction of plankton 2 million years ago that coincided with a nearby supernova. There is a difference in the orientation and motility of planktons when excess of UV rays reach earth. Researchers speculate that the extinction was caused by a significant weakening of the ozone layer at that time when the radiation from the supernova produced nitrogen oxides that catalyzed the destruction of ozone (plankton are particularly susceptible to effects of UV light, and are vitally important to marine food webs).

Public Policy in Response to the Ozone Hole

The full extent of the damage that CFCs have caused to the ozone layer is not known and will not be known for decades; however, marked decreases in column ozone have already been observed (as explained above).

After a 1976 report by the U.S. National Academy of Sciences concluded that credible scientific evidence supported the ozone depletion hypothesis, a few countries, including the United States, Canada, Sweden, and Norway, moved to eliminate the use of CFCs in aerosol spray cans. At the time this was widely regarded as a first step towards a more comprehensive regulation policy, but progress in this direction slowed in subsequent years, due to a combination of political factors (continued resistance from the halocarbon industry and a general change in attitude towards environmental regulation during the first two years of the Reagan administration) and scientific developments (subsequent National Academy assessments which indicated that the first estimates of the magnitude of ozone depletion had been overly large). The European Community rejected proposals to ban CFCs in aerosol sprays while even in the U.S., CFCs continued to be used as refrigerants and for cleaning circuit boards. Worldwide CFC production fell sharply after the U.S. aerosol ban, but by 1986 had returned nearly to its 1976 level. In 1980, DuPont closed down its research programme into halocarbon alternatives.

The US Government's attitude began to change again in 1983, when William Ruckelshaus replaced Anne M. Burford as Administrator of the United States Environmental Protection Agency. Under Ruckelshaus and his successor, Lee Thomas, the EPA pushed for an international approach to halocarbon regulations. In 1985 20 nations, including most of the major CFC producers, signed the Vienna Convention which established a framework for negotiating international regulations on ozone-depleting substances. That same year, the discovery of the Antarctic ozone hole was announced, causing a revival in public attention to the issue. In 1987, representatives from 43 nations signed the Montreal Protocol. Meanwhile, the halocarbon industry shifted its position and started supporting a protocol to limit CFC production. The reasons for this were in part explained by "Dr. Mostafa Tolba, former head of the UN Environment Programme, who was quoted in the June 30, 1990 edition of The New Scientist, '.the chemical industry supported the

Montreal Protocol in 1987 because it set up a worldwide schedule for phasing out CFCs, which [were] no longer protected by patents. This provided companies with an equal opportunity to market new, more profitable compounds.'"

At Montreal, the participants agreed to freeze production of CFCs at 1986 levels and to reduce production by 50 per cent by 1999. After a series of scientific expeditions to the Antarctic produced convincing evidence that the ozone hole was indeed caused by chlorine and bromine from manmade organohalogens, the Montreal Protocol was strengthened at a 1990 meeting in London. The participants agreed to phase out CFCs and halons entirely (aside from a very small amount marked for certain "essential" uses, such as asthma inhalers) by 2000. At a 1992 meeting in Copenhagen, the phase out date was moved up to 1996.

To some extent, CFCs have been replaced by the less damaging hydro-chloro-fluoro-carbons (HCFCs), although concerns remain regarding HCFCs also. In some applications, hydro-fluoro-carbons (HFCs) have been used to replace CFCs. HFCs, which contain no chlorine or bromine, do not contribute at all to ozone depletion although they are potent greenhouse gases. The best known of these compounds is probably HFC-134a (R-134a), which in the United States has largely replaced CFC-12 (R-12) in automobile air conditioners. In laboratory analytics (a former "essential" use) the ozone depleting substances can be replaced with various other solvents.

Ozone Diplomacy, by Richard Benedick (Harvard University Press, 1991) gives a detailed account of the negotiation process that led to the Montreal Protocol. Pielke and Betsill provide an extensive review of early US government responses to the emerging science of ozone depletion by CFCs.

Current Events and Future Prospects of Ozone Depletion

Since the adoption and strengthening of the Montreal Protocol has led to reductions in the emissions of CFCs, atmospheric concentrations of the most significant compounds have been declining. These substances are being gradually removed from the atmosphere. By 2015, the Antarctic ozone hole would have reduced by only 1 million km² out of 25 (Newman *et al.*, 2004); complete recovery of the Antarctic ozone layer will not occur until the year 2050 or later. Work has suggested that a detectable (and statistically significant) recovery will not occur until around 2024, with ozone levels recovering to 1980 levels by around 2068.

There is a slight caveat to this, however. Global warming from CO_2 is expected to cool the stratosphere. This, in turn, would lead to a relative increase in ozone depletion and the frequency of ozone holes. The effect may not be linear; ozone holes form because of polar stratospheric clouds; the formation of polar stratospheric clouds has a temperature threshold above which they will not form; cooling of the Arctic stratosphere might lead to Antarctic-ozone-hole-like conditions. But at the moment this is not clear.

Even though the stratosphere as a whole is cooling, high-latitude areas may become increasingly predisposed to springtime stratospheric warming events as weather patterns change in response to higher greenhouse gas loading. This would cause PSCs to disappear earlier in the season, and may explain why Antarctic ozone hole seasons have tended to end somewhat earlier since 2000 as compared with the most prolonged ozone holes of the 1990s.

The decrease in ozone-depleting chemicals has also been significantly affected by a decrease in bromine-containing chemicals. The data suggest that substantial natural sources exist for atmospheric methyl bromide (CH_3Br).

The 2004 ozone hole ended in November 2004, daily minimum stratospheric temperatures in the Antarctic lower stratosphere increased to levels that are too warm for the formation of polar stratospheric clouds (PSCs) about 2 to 3 weeks earlier than in most recent years.

The Arctic winter of 2005 was extremely cold in the stratosphere; PSCs were abundant over many high-latitude areas until dissipated by a big warming event, which started in the upper stratosphere during February and spread throughout the Arctic stratosphere in March. The size of the Arctic area of anomalously low total ozone in 2004-2005 was larger than in any year since 1997. The predominance of anomalously low total ozone values in the Arctic region in the winter of 2004-2005 is attributed to the very low stratospheric temperatures and meteorological conditions favourable for ozone destruction along with the continued presence of ozone destroying chemicals in the stratosphere.

A 2005 IPCC summary of ozone issues observed that observations and model calculations suggest that the global average amount of ozone depletion has now approximately stabilized. Although considerable variability in ozone is expected from year to year, including in polar regions where depletion is largest, the ozone layer is expected to begin to recover in coming decades due to declining ozone-depleting substance concentrations, assuming full compliance with the Montreal Protocol.

Temperatures during the Arctic winter of 2006 stayed fairly close to the long-term average until late January, with minimum readings frequently cold enough to produce PSCs. During the last week of January, however, a major warming event sent temperatures well above normal—much too warm to support PSCs. By the time temperatures dropped back to near normal in March, the seasonal norm was well above the PSC threshold. Preliminary satellite instrument-generated ozone maps show seasonal ozone buildup slightly below the long-term means for the Northern Hemisphere as a whole, although some high ozone events have occurred. During March 2006, the Arctic stratosphere poleward of 60 degrees North Latitude was free of anomalously low ozone areas except during the three-day period from March 17 to 19 when the total ozone cover fell below 300 DU over part of the North Atlantic region from Greenland to Scandinavia.

The area where total column ozone is less than 220 DU (the accepted definition of the boundary of the ozone hole) was relatively small until around 20 August 2006. Since then the ozone hole area increased rapidly, peaking at 29 million km² September 24. In October 2006, NASA reported that the year's ozone hole set a new area record with a daily average of 26 million km² between 7 September and 13 October 2006; total ozone thicknesses fell as low as 85 DU on October 8. The two factors combined, 2006 sees the worst level of depletion in recorded ozone history. The depletion is attributed to the temperatures above the Antarctic reaching the lowest recording since comprehensive records began in 1979.

The Antarctic ozone hole is expected to continue for decades. Ozone concentrations in the lower stratosphere over Antarctica will increase by 5 per cent-10 per cent by 2020 and return to pre-1980 levels by about 2060-2075, 10-25 years later than predicted in earlier assessments. This is because of revised estimates of atmospheric concentrations of Ozone Depleting Substances—and a larger predicted future usage in developing countries. Another factor which may aggravate ozone depletion is the draw-down of nitrogen oxides from above the stratosphere due to changing wind patterns.

History of the Research

The basic physical and chemical processes that lead to the formation of an ozone layer in the earth's stratosphere were discovered by Sydney Chapman in 1930. These are discussed in the article Ozone-oxygen cycle—briefly, short-wavelength UV radiation splits an oxygen (O_2) molecule into two oxygen (O) atoms, which then combine with other oxygen molecules to form ozone. Ozone is removed when an oxygen atom and an ozone molecule "recombine" to form two oxygen molecules, i.e. $O + O_3 \rightarrow 2O_2$. In the 1950s, David Bates and Marcel Nicolet presented evidence that various free radicals, in particular hydroxyl (OH) and nitric oxide (NO), could catalyze this recombination reaction, reducing the overall amount of ozone. These free radicals were known to be present in the stratosphere, and so were regarded as part of the natural balance—it was estimated that in their absence, the ozone layer would be about twice as thick as it currently is.

In 1970 Prof. Paul Crutzen pointed out that emissions of *nitrous* oxide (N_2O), a stable, long-lived gas produced by soil bacteria, from the earth's surface could affect the amount of *nitric* oxide (NO) in the stratosphere. Crutzen showed that nitrous oxide lives long enough to reach the stratosphere, where it is converted into NO. Crutzen then noted that increasing use of fertilizers might have led to an increase in nitrous oxide emissions over the natural background, which would in turn result in an increase in the amount of NO in the stratosphere. Thus human activity could have an impact on the stratospheric ozone layer. In the following year, Crutzen and (independently) Harold Johnston suggested that NO emissions from supersonic aircraft, which fly in the lower stratosphere, could also deplete the ozone layer.

The Rowland-Molina Hypothesis

In 1974 Frank Sherwood Rowland, Chemistry Professor at the University of California at Irvine, and his postdoctoral associate Mario J. Molina suggested that long-lived organic halogen compounds, such as CFCs, might behave in a similar fashion as Crutzen had proposed for nitrous oxide. James Lovelock (most popularly known as the creator of the Gaia hypothesis) had recently discovered, during a cruise in the South Atlantic in 1971, that almost all of the CFC compounds manufactured since their invention in 1930 were still present in the atmosphere. Molina and Rowland concluded that, like N_2O, the CFCs would reach the stratosphere where they would be dissociated by UV light, releasing Cl atoms. (A year earlier, Richard Stolarski and Ralph Cicerone at the University of Michigan had shown that Cl is even more efficient than NO at catalyzing the destruction of ozone. Similar conclusions were reached by Michael McElroy and Steven Wofsy at Harvard University. Neither group, however, had realized that CFC's were a potentially large source of stratospheric chlorine—instead, they had been investigating the possible effects of HCl emissions from the Space Shuttle, which are very much smaller.)

The Rowland-Molina hypothesis was strongly disputed by representatives of the aerosol and halocarbon industries. The Chair of the Board of DuPont was quoted as saying that ozone depletion theory is "a science fiction tale.a load of rubbish.utter nonsense". Robert Abplanalp, the President of Precision Valve Corporation (and inventor of the first practical aerosol spray can valve), wrote to the Chancellor of UC Irvine to complain about Rowland's public statements (Roan, p 56.) Nevertheless, within three years most of the basic assumptions made by Rowland and Molina were confirmed by laboratory measurements and by direct observation in the stratosphere. The concentrations of the source gases (CFC's and related compounds) and the chlorine reservoir species (HCl and $ClONO_2$) were measured throughout the stratosphere, and demonstrated that CFCs were indeed the major source of stratospheric chlorine, and that nearly all of the CFCs emitted would eventually reach the stratosphere. Even more convincing was the measurement, by James G. Anderson and collaborators, of chlorine monoxide (ClO) in the stratosphere. ClO is produced by the reaction of Cl with ozone— its observation thus demonstrated that Cl radicals not only were present in the stratosphere but also were actually involved in destroying ozone. McElroy and Wofsy extended the work of Rowland and Molina by showing that Bromine atoms were even more effective catalysts for ozone loss than chlorine atoms and argued that the brominated organic compounds known as halons, widely used in fire extinguishers, were a potentially large source of stratospheric bromine. In 1976 the U.S. National Academy of Sciences released a report which concluded that the ozone depletion hypothesis was strongly supported by the scientific evidence. Scientists calculated that if CFC production continued to increase at the going rate of 10 per cent per year until 1990 and then remain steady, CFCs would cause a global ozone loss of 5 to 7 per cent by 1995, and a 30 to 50 per cent loss by 2050. In response the United States, Canada, Sweden and Norway banned the use of CFCs in aerosol spray cans in 1978. However, subsequent research, summarized by the National Academy in reports issued between 1979 and 1984, appeared to show that the earlier estimates of global ozone loss had been too large.

Crutzen, Molina, and Rowland were awarded the 1995 Nobel Prize in Chemistry for their work on stratospheric ozone.

The Ozone Hole

The discovery of the Antarctic "ozone hole" by British Antarctic Survey scientists Farman, Gardiner and Shanklin (announced in a paper in *Nature* in May 1985) came as a shock to the scientific community, because the observed decline in polar ozone was far larger than anyone had anticipated. Satellite measurements showing massive depletion of ozone around the south pole were becoming available at the same time. However, these were initially rejected as unreasonable by data quality control algorithms (they were filtered out as errors since the values were unexpectedly low); the ozone hole was detected only in satellite data when the raw data was reprocessed following evidence of ozone depletion in *in situ* observations. When the software was rerun without the flags, the ozone hole was seen as far back as 1976.

Susan Solomon, an atmospheric chemist at the National Oceanic and Atmospheric Administration (NOAA), proposed that chemical reactions on polar stratospheric clouds (PSCs) in the cold Antarctic stratosphere caused a massive, though localized and seasonal, increase in the amount of chlorine present in active, ozone-destroying forms. The polar stratospheric clouds in Antarctica are only formed when there are very low temperatures, as low as −80 °C, and early spring conditions. In such conditions the ice crystals of the cloud provide a suitable surface for conversion of unreactive chlorine compounds into reactive chlorine compounds which can deplete ozone easily.

Moreover the polar vortex formed over Antarctica is very tight and the reaction which occurs on the surface of the cloud crystals is far different from when it occurs in atmosphere. These conditions have led to ozone hole formation in Antarctica. This hypothesis was decisively confirmed, first by laboratory measurements and subsequently by direct measurements, from the ground and from high-altitude airplanes, of very high concentrations of chlorine monoxide (ClO) in the Antarctic stratosphere.

Alternative hypotheses, which had attributed the ozone hole to variations in solar UV radiation or to changes in atmospheric circulation patterns, were also tested and shown to be untenable.

Meanwhile, analysis of ozone measurements from the worldwide network of ground-based Dobson spectrophotometers led an international panel to conclude that the ozone layer was in fact being depleted, at all latitudes outside of the tropics. These trends were confirmed by satellite measurements. As a consequence, the major halocarbon producing nations agreed to phase out production of CFCs, halons, and related compounds, a process that was completed in 1996.

Since 1981 the United Nations Environment Programme has sponsored a series of reports on scientific assessment of ozone depletion. The most recent is from 2007 where satellite measurements have shown the hole in the ozone layer is recovering and is now the smallest it has been for about a decade.

Controversy Regarding Ozone Science and Policy

That ozone depletion takes place is not seriously disputed in the scientific community. There is a consensus among atmospheric physicists and chemists that the scientific understanding has now reached a level where countermeasures to control CFC emissions are justified, although the decision is ultimately one for policy-makers.

Despite this consensus, the science behind ozone depletion remains complex, and some who oppose the enforcement of countermeasures point to some of the uncertainties. For example, although increased UVB has been shown to constitute a melanoma risk, it has been difficult for statistical studies to establish a direct link between ozone depletion and increased rates of melanoma. Although melanomas did increase significantly during the period 1970-1990, it is difficult to separate reliably the effect of ozone depletion from the effect of changes in lifestyle factors (e.g., increasing rates of air travel).

Ozone Depletion and Global Warming

Although they are often interlinked in the mass media, the connection between global warming and ozone depletion is not strong. There are four areas of linkage:

- ❖ The same CO_2 radiative forcing that produces near-surface global warming is expected to cool the stratosphere. This cooling, in turn, is expected to produce a relative *increase* in polar ozone (O_3) depletion and the frequency of ozone holes.
- ❖ Conversely, ozone depletion represents a radiative forcing of the climate system. There are two opposing effects: Reduced ozone causes the stratosphere to absorb less solar radiation, thus cooling the stratosphere while warming the troposphere; the resulting colder stratosphere emits less long-wave radiation downward, thus cooling the troposphere. Overall, the cooling dominates; the IPCC concludes that "*observed stratospheric O_3 losses over the past two decades have caused a negative forcing of the surface-troposphere system*" of about -0.15 ± 0.10 watts per square meter (W/m^2).
- ❖ One of the strongest predictions of the greenhouse effect is that the stratosphere will cool. Although this cooling has been observed, it is not trivial to separate the effects of changes in the concentration of greenhouse gases and ozone depletion since both will lead to cooling. However, this can be done by numerical stratospheric modeling. Results from the National Oceanic and Atmospheric Administration's Geophysical Fluid Dynamics Laboratory show that above 20 km (12.4 miles), the greenhouse gases dominate the cooling.
- ❖ Ozone depleting chemicals are also greenhouse gases. The increases

in concentrations of these chemicals have produced 0.34 ± 0.03 W/m² of radiative forcing, corresponding to about 14 per cent of the total radiative forcing from increases in the concentrations of well-mixed greenhouse gases.

❖ The long term modeling of the process, its measurement, study, design of theories and testing take decades to both document, gain wide acceptance, and ultimately become the dominant paradigm. Several theories about the destruction of ozone, were hyphtosized in the 1980s, published in the late 1990s, and are currently being proven. Dr. Drew Schindell, and Dr Paul Newman, NASA Goddard, proposed a theory in the late 1990s, using a SGI Origin 2000 supercomputer, that modeled ozone destruction, accounted for 78 per cent of the ozone destroyed. Further refinement of that model, accounted for 89 per cent of the ozone destroyed, but pushed back the estimated recovery of the ozone hole from 75 years to 150 years. (An important part of that model is the lack of staratospheric flight due to depletion of fossil fuels.)

Misconceptions about Ozone Depletion

A few of the more common misunderstandings about ozone depletion are addressed briefly here; more detailed discussions can be found in the ozone-depletion FAQ.

CFCs are "Too Heavy" to Reach the Stratosphere

It is sometimes stated that since CFC molecules are much heavier than nitrogen or oxygen, they cannot reach the stratosphere in significant quantities. But atmospheric gases are not sorted by weight; the forces of wind (turbulence) are strong enough to fully intermix gases in the atmosphere. CFCs are heavier than air, but just like argon, krypton and other heavy gases with a long lifetime, they are uniformly distributed throughout the turbosphere and reach the upper atmosphere.

Man-made Chlorine is Insignificant Compared to Natural Sources

Another objection occasionally voiced is that *It is generally agreed that natural sources of tropospheric chlorine (volcanoes, ocean spray, etc.) are four to five orders of magnitude larger than man-made sources.* While strictly true, *tropospheric* chlorine is irrelevant; it is *stratospheric* chlorine that matters to ozone depletion. Chlorine from ocean spray is soluble and thus is washed out by rainfall before it reaches the stratosphere. CFCs, in contrast, are insoluble and long-lived, which allows them to reach the stratosphere. Even in the lower atmosphere there is more chlorine present in the form of CFCs and related haloalkanes than there is in HCl from salt spray, and in the stratosphere the halocarbons dominate overwhelmingly. Only one of these halocarbons, methyl chloride, has a predominantly natural source, and it is responsible for about 20 percent of the chlorine in the stratosphere; the remaining 80 per cent comes from manmade compounds.

Very large volcanic eruptions can inject HCl directly into the stratosphere, but direct measurements have shown that their contribution is small compared to that of chlorine from CFCs. A similar erroneous assertion is that soluble halogen compounds from the volcanic plume of Mount Erebus on Ross Island, Antarctica are a major contributor to the Antarctic ozone hole.

An Ozone Hole was First Observed in 1956

G.M.B. Dobson (Exploring the Atmosphere, 2nd Edition, Oxford, 1968) mentioned that when springtime ozone levels over Halley Bay were first measured, he was surprised to find that they were ~320 DU, about 150 DU below spring levels, ~450 DU, in the Arctic. These, however, were the pre-ozone hole normal climatological values. What Dobson describes is essentially the *baseline* from which the ozone hole is measured: actual ozone hole values are in the 150-100 DU range.

The discrepancy between the Arctic and Antarctic noted by Dobson was primarily a matter of timing: during the Arctic spring ozone levels rose smoothly, peaking in April, whereas in the Antarctic they stayed approximately constant during early spring, rising abruptly in November when the polar vortex broke down.

The behaviour seen in the Antarctic ozone hole is completely different. Instead of staying constant, early springtime ozone levels suddenly drop from their already low winter values, by as much as 50 per cent, and normal values are not reached again until December.

If the theory were correct, the ozone hole should be above the sources of CFCs

CFCs are well mixed in the troposphere and the stratosphere. The reason the ozone hole occurs above Antarctica is not because there are more CFCs there but because the low temperatures allow polar stratospheric clouds to form. There have been anomalous discoveries of significant, serious, localized "holes" above other parts of the globe.

The "Ozone Hole" is a Hole in the Ozone Layer

When the "ozone hole" forms, essentially all of the ozone in the lower stratosphere is destroyed. The upper stratosphere is much less affected, however, so that the overall amount of ozone over the continent declines by 50 percent or more. The ozone hole does not go all the way through the layer; on the other hand, it is not a uniform 'thinning' of the layer either. It's a "hole" in the sense of "a hole in the ground", a depression, not in the sense of "a hole in the windshield."

World Ozone Day

In 1994, the United Nations General Assembly voted to designate September 16 as "World Ozone Day", to commemorate the signing of the Montreal Protocol on that date in 1987.

5.5 Montreal Protocol

The *Montreal Protocol on Substances That Deplete the Ozone Layer* is an international treaty designed to protect the ozone layer by phasing out the production of a number of substances believed to be responsible for ozone depletion. The treaty was opened for signature on September 16, 1987 and entered into force on January 1, 1989 followed by a first meeting in Helsinki, May 1989. Since then, it has undergone seven revisions, in 1990 (London), 1991 (Nairobi), 1992 (Copenhagen), 1993 (Bangkok), 1995 (Vienna), 1997 (Montreal), and 1999 (Beijing). Due to its widespread adoption and implementation it has been hailed as an example of exceptional international co-operation with Kofi Annan quoted as saying it is "Perhaps the single most successful international agreement to date.".

Terms and Purpose

The treaty is structured around several groups of halogenated hydrocarbons that have been shown to play a role in ozone depletion. All of these ozone depleting substances contain either chlorine or bromine (substances containing fluorine-only do not harm the ozone layer). For a table of ozone-depleting substances see:

For each group, the treaty provides a timetable on which the production of those substances must be phased out and eventually eliminated.

The stated purpose of the treaty is that the signatory states:

....Recognizing that world-wide emissions of certain substances can significantly deplete and otherwise modify the ozone layer in a manner that is likely to result in adverse effects on human health and the environment,. Determined to protect the ozone layer by taking precautionary measures to control equitably total global emissions of substances that deplete it, with the ultimate objective of their elimination on the basis of developments in scientific knowledge. Acknowledging that special provision is required to meet the needs of developing countries....

Shall accept a series of stepped limits on CFC use and production, including:

- from 1991 to 1992 its levels of consumption and production of the controlled substances in Group I of Annex A do not exceed 150 percent of its calculated levels of production and consumption of those substances in 1986;
- from 1994 its calculated level of consumption and production of the controlled substances in Group I of Annex A does not exceed, annually, twenty-five percent of its calculated level of consumption and production in 1986; and
- from 1996 its calculated level of consumption and production of the controlled substances in Group I of Annex A does not exceed zero.

There is a slower phase-out (to zero by 2010) of other substances (halon 1211, 1301, 2402; CFCs 13, 111, 112, etc) and some chemicals get individual attention (Carbon tetrachloride; 1,1,1-trichloroethane). The phasing-out of the less active HCFCs started only in 1996 and will go on until a complete phasing-out is achieved in 2030.

There are a few exceptions for "essential uses", where no acceptable substitutes have been found (for example, in the metered dose inhalers commonly used to treat asthma and other respiratory problems) or Halon fire suppression systems used in submarines and aircraft (but not in general industry).

The substances in Group I of Annex A are:

- $CFCl_3$ (CFC-11)
- CF_2Cl_2 (CFC-12)
- $C_2F_3Cl_3$ (CFC-113)
- $C_2F_4Cl_2$ (CFC-114)
- C_2F_5Cl (CFC-115)

The provisions of the Protocol include the requirement that the Parties to the Protocol base their future decisions on the current scientific, environmental, technical, and economic information that is assessed through panels drawn from the worldwide expert communities. To provide that input to the decision-making process, advances in understanding on these topics were assessed in 1989, 1991, 1994, 1998 and 2002 in a series of reports entitled Scientific assessment of ozone depletion.

Several reports have been published by various governmental and non-governmental organizations to present alternatives to the ozone depleting substances, since the substances have been used in various technical sectors, like in refrigerating, agriculture, energy production, and laboratory measurements.

History

In 1973 Chemists Frank Sherwood Rowland and Mario Molina, then at the University of California, Irvine, began studying the impacts of CFCs in the earth's atmosphere. They discovered that CFC molecules were stable enough to remain in the atmosphere until they got up into the middle of the stratosphere where they would finally (after an average of 50-100 years for two common CFCs) be broken down by ultraviolet radiation releasing a chlorine atom. Rowland and Molina then proposed that these chlorine atoms might be expected to cause the breakdown of large amounts of ozone (O_3) in the stratosphere. Their argument was based upon an analogy to contemporary work by Paul J. Crutzen and Harold Johnston, which had shown that nitric oxide (NO) could catalyze the destruction of ozone. (Several other scientists, including Ralph Cicerone, Richard Stolarski, Michael McElroy, and Steven Wofsy had independently proposed that chlorine could catalyze ozone loss, but none had realized that CFCs were a potentially large source of chlorine.) Crutzen, Molina and Rowland were awarded the 1995 Nobel Prize for Chemistry for their work on this problem.

The environmental consequence of this discovery was that, since stratospheric ozone absorbs most of the ultraviolet-B (UV-B) radiation reaching the surface of the planet, depletion of the ozone layer by CFCs would lead to an in increase in UV-B radiation at the surface, resulting in an increase in skin cancer and other impacts such as damage to crops and to marine phytoplankton.

But the Rowland-Molina hypothesis was strongly disputed by representatives of the aerosol and halocarbon industries. The chair of the board of DuPont was quoted as saying that ozone depletion theory is "a science fiction tale.a load of rubbish.utter nonsense". Robert Abplanalp, the president of Precision Valve Corporation (and inventor of the first practical aerosol spray can valve), wrote to the Chancellor of UC Irvine to complain about Rowland's public statements (Roan, p. 56.)

After publishing their pivotal paper in June 1974, Rowland and Molina testified at a hearing before the U.S. House of Representatives in December, 1974. As a result significant funding was made available to study various aspects of the problem and to confirm the initial findings. In 1976 the U.S. National Academy of Sciences (NAS) released a report that confirmed the scientific credibility of the ozone depletion hypothesis. NAS continued to publish assessments of related science for the next decade.

Then, in 1985, British Antarctic Survey scientists Farman, Gardiner and Shanklin shocked the scientific community when they published results of a study showing an ozone "hole" in the journal Nature—showing a decline in polar ozone far larger than anyone had anticipated. That same year, 20 nations, including most of the major CFC producers, signed the Vienna Convention, which established a framework for negotiating international regulations on ozone-depleting substances. But the CFC industry did not give up that easily. As late as 1986, the Alliance for Responsible CFC Policy (an association representing the CFC industry founded by DuPont) was still arguing that the science was too uncertain to justify any action. In 1987, DuPont testified before the US Congress that "we believe that there is no immediate crisis that demands unilateral regulation."

Multilateral Fund

The *Multilateral Fund for the Implementation of the Montreal Protocol* provides funds to help developing countries to phase out the use of ozone-depleting substances. The Multilateral Fund was the first financial mechanism to be created under an international treaty. It embodies the principle agreed at the United Nations Conference on Environment and Development in 1992 that countries have a common but differentiated responsibility to protect and manage the global commons.

The Fund is managed by an executive committee with an equal representation of seven industrialized and seven Article 5 countries, which are elected annually by a Meeting of the Parties. The Committee reports annually to the Meeting of the Parties on its operations.

Up to 20 percent of the contributions of contributing parties can also be delivered through their bilateral agencies in the form of eligible projects and activities. The fund is replenished on a three-year basis by the donors. Pledges amount to US$ 2.1 billion over the period 1991 to 2005. Funds are used, for example, to finance the conversion of existing manufacturing processes, train personnel, pay royalties and patent rights on new technologies, and establish national ozone offices.

Parties

At present, 191 nations have become party to the Montreal Protocol (see external link below). Those 5 that are not as of September 2007 are Andorra, Iraq, San Marino, Timor-Leste and Vatican City.

Impact

Since the Montreal Protocol came into effect, the atmospheric concentrations of the most important chlorofluorocarbons and related chlorinated hydrocarbons have either leveled off or decreased . Halon concentrations have continued to increase, as the halons presently stored in fire extinguishers are released, but their rate of increase has slowed and their abundances are expected to begin to decline by about 2020. Also, the concentration of the HCFCs increased drastically at least partly because for many uses CFCs (e.g. used as solvents or refrigerating agents) were substituted with HCFCs. While there have been reports of attempts by individuals to circumvent the ban, e.g. by smuggling CFCs from undeveloped to developed nations, the overall level of compliance has been high. In consequence, the Montreal Protocol has often been called the most successful international environmental agreement to date. In a 2001 report, NASA found the ozone hole over Antarctica had remained the same size for the previous three years, however in 2003 the ozone hole grew to its second largest size.

Unfortunately, the hydrochlorofluorocarbons, or HCFCs, and hydrofluorocarbons, or HFCs, are now thought to contribute to anthropogenic global warming. On a molecule-for-molecule basis, these compounds are up to 10,000 times more potent greenhouse gases than carbon dioxide. The Montreal Protocol currently calls for a complete phase-out of HCFCs by 2030, but does not place any restriction on HFCs. Since the CFCs themselves are equally powerful as greenhouse gases, the mere substitution of HFCs for CFCs does not significantly increase the rate of anthropogenic global warming, but over time a steady increase in their use could increase the danger that human activity will change the climate.

5.6 Convention on Long-Range Transboundary Air Pollution

The *Convention on Long-Range Transboundary Air Pollution,* often abbreviated as *Air Pollution* or CLRTAP, is intended to protect the human environment against air pollution and to gradually reduce and prevent air pollution, including long-range transboundary air pollution. The convention opened for signature on 1979-11-13 and entered into force on 1983-03-16. The current parties to the Convention are shown on the right.

The Convention, which now has 51 Parties, identifies the Executive Secretary of the United Nations Economic Commission for Europe (UNECE) as its secretariat. Since 1979 the Convention on Long-range Transboundary Air Pollution has addressed some of the major environmental problems of the UNECE region through scientific collaboration and policy negotiation. The Convention has been extended by eight protocols that identify specific measures to be taken by Parties to cut their emissions of air pollutants. The aim of the Convention is that Parties shall endeavour to limit and, as far as possible, gradually reduce and prevent air pollution including long-range transboundary air pollution. Parties develop policies and strategies to combat the discharge of air pollutants through exchanges of information, consultation, research and monitoring.

The Parties meet annually at sessions of the Executive Body to review ongoing work and plan future activities including a workplan for the coming year. The three main subsidiary bodies—the Working Group on Effects, the Steering Body to EMEP and the Working Group on Strategies and Review— as well as the Convention's Implementation Committee, report to the Executive Body each year. Currently, the Convention's priority activities include review and possible revision of its most recent protocols, implementation of the Convention and its protocols across the entire UNECE region (with special focus on Eastern Europe, the Caucasus and Central Asia and South-East Europe) and sharing its knowledge and information with other regions of the world.

5.7 Nitrogen Oxide Protocol

Protocol to the 1979 Convention on Long-Range Transboundary Air Pollution Concerning the Control of Emissions of Nitrogen Oxides or Their Transboundary Fluxes, opened for signature on 31 October 1988 and entered into force on 14 February 1991, was to provide for the control or reduction of nitrogen oxides and their transboundary fluxes.

Parties: (28) Austria, Belarus, Belgium, Bulgaria, Canada, Czech Republic,

Denmark, Estonia, European Union, Finland, France, Germany, Greece, Hungary, Ireland, Italy, Liechtenstein, Luxembourg, Netherlands, Norway, Russia, Slovakia, Spain, Sweden, Switzerland, Ukraine, United Kingdom, United States

Countries that have signed, but not yet ratified: Poland.

5.8 Climate Change

Climate change is any long-term significant change in the "average weather" that a given region experiences. Average weather may include average temperature, precipitation and wind patterns. It involves changes in the variability or average state of the atmosphere over durations ranging from decades to millions of years. These changes can be caused by dynamic process on Earth, external forces including variations in sunlight intensity, and more recently by human activities.

In recent usage, especially in the context of environmental policy, the term "climate change" often refers to changes in modern climate (see global warming). For information on temperature measurements over various periods, and the data sources available, see temperature record. For attribution of climate change over the past century, see attribution of recent climate change.

Climate Change Factors

Climate changes reflect variations within the Earth's atmosphere, processes in other parts of the Earth such as oceans and ice caps, and the effects of human activity. The external factors that can shape climate are often called climate forcings and include such processes as variations in solar radiation, the Earth's orbit, and greenhouse gas concentrations.

Variations within the Earth's Climate

Weather is the day-to-day state of the atmosphere, and is a chaotic non-linear dynamical system. On the other hand, *climate*—the average state of weather—is fairly stable and predictable. Climate includes the average temperature, amount of precipitation, days of sunlight, and other variables that might be measured at any given site. However, there are also changes within the Earth's environment that can affect the climate.

Glaciation

Glaciers are recognized as being among the most sensitive indicators of climate change, advancing substantially during climate cooling (e.g., the Little Ice Age) and retreating during climate warming on moderate time scales. Glaciers grow and collapse, both contributing to natural variability and greatly amplifying externally forced changes. For the last century, however, glaciers have been unable to regenerate enough ice during the winters to make up for the ice lost during the summer months (see glacier retreat).

The most significant climate processes of the last several million years are the glacial and interglacial cycles of the present ice age. Though shaped by orbital variations, the internal responses involving continental ice sheets and 130 m sea-level change certainly played a key role in deciding what climate response would be observed in most regions. Other changes, including Heinrich events, Dansgaard—Oeschger events and the Younger Dryas show the potential for glacial variations to influence climate even in the absence of specific orbital changes.

Ocean Variability

On the scale of decades, climate changes can also result from interaction of the atmosphere and oceans. Many climate fluctuations—including not only the El Niño Southern oscillation (the best known) but also the Pacific decadal oscillation, the North Atlantic oscillation, and the Arctic oscillation—owe their existence at least in part to different ways that heat can be stored in the oceans and move between different reservoirs. On longer time scales ocean processes such as thermohaline circulation play a key role in redistributing heat, and can dramatically affect climate.

The Memory of Climate

More generally, most forms of internal variability in the climate system can be recognized as a form of hysteresis, meaning that the current state of climate reflects not only the inputs, but also the history of how it got there. For example, a decade of dry conditions may cause lakes to shrink, plains to dry up and deserts to expand. In turn, these conditions may lead to less rainfall in the following years. In short, climate change can be a self-perpetuating process because different aspects of the environment respond at different rates and in different ways to the fluctuations that inevitably occur.

Non-climate Factors Driving Climate Change

Current studies indicate that radiative forcing by greenhouse gases is the primary cause of global warming. Greenhouse gases are also important in understanding Earth's climate history. According to these studies, the greenhouse effect, which is the warming produced as greenhouse gases trap heat, plays a key role in regulating Earth's temperature.

Over the last 600 million years, carbon dioxide concentrations have varied from perhaps >5000 ppm to less than 200 ppm, due primarily to the effect of geological processes and biological innovations. It has been argued by Veizer *et al.*, 1999, that variations in greenhouse gas concentrations over tens of millions of years have not been well correlated to climate change, with plate tectonics perhaps playing a more dominant role. More recently Royer *et al.* have used the CO_2—climate correlation to derive a value for the climate sensitivity. There are several examples of rapid changes in the concentrations of greenhouse gases in the Earth's atmosphere that do appear to correlate to strong warming, including the Paleocene-Eocene thermal maximum, the Permian-Triassic extinction event, and the end of the Varangian snowball earth event.

During the modern era, the naturally rising carbon dioxide levels are implicated as the primary cause of global warming since 1950. According to the Intergovernmental Panel on Climate Change (IPCC), 2007, the atmospheric concentration of CO_2 in 2005 was 379 ppm^3 compared to the pre-industrial levels of 280 ppm^3. Thermodynamics and Le Chatelier's principle explain the characteristics of the dynamic equilibrium of a gas in solution such as the vast amount of CO_2 held in solution in the world's oceans moving into and returning from the atmosphere. These principles can be observed as bubbles which rise in a pot of water heated on a stove, or in a glass of cold beer allowed to sit at room temperature; gases dissolved in liquids are released under certain circumstances.

Plate Tectonics

On the longest time scales, plate tectonics will reposition continents, shape oceans, build and tear down mountains and generally serve to define the stage upon which climate exists. More recently, plate motions have been implicated in the intensification of the present ice age when, approximately 3 million years ago, the North and South American plates collided to form the Isthmus of Panama and shut off direct mixing between the Atlantic and Pacific Oceans.

Solar Variation

The sun is the ultimate source of essentially all heat in the climate system. The energy output of the sun, which is converted to heat at the Earth's surface, is an integral part of shaping the Earth's climate. On the longest time scales, the sun itself is getting brighter with higher energy output; as it continues its main sequence, this slow change or evolution affects the Earth's atmosphere. It is thought that, early in Earth's history, the sun was too cold to support liquid water at the Earth's surface, leading to what is known as the Faint young sun paradox.

On more modern time scales, there are also a variety of forms of solar variation, including the 11-year solar cycle and longer-term modulations. However, the 11-year sunspot cycle does not manifest itself clearly in the climatological data. Solar intensity variations are considered to have been influential in triggering the Little Ice Age, and for some of the warming observed from 1900 to 1950. The cyclical nature of the sun's energy output is not yet fully understood; it differs from the very slow change that is happening within the sun as it ages and evolves.

Orbital Variations

In their effect on climate, orbital variations are in some sense an extension of solar variability, because slight variations in the Earth's orbit lead to changes in the distribution and abundance of sunlight reaching the Earth's surface. Such orbital variations, known as Milankovitch cycles, are a highly predictable consequence of basic physics due to the mutual interactions of the Earth, its moon, and the other planets. These variations are considered the driving factors underlying the glacial and interglacial cycles of the present ice age. Subtler variations are also present, such as the repeated advance and retreat of the Sahara desert in response to orbital precession.

Volcanism

A single eruption of the kind that occurs several times per century can affect climate, causing cooling for a period of a few years. For example, the eruption of Mount Pinatubo in 1991 affected climate substantially. Huge eruptions, known as large igneous provinces, occur only a few times every hundred million years, but can reshape climate for millions of years and cause mass extinctions. Initially, scientists thought that the dust emitted into the atmosphere from large volcanic eruptions was responsible for the cooling by partially blocking the transmission of solar radiation to the Earth's surface. However, measurements indicate that most of the dust thrown in the atmosphere returns to the Earth's surface within six months.

Volcanoes are also part of the extended carbon cycle. Over very long (geological) time periods, they release carbon dioxide from the earth's interior, counteracting the uptake by sedimentary rocks and other geological carbon dioxide sinks. However, this contribution is insignificant compared to the current anthropogenic emissions. The US Geological Survey estimates that human activities generate more than 130 times the amount of carbon dioxide emitted by volcanoes.

Human Influences on Climate Change

Anthropogenic factors are human activities that change the environment and influence climate. In some cases the chain of causality is direct and unambiguous (e.g., by the effects of irrigation on temperature and humidity), while in others it is less clear. Various hypotheses for human-induced climate change have been debated for many years.

The biggest factor of present concern is the increase in CO_2 levels due to emissions from fossil fuel combustion, followed by aerosols (particulate matter in the atmosphere), which exert a cooling effect, and cement manufacture. Other factors, including land use, ozone depletion, animal agriculture and deforestation, also affect climate.

Fossil Fuels

Beginning with the industrial revolution in the 1850s and accelerating ever since, the human consumption of fossil fuels has elevated CO_2 levels from a concentration of ~280 ppm to more than 380 ppm today. These increases are projected to reach more than 560 ppm before the end of the 21st century. It is known that carbon dioxide levels are substantially higher now than at any time in the last 750,000 years. Along with rising methane levels, these changes are anticipated to cause an increase of 1.4-5.6 °C between 1990 and 2100 (see global warming).

Aerosols

Anthropogenic aerosols, particularly sulphate aerosols from fossil fuel combustion, exert a cooling influence. This, together with natural variability, is believed to account for the relative "plateau" in the graph of 20th-century temperatures in the middle of the century.

Cement Manufacture

Cement manufacturing is the third largest cause of man-made carbon dioxide emissions. Carbon dioxide is produced when calcium carbonate ($CaCO_3$) is heated to produce the cement ingredient calcium oxide (CaO, also called *quicklime*). While fossil fuel combustion and deforestation each produce significantly more carbon dioxide (CO_2), cement-making is responsible for approximately 2.5 per cent of total worldwide emissions from industrial sources (energy plus manufacturing sectors).

Land Use

Prior to widespread fossil fuel use, humanity's largest effect on local climate is likely to have resulted from land use. Irrigation, deforestation, and agriculture fundamentally change the environment. For example, they change the amount of water going into and out of a given location. They also may change the local albedo by influencing the ground cover and altering the amount of sunlight that is absorbed. For example, there is evidence to suggest that the climate of Greece and other Mediterranean countries was permanently changed by widespread deforestation between 700 BC and 1 AD (the wood being used for shipbuilding, construction and fuel), with the result that the modern climate in the region is significantly hotter and drier, and the species of trees that were used for shipbuilding in the ancient world can no longer be found in the area.

A controversial hypothesis by William Ruddiman called the early anthropocene hypothesis suggests that the rise of agriculture and the accompanying deforestation led to the increases in carbon dioxide and methane during the period 5000-8000 years ago. These increases, which reversed previous declines, may have been responsible for delaying the onset of the next glacial period, according to Ruddimann's overdue-glaciation hypothesis.

In modern times, a 2007 Jet Propulsion Laboratory study found that the average temperature of California has risen about 2 degrees over the past 50 years, with a much higher increase in urban areas. The change was attributed mostly to extensive human development of the landscape.

Livestock

According to a 2006 United Nations report, Livestock's Long Shadow, livestock is responsible for 18 per cent of the world's greenhouse gas emissions as measured in CO_2 equivalents. This however includes land usage change, meaning deforestation in order to create grazing land. In the Amazon Rainforest, 70 per cent of deforestation is to make way for grazing land, so this is the major factor in the 2006 UN FAO report, which was the first agricultural report to include land usage change into the radiative forcing of livestock. In addition to CO_2 emissions, livestock produces 65 per cent of human-induced nitrous oxide (which has 296 times the global warming potential of CO_2) and 37 per cent of human-induced methane (which has 23 times the global warming potential of CO_2).

Interplay of Factors

If a certain forcing (for example, solar variation) acts to change the climate, then there may be mechanisms that act to amplify or reduce the effects. These are called positive and negative feedbacks. As far as is known, the climate system is generally stable with respect to these feedbacks: positive feedbacks do not "run away". Part of the reason for this is the existence of a powerful negative feedback between temperature and emitted radiation: radiation increases as the fourth power of absolute temperature.

However, a number of important positive feedbacks do exist. The glacial and interglacial cycles of the present ice age provide an important example. It is believed that orbital variations provide the timing for the growth and retreat of ice sheets. However, the ice sheets themselves reflect sunlight back into space and hence promote cooling and their own growth, known as the ice-albedo feedback. Further, falling sea levels and expanding ice decrease plant growth and indirectly lead to declines in carbon dioxide and methane. This leads to further cooling. Conversely, rising temperatures caused, for example, by anthropogenic emissions of greenhouse gases could lead to decreased snow and ice cover, revealing darker ground underneath, and consequently result in more absorption of sunlight.

Water vapor, methane, and carbon dioxide can also act as significant positive feedbacks, their levels rising in response to a warming trend, thereby accelerating that trend. Water vapor acts strictly as a feedback (excepting small amounts in the stratosphere), unlike the other major greenhouse gases, which can also act as forcings.

More complex feedbacks include heat movement from the equatorial regions to the northern latitudes and involve the possibility of altered water currents with in the oceans or air currents with in the atmosphere. A significant concern is that melting glacial ice from Greenland may interfere and change the thermohaline circulation of water in the North Atlantic, affecting the Gulf Stream which brings warmer water to replace sinking colder water; which would change the distribution of heat to Europe and the east coast of the United States.

Other potential feedbacks are not well understood and may either inhibit or promote warming. For example, it is unclear whether rising temperatures promote or inhibit vegetative growth, which could in turn draw down either more or less carbon dioxide. Similarly, increasing temperatures may lead to either more or less cloud cover. Since on balance cloud cover has a strong cooling effect, any change to the abundance of clouds also affects climate.

Monitoring the Current Status of Climate

Testing for spatial dependence between independently measured values in an ordered set is based on applying Fisher's F-test to the variance of a set and the first variance term of the ordered set. Charting statistically significant variance terms gives a sampling variogram that shows where spatial dependence in our sample space of time dissipates into randomness. The lag of a sampling variogram is a statistically robust measure for a change in a climate statistic.

Scientists use "Indicator time series" that represent the many aspects of climate and ecosystem status. The time history provides a historical context. Current status of the climate is also monitored with climate indices.

Evidence for Climatic Change

Evidence for climatic change is taken from a variety of sources that can be used to reconstruct past climates. Most of the evidence is indirect—climatic changes are inferred from changes in indicators that reflect climate, such as vegetation, dendrochronology, ice cores, sea level change, and glacial retreat.

Pollen Analysis

Palynology is the science that studies contemporary and fossil palynomorphs, including pollen. Palynology is used to infer the geographical distribution of plant species, which vary under different climate conditions. Different groups of plants have pollen with distinctive shapes and surface textures, and since the outer surface of pollen is composed of a very resilient material, they resist decay. Changes in the type of pollen found in different sedimentation levels in lakes, bogs or river deltas indicate changes in plant communities; which are dependent on climate conditions.

Beetles

Remains of beetles are common in freshwater and land sediments. Different species of beetles tend to be found under different climatic conditions. Knowledge of the present climatic range of the different species, and of the age of the sediments in which remains are found, allows past climatic conditions to be inferred.

Glacial Geology

Advancing glaciers leave behind moraines and other features that often have datable material in them, recording the time when a glacier advanced and deposited a feature. Similarly, by tephrochronological techniques, the lack of glacier cover can be identified by the presence of datable soil or volcanic tephra horizons. Glaciers are considered one of the most sensitive climate indicators by the IPCC, and their recent observed variations provide a global signal of climate change. See Retreat of glaciers since 1850.

Examples of Climate Change

Climate change has continued throughout the entire history of Earth. The field of paleoclimatology has provided information of climate change in the ancient past, supplementing modern observations of climate.

1. Climate of the deep past
    * Faint young sun paradox
    * Snowball earth
    * Oxygen Catastrophe
2. Climate of the last 500 million years
    * Phanerozoic overview
    * Paleocene-Eocene Thermal Maximum
    * Cretaceous Thermal Maximum
    * Permo-Carboniferous Glaciation
    * Ice ages
3. Climate of recent glaciations
    * Dansgaard-Oeschger event
    * Younger Dryas
    * Ice age temperatures
4. Recent climate
    * Holocene Climatic Optimum
    * Medieval Warm Period
    * Little Ice Age
    * Year Without a Summer
    * Temperature record of the past 1000 years
    * Global warming
    * Hardiness Zone Migration

Climate Change and Biodiversity

The life cycles of many wild plants and animals are closely linked to the passing of the seasons; climatic changes can lead to interdependent pairs of species (e.g. a wild flower and its pollinating insect) losing synchronization, if, for example, one has a cycle dependent on day length and the other on temperature or precipitation. In principle, at least, this could lead to extinctions or changes in the distribution and abundance of species. One phenomenon is the movement of species northwards in Europe. A recent study by Butterfly Conservation in the UK, has shown that relatively common species with a southerly distribution have moved north, whilst scarce upland species have become rarer and lost territory towards the south. This picture has been mirrored across several invertebrate groups. Drier summers could lead to more periods of drought, potentially affecting many species of animal and plant. For example, in the UK during the drought year of 2006 significant numbers of trees died or showed dieback on light sandy soils. In Australia, since the early 90s, tens of thousands of flying foxes (Pteropus) have died as a direct result of extreme heat. Wetter, milder winters might affect temperate mammals or insects by preventing them hibernating or entering torpor during periods when food is scarce. One predicted change is the ascendancy of 'weedy' or opportunistic species at the expense of scarcer species with narrower or more specialized ecological requirements. One example could be the expanses of bluebell seen in many woodlands in the UK. These have an early growing and flowering season before competing weeds can develop and the tree canopy closes. Milder winters can allow weeds to overwinter as adult plants or germinate sooner, whilst trees leaf earlier, reducing the length of the window for bluebells to complete their life cycle. Organisations such as Wildlife Trust, World Wide Fund for Nature, Birdlife International and the Audubon Society are actively monitoring and research the effects of climate change on biodiversity and advance policies in areas such as landscape scale conservation to promote adaptation to climate change.

5.9 Kyoto Protocol

The *Kyoto Protocol* is a protocol to the international Framework Convention on Climate Change with the objective of reducing greenhouse gases that cause climate change. It was adopted on 11 December 1997 by the 3rd Conference of the Parties, which was meeting in Kyoto, and it entered into force on 16 February 2005. As of June 2008, 182 parties have ratified the protocol. Of these, 36 developed cg countries (plus the EU as a party in its own right) are required to reduce greenhouse gas emissions to the levels specified for each of them in the treaty (representing over 61.6 per cent of emissions from Annex I countries), with three more countries intending to participate. One hundred thirty-seven (137) developing countries have ratified the protocol, including Brazil, China and India, but have no obligation beyond monitoring and reporting emissions. The United States has not ratified the treaty. Among various experts, scientists, and critics, there is debate about the usefulness of the protocol, and there have been cost-benefit studies performed on its usefulness.

Description

The Kyoto Protocol is an agreement made under the United Nations Framework Convention on Climate Change (UNFCCC). Countries that ratify this protocol commit to reducing their emissions of carbon dioxide and five other greenhouse gases (GHG), or engaging in emissions trading if they maintain or increase emissions of these green house gases.

The Kyoto Protocol now covers more than 170 countries globally but only 60 per cent of countries in terms of global greenhouse gas emissions. As of December 2007, the US and Kazakhstan are the only signatory nations not to have ratified the act. The first commitment period of the Kyoto Protocol ends in 2012, and international talks began in May 2007 on a subsequent commitment period.

At its heart, the Kyoto Protocol establishes the following principles:

- ❖ Kyoto is underwritten by governments and is governed by global legislation enacted under the UN's aegis.
- ❖ Governments are separated into two general categories: developed countries, referred to as Annex I countries (who have accepted greenhouse gas emission reduction obligations and must submit an annual greenhouse gas inventory), and developing countries, referred to as Non-Annex I countries (who have no greenhouse gas emission reduction obligations but may participate in the Clean Development Mechanism).
- ❖ Any Annex I country that fails to meet its Kyoto obligation will be penalized by having to submit 1.3 emission allowances in a second commitment period for every ton of greenhouse gas emissions they exceed their cap in the first commitment period (i.e., 2008-2012).
- ❖ As of January 2008, and running through 2012, Annex I countries have to reduce their greenhouse gas emissions by a collective average

of 5 per cent below their 1990 levels (for many countries, such as the EU member states, this corresponds to some 15 per cent below their expected greenhouse gas emissions in 2008). While the average emissions reduction is 5 per cent, national limitations range from an 8 per cent average reduction across the European Union to a 10 per cent emissions increase for Iceland; but, since the EU's member states each have individual obligations, much larger increases (up to 27%) are allowed for some of the less developed EU countries (see below Increase in greenhouse gas emission since 1990). Reduction limitations expire in 2013.

❖ Kyoto includes "flexible mechanisms" which allow Annex I economies to meet their greenhouse gas emission limitation by purchasing GHG emission reductions from elsewhere. These can be bought either from financial exchanges, from projects which reduce emissions in non-Annex I economies under the Clean Development Mechanism (CDM), from other Annex 1 countries under the JI, or from Annex I countries with excess allowances. Only CDM Executive Board-accredited Certified Emission Reductions (CER) can be bought and sold in this manner. Under the aegis of the UN, Kyoto established this Bonn-based Clean Development Mechanism Executive Board to assess and approve projects ("CDM Projects") in Non-Annex I economies prior to awarding CERs. (A similar scheme called "Joint Implementation" or "JI" applies in transitional economies mainly covering the former Soviet Union and Eastern Europe).

In practice this means that Non-Annex I economies have no GHG emission restrictions, but when a greenhouse gas emission reduction project (a "Greenhouse Gas Project") is implemented in these countries the project will receive Carbon Credits, which can then be sold to Annex I buyers.

These Kyoto mechanisms are in place for two main reasons:

❖ there were fears that the cost of complying with Kyoto would be expensive for many Annex I countries, especially those countries already home to efficient, low greenhouse gas emitting industries, and high prevailing environmental standards. Kyoto therefore allows these countries to purchase (cheaper) carbon credits on the world market instead of reducing greenhouse gas emissions domestically, and

❖ this is seen as a means of encouraging Non-Annex I developing economies to reduce greenhouse gas emissions through sustainable development, since doing so is now economically viable because of the investment flows from the sale of Carbon Credits.

All the Annex I economies have established Designated National Authorities to manage their greenhouse gas portfolios under Kyoto. Countries including Japan, Canada, Italy, the Netherlands, Germany, France, Spain and many more are actively promoting government carbon funds and supporting multilateral carbon funds intent on purchasing Carbon Credits from Non-Annex I countries. These government organizations are working closely with their major utility, energy, oil & gas and chemicals conglomerates to try to acquire as many Greenhouse Gas Certificates as cheaply as possible.

Virtually all of the Non-Annex I countries have also set up their own Designated National Authorities to manage the Kyoto process (and specifically the "CDM process" whereby these host government entities decide which Greenhouse Gas Projects they do or do not wish to support for accreditation by the CDM Executive Board).

The objectives of these opposing groups are quite different. Annex I entities want Carbon Credits as cheaply as possible, whilst Non-Annex I entities want to maximize the value of Carbon Credits generated from their domestic Greenhouse Gas Projects.

Objectives

The objective is to achieve "stabilization of greenhouse gas concentrations in the atmosphere at a level that would prevent dangerous anthropogenic interference with the climate system." The Intergovernmental Panel on Climate Change (IPCC) has predicted an average global rise in temperature of 1.4°C (2.5°F) to 5.8°C (10.4°F) between 1990 and 2100. Proponents also note that Kyoto is a first step as requirements to meet the UNFCCC will be modified until the objective is met, as required by UNFCCC Article 4.2(d).

Status of the Agreement

The treaty was negotiated in Kyoto, Japan in December 1997, opened for signature on March 16, 1998, and closed on March 15, 1999. The agreement came into force on February 16, 2005 following ratification by Russia on November 18, 2004. As of April 2008, a total of 178 countries and other governmental entities have ratified the agreement (representing over 61.6% of emissions from Annex I countries).

According to article 25 of the protocol, it enters into force "on the ninetieth day after the date on which not less than 55 Parties to the Convention, incorporating Parties included in Annex I which accounted in total for at least 55 per cent of the total carbon dioxide emissions for 1990 of the Parties included in Annex I, have deposited their instruments of ratification, acceptance, approval or accession." Of the two conditions, the "55 parties" clause was reached on May 23, 2002 when Iceland ratified. The ratification by Russia on 18 November 2004 satisfied the "55 per cent" clause and brought the treaty into force, effective February 16, 2005.

Details of the Agreement

According to a press release from the United Nations Environment Programme:

> *"The Kyoto Protocol is an agreement under which industrialized countries will reduce their collective emissions of greenhouse gases by 5.2 per cent compared to the year 1990 (but note that, compared to the emissions levels that would be expected by 2010 without the Protocol, this limitation represents a 29% cut). The goal is to lower overall emissions of six greenhouse gases—carbon dioxide, methane, nitrous oxide, sulfur hexafluoride, hydrofluorocarbons, and perfluorocarbons—averaged over the period of 2008-2012. National limitations range from 8 per cent reductions for the European Union and some others to 7 per cent for the US, 6 per cent for Japan, 0 per cent for Russia, and permitted increases of 8 per cent for Australia and 10 per cent for Iceland.*

It is an agreement negotiated as an amendment to the United Nations Framework Convention on Climate Change (UNFCCC, which was adopted at the Earth Summit in Rio de Janeiro in 1992). All parties to the UNFCCC can sign or ratify the Kyoto Protocol, while non-parties to the UNFCCC cannot. The Kyoto Protocol was adopted at the third session of the Conference of Parties to the UNFCCC (COP3) in 1997 in Kyoto, Japan. Most provisions of the Kyoto Protocol apply to developed countries, listed in Annex I to the UNFCCC. Emission figures exclude international aviation and shipping.

Common but Differentiated Responsibility

The United Nations Framework Convention on Climate Change agreed to a set of a "common but differentiated responsibilities." The parties agreed that:

1. the largest share of historical and current global emissions of greenhouse gases has originated in developed countries,
2. per capita emissions in developing countries are still relatively low, and
3. the share of global emissions originating in developing countries will grow to meet their social and development needs.

In other words, China, India, and other developing countries were not included in any numerical limitation of the Kyoto Protocol because they were not the main contributors to the greenhouse gas emissions during the pre-treaty industrialization period. However, even without the commitment to reduce according to the Kyoto target, developing countries do share the common responsibility that all countries have in reducing emissions.

Financial Commitments

The Protocol also reaffirms the principle that developed countries have to pay billions of dollars, and supply technology to other countries for climate-related studies and projects. This was originally agreed in the UNFCCC.

Emissions Trading

Kyoto is a 'cap and trade' system that imposes national caps on the emissions of Annex I countries. On average, this cap requires countries to reduce their emissions 5.2 per cent below their 1990 baseline over the 2008 to 2012 period. Although these caps are national-level commitments, in practice most countries will devolve their emissions targets to individual industrial entities, such as a power plant or paper factory. One example of a 'cap and trade' system is the 'EU ETS'. Other schemes may follow suit in time.

This means that the ultimate buyers of credits are often individual companies that expect their emissions to exceed their quota (their Assigned Allocation Units, AAUs or 'allowances' for short). Typically, they will purchase credits directly from another party with excess allowances, from a broker, from a JI/CDM developer, or on an exchange.

National governments, some of whom may not have devolved responsibility for meeting Kyoto obligations to industry, and that have a net deficit of allowances, will buy credits for their own account, mainly from JI/CDM developers. These deals are occasionally done directly through a national fund or agency, as in the case of the Dutch government's ERUPT programme, or via collective funds such as the World Bank's Prototype Carbon

Fund (PCF). The PCF, for example, represents a consortium of six governments and 17 major utility and energy companies on whose behalf it purchases Credits.

Since allowances and carbon credits are tradeable instruments with a transparent price, financial investors can buy them on the spot market for speculation purposes, or link them to futures contracts. A high volume of trading in this secondary market helps price discovery and liquidity, and in this way helps to keep down costs and set a clear price signal in CO2 which helps businesses to plan investments. This market has grown substantially, with banks, brokers, funds, arbitrageurs and private traders now participating in a market valued at about $60 billion in 2007. Emissions Trading PLC, for example, was floated on the London Stock Exchange's AIM market in 2005 with the specific remit of investing in emissions instruments.

Although Kyoto created a framework and a set of rules for a global carbon market, there are in practice several distinct schemes or markets in operation today, with varying degrees of linkages among them.

Kyoto enables a group of several Annex I countries to join together to create a market-within-a-market. The EU elected to be treated as such a group, and created the EU Emissions Trading Scheme (ETS). The EU ETS uses EAUs (EU Allowance Units), each equivalent to a Kyoto AAU. The scheme went into operation on 1 January 2005, although a forward market has existed since 2003.

The UK established its own learning-by-doing voluntary scheme, the UK ETS, which ran from 2002 through 2006. This market existed alongside the EU's scheme, and participants in the UK scheme have the option of applying to opt out of the first phase of the EU ETS, which lasts through 2007.

The sources of Kyoto credits are the Clean Development Mechanism (CDM) and Joint Implementation (JI) projects. The CDM allows the creation of new carbon credits by developing emission reduction projects in Non-Annex I countries, while JI allows project-specific credits to be converted from existing credits within Annex I countries. CDM projects produce Certified Emission Reductions (CERs), and JI projects produce Emission Reduction Units (ERUs), each equivalent to one AAU. Kyoto CERs are also accepted for meeting EU ETS obligations, and ERUs will become similarly valid from 2008 for meeting ETS obligations (although individual countries may choose to limit the number and source of CER/JIs they will allow for compliance purposes starting from 2008). CERs/ERUs are overwhelmingly bought from project developers by funds or individual entities, rather than being exchange-traded like allowances.

Since the creation of Kyoto instruments is subject to a lengthy process of registration and certification by the UNFCCC, and the projects themselves require several years to develop, this market is at this point largely a forward market where purchases are made at a discount to their equivalent currency, the EUA, and are almost always subject to certification and delivery (although up-front payments are sometimes made). According to IETA, the market value of CDM/JI credits transacted in 2004 was EUR 245 m; it is estimated that more than EUR 620 m worth of credits were transacted in 2005.

Several non-Kyoto carbon markets are in existence or being planned, and these are likely to grow in importance and numbers in the coming years. These include the New South Wales Greenhouse Gas Abatement Scheme, the Regional Greenhouse Gas Initiative and Western Climate Initiative in the United States, the Chicago Climate Exchange and the State of California's recent initiative to reduce emissions.

These initiatives, taken together may create a series of partly-linked markets, rather than a single carbon market. The common theme across most of them is the adoption of market-based mechanisms centered on carbon credits that represent a reduction of CO_2 emissions. The fact that some of these initiatives have similar approaches to certifying their credits makes it conceivable that carbon credits in one market may in the long run be tradeable in other schemes. This would broaden the current carbon market far more than the current focus on the CDM/JI and EU ETS domains. An obvious precondition, however, is a realignment of penalties and fines to similar levelsince these create an effective ceiling for each market.

Revisions

The protocol left several issues open to be decided later by the sixth Conference of Parties (COP). COP6 attempted to resolve these issues at its meeting in the Hague in late 2000, but was unable to reach an agreement due to disputes between the European Union on the one hand (which favoured a tougher agreement) and the United States, Canada, Japan and Australia on the other (which wanted the agreement to be less demanding and more flexible).

In 2001, a continuation of the previous meeting (COP6bis) was held in Bonn where the required decisions were adopted. After some concessions, the supporters of the protocol (led by the European Union) managed to get Japan and Russia in as well by allowing more use of carbon dioxide sinks. COP7 was held from 29 October 2001 through 9 November 2001 in Marrakech to establish the final details of the protocol.

The first Meeting of the Parties to the Kyoto Protocol (MOP1) was held in Montreal from November 28 to December 9, 2005, along with the 11th conference of the Parties to the UNFCCC (COP11). See United Nations Climate Change Conference.

The 3rd of December 2007, Australia ratified the protocol during the first day of the COP13 in Bali.

Enforcement

If the Enforcement Branch determines that an Annex I country is not in compliance with its emissions limitation, then that country is required to make up the difference plus an additional 30 per cent. In addition, that country will be suspended from making transfers under an emissions trading programme.

Current Positions of Governments

Australia

Despite being one of the biggest emitters on a per capita basis' the country was granted a limitation of an 8 per cent increase. This was because of considerations specified in Article 4, section 8(h) of the Convention.

The Australian Prime Minister at the time, John Howard (Liberal Party), declined to ratify the Agreement, arguing that the protocol would cost Australians jobs, due to countries with booming economies and massive populations such as China and India not having any reduction obligations. Further, it was claimed that Australia was already doing enough to cut emissions; having pledged $300 million to reduce Greenhouse gas emissions over three years.

Australia's new government formed by the Australian Labor Party after the November 2007 election fully supports the protocol and Prime Minister Kevin Rudd signed the instrument of ratification immediately after assuming office on 3 December 2007, just before the meeting of the UN Framework Convention on Climate Change ; it took effect in March, 2008. When still in Opposition, Kevin Rudd commissioned Professor Ross Garnaut to report into the economic issues of reducing greenhouse gas emissions. Garnaut's report is due to be handed to the Australian Government in September 2008, with a draft in June 2008.

Analysis has projected Australia's greenhouse gas emissions at 109 per cent of the 1990 emissions level over the period 2008-12, calculated including the effects of Land use, land-use change and forestry (LULUCF). This is slightly above its 108 per cent Kyoto Protocol limitation. As of 2007, the UNFCCC is reporting that Australia's 2004 greenhouse gas emissions were at 125.6 per cent of 1990 levels, calculated without the LULUCF correction. http://unfccc.int/files/inc/graphics/image/gif/graph3_2007_ori.gif

The previous Australian Government, along with the United States, agreed to sign the Asia Pacific Partnership on Clean Development and Climate at the ASEAN regional forum on 28 July 2005. Furthermore, the Australian state of New South Wales (NSW) commenced The NSW Greenhouse Gas Abatement Scheme (GGAS). This mandatory greenhouse gas emissions trading scheme commenced on 1 January 2003 and is currently being trialled by the state government in NSW alone. Uniquely this scheme allows Accredited Certificate Providers (ACP) to trade emissions from householders in the state. As of 2006 the scheme is still in place despite the outgoing Prime Minister's clear dismissal of emissions trading as a credible solution to climate change. Following the example of NSW, the National Emissions Trading Scheme (NETS) has been established as an initiative of State and Territory Governments of Australia, all of which have Labor Party governments. The focus of NETS is to bring into existence an intra-Australian carbon trading scheme and to coordinate policy developments to this end. According to the Constitution of Australia, environmental matters are under the jurisdiction of the States, and the NETS is intended to facilitate ratification of the Kyoto Protocol by the incoming Labor Government.

Greenpeace have called Clause 3.7 of the Kyoto Protocol the *"Australia Clause"*, as Australia was the major beneficiary. The clause allows for Annex 1 countries with high rates of land clearing in 1990 to consider that year a base level. Greenpeace argues that Australia had extremely high levels of land clearing in 1990, and that this meant that Australia's "baseline" was unusually high compared to other countries.

Canada

On December 17, 2002, Canada ratified the treaty that came into force in February 2005, requiring it to reduce emissions to 6 per cent below 1990 levels during the 2008-2012 commitment period. At that time, numerous polls showed support for the Kyoto protocol at around 70 per cent. Despite strong public support, there was still some opposition, particularly by the Canadian Alliance, precursor to the governing Conservative Party, some business groups, and energy concerns, using arguments similar to those being used in the US. In particular, there was a fear that since US companies would not be affected by the Kyoto Protocol that Canadian companies would be at a disadvantage in terms of trade. In 2005, the result was limited to an ongoing "war of words", primarily between the government of Alberta (Canada's primary oil and gas producer) and the federal government. As of 2003, the federal government claimed to have spent or committed 3.7 billion dollars on climate change programmes. By 2004, CO_2 emissions had risen to 27 per cent above 1990 levels (which compares unfavourably to the 16 per cent increase in emissions by the United States during that time).

In January 2006, a Conservative minority government under Stephen Harper was elected, who previously has expressed opposition to Kyoto, and in particular to the plan to participate in international emission trading. Rona Ambrose, who replaced Stéphane Dion as the environment minister, has since endorsed some types of emission trading, and indicated interest in international trading. On April 25, 2006, Ambrose announced that Canada would have no chance of meeting its targets under Kyoto, and would look to participate in U.S. sponsored Asia-Pacific Partnership on Clean Development and Climate. "We've been looking at the Asia-Pacific Partnership for a number of months now because the key principles around [it] are very much in line with where our government wants to go," Ambrose told reporters. On May 2, 2006, it was reported that environmental funding designed to meet the Kyoto standards had been cut, while the Harper government develops a new plan to take its place. As the co-chair of UN Climate Change Conference in Nairobi in November 2006, Canada and its government received criticism from environmental groups and from other governments for its climate change positions. On January 4, 2007, Rona Ambrose moved from the Ministry of the Environment to become Minister of Intergovernmental Affairs. The Environment portfolio went to John Baird, the former President of the Treasury Board.

Canada's federal government has introduced legislation to set mandatory emissions targets for industry, but it will not take effect until an estimated 2050. The government has since begun working with opposition parties to improve the legislation.

A private member's bill, was put forth by Pablo Rodriguez, Liberal, aiming to force the government to "ensure that Canada meets its global climate change obligations under the Kyoto Protocol." With the support of the Liberals, the New Democratic Party and the Bloc Québécois, and with the current minority situation, the bill passed the House of Commons on 14 February, 2007 with a vote of 161-113, and is now being considered by the Senate. If passed, the bill would give the government 60 days to form a detailed plan of action. The government has flatly refused to abide by the bill, which may spark a constitutional crisis, lawsuit, or non-confidence motion once the bill becomes law, as is expected.

In May 2007 Friends of the Earth sued the Canadian federal government for failing to meet its Kyoto Protocol obligations to cut greenhouse gas emissions linked to global warming. This was based on a clause in the Canadian Environmental Protection Act that requires Ottawa to "prevent air pollution that violates an international agreement binding on Canada". Canada's obligation to the treaty began in 2008.

Regardless of the national position, some individual provinces are pursuing policies to restrain emissions, including Quebec and British Columbia and Manitoba as part of the Western Climate Initiative.

People's Republic of China

In 2004 the total greenhouse gas emissions from the People's Republic of China were about 54 per cent of the USA emissions. However, China is now building on average one coal-fired power plant every week, and plans to continue doing so for years. Various predictions see China overtaking the US in total greenhouse emissions between late 2007 and 2010, and according to many other estimates, this already occurred in 2006.

The Chinese government insists that the gas emissions level of any given country is a multiplication of its per capita emission and its population. Because China has put into place population control measures while maintaining low emissions per capita, it claims it should therefore in both of the above aspects be considered a contributor to the world's environment. In addition, the country's energy intensity—measured as energy consumption per unit of GDP—was lowered by 47 per cent between 1991 and 2005; from 1950 to 2002, China's carbon dioxide emissions from fossil sources accounted for only 9.33 per cent of the global total in the same period, and in 2004, its per capita emission of carbon dioxide from fossil sources was 3.65 tons, which is 87 per cent of the world average and 33 per cent of that of Organization for Economic Co-operation and Development countries.

In June of 2007, China unveiled a 62-page climate change plan and promised to put climate change at the heart of its energy policies but insisted that developed countries had an "unshirkable responsibility" to take the lead on cutting greenhouse gas emissions and that the "common but differentiated responsibility" principle, as agreed up in the UNFCCC should be applied.

In response to critics of the nation's energy policy, China responded that those criticisms were unjust , while studies of carbon leakage suggest that nearly a quarter of China's emissions result from exports for consumption by developed countries.

European Union

On May 31, 2002, all fifteen then-members of the European Union deposited the relevant ratification paperwork at the UN. The EU produces around 22 per cent of global greenhouse gas emissions, and has agreed to a cut, on average, by 8 per cent from 1990 emission levels. On 10 January 2007, the European Commission announced plans for a European Union energy policy that included a unilateral 20 per cent reduction in GHG emissions by 2020. The EU has consistently been one of the major nominal supporters of the Kyoto Protocol, negotiating hard to get wavering countries on board.

In December 2002, the EU created an emissions trading system in an effort to meet these tough targets. Quotas were introduced in six key industries: energy, steel, cement, glass, brick making, and paper/cardboard. There are also fines for member nations that fail to meet their obligations, starting at $\in 40$/ton of carbon dioxide in 2005, and rising to $\in 100$/ton in 2008. Current EU projections suggest that by 2008 the EU will be at 4.7 per cent below 1990 levels.

Transport CO_2 emissions in the EU grew by 32 per cent between 1990 and 2004. The share of transport in CO_2 emissions was 21 per cent in 1990, but by 2004 this had grown to 28 per cent.

The position of the EU is not without controversy in Protocol negotiations, however. One criticism is that, rather than reducing 8 per cent, all the EU member countries should cut 15 per cent as the EU insisted a uniform target of 15 per cent for other developed countries during the negotiation while allowing itself to share a big reduction in the former East Germany to meet the 15 per cent goal for the entire EU. Also, emission levels of former Warsaw Pact countries who now are members of the EU have already been reduced as a result of their economic restructuring. This may mean that the region's 1990 baseline level is inflated compared to that of other developed countries, thus giving European economies a potential competitive advantage over the U.S.

Both the EU (as the European Community) and its member states are signatories to the Kyoto treaty. Greece, however was excluded from the Kyoto Protocol on Earth Day (April 22, 2008) due to unfulfilled commitment of creating the adequate mechanisms of monitoring and reporting emissions, which is the minimum obligation, and delivering false reports by having no other data to report.

Germany

Germany has reduced greenhouse gas emissions by 17.2 per cent between 1990 and 2004. On June 28, 2006, the German government announced it would exempt its coal industry from requirements under the EU internal emission trading system. Claudia Kemfert, an energy professor at the German Institute for Economic Research in Berlin said, "For all its support for a clean environment and the Kyoto Protocol, the cabinet decision is very disappointing. The energy lobbies have played a big role in this decision."

United Kingdom

The energy policy of the United Kingdom fully endorses goals for carbon dioxide emissions reduction and has committed to proportionate reduction in national emissions on a phased basis. The United Kingdom is a signatory to the Kyoto Protocol. On March 13, 2007, a draft Climate Change Bill was published after cross-party pressure over several years, led by environmental groups. Informed by the Energy White Paper 2003, The Bill aims to put in place a framework to achieve a mandatory 60 per cent cut in the UK's carbon emissions by 2050 (compared to 1990 levels), with an intermediate target of between 26 per cent and 32 per cent by 2020. If approved, the United Kingdom is likely to become the first country to set such a long-range and significant carbon reduction target into law.

The UK currently appears on course to meet its Kyoto limitation for the basket of greenhouse gases, assuming the Government is able to curb rising carbon dioxide emissions between now (2007) and the period 2008-2012. Although the UK's overall greenhouse gas emissions have fallen, annual net carbon dioxide emissions have risen by around 2 per cent since The Labour Party came to power in 1997. As a result it now seems highly unlikely that the Government will be able to honour its manifesto pledge to cut carbon dioxide emissions by 20 per cent from 1990 levels by the year 2010, unless immediate and drastic action is taken under after the passing of the Climate Change Bill.

France

In 2004, France shut down its last coal mine, and now gets 80 per cent of its electricity from nuclear power and therefore has relatively low CO_2 emissions

Norway

Between 1990 and 2006, Norway's carbon emissions increased by almost 8 per cent. As well as directly reducing their own greenhouse gas emissions, Norway's idea for carbon neutrality, is that they will finance reforestation in China, which is still legal under the Kyoto protocol.

India

India signed and ratified the Protocol in August, 2002. Since India is exempted from the framework of the treaty, it is expected to gain from the protocol in terms of transfer of technology and related foreign investments. At the G-8 meeting in June 2005, Indian Prime Minister Manmohan Singh pointed out that the per-capita emission rates of the developing countries are a tiny fraction of those in the developed world. Following the principle of *common but differentiated responsibility*, India maintains that the major responsibility of curbing emission rests with the developed countries, which have accumulated emissions over a long period of time. However, the U.S. and other Western nations assert that India, along with China, will account for most of the emissions in the coming decades, owing to their rapid industrialization and economic growth.

Russia

Vladimir Putin approved the treaty on November 4, 2004 and Russia officially notified the United Nations of its ratification on November 18, 2004. The issue of Russian ratification was particularly closely watched in the international community, as the accord was brought into force 90 days after Russian ratification (February 16, 2005).

President Putin had earlier decided in favour of the protocol in September 2004, along with the Russian cabinet, against the opinion of the Russian Academy of Sciences, of the Ministry for Industry and Energy and of the then president's economic adviser, Andrey Illarionov, and in exchange to EU's support for Russia's admission in the WTO. As anticipated after this, ratification by the lower (22 October 2004) and upper house of parliament did not encounter any obstacles.

The Kyoto Protocol limits emissions to a percentage increase or decrease from their 1990 levels. Since 1990 the economies of most countries in the former Soviet Union have collapsed, as have their greenhouse gas emissions. Because of this, Russia should have no problem meeting its commitments under Kyoto, as its current emission levels are substantially below its limitations. It is debatable whether Russia will benefit from selling emissions credits to other countries in the Kyoto Protocol.

United States

The United States (U.S.), although a signatory to the Kyoto Protocol, has neither ratified nor withdrawn from the Protocol. The signature alone is symbolic, as the Kyoto Protocol is non-binding on the United States unless ratified. The United States was, as of 2005, the largest single emitter of carbon dioxide from the burning of fossil fuels. The "Lieberman-Warner Climate Security Act of 2008"lso more commonly referred to in the U.S. as the "Cap and Trade Bill", was proposed for greater U.S. allignment with the Kyoto standards and goals. The current bill is almost 500 pages long, and provides for establishment of a federal bureau of Carbon Trading, Regulation, and Enforcement with mandates which some authorities suggest will amount to the largest tax increase in the history of the United States.

On July 25, 1997, before the Kyoto Protocol was finalized (although it had been fully negotiated, and a penultimate draft was finished), the U.S. Senate unanimously passed by a 95-0 vote the Byrd-Hagel Resolution (S. Res. 98), which stated the sense of the Senate was that the United States should not be a signatory to any protocol that did not include binding targets and timetables for developing as well as industrialized nations or "would result in serious harm to the economy of the United States". On November 12, 1998, Vice President Al Gore symbolically signed the protocol. Both Gore and Senator Joseph Lieberman indicated that the protocol would not be acted upon in the Senate until there was participation by the developing nations. The Clinton Administration never submitted the protocol to the Senate for ratification.

The Clinton Administration released an economic analysis in July 1998, prepared by the Council of Economic Advisors, which concluded that with emissions trading among the Annex B/Annex I countries, and participation of key developing countries in the "Clean Development Mechanism"—which grants the latter business-as-usual emissions rates through 2012—the costs of implementing the Kyoto Protocol could be reduced as much as 60 per cent from many estimates. Other economic analyses, however, prepared by the Congressional Budget Office and the Department of Energy· Energy Information Administration (EIA) , demonstrated a potentially large loss to GDP from implementing the Protocol of up to 4.2 per cent (EIA).

The current President, George W. Bush, has indicated that he does not intend to submit the treaty for ratification, not because he does not support the Kyoto principles, but because of the exemption granted to China (the world's largest emitter of carbon dioxide). Bush also opposes the treaty because of the strain he believes the treaty would put on the economy; he emphasizes the uncertainties which he believes are present in the climate change issue. Furthermore, the U.S. is concerned with broader exemptions of the treaty. For example, the U.S. does not support the split between Annex I countries and others. Bush said of the treaty:

> This is a challenge that requires a 100 per cent effort; ours, and the rest of the world's. The world's second-largest emitter of greenhouse gases is the People's Republic of China. Yet, China was entirely exempted from the requirements of the Kyoto Protocol. India and Germany are among the top emitters. Yet, India was also exempt from Kyoto... America's unwillingness to embrace a flawed treaty should not be read by our friends and allies as any abdication of responsibility. To the contrary, my administration is committed to a leadership role on the issue of climate change ... Our approach must be consistent with the long-term goal of stabilizing greenhouse gas concentrations in the atmosphere."

In June 2002, the United States Environmental Protection Agency (EPA) released the "Climate Action Report 2002". Some observers have interpreted this report as being supportive of the protocol, although the report itself does not explicitly endorse the protocol. At the G-8 meeting in June 2005 administration officials expressed a desire for "practical commitments industrialized countries can meet without damaging their economies". According to those same officials, the United States is on track to fulfill its pledge to reduce its carbon intensity 18 per cent by 2012. The United States has signed the Asia Pacific Partnership on Clean Development and Climate, a pact that allows those countries to set their goals for reducing greenhouse gas emissions individually, but with no enforcement mechanism. Supporters of the pact see it as complementing the Kyoto Protocol while being more flexible, but critics have said the pact will be ineffective without any enforcement measures.

The Administration's position is not uniformly accepted in the U.S. For example, Paul Krugman notes that the target 18 per cent reduction in carbon intensity is still actually an increase in overall emissions. The White House has also come under criticism for downplaying reports that link human activity and greenhouse gas emissions to climate change and that a White House official, former oil industry advocate and current Exxon Mobil officer, Philip Cooney, watered down descriptions of climate research that had already been approved by government scientists, charges the White House denies. Critics point to the Bush administration's close ties to the oil and gas industries. In June 2005, State Department papers showed the administration thanking Exxon executives for the company's "active involvement" in helping to determine climate change policy, including the U.S. stance on Kyoto. Input from the business lobby group Global Climate Coalition was also a factor.

In 2002, Congressional researchers who examined the legal status of the Protocol advised that signature of the UNFCCC imposes an obligation to refrain from undermining the Protocol's object and purpose, and that while the President probably cannot implement the Protocol alone; Congress can create compatible laws on its own initiative.

Local Governments

As of January 18, 2007, eight Northeastern US states are involved in the Regional Greenhouse Gas Initiative (RGGI), which is a state level emissions capping and trading programme. It is believed that the state-level programme will indirectly apply pressure on the federal government by demonstrating that reductions can be achieved without being a signatory of the Kyoto Protocol.

- **Participating states**: Maine, New Hampshire, Vermont, Connecticut, New York, New Jersey, Delaware, Massachusetts, and Maryland (these states represent over 46 million people).
- **Observer states and regions**: Pennsylvania, District of Columbia, Rhode Island.

On August 31, 2006, the California Legislature (representing over 33 million Californians) reached an agreement with Governor Arnold Schwarzenegger to reduce the state's greenhouse-gas emissions, which rank at 12th-largest in the world, by 25 per cent by the year 2020. This resulted in the Global Warming Solutions Act which effectively puts California in line with the Kyoto limitations, but at a date later than the 2008-2012 Kyoto commitment period.

As of December 4, 2007, 740 US cities in 50 states, the District of Columbia and Puerto Rico, representing over 76 million Americans support Kyoto after Mayor Greg Nickels of Seattle started a nationwide effort to get cities to agree to the protocol. On October 29, 2007, it was reported that Seattle met their target reduction in 2005, reducing their greenhouse gas emissions by 8 percent since 1990.

- *Large participating cities*: Albany, New York; Albuquerque, New Mexico; Alexandria, Virginia; Ann Arbor, Michigan; Arlington, Texas; Atlanta, Georgia; Austin, Texas; Baltimore, Maryland; Berkeley, California; Bristol, Connecticut; Boston, Massachusetts; Charleston, South Carolina; Chattanooga, Tennessee; Chicago, Illinois; Cincinnati, Ohio; ; Dallas, Texas; Denver, Colorado; Des Moines, Iowa; Fairfield,

Connecticut; Fayetteville, Arkansas; Hartford, Connecticut; Honolulu, Hawaii; Indianapolis, Indiana; Jersey City, New Jersey; Lansing, Michigan; Las Vegas, Nevada; Lawrence, Kansas; Lexington, Kentucky; Lincoln, Nebraska; Little Rock, Arkansas; Los Angeles, California; Louisville, Kentucky; Madison, Wisconsin; Miami, Florida; Milwaukee, Wisconsin; Minneapolis, Minnesota; Nashville, Tennessee; New Orleans, Louisiana; New York, New York; Oakland, California; Omaha, Nebraska; Pasadena, California; Philadelphia, Pennsylvania; Phoenix, Arizona; Pittsburgh, Pennsylvania; Portland, Oregon; Providence, Rhode Island; Richmond, Virginia; Sacramento, California; Salt Lake City, Utah; San Antonio, Texas; San Francisco, California; San Jose, California; Santa Ana, California; Santa Fe, New Mexico; Seattle, Washington; Sioux City, Iowa; St. Louis, Missouri; Tacoma, Washington; Tallahassee, Florida; Tampa, Florida; Topeka, Kansas; Tulsa, Oklahoma; Vancouver, Washington; Virginia Beach, Virginia; Washington, D.C.; West Palm Beach, Florida; Wilmington, North Carolina.

Support

Advocates of the Kyoto Protocol state that reducing these emissions is crucially important, as carbon dioxide is causing the earth's atmosphere to heat up. This is supported by attribution analysis. No country has passed national legislation requiring compliance with their treaty obligation. The governments of all of the countries whose parliaments have ratified the Protocol are supporting it. Most prominent among advocates of Kyoto have been the European Union and many environmentalist organizations. The United Nations and some individual nations' scientific advisory bodies (including the G8 national science academies) have also issued reports favouring the Kyoto Protocol.

An international day of action was planned for 3 December 2005, to coincide with the Meeting of the Parties in Montreal. The planned demonstrations were endorsed by the Assembly of Movements of the World Social Forum.

A group of major Canadian corporations also called for urgent action regarding climate change, and have suggested that Kyoto is only a first step.

In the United States, there is at least one student group, Kyoto Now!, which aims to use student interest to support pressure towards reducing emissions as targeted by the Kyoto Protocol compliance.

Opposition

Some public policy experts who are skeptical of global warming, although few, see Kyoto as a scheme to either slow the growth of the world's industrial democracies or to transfer wealth to the third world in what they claim is a global socialism initiative. Others argue the protocol does not go far enough to curb greenhouse emissions (Niue, The Cook Islands, and Nauru added notes to this effect when signing the protocol).

Some environmental economists have been critical of the Kyoto Protocol. Many see the costs of the Kyoto Protocol as outweighing the benefits, some believing the standards which Kyoto sets to be too optimistic, others seeing a highly inequitable and inefficient agreement which would do little to curb greenhouse gas emissions. Finally, some economists as Gwyn Prins and Steve Rayner think that an entirely different approach needs to be followed than the approach suggested by the Kyoto-protocol.

Further, there is controversy surrounding the use of 1990 as a base year as well as not using per capita emissions as a basis. Countries had different achievements in energy efficiency in 1990. For example, the former Soviet Union and eastern European countries did little to tackle the problem and their energy efficiency was at its worst level in 1990; the year just before their communist regimes fell. On the other hand, Japan, as a big importer of natural resources, had to improve its efficiency after the 1973 oil crisis and its emissions level in 1990 was better than most developed countries. However, such efforts were set aside, and the inactivity of the former Soviet Union was overlooked and could even generate big income due to the emission trade. There is an argument that the use of per capita emissions as a basis in the following Kyoto-type treaties can reduce the sense of inequality among developed and developing countries alike, as it can reveal inactivities and responsibilities among countries.

Cost-benefit Analysis

Economists have been trying to analyze the overall net benefit of Kyoto Protocol through cost-benefit analysis. There is disagreement due to large uncertainties in economic variables. Some of the estimates indicate either that observing the Kyoto Protocol is more expensive than not observing the Kyoto Protocol or that the Kyoto Protocol has a marginal net benefit which exceeds the cost of simply adjusting to global warming. However, a study by De Leo *et al.* found that *"accounting only for local external costs, together with production costs, to identify energy strategies, compliance with the Kyoto Protocol would imply lower, not higher, overall costs."*

The recent Copenhagen consensus project found that the Kyoto Protocol would slow down the process of global warming, but have a superficial overall benefit. Defenders of the Kyoto Protocol argue, however, that while the initial greenhouse gas cuts may have little effect, they set the political precedent for bigger (and more effective) cuts in the future. They also advocate commitment to the precautionary principle. Critics point out that additional higher curbs on carbon emission are likely to cause significantly higher increase in cost, making such defense moot. Moreover, the precautionary principle could apply to any political, social, economic or environmental consequence, which might have equally devastating effect in terms of poverty and environment, making the precautionary argument irrelevant. The Stern Review (a UK government sponsored report into the economic impacts of climate change) concluded that one percent of global GDP is required to be invested in order to mitigate the effects of climate change, and that failure to do so could risk a recession worth up to twenty percent of global GDP.

Discount Rates

One problem in attempting to measure the "absolute" costs and benefits of different policies to global warming is choosing a proper discount rate. Over a long time horizon such as that in which benefits accrue under Kyoto, small changes in the discount rate create very large discrepancies between net benefits in various studies. However, this difficulty is generally not applicable to "relative" comparison of alternative policies under a long time horizon. This is because changes in discount rates tend to equally adjust the net cost/benefit of different policies unless there are significant discrepancies of cost and benefit over time horizon.

While it has been difficult to arrive at a scenario under which the net benefits of Kyoto are positive using traditional discounting methods such as the Shadow Price of Capital approach, there is an argument that a much lower discount rate should be utilized; that high rates are biased toward the current generation. This may appear to be a philosophical value judgment, outside the realm of economics, but it could be equally argued that the study of the allocation of resources does include how those resources are allocated over time.

Increase in Greenhouse Gas Emission Since 1990

Below is a list of the change in greenhouse gas emissions from 1990 to 2004 for some countries that are part of the Climate Change Convention as reported by the United Nations.

Value in Percent

Country	Change in greenhouse gas Emissions (1990-2004) excluding LULUCF	Change in greenhouse gas Emissions (1990-2004) including LULUCF	EU Assigned Objective for 2012	Treaty Obligation 2008-2012
Germany	-17	-18.2	-21	-8
Canada	+27	+26.6	N/A	-6
Australia	+25	+5.2	N/A	+8
Spain	+49	+50.4	+15	-8
Norway	+10	-18.7	N/A	+1
New Zealand	+21	+17.9	N/A	0
France	-0.8	-6.1	0	-8
Greece	+27	+25.3	+25	-8
Ireland	+23	+22.7	+13	-8
Japan	+6.5	+5.2	N/A	-6
United Kingdom	-14	-58.8	-12.5	-8
Portugal	+41	+28.9	+27	-8
EU-15	-0.8	-2.6	N/A	-8

Below is a table of the changes in CO_2 emission of some other countries which are large contributors, but are not required to meet numerical limitations.

Country	Change in greenhouse gas Emissions (1990-2004)
China	+47%
India	+55%

Comparing total greenhouse gas emissions in 2004 to 1990 levels, the US emissions were up by 16 per cent, with irregular fluctuations from one year to another but a general trend to increase. At the same time, the EU group of 23 (EU-23) Nations had reduced their emissions by 5 per cent. In addition, the EU-15 group of nations (a large subset of EU-23) reduced their emissions by 0.8 per cent between 1990 and 2004, while emission rose 2.5 per cent from 1999 to 2004. Part of the increases for some of the European Union countries are still in line with the treaty, being part of the cluster of countries implementation (see objectives in the list above).

As of year-end 2006, the United Kingdom and Sweden were the only EU countries on pace to meet their Kyoto emissions commitments by 2010. While UN statistics indicate that, as a group, the 36 Kyoto signatory countries can meet the 5 per cent reduction target by 2012, most of the progress in greenhouse gas reduction has come from the stark decline in Eastern European countries' emissions after the fall of communism in the 1990s.

Successor

In the non-binding 'Washington Declaration' agreed on February 16, 2007, Heads of governments from Canada, France, Germany, Italy, Japan, Russia, United Kingdom, the United States, Brazil, China, India, Mexico and South Africa agreed in principle on the outline of a successor to the Kyoto Protocol. They envisage a global cap-and-trade system that would apply to both industrialized nations and developing countries, and hoped that this would be in place by 2009.

On June 7, 2007, leaders at the 33rd G8 summit agreed that the G8 nations would 'aim to at least halve global CO_2 emissions by 2050'. The details enabling this to be achieved would be negotiated by environment ministers within the United Nations Framework Convention on Climate Change in a process that would also include the major emerging economies.

A round of climate change talks under the auspices of the United Nations Framework Convention on Climate Change (UNFCCC) (Vienna Climate Change Talks 2007) concluded in 31 August 2007 with agreement on key elements for an effective international response to climate change.

A key feature of the talks was a United Nations report that showed how energy efficiency could yield significant cuts in emissions at low cost.

The talks are meant to set the stage for a major international meeting to be held in Nusa Dua, Bali, which started on 3 December, 2007.

Asia Pacific Partnership on Clean Development and Climate

The Asia Pacific Partnership on Clean Development and Climate is an agreement between seven nations: Australia, Canada, China, India, Japan, South Korea, and the United States. Between them, these seven countries are responsible for more than half of the world's carbon dioxide emissions.

The partnership had its official launch in January 2006 at a ceremony in Sydney, Australia. The alliance states that member nations have initiated nearly 100 projects aimed at clean energy capacity building and market formation since then. Building on these activities, long-term projects are scheduled to deploy clean energy and environment technologies and services. The pact allows those countries to set arbitrary goals for reducing greenhouse gas emissions individually, without any enforcement mechanism for these goals.

Supporters of the pact see it as "complementing the Kyoto Protocol" whilst being more flexible. Critics have said the pact will be ineffective without any enforcement measures and is a means to undermine the negotiations leading to the Protocol scheduled to replace the current Kyoto Protocol (negotiations started in Montreal in December 2005). U.S. Senator John McCain said the partnership "[amounted] to nothing more than a nice little public relations ploy." , while the Economist described the partnership as "patent fig-leaf for the refusal of America and Australia to ratify Kyoto" .

5.10 Climate change and agriculture

Climate change and agriculture are interrelated processes, both of which take place on a global scale. Global warming is projected to have significant impacts on conditions affecting agriculture, including temperature, precipitation and glacial run-off. These conditions determine the carrying capacity of the biosphere to produce enough food for the human population and domesticated animals. Rising carbon dioxide levels would also have effects, both detrimental and beneficial, on crop yields. The overall effect of climate change on agriculture will depend on the balance of these effects. Assessment of the effects of global climate changes on agriculture might help to properly anticipate and adapt farming to maximize agricultural production.

At the same time, agriculture has been shown to produce significant effects on climate change, primarily through the production and release of greenhouse gases such as carbon dioxide, methane, and nitrous oxide, but also by altering the earth's land cover, which can change its ability to absorb or reflect heat and light, thus contributing to radiative forcing. Land use change such as deforestation and desertification, together with use of fossil fuels, are the major anthropogenic sources of carbon dioxide; agriculture itself is the major contributor to increasing methane and nitrous oxide concentrations in earth's atmosphere.

Impact of climate change on agriculture

Despite technological advances, such as improved varieties, genetically modified organisms, and irrigation systems, weather is still a key factor in agricultural productivity, as well as soil properties and natural communities. The effect of climate on agriculture is related to variabilities in local climates rather than in global climate patterns. Consequently, agronomists consider any assessment has to be individually consider each local area.

On the other hand, agricultural trade has grown in recent years, and now provides significant amounts of food, on a national level to major importing countries, as well as comfortable income to exporting ones. The international aspect of trade and security in terms of food implies the need to also consider the effects of climate change on a global scale.

A study published in *Science* suggest that, due to climate change, "southern Africa could lose more than 30 per cent of its main crop, maize, by 2030. In South Asia losses of many regional staples, such as rice, millet and maize could top 10 per cent".

The 2001 IPCC Third Assessment Report concluded that the poorest countries would be hardest hit, with reductions in crop yields in most tropical and sub-tropical regions due to decreased water availability, and new or changed insect pest incidence. In Africa and Latin America many rainfed crops are near their maximum temperature tolerance, so that yields are likely to fall sharply for even small climate changes; falls in agricultural productivity of up to 30 per cent over the 21st century are projected. Marine life and the fishing industry will also be severely affected in some places.

Climate change induced by increasing greenhouse gases is likely to affect crops differently from region to region. For example, average crop yield is expected to drop down to 50 per cent in Pakistan according to the UKMO scenario whereas corn production in Europe is expected to grow up to 25 per cent in optimum hydrologic conditions.

More favourable effects on yield tend to depend to a large extent on realization of the potentially beneficial effects of carbon dioxide on crop growth and increase of efficiency in water use. Decrease in potential yields is likely to be caused by shortening of the growing period, decrease in water availability and poor vernalization.

In the long run, the climatic change could affect agriculture in several ways:

* *productivity*, in terms of quantity and quality of crops,
* *agricultural practices*, through changes of water use (irrigation) and agricultural inputs such as herbicides, insecticides and fertilizers,
* *environmental effects*, in particular in relation of frequency and intensity of soil drainage (leading to nitrogen leaching), soil erosion, reduction of crop diversity,
* *rural space*, through the loss and gain of cultivated lands, land speculation, land renunciation, and hydraulic amenities, and
* *adaptation*, organisms may become more or less competitive, as well as humans may develop urgency to develop more competitive organisms, such as flood resistant or salt resistant varieties of rice.

They are large uncertainties to uncover, particularly because there is lack of information on many specific local regions, and include the uncertainties on magnitude of climate change, the effects of technological changes on productivity, global food demands, and the numerous possibilities of adaptation.

Most agronomists believe that agricultural production will be mostly affected by the severity and pace of climate change, not so much by gradual trends in climate. If change is gradual, there may be enough time for biota adjustment. Rapid climate change, however, could harm agriculture in many countries, especially those that are already suffering from rather poor soil and climate conditions, because there is less time for optimum natural selection and adaption.

Shortage in Grain Production

Between 1996 and 2003, grain production has stabilized slightly over 1800 millions of tons. In 2000, 2001, 2002 and 2003, grain stocks have been dropping, resulting in a global grain harvest that was short of consumption by 93 millions of tons in 2003.

The earth's average temperature has been rising since the late 1970s, with nine of the 10 warmest years on record occurring since 1995 . In 2002, India and the United States suffered sharp harvest reductions because of record temperatures and drought. In 2003 Europe suffered very low rainfall throughout spring and summer, and a record level of heat damaged most crops from the United Kingdom and France in the Western Europe through Ukraine in the East. Bread prices have been rising in several countries in the region.

Poverty Impacts

Researchers at the Overseas Development Institute (ODI) have investigated the potential impacts climate change could have on agriculture, and how this would affect attempts at alleviating poverty in the developing world. They argued that the effects from moderate climate change are likely to be mixed for developing countries. However, the vulnerability of the poor in developing countries to short term impacts from climate change, notably the increased frequency and severity of adverse weather events is likely to have a negative impact. This, they say, should be taken into account when defining agricultural policy.

Crop Development Models

Models for climate behaviour are frequently inconclusive. In order to further study effects of global warming on agriculture, other types of models, such as *crop development models, yield prediction*, quantities of *water or fertilizer consumed*, can be used. Such models condense the knowledge accumulated of the climate, soil, and effects observed of the results of various agricultural practices. They thus could make it possible to test strategies of adaptation to modifications of the environment.

Because these models are necessarily simplifying natural conditions (often based on the assumption that weeds, disease and insect pests are controlled), it is not clear whether the results they give will have an *in-field* reality. However, some results are partly validated with an increasing number of experimental results.

Other models, such as *insect and disease development* models based on climate projections are also used (for example simulation of aphid reproduction or septoria (cereal fungal disease) development).

Scenarios are used in order to estimate climate changes effects on crop development and yield. Each scenario is defined as a set of meteorological variables, based on generally accepted projections. For example, many models are running simulations based on doubled carbon dioxide projections, temperatures raise ranging from 1°C up to 5°C, and with rainfall levels an increase or decrease of 20 per cent. Other parameters may include humidity, wind, and solar activity. Scenarios of crop models are testing farm-level adaptation, such as sowing date shift, climate adapted species (vernalisation need, heat and cold resistance), irrigation and fertilizer adaptation, resistance to disease. Most developed models are about wheat, maize, rice and soybean.

Temperature Potential Effect on Growing Period

Duration of crop growth cycles are above all, related to temperature. An increase in temperature will speed up development. In the case of an annual crop, the duration between sowing and harvesting will shorten (for example, the duration in order to harvest corn could shorten between one and four weeks). The shortening of such a cycle could have an adverse effect on productivity because senescence would occur sooner.

Potential Effect of Atmospheric Carbon Dioxide on Yield

Carbon dioxide is essential to plant growth. Rising CO_2 concentration in the atmosphere can have both positive and negative consequences.

Increased CO_2 is expected to have positive physiological effects by increasing the rate of photosynthesis. Currently, the amount of carbon dioxide in the atmosphere is 380 parts per million. In comparison, the amount of oxygen is 210,000 ppm. This means that often plants may be starved of carbon dioxide, being outnumbered by the photosynthetic pollutant oxygen. The effects of an increase in carbon dioxide would be higher on C3 crops (such as wheat) than on C4 crops (such as maize), because the former is more susceptible to carbon dioxide shortage. Under optimum conditions of temperature and humidity, the yield increase could reach 36 per cent, if the levels of carbon dioxide are doubled.

However, other studies also show a change in harvest quality. The growth improvement in C3 plants could favour vegetative biomass on grain biomass; thus leading to a decrease in grain production yield.

Further, few studies have looked at the impact of elevated carbon dioxide concentrations on whole farming systems. Most models study the relationship between CO_2 and productivity in isolation from other factors associated with climate change, such as an increased frequency of extreme weather events, seasonal shifts, and so on. Moreover, it is reasonable to expect that weed productivity will increase in parallel with that of crop and pasture plants; potentially raising the cost of defensive expenditures like herbicides.

In 2005, the Royal Society in London concluded that the purported benefits of CO_2 fertilization are "likely to be far lower than previously estimated" when factors such as increasing ground-level ozone are taken into account."

Effect on Quality

According to the IPCC's TAR, "The importance of climate change impacts on grain and forage quality emerges from new research. For rice, the amylose content of the grain—a major determinant of cooking quality—is increased under elevated CO_2" (Conroy *et al.*, 1994). Cooked rice grain from plants grown in high-CO2 environments would be firmer than that from today's plants. However, concentrations of iron and zinc, which are important for human nutrition, would be lower (Seneweera and Conroy, 1997). Moreover, the protein content of the grain decreases under combined increases of temperature and CO_2 (Ziska *et al.*, 1997)."

Studies have shown that higher CO_2 levels lead to reduced plant uptake of nitrogen (and a smaller number showing the same for trace elements such as zinc) resulting in crops with lower nutritional value. This would primarily impact on populations in poorer countries less able to compensate by eating more food, more varied diets, or possibly taking supplements.

Reduced nitrogen content in grazing plants has also been shown to reduce animal productivity in sheep, which depend on microbes in their gut to digest plants, which in turn depend on nitrogen intake.

Agricultural Surfaces and Climate Changes

Climate change is likely to increase the amount of arable land in high-lattitude region by reduction of the amount of frozen lands. A 2005 study reports that temperature in siberia has increased three degree celcius in average since 1960 (much more than the rest of the world). However, reports about the impact of global warming on russian agriculture indicate conflicting probable effects: while they expect an northward extension of farmable lands, they also warn of possible productivity losses and increased risk of drought.

Sea levels are expected to get up to one meter higher by 2100, though this projection is disputed. A rise in the sea level would result in an agricultural land loss, in particular in areas such as South East Asia. Erosion, submergence of shorelines, salinity of the water table due to the increased sea levels, could mainly affect agriculture through inundation of low-lying lands.

Erosion and Fertility

With global warming, soil degradation is more likely to occur, and soil fertility would probably be affected by global warming. However, because the ratio of carbon to nitrogen is a constant, a doubling of carbon is likely to imply a higher storage of nitrogen in soils as nitrates, thus providing higher fertilizing elements for plants, providing better yields. The average needs for nitrogen could decrease, and give the opportunity of changing often costly fertilisation strategies.

Due to the extremes of climate that would result, the increase in precipitations would probably result in greater risks of erosion, whilst at the same time providing soil with better hydration, according to the intensity of the rain. The possible evolution of the organic matter in the soil is a highly contested issue: while the increase in the temperature would induce a greater rate in the production of minerals, lessening the soil organic matter content, the atmospheric CO_2 concentration would tend to increase it.

Potential Effects of Global Climate Change on Pests, Diseases and Weeds

A very important point to consider is that weeds would undergo the same acceleration of cycle as cultivated crops, and would also benefit from carbonaceous fertilization. Since most weeds are C3 plants, they are likely to compete even more than now against C4 crops such as corn. However, on the other hand, some results make it possible to think that weedkillers could gain in effectiveness with the temperature increase.

Global warming would cause an increase in rainfall in some areas, which would lead to an increase of atmospheric humidity and the duration of the wet seasons. Combined with higher temperatures, these could favour the development of fungal diseases. Similarly, because of higher temperatures and humidity, there could be an increased pressure from insects and disease vectors.

Glacier Retreat and Disappearance

The continued retreat of glaciers will have a number of different quantitative impacts. In areas that are heavily dependent on water runoff from glaciers that melt during the warmer summer months, a continuation of the current retreat will eventually deplete the glacial ice and substantially reduce or eliminate runoff. A reduction in runoff will affect the ability to irrigate crops and will reduce summer stream flows necessary to keep dams and reservoirs replenished.

According to a UN climate report, the Himalayan glaciers that are the principal dry-season water sources of Asia's biggest rivers—Ganges, Indus, Brahmaputra, Yangtze, Mekong, Salween and Yellow—could disappear by 2035 as temperatures rise. Approximately 2.4 billion people live in the drainage basin of the Himalayan rivers. India, China, Pakistan, Afghanistan, Bangladesh, Nepal and Myanmar could experience floods followed by severe droughts in coming decades. In India alone, the Ganges provides water for drinking and farming for more than 500 million people.

Ozone and UV-B

Some scientists think agriculture could be affected by any decrease in stratospheric ozone, which could increase biologically dangerous ultraviolet radiation B. Excess ultraviolet radiation B can directly effect plant physiology and cause massive amounts of mutations, and indirectly through changed pollinator behaviour, though such changes are difficult to quantify. However, it has not yet been ascertained whether an increase in greenhouse gases would decrease stratospheric ozone levels.

In addition, a possible effect of rising temperatures is significantly higher levels of ground-level ozone, which would substantially lower yields.

Impact of Agriculture on Climate Change

The agricultural sector is a driving force in the gas emissions and land use effects thought to cause climate change. In addition to being a significant user of land and consumer of fossil fuel, agriculture contributes directly to greenhouse gas emissions through practices such as rice production and the raising of livestock ; according to the Intergovernmental Panel on Climate Change, the three main causes of the increase in greenhouse gases observed over the past 250 years have been fossil fuels, land use, and agriculture.

Land Use

Agriculture contributes to greenhouse gas increases through land use in four main ways:

- ❖ CO_2 releases linked to deforestation,
- ❖ Methane releases from rice cultivation,
- ❖ Methane releases from enteric fermentation in cattle,
- ❖ Nitrous oxide releases from fertilizer application.

Together, these agricultural processes comprise 54 per cent of methane emissions, roughly 80 per cent of nitrous oxide emissions, and virtually all carbon dioxide emissions tied to land use.

The planet's major changes to land cover since 1750 have resulted from deforestation in temperate regions: when forests and woodlands are cleared to make room for fields and pastures, the albedo of the affected area increases, which can result in either warming or cooling effects, depending on local conditions. Deforestation also affects regional carbon reuptake, which can result in increased concentrations of CO_2, the dominant greenhouse gas. Land-clearing methods such as slash and burn compound these effects by burning biomatter, which directly releases greenhouse gases and particulate matter such as soot into the air.

Livestock

Livestock and livestock-related activities such as deforestation and increasingly fuel-intensive farming practices are responsible for over 18 per cent of human-made greenhouse gas emissions, including:

- ❖ 9 per cent of global carbon dioxide emissions,
- ❖ 35-40 per cent of global methane emissions (chiefly due to enteric fermentation and manure), and
- ❖ 64 per cent of global nitrous oxide emissions (chiefly due to fertilizer use.).

Livestock activities also contribute disproportionately to land-use effects, since crops such as corn and alfalfa are cultivated in order to feed the animals.

Worldwide, livestock production occupies 70 per cent of all land used for agriculture, or 30 per cent of the land surface of the Earth.

5.11 Environmental Effects of Fishing

The environmental effects of fishing can be divided into issues that involve the availability of fish to be caught, such as overfishing, sustainable fisheries, and fisheries management; and issues that involve the impact of fishing on the environment, such as by-catch.

These conservation issues are part of marine conservation, and are addressed in fisheries science programmes. There is a growing gap between how many fish are available to be caught and humanity's desire to catch them, a problem that gets worse as the world population grows.

Similar to other environmental issues, there can be conflict between the fishermen who depend on fishing for their livelihoods and fishery scientists who realise that if future fish populations are to be sustainable then some fisheries must reduce or even close.

The journal *Science* published a four-year study in November 2006, which predicted that, at prevailing trends, the world would run out of wild-caught seafood in 2048. The scientists stated that the decline was a result of overfishing, pollution and other environmental factors that were reducing the population of fisheries at the same time as their ecosystems were being degraded. Yet again the analysis has met criticism as being fundamentally flawed, and many fishery management officials, industry representatives and scientists challenge the findings, although the debate continues. Many countries, such as Tonga, the United States, Australia and New Zealand, and international management bodies have taken steps to appropriately manage marine resources.

Effects on Habitat

Some fishing techniques also may cause habitat destruction. Dynamite fishing and cyanide fishing, which are illegal in many places, harm surrounding habitat. Bottom trawling, the practice of pulling a fishing net along the sea bottom behind trawlers, removes around 5 to 25 per cent of an area's seabed life on a single run. A 2005 report of the UN Millennium Project, commissioned by UN Secretary-General Kofi Annan, recommended the elimination of bottom trawling on the high seas by 2006 to protect seamounts and other ecologically sensitive habitats.

In mid October 2006, U.S. President Bush joined other world leaders calling for a moratorium on deep-sea trawling, a practice shown to often have harmful effects on sea habitat and, hence, on fish populations.

Overfishing

Overfishing has also been widely reported due to increases in the volume of fishing hauls to feed a quickly growing number of consumers. This has led to the breakdown of some sea ecosystems and several fishing industries whose catch has been greatly diminished. The extinction of many species has also been reported. According to an FAO estimate, over 70 per cent of the world's fish species are either fully exploited or depleted. According to Nitin Desai, Secretary General of the 2002 World Summit on Sustainable Development, "Overfishing cannot continue, the depletion of fisheries poses a major threat to the food supply of millions of people."

The cover story of the May 15, 2003 issue of the science journal *Nature*—with Dr. Ransom A. Myers, an internationally prominent fisheries biologist (Dalhousie University, Halifax, Canada) as the lead author—was devoted to a summary of the scientific information. The story asserted that, as compared with 1950 levels, only a remnant (in some instances, as little as 10 per cent) of all large ocean-fish stocks are left in the seas. These large ocean fish are the species at the top of the food chains (e.g., tuna, cod, among others). However, this article was subsequently criticized as being fundamentally flawed, although much debate still exists (Walters 2003; Hampton *et al.* 2005; Maunder *et al.* 2006; Polacheck 2006; Sibert *et al.* 2006) and the majority of fisheries scientists now consider the results irrelevant with respect to large pelagics (the open seas).

The environmental impact of recreational fishing may be alleviated to some extent by catch and release fishing.

Ecological Disruption

Fishing may disrupt food webs by targeting specific, in-demand species. There might be too much fishing of prey species such as sardines and anchovies, thus reducing the food supply for the predators. It may also cause the increase of prey species when the target fishes are predator species such as salmon and tuna.

By-catch

By-catch is the portion of the catch that is not the target species. These are either kept to be sold or discarded. In some instances the discarded portion is known as discards.

Possible Remedies

Many governments have implemented fisheries management policies designed to curb the environmental impact of fishing. Fishing conservation aims to control the human activities that may completely decrease a fish stock or washout an entire aquatic environment. These laws include the quotas on the total catch of particular species in a fishery, limits on the number of vessels allowed in specific areas, and the imposition of seasonal restrictions on fishing.

Fish farming has also been proposed as a more sustainable alternative to traditional capture of wild fish. However, fish farming has been found to have negative impacts on nearby wild fish.

5.12 Forest Protection

Forest protection is a general term describing methods purported to preserve or improve a forest threatened or affected by abuse. There is considerable debate over the effectiveness of forest protection methods. One simple type of forest protection is the purchasing of land in order to secure it, or in order to plant trees (afforestation). It can also mean forest management or the designation of areas such as natural reservoirs which are intended to be left to themselves. However, merely purchasing a piece of land does not prevent it from being used by others for poaching and illegal logging. A better way to protect a forest, particularly old growth forests in remote areas, is to obtain a part of it and to live on and monitor the purchased land. Even in the USA, these measures sometimes don't suffice because arson can burn a forest to the ground, leaving burnt areas free for different use.

Enforcement of laws regarding purchased forest land is weak or non-existent in most parts of the world. In the increasingly dangerous South America, home of major rainforests, officials of the Brazilian National Agency for the Environment (IBAMA) have recently been shot during their routine duties. Another issue about living on purchased forest-land is that there may not be a suitable site for a standard home without clearing land, which defies the purpose of protection. Alternatives include building a treehouse or an earthhouse. This is being done currently by indigenous people in South America to protect large reservoirs. In former times, North American Native Americans used to live in tipies or mandan earthhouses, which also require less land. An undertaking to develop modern treehouses is being taken by a company from Germany called "True School treehouses"

A compromise is to conduct agriculture and stock farming, or sustainable wood management. This ascribes different values to forest land and farmland, for which many areas are clear felled. A number of less successful methods of forest protection have been tried, such as the trade in certified wood. The types of abuse that forest protection seeks to prevent include:

❖ Aggressive or unsustainable farming and logging, and
❖ Expanding city development caused by population explosion and the resulting urban sprawl.

Protecting a small section of land in a larger forest may also have limited value. For example, tropical rainforests can die if they decrease in size, since they are dependent on the moist microclimate which they create.

A recent discovery in Europe relating to forest protection is that urban areas have forests of their own. Many cities have tens of thousands of trees which constitute forests. In addition the air in the cities is lately becoming better, providing conditions favourable for small associated species such as mosses and lichens.

5.13 Environmental Effects of Pesticides

Over 98 per cent of sprayed insecticides and 95 per cent of herbicides reach a destination other than their target species, including nontarget species, air, water, bottom sediments, and food. Pesticide contaminates land and water when it escapes from production sites and storage tanks, when it runs off from fields, when it is discarded, when it is sprayed aerially, and when it is sprayed into water to kill algae. The amount of pesticide that migrates from the intended application area is influenced by the particular chemical's properties: its propensity for binding to soil, its vapor pressure, its water solubility, and its resistance to being broken down over time. Factors in the soil, such as its texture, its ability to retain water, and the amount of organic matter contained in it, also affect the amount of pesticide that will leave the area. Some pesticides contribute to global warming and the depletion of the ozone layer.

Air

Pesticides can contribute to air pollution. Pesticide drift occurs when pesticides suspended in the air as particles are carried by wind to other areas, potentially contaminating them. Pesticides that are applied to crops can volatilize and may be blown by winds into nearby areas, potentially posing a threat to wildlife. Also, droplets of sprayed pesticides or particles from pesticides applied as dusts may travel on the wind to other areas, or pesticides may adhere to particles that blow in the wind, such as dust particles. Ground spraying produces less pesticide drift than aerial spraying does. Farmers can employ a buffer zone around their crop, consisting of empty land or non-crop plants such as evergreen trees to serve as windbreaks and absorb the pesticides, preventing drift into other areas. Such windbreaks are legally required in the Netherlands.

Pesticides that are sprayed onto fields and used to fumigate soil can give off chemicals called volatile organic compounds, which can react with other chemicals and form a pollutant called ozone, accounting for an estimated 6 per cent of the total ozone production.

Water

In the United States, pesticides were found to pollute every stream and over 90 per cent of wells sampled in a study by the US Geological Survey. Pesticide residues have also been found in rain and groundwater. Studies by the UK government showed that pesticide concentrations exceeded those allowable for drinking water in some samples of river water and groundwater.

Pesticide impacts on aquatic systems are often studied using a hydrology transport model to study movement and fate of chemicals in rivers and streams. As early as the 1970s quantitative analysis of pesticide runoff was conducted in order to predict amounts of pesticide that would reach surface waters.

There are four major routes through which pesticides reach the water: it may drift outside of the intended area when it is sprayed, it may percolate, or leach, through the soil, it may be carried to the water as runoff, or it may be spilled, for example accidentally or through neglect. They may also be carried to water by eroding soil. Factors that affect a pesticide's ability to contaminate water include its water solubility, the distance from an application site to a body of water, weather, soil type, presence of a growing crop, and the method used to apply the chemical.

Maximum limits of allowable concentrations for individual pesticides in public bodies of water are set by the Environmental Protection Agency in the US. Similarly, the government of the United Kingdom sets Environmental Quality Standards (EQS), or maximum allowable concentrations of some pesticides in bodies of water above which toxicity may occur. The European Union also regulates maximum concentrations of pesticides in water.

Soil

Many of the chemicals used in pesticides are persistent soil contaminants, whose impact may endure for decades and adversely affect soil conservation. The use of pesticides decreases the general biodiversity in the soil. Not using the chemicals results in higher soil quality, with the additional effect that more organic matter in the soil allows for higher water retention. This helps increase yields for farms in drought years, when organic farms have had yields 20-40 per cent higher than their conventional counterparts. A smaller content of organic matter in the soil increases the amount of pesticide that will leave the area of application, because organic matter binds to and helps break down pesticides.

Plants

Nitrogen fixation, which is required for the growth of higher plants, is hindered by pesticides in soil. The insecticides DDT, methyl parathion, and especially pentachlorophenol have been shown to interfere with legume-rhizobium chemical signaling. Reduction of this symbiotic chemical signaling results in reduced nitrogen fixation and thus reduced crop yields. Root nodule formation in these plants saves the world economy $10 billion in synthetic nitrogen fertilizer every year.

Pesticides can kill bees and are strongly implicated in pollinator decline, the loss of species that pollinate plants, including through the mechanism of Colony Collapse Disorder, in which worker bees from a beehive or Western honey bee colony abruptly disappear. Application of pesticides to crops that are in bloom can kill honeybees, which act as pollinators. The USDA and USFWS estimate that US farmers lose at least $200 million a year from reduced crop pollination because pesticides applied to fields eliminate about a fifth of honeybee colonies in the US and harm an additional 15 per cent.

Persistent Organic Pollutants

Persistent organic pollutants (POPs) are compounds that resist degradation and thus remain in the environment for years. Some pesticides, including aldrin, chlordane, DDT, dieldrin, endrin, heptachlor, hexachlorobenzene, mirex, and toxaphene, are considered POPs. POPs have the ability to volatilize and travel great distances through the atmosphere to become deposited in remote regions. The chemicals also have the ability to bioaccumulate and biomagnify, and can bioconcentrate (i.e. become more concentrated) up to 70,000 times their original concentrations. POPs may continue to poison non-target organisms in the environment and increase risk to humans by disruption in the endocrine, reproductive, and immune systems; cancer; neurobehavioural disorders, infertility and mutagenic effects, although very little is currently known about these chronic effects. Some POPs have been banned, while others continue to be used.

Animals

Pesticides inflict ext]remely widespread damage to biota, and many countries have acted to discourage pesticide usage through their Biodiversity Action Plans.

Animals may be poisoned by pesticide residues that remain on food after spraying, for example when wild animals enter sprayed fields or nearby areas shortly after spraying.

Widespread application of pesticides can eliminate food sources that certain types of animals need, causing the animals to relocate, change their diet, or starve. Poisoning from pesticides can travel up the food chain; for example, birds can be harmed when they eat insects and worms that have consumed pesticides. Some pesticides can bioaccumulate, or build up to toxic levels in the bodies of organisms that consume them over time, a phenomenon that impacts species high on the food chain especially hard.

The USDA and USFWS estimate that about 20 per cent of the endangered and threatened species in the US are jeopardized by use of pesticides.

Birds

Birds are common examples of nontarget organisms that are impacted by pesticide use. Rachel Carson's landmark book *Silent Spring* dealt with the topic of loss of bird species due to bioaccumulation of pesticides in their tissues. There is evidence that birds are continuing to be harmed by pesticide use. In the farmland of Britain, populations of ten different species of birds have declined by 10 million breeding individuals between 1979 and 1999, a phenomenon thought to have resulted from loss of plant and invertebrate species on which the birds feed. Throughout Europe, 116 species of birds are now threatened. Reductions in bird populations have been found to be associated with times and areas in which pesticides are used. In another example, some types of fungicides used in peanut farming are only slightly toxic to birds and mammals, but may kill off earthworms, which can in turn reduce populations of the birds and mammals that feed on them.

Some pesticides come in granular form, and birds and other wildlife may eat the granules, mistaking them for grains of food. A few granules of a pesticide is enough to kill a small bird.

The herbicide paraquat, when sprayed onto bird eggs, causes growth abnormalities in embryos and reduces the number of chicks that hatch successfully, but most herbicides do not directly cause much harm to birds. Herbicides may endanger bird populations by reducing their habitat.

The USDA and USFWS estimate that over 67 million birds are killed by pesticides each year in the US.

Aquatic Life

Fish and other aquatic biota may be harmed by pesticide-contaminated water. Pesticide surface runoff into rivers and streams can be highly lethal to aquatic life, sometimes killing all the fish in a particular stream. For example, in Montague P.E.I., nine "fish kills" happened in one year: every fish, snake, and snail was killed in a river called Sutherland's Hole near potato farms from which herbicides, insecticides, and fungicides ran off after heavy rains. Pesticide-related fish kills are frequently unreported and likely underestimated.

Application of herbicides to bodies of water can cause fish kills when the dead plants rot and use up the water's oxygen, suffocating the fish. Some herbicides, such as copper sulfite, that are applied to water to kill plants are toxic to fish and other water animals at concentrations similar to those used to kill the plants. Repeated exposure to sublethal doses of some pesticides can cause physiological and behavioural changes in fish that reduce populations, such as abandonment of nests and broods, decreased immunity to disease, and increased failure to avoid predators. Application of herbicides to bodies of water can kill off plants on which fish depend for their habitat.

Pesticides can accumulate in bodies of water to levels that kill off zooplankton, the main source of food for young fish. Pesticides can kill off the insects on which some fish feed, causing the fish to travel farther in search of food and exposing them to greater risk from predators.

The USDA and USFWS estimate that between 6 and 14 million fish are killed by pesticides each year in the US.

The faster a given pesticide breaks down in the environment, the less threat it poses to aquatic life. Insecticides are more toxic to aquatic life than herbicides and fungicides.

Amphibians

Some scientists believe that certain common pesticides already exist at levels capable of killing amphibians in California. They warn that the breakdown products of these pesticides can be 10 to 100 times more toxic to amphibians than the original pesticides. Direct contact of sprays of some pesticides (either by drift from nearby applications or accidental or deliberate sprays) can be highly lethal to amphibians. A Canadian study showed that exposing tadpoles to endosulfan, an organochloride pesticide at levels that are likely to be found in habitats near fields sprayed with the chemical kills the tadpoles and causes behavioural and growth abnormalities.

US scientists have found that some pesticides used in farming disrupt the nervous systems of frogs, and that use of these pesticides is correlated with a decline in the population of frogs in the Sierra Nevada. In the past several decades, decline in amphibian populations has been occurring all over the world, for unexplained reasons which are thought to be varied but of which pesticides may be a part. Being downwind from agricultural land on which pesticides are used has been linked to the decline in population of threatened frog species in California.

Mixtures of multiple pesticides appear to have a cumulative toxic effect on frogs. Tadpoles from ponds with multiple pesticides present in the water take longer to metamorphose into frogs and are smaller when they do, decreasing their ability to catch prey and avoid predators.

In Minnesota, pesticide use has been causally linked to congenital deformities in frogs such as eye, mouth, and limb malformations. Researchers in California found that similar deformities in frogs in the US and Canada may have been caused by breakdown products from pesticides which themselves did not pose a threat.

The herbicide atrazine has been shown to turn male frogs into hermaphrodites, decreasing their ability to reproduce.

Pest Resistance

Pests may evolve to become resistant to pesticides. Many pests will initially be very susceptible to pesticides, but some with slight variations in their genetic makeup are resistant and therefore survive to reproduce. Through natural selection, the pests may eventually become very resistant to the pesticide.

Pest resistance to a pesticide is commonly managed through pesticide rotation, which involves alternating among pesticide classes with different modes of action to delay the onset of or mitigate existing pest resistance.

Tank mixing pesticides is the combination of two or more pesticides with different modes of action in order to improve individual pesticide application results and delay the onset of or mitigate existing pest resistance.

Pest Rebound and Secondary Pest Outbreaks

Non-target organisms, organisms that the pesticides are not intended to kill, can be severely impacted by use of the chemicals. In some cases, where a pest insect has some controls from a beneficial predator or parasite, an insecticide application can kill both pest and beneficial populations. A study comparing biological pest control and use of pyrethroid insecticide for diamondback moths, a major cabbage family insect pest, showed that the insecticide application created a rebounded pest population due to loss of insect predators, whereas the biocontrol did not show the same effect. Likewise, pesticides sprayed in an effort to control adult mosquitoes, may temporarily depress mosquito populations, however they may result in a larger population in the long run by damaging the natural controlling factors. This phenomenon, wherein the population of a pest species rebounds to equal or greater numbers than it had before pesticide use, is called pest resurgence and can be linked to elimination of predators and other natural enemies of the pest.

Loss of predator species can also lead to a related phenomenon called secondary pest outbreaks, an increase in problems from species which were not originally very damaging pests due to loss of their predators or parasites. An estimated third of the 300 most damaging insects in the US were originally secondary pests and only became a major problem after the use of pesticides. In both pest resurgence and secondary pest outbreaks, the natural enemies have been found to be more susceptible to the pesticides than the pests themselves, in some cases causing the pest population to be higher than it was before the use of pesticide.

5.14 Radioactive Contamination

Radioactive contamination is the uncontrolled distribution of radioactive material in a given environment. The amount of radioactive material released in an accident is called the source term.

Sources of Contamination

Radioactive contamination is typically the result of a spill or accident during the production or use of radionuclides (radioisotopes), an unstable nucleus which has excessive energy. Contamination may occur from radioactive gases, liquids or particles. For example, if a radionuclide used in nuclear medicine is accidentally spilled, the material could be spread by people as they walk around. Radioactive contamination may also be an inevitable result of certain processes, such as the release of radioactive xenon in nuclear fuel reprocessing.

In cases that radioactive material cannot be contained, it may be diluted to safe concentrations. Nuclear fallout is the distribution of radioactive contamination by a nuclear explosion. For a discussion of environmental contamination by alpha emitters please see actinides in the environment. Containment is what differentiates radioactive material from radioactive contamination. Therefore, radioactive material in sealed and designated containers is not properly referred to as contamination, although the units of measurement might be the same.

Measurement

Radioactive contamination may exist on surfaces or in volumes of material or air. In a nuclear power plant, detection and measurement of radioactivity and contamination is often the job of a Certified Health Physicist.

Surface Contamination

Surface contamination is usually expressed in units of radioactivity per unit of area. For SI, this is becquerels per square meter (or Bq/m^2). Other units such as picoCuries per 100 cm^2 or disintegrations per minute per square centimeter (1 dpm/cm^2 = 166 2/3 Bq/m^2) may be used. Surface contamination may either be fixed or removable. In the case of fixed contamination, the radioactive material cannot by definition be spread, but it is still measurable.

Hazards

In practice there is no such thing as zero radioactivity. Not only is the entire world constantly bombarded by cosmic rays, but every living creature on earth contains significant quantities of carbon-14 and most (including humans) contain significant quantities of potassium-40. These tiny levels of radiation are not any more harmful than sunlight, but just as excessive quantities of sunlight can be dangerous, so too can excessive levels of radiation.

Low Level Contamination

The hazards to people and the environment from radioactive contamination depend on the nature of the radioactive contaminant, the level of contamination, and the extent of the spread of contamination. Low levels of radioactive contamination pose little risk, but can still be detected by radiation instrumentation. In the case of low-level contamination by isotopes with a short half-life, the best course of action may be to simply allow the material to naturally decay. Longer-lived isotopes should be cleaned up and properly disposed off, because even a very low level of radiation can be life-threatening when in long exposure to it. Therefore, whenever there's any radiation in an area, many people take extreme caution when approaching.

High Level Contamination

High levels of contamination may pose major risks to people and the environment. People can be exposed to potentially lethal radiation levels, both externally and internally, from the spread of contamination following an accident (or a deliberate initiation) involving large quantities of radioactive material. The biological effects of external exposure to radioactive contamination are generally the same as those from an external radiation source not involving radioactive materials, such as x-ray machines, and are dependent on the absorbed dose.

Biological Effects

The biological effects of internally deposited radionuclides depend greatly on the activity and the biodistribution and removal rates of the radionuclide, which in turn depends on its chemical form. The biological effects may also depend on the chemical toxicity of the deposited material, independent of its radioactivity. Some radionuclides may be generally distributed throughout the body and rapidly removed, as is the case with tritiated water. Some radionuclides may target specific organs and have much lower removal rates. For instance, the thyroid gland takes up a large percentage of any iodine that enters the body. If large quantities of radioactive iodine are inhaled or ingested, the thyroid may be impaired or destroyed, while other tissues are affected to a lesser extent. Radioactive iodine is a common fission product; it was a major component of the radiation released from the Chernobyl disaster, leading to many cases of pediatric thyroid cancer and hypothyroidism. On the other hand, radioactive iodine is used in the diagnosis and treatment of many diseases of the thyroid precisely because of the thyroid's selective uptake of iodine.

Means of Contamination

Radioactive contamination can enter the body through ingestion, inhalation, absorption, or injection. For this reason, it is important to use personal protective equipment when working with radioactive materials. Radioactive contamination may also be ingested as the result of eating contaminated plants and animals or drinking contaminated water or milk from exposed animals. Following a major contamination incident, all potential pathways of internal exposure should be considered.

5.15 Radioactive Waste

Radioactive wastes are waste types containing radioactive chemical elements that do not have a practical purpose. They are sometimes the products of nuclear processes, such as nuclear fission. However, industries not directly connected to the nuclear industry can produce large quantities of radioactive waste. It has been estimated, for instance, that the past 20 years the oil-producing endeavors of the United States have accumulated eight million tons of radioactive wastes. The majority of radioactive waste is "low-level waste", meaning it contains low levels of radioactivity per mass or volume. This type of waste often consists of used protective clothing, which is only slightly contaminated but still dangerous in case of radioactive contamination of a human body through ingestion, inhalation, absorption, or injection. In the United States alone, the Department of Energy states that there are "millions of gallons of radioactive waste" as well as "thousands of tons of spent nuclear fuel and material" and also "huge quantities of contaminated soil and water". Despite these copious quantities of waste, the DOE has a goal of cleaning all presently contaminated sites successfully by 2025. The Fernald, Ohio site for example had "31 million pounds of uranium product", "2.5 billion pounds of waste", "2.75 million cubic yards of contaminated soil and debris", and a "223 acre portion of the underlying Great Miami Aquifer had uranium levels above drinking standards". The United States currently has at least 108 sites it currently designates as areas that are contaminated and unusable, sometimes many thousands of acres The DOE wishes to try and clean or mitigate many or all by 2025, however the task can be difficult and it acknowledges that some will never be completely remediated, and just in one of these 108 larger designations, Oak Ridge National Laboratory, there were for example at least "167 known contaminant release sites" in one of the three subdivisions of the 37,000-acre (150 km²) site. Some of the U.S. sites were smaller in nature, however, and cleanup issues were simpler to address, and the DOE has successfully completed cleanup, or at least closure, of several sites.

The issue of disposal methods for nuclear waste was one of the most pressing current problems the international nuclear industry faced when trying to establish a long term energy production plan, yet there was hope it could be safely solved. A recent research report on the Nuclear Industry perspective of the current state of scientific knowledge in predicting the extent that waste would find its way from the deep burial facility—back to soil and drinking water (such that it presents a direct threat to the health of human beings—as well as to other forms of life) is presented in a document from the IAEA (The International Atomic Energy Agency)—which was published in October 2007 This document states "The capacity to model all the effects involved in the dissolution of the waste form, in conditions similar to the disposal site, is the final goal of all the research undertaken by many research groups over many years. As we will see in this report, this kind of investigation is far from being finished" . In the United States, the DOE acknowledges much progress in addressing the waste problems of the industry, and successful remediation of some contaminated sites, yet also major uncertainties and sometimes complications and setbacks in handling the issue properly, cost effectively, and in the projected time frame. In other countries with lower ability or will to maintain environmental integrity the issue would be even more problematic.

The Nature and Significance of Radioactive Waste

Radioactive waste typically comprises a number of radioisotopes: unstable configurations of elements that decay, emitting ionizing radiation which can be harmful to human health and to the environment. Those isotopes emit different types and levels of radiation, which last for different periods of time.

Physics

The radioactivity of all nuclear waste diminishes with time. All radioisotopes contained in the waste have a half-life—the time it takes for any radionuclide to lose half of its radioactivity and eventually all radioactive waste decays into non-radioactive elements. Certain radioactive elements (such as plutonium-239) in "spent" fuel will remain hazardous to humans and other living beings for hundreds of thousands of years. Other radioisotopes will remain hazardous for millions of years. Thus, these wastes must be shielded for centuries and isolated from the living environment for hundreds of millennia. Some elements, such as Iodine-131, have a short half-life (around 8 days in this case) and thus they will cease to be a problem much more quickly than other, longer-lived, decay products but their activity is much greater initially. The two tables show some of the major radioisotopes, their half-lives, and their radiation yield as a proportion of the yield of fission of Uranium-235.

The faster a radioisotope decays, the more radioactive it will be. The energy and the type of the ionizing radiation emitted by a pure radioactive substance are important factors in deciding how dangerous it will be. The chemical properties of the radioactive element will determine how mobile the substance is and how likely it is to spread into the environment and contaminate human bodies. This is further complicated by the fact that many radioisotopes do not decay immediately to a stable state but rather to a radioactive decay product leading to decay chains.

Chemistry

The chemical properties of the radioactive substance and the other substances found within (and near) the waste store has a great effect upon the ability of the waste to cause harm to humans or other organisms. For instance TcO_4^- tends to adsorb on the surfaces of steel objects which reduces its ability to move out of the waste store in water.

Pharmacokinetics

Exposure to high levels of radioactive waste may cause serious harm or death. Treatment of an adult animal with radiation or some other mutation-causing effect, such as a cytotoxic anti-cancer drug, may cause cancer in the animal. In humans it has been calculated that a 1 sievert dose has a 5 per cent chance of causing cancer and a 1 per cent chance of causing a mutation in a gamete which can be passed to the next generation. If a developing organism such as an unborn child is irradiated, then it is possible to induce a birth defect but it is unlikely that this defect will be in a gamete or a gamete forming cell.

Depending on the decay mode and the pharmacokinetics of an element (how the body processes it and how quickly), the threat due to exposure to a given activity of a radioisotope will differ. For instance Iodine-131 is a short-lived beta and gamma emitter but because it concentrates in the thyroid gland, it is more able to cause injury than cesium-137 which, being water soluble, is rapidly excreted in urine. In a similar way, the alpha emitting actinides and radium are considered very harmful as they tend to have long biological half-lives and their radiation has a high linear energy transfer value. Because of such differences, the rules determining biological injury differ widely according to the radioisotope, and sometimes also the nature of the chemical compound which contains the radioisotope.

Philosophy

The main objective in managing and disposing of radioactive (or other) waste is to protect people and the environment. This means isolating or diluting the waste so that the rate or concentration of any radionuclide returned to the biosphere is harmless. To achieve this the preferred technology to date has been deep and secure burial for the more dangerous wastes; transmutation, long-term retrievable storage, and removal to space have also been suggested. Management options for waste are discussed below.

Radioactivity by definition reduces over time, so in principle the waste needs to be isolated for a particular period of time until its components have decayed such that it no longer poses a threat. In practice this can mean periods of hundreds of thousands of years, depending on the nature of the waste involved.

Though an affirmative answer is often taken for granted, the question as to whether or not we should endeavor to avoid causing harm to remote future generations, perhaps thousands upon thousands of years hence, is essentially one which must be dealt with by philosophy.

Sources of Waste

Radioactive waste comes from a number of sources. The majority originates from the nuclear fuel cycle and nuclear weapon reprocessing. However, other sources include medical and industrial wastes, as well as naturally occurring radioactive materials (NORM) that can be concentrated as a result of the processing or consumption of coal, oil and gas, and some minerals.

Nuclear Fuel Cycle

Front End

Waste from the front end of the nuclear fuel cycle is usually alpha emitting waste from the extraction of uranium. It often contains radium and its decay products.

Uranium dioxide (UO_2) concentrate from mining is not very radioactive—only a thousand or so times as radioactive as the granite used in buildings. It is refined from yellowcake (U_3O_8), then converted to uranium hexafluoride gas (UF_6). As a gas, it undergoes enrichment to increase the U-235 content from 0.7 per cent to about 4.4 per cent (LEU). It is then turned into a hard ceramic oxide (UO_2) for assembly as reactor fuel elements.

The main by-product of enrichment is depleted uranium (DU), principally the U-238 isotope, with a U-235 content of ~0.3 per cent. It is stored, either as UF_6 or as U_3O_8. Some is used in applications where its extremely high density makes it valuable, such as the keels of yachts, and anti-tank shells. It is also used (with recycled plutonium) for making mixed oxide fuel (MOX) and to dilute highly enriched uranium from weapons stockpiles which is now being redirected to become reactor fuel. This dilution, also called downblending, means that any nation or group that acquired the finished fuel would have to repeat the (very expensive and complex) enrichment process before assembling a weapon.

Back End

The back end of the nuclear fuel cycle, mostly spent fuel rods, contains fission products that emit beta and gamma radiation, and that emit alpha particles, such as uranium-234, neptunium-237, plutonium-238 and americium-241, and even sometimes some neutron emitters such as californium (Cf). These isotopes are formed in nuclear reactors.

It is important to distinguish the processing of uranium to make fuel from the reprocessing of used fuel. Used fuel contains the highly radioactive products of fission (see high level waste below). Many of these are neutron absorbers, called neutron poisons in this context. These eventually build up to a level where they absorb so many neutrons that the chain reaction stops, even with the control rods completely removed. At that point the fuel has to be replaced in the reactor with fresh fuel, even though there is still a substantial quantity of uranium-235 and plutonium present. In the United States, this used fuel is stored, while in countries such as the United Kingdom, France, and Japan, the fuel is reprocessed to remove the fission products, and the fuel can then be re-used. This reprocessing involves handling highly radioactive materials, and the fission products removed from the fuel are a concentrated form of high-level waste as are the chemicals used in the process.

Proliferation Concerns

When dealing with uranium and plutonium, the possibility that they may be used to build nuclear weapons is often a concern. Active nuclear reactors and nuclear weapons stockpiles are very carefully safeguarded and controlled. However, high-level waste from nuclear reactors may contain plutonium. Ordinarily, this plutonium is reactor-grade plutonium, containing a mixture of plutonium-239 (highly suitable for building nuclear weapons), plutonium-240 (an undesirable contaminant and highly radioactive), plutonium-241, and plutonium-238; these isotopes are difficult to separate. Moreover, high-level waste is full of highly radioactive fission products. However, most fission products are relatively short-lived. This is a concern since if the waste is stored, perhaps in deep geological storage, over many years the fission products decay, decreasing the radioactivity of the waste and making the plutonium easier to access. Moreover, the undesirable contaminant Pu-240 decays faster than the Pu-239, and thus the quality of the bomb material increases with time (although its quantity decreases during that time as well). Thus, some have argued, as time passes, these deep storage areas have the potential to become "plutonium mines", from which material for nuclear weapons can be acquired with relatively little difficulty. Critics of the latter idea point out that the half-life of Pu-240 is 6,560 years and Pu-239 is 24,110 years, and thus the relative enrichment of one isotope to the other with time occurs with a half-life of 9,000 years (that is, it takes 9000 years for the *fraction* of Pu-240 in a sample of mixed plutonium isotopes, to spontaneously decrease by half— a typical enrichment needed to turn reactor-grade into weapons-grade Pu). Thus "weapons grade plutonium mines" would be a problem for the very far future (>9,000 years from now), so that there remains a great deal of time for technology to advance to solve this problem, before it becomes acute.

Pu-239 decays to U-235 which is suitable for weapons and which has a very long half life (roughly 10^9 years). Thus plutonium may decay and leave uranium-235. However, modern reactors are only moderately enriched with U-235 relative to U-238, so the U-238 continues to serve as denaturation agent for any U-235 produced by plutonium decay.

One solution to this problem is to recycle the plutonium and use it as a fuel e.g. in fast reactors. But in the minds of some, the very existence of the nuclear fuel reprocessing plant needed to separate the plutonium from the other elements represents a proliferation concern. In pyrometallurgical fast reactors, the waste generated is an actinide compound that cannot be used for nuclear weapons.

Nuclear Weapons Reprocessing

Waste from nuclear weapons reprocessing (as opposed to production, which requires primary processing from reactor fuel) is unlikely to contain much beta or gamma activity other than tritium and americium. It is more likely to contain alpha emitting actinides such as Pu-239 which is a fissile material used in bombs, plus some material with much higher specific activities, such as Pu-238 or Po.

In the past the neutron trigger for a bomb tended to be beryllium and a high activity alpha emitter such as polonium; an alternative to polonium is Pu-238. For reasons of national security, details of the design of modern bombs are normally not released to the open literature. It is likely however that a D-T fusion reaction in either an electrically driven device or a D-T fusion reaction driven by the chemical explosives would be used to start up a modern device.

Some designs might well contain a radioisotope thermoelectric generator using Pu-238 to provide a longlasting source of electrical power for the electronics in the device.

It is likely that the fissile material of an old bomb which is due for refitting will contain decay products of the plutonium isotopes used in it, these are likely to include alpha-emitting Np-236 from Pu-240 impurities, plus some U-235 from decay of the Pu-239; however, due to the relatively long half-life of these Pu isotopes, these wastes from radioactive decay of bomb core material would be very small, and in any case, far less dangerous (even in terms of simple radioactivity) than the Pu-239 itself.

The beta decay of Pu-241 forms Am-241; the in-growth of americium is likely to be a greater problem than the decay of Pu-239 and Pu-240 as the americium is a gamma emitter (increasing external-exposure to workers) and is an alpha emitter which can cause the generation of heat. The plutonium could be separated from the americium by several different processes; these would include pyrochemical processes and aqueous/organic solvent extraction. A truncated PUREX type extraction process would be one possible method of making the separation.

Medical

Radioactive medical waste tends to contain beta particle and gamma ray emitters. It can be divided into two main classes. In diagnostic nuclear medicine a number of short-lived gamma emitters such as technetium-99m are used. Many of these can be disposed of by leaving it to decay for a short time before disposal as normal trash. Other isotopes used in medicine, with half-lives in parentheses:

- ❖ Y-90, used for treating lymphoma (2.7 days),
- ❖ I-131, used for thyroid function tests and for treating thyroid cancer (8.0 days),
- ❖ Sr-89, used for treating bone cancer, intravenous injection (52 days),
- ❖ Ir-192, used for brachytherapy (74 days),
- ❖ Co-60, used for brachytherapy and external radiotherapy (5.3 years),
- ❖ Cs-137, used for brachytherapy, external radiotherapy (30 years).

Industrial

Industrial source waste can contain alpha, beta, neutron or gamma emitters.

Gamma emitters are used in radiography while neutron emitting sources are used in a range of applications, such as oil well logging.

Naturally Occurring Radioactive Material (NORM)

Processing of substances containing natural radioactivity; this is often known as NORM. A lot of this waste is alpha particle-emitting matter from the decay chains of uranium and thorium. The main source of radiation in the human body is potassium-40 (^{40}K).

Coal

Coal contains a small amount of radioactive uranium, barium, thorium and potassium, but, in the case of pure coal, this is significantly less than the average concentration of those elements in the Earth's crust. However, the surrounding strata, if shale or mudstone, often contains slightly more than average and this may also be reflected in the ash content of 'dirty' coals. The more active ash minerals become concentrated in the fly ash precisely because they do not burn well . However, the radioactivity of fly ash is still very low. It is about the same as black shale and is less than phosphate rocks, but is more of a concern because a small amount of the fly ash ends up in the atmosphere where it can be inhaled.

Oil and Gas

Residues from the oil and gas industry often contain radium and its daughters. The sulphate scale from an oil well can be very radium rich, while the water, oil and gas from a well often contains radon. The radon decays to form solid radioisotopes which form coatings on the inside of pipework. In an oil processing plant the area of the plant where propane is processed is often one of the more contaminated areas of the plant as radon has a similar boiling point as propane.

Types of Radioactive Waste

Although not significantly radioactive, *uranium mill tailings* are waste. They are byproduct material from the rough processing of uranium-bearing ore. They are sometimes referred to as 11(e) wastes, from the section of the U.S. Atomic Energy Act that defines them. Uranium mill tailings typically also contain chemically-hazardous heavy metals such as lead and arsenic. Vast mounds of uranium mill tailings are left at many old mining sites, especially in Colorado, New Mexico, and Utah.

Low level waste (LLW) is generated from hospitals and industry, as well as the nuclear fuel cycle. It comprises paper, rags, tools, clothing, filters, etc., which contain small amounts of mostly short-lived radioactivity. Commonly, LLW is designated as such as a precautionary measure if it originated from any region of an 'Active Area', which frequently includes offices with only a remote possibility of being contaminated with radioactive materials. Such LLW typically exhibits no higher radioactivity than one would expect from the same material disposed of in a non-active area, such as a normal office block. Some high activity LLW requires shielding during handling and transport but most LLW is suitable for shallow land burial. To reduce its volume, it is often compacted or incinerated before disposal. Low level waste is divided into four classes, class A, B, C and GTCC, which means "Greater Than Class C".

Intermediate level waste (ILW) contains higher amounts of radioactivity and in some cases requires shielding. ILW includes resins, chemical sludge and metal reactor fuel cladding, as well as contaminated materials from reactor decommissioning. It may be solidified in concrete or bitumen for disposal. As a general rule, short-lived waste (mainly non-fuel materials from reactors) is buried in shallow repositories, while long-lived waste (from fuel and fuel-reprocessing) is deposited in deep underground facilities. U.S. regulations do not define this category of waste; the term is used in Europe and elsewhere.

High level waste (HLW) is produced by nuclear reactors. It contains fission products and transuranic elements generated in the reactor core. It is highly radioactive and often thermally hot. LLW and ILW accounts for over 95 per cent of the total radioactivity produced in the process of nuclear electricity generation. The amount of HLW worldwide is currently increasing by about 12,000 metric tons every year, which is the equival to about 100 double-decker busses or a two-story structure built on top of a basketball court.

Transuranic waste (TRUW) as defined by U.S. regulations is, without regard to form or origin, waste that is contaminated with alpha-emitting transuranic radionuclides with half-lives greater than 20 years, and concentrations greater than 100 nCi/g (3.7 MBq/kg), excluding High Level Waste. Elements that have an atomic number greater than uranium are called transuranic ("beyond uranium"). Because of their long half-lives, TRUW is disposed more cautiously than either low level or intermediate level waste. In the U.S. it arises mainly from weapons production, and consists of clothing, tools, rags, residues, debris and other items contaminated with small amounts of radioactive elements (mainly plutonium).

Under U.S. law, TRUW is further categorized into "contact-handled" (CH) and "remote-handled" (RH) on the basis of radiation dose measured at the surface of the waste container. CH TRUW has a surface dose rate not greater than 200 mrem per hour (2 mSv/h), whereas RH TRUW has a surface dose rate of 200 mrem per hour (2 mSv/h) or greater. CH TRUW does not have the very high radioactivity of high level waste, nor its high heat generation, but RH TRUW can be highly radioactive, with surface dose rates up to 1000000 mrem per hour (10000 mSv/h). The United States currently permanently disposes of TRUW generated from nuclear power plants and military facilities at the Waste Isolation Pilot Plant.

Management of Waste

Nuclear waste requires sophisticated treatment and management in order to successfully isolate it from interacting with the biosphere. This usually necessitates treatment, followed by a long-term management strategy involving storage, disposal or transformation of the waste into a non-toxic form .

Initial Treatment of Waste

Vitrification

Long-term storage of radioactive waste requires the stabilization of the waste into a form which will not react, nor degrade, for extended periods of time. One way to do this is through vitrification. Currently at Sellafield the high-level waste (PUREX first cycle raffinate) is mixed with sugar and then calcined. Calcination involves passing the waste through a heated, rotating tube. The purposes of calcination are to evaporate the water from the waste, and de-nitrate the fission products to assist the stability of the glass produced.

The 'calcine' generated is fed continuously into an induction heated furnace with fragmented glass. The resulting glass is a new substance in which the waste products are bonded into the glass matrix when it solidifies. This product, as a molten fluid, is poured into stainless steel cylindrical containers ("cylinders") in a batch process. When cooled, the fluid solidifies ("vitrifies") into the glass. Such glass, after being formed, is very highly resistant to water. According to the ITU, it will require about 1 million years for 10 per cent of such glass to dissolve in water.

After filling a cylinder, a seal is welded onto the cylinder. The cylinder is then washed. After being inspected for external contamination, the steel cylinder is stored, usually in an underground repository. In this form, the waste products are expected to be immobilized for a very long period of time (many thousands of years).

The glass inside a cylinder is usually a black glossy substance. All this work (in the United Kingdom) is done using hot cell systems. The sugar is added to control the ruthenium chemistry and to stop the formation of the volatile RuO_4 containing radio ruthenium. In the west, the glass is normally a borosilicate glass (similar to Pyrex), while in the former Soviet bloc it is normal to use a phosphate glass. The amount of fission products in the glass must be limited because some (palladium, the other Pt group metals, and tellurium) tend to form metallic phases which separate from the glass. In Germany a vitrification plant is in use; this is treating the waste from a small demonstration reprocessing plant which has since been closed down.

Ion Exchange

It is common for medium active wastes in the nuclear industry to be treated with ion exchange or other means to concentrate the radioactivity into a small volume. The much less radioactive bulk (after treatment) is often then discharged. For instance, it is possible to use a ferric hydroxide floc to remove radioactive metals from aqueous mixtures . After the radioisotopes are absorbed onto the ferric hydroxide, the resulting sludge can be placed in a metal drum before being mixed with cement to form a solid waste form. In order to get better long-term performance (mechanical stability) from such forms, they may be made from a mixture of fly ash, or blast furnace slag, and portland cement, instead of normal concrete (made with portland cement, gravel and sand).

Synroc

The Australian Synroc (synthetic rock) is a more sophisticated way to immobilize such waste, and this process may eventually come into commercial use for civil wastes (it is currently being developed for U.S. military wastes). Synroc was invented by the late Prof Ted Ringwood (a geochemist) at the Australian National University. The Synroc contains pyrochlore and cryptomelane type minerals. The original form of Synroc (Synroc C) was designed for the liquid high level waste (PUREX raffinate) from a light water reactor. The main minerals in this Synroc are hollandite $(BaAl_2Ti_6O_{16})$, zirconolite $(CaZrTi_2O_7)$ and perovskite $(CaTiO_3)$. The zirconolite and perovskite are hosts for the actinides. The strontium and barium will be fixed in the perovskite. The caesium will be fixed in the hollandite.

Long Term Management of Waste

We are talking here of durations ranging from 10,000 to 1,000,000 years, according to studies based on the effect of estimated radiation doses. It is worthwhile noting that state of the art only allows geological considerations for such long periods. Researchers suggest that forecasts of health detriment for such periods *should be examined critically*. Practical studies only consider up to 100 years as far as effective planning and cost evaluations are concerned.

Storage

High-level radioactive waste is stored temporarily in spent fuel pools and in dry cask storage facilities. This allows the shorter-lived isotopes to decay before further handling.

In 1997, in the 20 countries which account for most of the world's nuclear power generation, spent fuel storage capacity at the reactors was 148,000 tonnes, with 59 per cent of this utilized. However, a number of nuclear power plants in countries that do not reprocess had nearly filled their spent fuel pools, and resorted to Away-from-reactor storage (AFRS). AFRS capacity in 1997 was 78,000 tonnes, with 44 per cent utilized, and annual additions of about 12,000 tonnes. AFRS cannot be expanded forever, and the lead times for final disposal sites have proven to be unpredictable.

In 1989 and 1992, France commissioned commercial plants to vitrify HLW left over from reprocessing oxide fuel, although there are adequate facilities elsewhere, notably in the United Kingdom and Belgium. The capacity of these western European plants is 2,500 canisters (1000 t) a year, and some have been operating for 18 years.

Geological Disposal

The process of selecting appropriate deep final repositories for high level waste and spent fuel is now under way in several countries with the first expected to be commissioned some time after 2010. However, many people remain uncomfortable with the immediate stewardship cessation of this management system. In Switzerland, the Grimsel Test Site is an international research facility investigating the open questions in radioactive waste disposal (). Sweden is well advanced with plans for direct disposal of spent fuel, since its Parliament decided that this is acceptably safe, using the KBS-3 technology. In Germany, there is a political discussion about the search for an *Endlager* (final repository) for radioactive waste, accompanied by loud protests especially in the Gorleben village in the Wendland area, which was seen ideal for the final repository until 1990 because of its location next to the border to the former German Democratic Republic. Gorleben is presently being used to store radioactive waste non-permanently, with a decision on final disposal to be made at some future time. The U.S. has opted for a final repository at Yucca Mountain in Nevada, but this project is widely opposed and is a hotly debated topic, with some of the main concerns being the long distance transportation of the waste from across the United States to this area, and the possibility of accidents over time that could occur. The Waste Isolation Pilot Plant in the United States is the world's first underground repository for transuranic waste. There is also a proposal for an international HLW repository in optimum geology, with Australia or Russia as possible locations, although the proposal for a global repository for Australia has raised fierce domestic political objections.

The Canadian government, for example, is seriously considering this method of disposal, known as the *Deep Geological Disposal* concept. Under the current plan, a vault is to be dug 500 to 1000 meters below ground, under the Canadian Shield, one of the most stable landforms on the planet. The vaults are to be dug inside geological formations known as *batholiths*, formed about a billion years ago. The used fuel bundles will be encased in a corrosion-resistant container, and further surrounded by a layer of *buffer material*, possibly of a special kind of clay (bentonite clay). The case itself is designed to last for thousands of years, while the clay would further slow the corrosion rates of the container. The batholiths themselves are chosen for their low ground-water movement rates, geological stability, and low economic value.

The Finnish government has already started building a vault to store nuclear waste 500 to 1000 meters below ground, not far from the Olkiluoto Nuclear Power Plant.

In the EU, Covra is negotiating about a European-wide waste disposal system with single disposal sites that can be used by several EU-countries. This EU-wide storage possibility is being researched under the SAPIERR-2 programme.

Storing high level nuclear waste above ground for a century or so is considered appropriate by many scientists. This allows for the material to be more easily observed and any problems detected and managed, while the decay over this time period significantly reduces the level of radioactivity and the associated harmful effects to the container material. It is also considered likely that over the next century newer materials will be developed which will not break down as quickly when exposed to a high neutron flux thus increasing the longevity of the container once it is permanently buried.

Sea-based options for disposal of radioactive waste include burial beneath a stable abyssal plain, burial in a subduction zone that would slowly carry the waste downward into the Earth's mantle, and burial beneath a remote natural or human-made island. While these approaches all have merit and would facilitate an international solution to the vexing problem of disposal of radioactive waste, they are currently not being seriously considered because of the legal barrier of the Law of the Sea and because in North America and Europe sea-based burial has become taboo from fear that such a repository could leak and cause widespread damage. Dumping of radioactive waste from ships has reinforced this concern, as has contamination of islands in the Pacific. However, sea-based approaches might come under consideration in the future by individual countries or groups of countries that cannot find other acceptable solutions.

Article 1 (Definitions), 7., of the 1996 Protocol to the Convention on the Prevention of Marine Pollution by Dumping of Wastes and Other Matter, (the London Dumping Convention) states:

> "Sea" means all marine waters other than the internal waters of States, as well as the seabed and the subsoil thereof; it does no include sub-seabed repositories accessed only from land."

A subduction zone accessed from land as is the embodiment of the subductive waste disposal method http://www3.telus.net/subductionservices/ is not nor ever has been prohibited by international agreement.

This method has been described as the most viable means of disposing of radioactive and the state-of-the-art in nuclear waste disposal technology.

Another approach termed Remix & Return would blend high-level waste with uranium mine and mill tailings down to the level of the original radioactivity of the uranium ore, then replace it in empty uranium mines. This approach has the merits of totally eliminating the problem of high-level waste, of providing jobs for miners who would double as disposal staff, and of facilitating a cradle-to-grave cycle for all radioactive materials.

Transmutation

There have been proposals for reactors that consume nuclear waste and transmute it to other, less-harmful nuclear waste. In particular, the was a proposed nuclear reactor with a nuclear fuel cycle that produced no transuranic waste and in fact, could consume transuranic waste. It proceeded as far as large-scale tests but was then canceled by the U.S. Government. Another approach, considered safer but requiring more development, is to dedicate subcritical reactors to the transmutation of the left-over transuranic elements.

Transmutation was banned in the US on April 1977 by President Carter due to the danger of plutonium proliferation however, President Reagan rescinded the ban in 1981 . Due to the economic losses and risks, construction of reprocessing plants during this time did not resume. Due to high energy demand, work on the method has continued in the EU. This has resulted in a practical nuclear research reactor called Myrrha in which transmutation is possible. Additionally, a new research programme called ACTINET has been started in the EU to make transmutation possible on a large, industrial scale. According to President Bush's Global Nuclear Energy Partnership (GNEP) of 2007, the US is now actively promoting research on transmutation technologies needed to markedly reduce the problem of nuclear waste treatment.

There have also been theoretical studies involving the use of fusion reactors as so called "actinide burners" where a fusion reactor plasma such as in a tokamak, could be "doped" with a small amount of the "minor" transuranic atoms which would be transmuted (meaning fissioned in the actinide case) to lighter elements upon their successive bombardment by the very high energy neutrons produced by the fusion of deuterium and tritium in the reactor. It was recently found by a study done at MIT, that only 2 or 3 fusion reactors with parameters similar to that of the International Thermonuclear Experimental Reactor (ITER) could transmute the entire annual minor actinide production from all of the light water reactors presently operating in the United States fleet while simultaneously generating approximately 1 gigawatt of power from each reactor.

Reuse of Waste

Another option is to find applications of the isotopes in nuclear waste so as to reuse them. . Already, caesium-137, strontium-90 and a few other isotopes are extracted for certain industrial applications such as food irradiation and radioisotope thermoelectric generators. While re-use does not eliminate the need to manage radioisotopes, it may reduce the quantity of waste produced.

Space Disposal

Space disposal is an attractive notion because it permanently removes nuclear waste from the environment. However, it has significant disadvantages, not least of which is the potential for catastrophic failure of a launch vehicle. Furthermore, the high number of launches that would be required—due to the fact that no individual rocket would be able to carry very much of the material relative to the material needed to be disposed of—makes the proposal impractical (for both economic and risk-based reasons). To further complicate matters, international agreements on the regulation of such a programme would need to be established.

It has been suggested that through the use of a stationary launch system many of the risks of catastrophic launch failure could be avoided. A promising concept is the use of high power lasers to launch "indestructible" containers from the ground into space. Such a system would require no rocket propellant, with the launch vehicle's payload making up a near entirety of the vehicle's mass. Without the use of rocket fuel on board there would be little chance of the vehicle exploding.

Another form of safe removal would possibly be the space elevator. Encasing the waste in glassified form inside a steel shell 9 inches (230 mm) thick, which in turn is tiled with shuttle tile to its exterior. If the launch vehicle fails just before reaching orbit, the waste ball will safely re-enter the earth's atmosphere. The steel shell would deform on impact, but would not rupture due to the density of the shell. Also, this would potentially allow the waste to be shot into the Sun.

Accidents Involving Radioactive Waste

A number of incidents have occurred when radioactive material was disposed of improperly, shielding during transport was defective, or when it was simply abandoned or even stolen from a waste store. In the former Soviet Union, waste stored in Lake Karachay was blown over the area during a dust storm after the lake had partly dried out. At Maxey Flat, a low-level radioactive waste facility located in Kentucky, containment trenches covered with dirt, instead of steel or cement, collapsed under heavy rainfall into the trenches and filled with water. The water that invaded the trenches became radioactive and had to be disposed of at the Maxey Flat facility itself. In other cases of radioactive waste accidents, lakes or ponds with radioactive waste accidentally overflowed into the rivers during exceptional storms.

Scavenging of abandoned radioactive material has been the cause of several other cases of radiation exposure, mostly in developing nations, which may have less regulation of dangerous substances (and sometimes less general education about radioactivity and its hazards) and a market for scavenged goods and scrap metal. The scavengers and those who buy the material are almost always unaware that the material is radioactive and it is selected for its aesthetics or scrap value. Irresponsibility on the part of the radioactive material's owners, usually a hospital, university or military, and the absence of regulation concerning radioactive waste, or a lack of enforcement of such regulations, have been significant factors in radiation exposures. For an example of an accident involving radioactive scrap originating from a hospital see the Goiânia accident.

Transportation accidents involving spent nuclear fuel from power plants are unlikely to have serious consequences due to the strength of the spent nuclear fuel shipping casks.

Radioactive Waste in Fiction and Popular Culture

In fiction, radioactive waste is often cited as the reason for gaining super-human powers and abilities. An example of this fictional scenario is the 1981 movie "Modern Problems" in which actor Chevy Chase portrays a jealous, harried air traffic controller Max Fiedler; Max Fiedler, recently dumped by his girlfriend, comes into contact with nuclear waste and is granted the power of telekinesis, which he uses to not only win her back, but to gain a little revenge.

In reality, of course, exposure to radioactive waste instead would lead to illness and/or death. In the science fiction television series, "Space: 1999," a massive nuclear waste dump on the Moon explodes, hurtling the Moon, and the inhabitants of "Moonbase Alpha" out of the Solar System at interstellar speeds. In the television comedy series Family Guy, the Griffin family all get super-human powers from toxic waste. When the local mayor Adam West tries to do the same thing, he gets lymphoma. In The Simpsons, many mutant three-eyed fish live near the Springfield Nuclear Power Plant.

6

Structural and Functional Dynamics of Microbial Life

6.1 Microbe

Microbes are single-cell organisms so tiny that millions can fit into the eye of a needle. They are the oldest form of life on earth. Microbe fossils date back more than 3.5 billion years to a time when the Earth was covered with oceans that regularly reached the boiling point, hundreds of millions of years before dinosaurs roamed the earth.

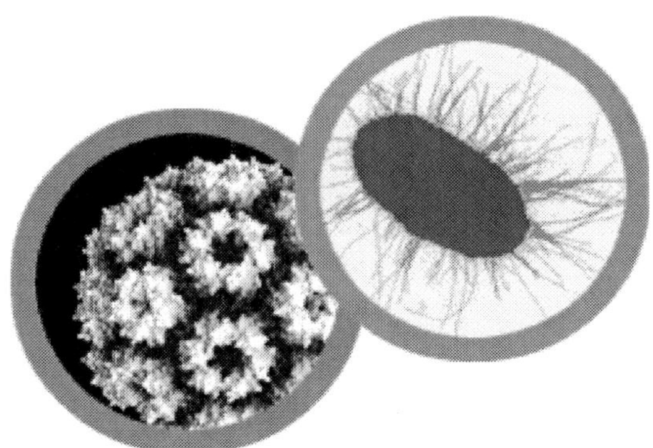

Without microbes, we couldn't eat or breathe.

Without us, they'd probably be just fine.

Understanding microbes is vital to understanding the past and the future of ourselves and our planet.

Microbes <my-crobes> are everywhere. There are more of them on a person's hand than there are people on the entire planet!

Microbes are in the air we breathe, the ground we walk on, the food we eat—they're even inside us!

We couldn't digest food without them—animals couldn't, either. Without microbes, plants couldn't grow, garbage wouldn't decay and there would be a lot less oxygen to breathe.

In fact, without these invisible companions, our planet wouldn't survive as we know it!

Microbe is a term for tiny creatures that individually are too small to be seen with the unaided eye. Microbes include *bacteria* (*back-tear-ee-uh*), *archaea* (*are-key-uh*), *fungi* (*fun-jeye*) and *protists* (*pro-tists*). You've probably heard of bacteria and fungi before. Archaea are bacteria-like creatures that have some traits not found in any true bacteria. Protists include primitive algae (*al-gee*), amoebas (*ah-me-buhs*), slime molds and protozoa (*pro-toe-zoh-uh*). We can also include *viruses* (*vye-rus-is*) as a major type of microbe, though there is a debate as to whether viruses can be considered living creatures or not.

To solve the case of what a microbe is, we have to use tools such as high-power microscopes. Let's zoom in on some microbes and see what a few of these strange creatures look like.

As you can see, microbes come in many varieties. They may live as individuals or cluster together in communities.

Well, let's say we could enlarge an average virus, the smallest of all microbes, to the size of a baseball.

An average bacterium would then be the size of the pitcher's mound.

And just one of the millions of cells that make up your body would be the size of the ballpark!

Continue reading the other parts of this section of the site to find out more about the different types of microbes and what they do.

6.2 Microorganism

A *microorganism* (also can be spelled as micro organism) or microbe is an organism that is microscopic (too small to be seen by the naked human eye). The study of microorganisms is called microbiology, a subject that began with Anton van Leeuwenhoek's discovery of microorganisms in 1675, using a microscope of his own design. Microorganisms are incredibly diverse and include bacteria, fungi, archaea, and protists, as well as some microscopic plants and animals such as plankton, and popularly-known animals such as the planarian and the amoeba. They do not include viruses and prions, which are generally classified as non-living. Most microorganisms are single-celled, or *unicellular*, but some multicellular organisms are microscopic, while some unicellular protists, and a bacteria called *Thiomargarita namibiensis* are visible to the naked eye.

Microorganisms live in all parts of the biosphere where there is liquid water, including hot springs, on the ocean floor, high in the atmosphere and deep inside rocks within the Earth's crust. Microorganisms are critical to nutrient recycling in ecosystems as they act as decomposers. As some microorganisms can fix nitrogen, they are a vital part of the nitrogen cycle, and recent studies indicate that airborne microbes may play a role in precipitation and weather.

Microbes are also exploited by people in biotechnology, both in traditional food and beverage preparation, as well as modern technologies based on genetic engineering. However, pathogenic microbes are harmful, since they invade and grow within other organisms, causing diseases that kill millions of people, other animals, and plants.

History

Evolution

Single-celled microorganisms were the first forms of life to develop on earth, approximately 3-4 billion years ago. Further evolution was slow, and for about 3 billion years in the Precambrian eon, all organisms were microscopic. So, for most of the history of life on Earth the only form of life were microorganisms. Bacteria, algae and fungi have been identified in amber that is 220 million years old, which shows that the morphology of microorganisms has changed little since the triassic period.

Most microorganisms can reproduce rapidly and microbes such as bacteria can also freely exchange genes by conjugation, transformation and transduction between widely-divergent species. This horizontal gene transfer, coupled with a high mutation rate and many other means of genetic variation, allows microorganisms to swiftly evolve (via natural selection) to survive in new environments and respond to environmental stresses. This rapid evolution is important in medicine, as it has led to the recent development of 'super-bugs'—pathogenic bacteria that are resistant to modern antibiotics.

Pre-Microbiology

The possibility that microorganisms might exist was discussed for many centuries before their actual discovery in the 17th century. The first ideas about microorganisms were those of the Roman scholar Marcus Terentius Varro in a book titled *On Agriculture* in which he warns against locating a homestead near swamps:

> "…and because there are bred certain minute creatures which cannot be seen by the eyes, which float in the air and enter the body through the mouth and nose and there cause serious diseases. "

This passage seems to indicate that the ancients were aware of the possibility that diseases could be spread by yet unseen organisms.

In *The Canon of Medicine* (1020), Abû Alî ibn Sînâ (Avicenna) stated that bodily secretion is contaminated by foul foreign earthly bodies before being infected. He also hypothesized that tuberculosis and other diseases might be contagious, *i.e.* that they were infectious diseases, and used quarantine to limit their spread.

When the Black Death bubonic plague reached al-Andalus in the 14th century, Ibn Khatima wrote that infectious diseases were caused by "contagious entities" that enter the human body. Later, in 1546, Girolamo Fracastoro proposed that epidemic diseases were caused by transferable seedlike entities that could transmit infection by direct or indirect contact, or even without contact over long distances.

All these early claims about the existence of microorganisms were speculative in nature and not based on any data or science. Microorganisms were neither proven, observed, nor correctly and accurately described until the 17th century. The reason for this was that all these early inquiries lacked the most fundamental tool in order for microbiology and bacteriology to exist as a science, and that was the microscope.

Discovery

Anton van Leeuwenhoek was the first person to observe microorganisms, using a microscope of his own design, thereby making him the first microbiologist. In doing so Leeuwenhoek would make one of the most important contributions to biology and open up the fields of microbiology and bacteriology. Prior to Leeuwenhoek's discovery of microorganisms in 1675, it had been a mystery as to why grapes could be turned into wine, milk into cheese, or why food would spoil. Leeuwenhoek did not make the connection between these processes and microorganisms, but using a microscope, he did establish that there were forms of life that were not visible to the naked eye. Leeuwenhoek's discovery, along with subsequent observations by Lazzaro Spallanzani and Louis Pasteur, ended the long-held belief that life spontaneously appeared from non-living substances during the process of spoilage.

Lazzarro Spallanzani found that microorganisms could only settle in a broth if the broth was exposed to the air. He also found that boiling the broth would sterilise it and kill the microorganisms. Louis Pasteur expanded upon Spallanzani's findings by exposing boiled broths to the air, in vessels that contained a filter to prevent all particles from passing through to the growth medium, and also in vessels with no filter at all, with air being admitted via a curved tube that would not allow dust particles to come in contact with the broth. By boiling the broth beforehand, Pasteur ensured that no microorganisms survived within the broths at the beginning of his experiment. Nothing grew in the broths in the course of Pasteur's experiment. This meant that the living organisms that grew in such broths came from outside, as spores on dust, rather than spontaneously generated within the broth. Thus, Pasteur dealt the death blow to the theory of spontaneous generation and supported germ theory.

In 1876, Robert Koch established that microbes can cause disease. He did this by finding that the blood of cattle who were infected with anthrax always had large numbers of *Bacillus anthracis*. Koch also found that he could transmit anthrax from one animal to another by taking a small sample of blood from the infected animal and injecting it into a healthy one, causing the healthy animal to become sick. He also found that he could grow the bacteria in a nutrient broth, inject it into a healthy animal, and cause illness. Based upon these experiments, he devised criteria for establishing a causal link between a microbe and a disease in what are now known as Koch's postulates. Though these postulates cannot be applied in all cases, they do retain historical importance in the development of scientific thought and can still be used today.

Classification and Structure

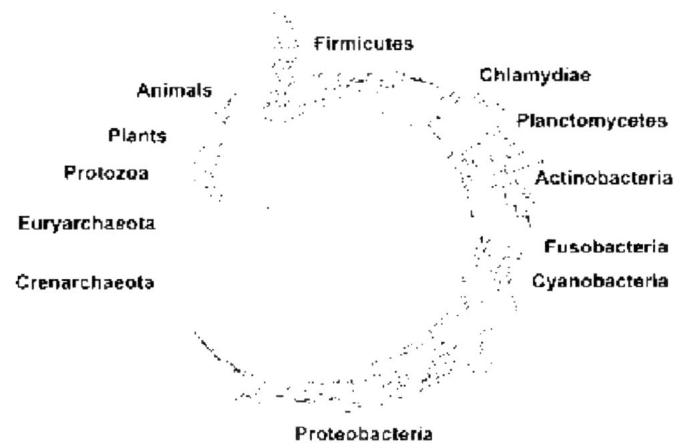

Microorganisms can be found almost anywhere in the taxonomic organization of life on the planet. Bacteria and archaea are almost always microscopic, while a number of eukaryotes are also microscopic, including most protists, some fungi, as well as some animals and plants. Viruses are generally regarded as not living and therefore are not microbes, although the field of microbiology also encompasses the study of viruses.

Prokaryotes

Prokaryotes are organisms that lack a cell nucleus and the other organelles found in eukaryotes. Prokaryotes are almost always unicellular, although some species such as myxobacteria can aggregate into complex structures as part of their life cycle. These organisms are divided into two groups, the archaea and the bacteria.

Bacteria

Bacteria are the most diverse and abundant group of organisms on Earth. Bacteria inhabit practically all environments where some liquid water is available and the temperature is below +140 °C. They are found in sea water, soil, air, animals' gastrointestinal tracts, hot springs and even deep beneath the Earth's crust in rocks. Practically all surfaces which have not been specially sterilized are covered in bacteria. The number of bacteria in the world is estimated to be around five million trillion trillion, or 5×10^{30}.

Bacteria are practically all invisible to the naked eye, with a few extremely rare exceptions, such as *Thiomargarita namibiensis*. They are unicellular organisms and lack membrane-bound organelles. Their genome is usually a single loop of DNA, although they can also harbor small pieces of DNA called plasmids. These plasmids can be transferred between cells through bacterial conjugation. Bacteria are surrounded by a cell wall, which provides strength and rigidity to their cells. They reproduce by binary fission or sometimes by budding, but do not undergo sexual reproduction. Some species form extraordinarily resilient spores, but for bacteria this is a mechanism for survival, not reproduction. Under optimal conditions bacteria can grow extremely rapidly and can double as quickly as every 10 minutes.

Archaea

Archaea are also single-celled organisms that lack nuclei. In the past, the differences between bacteria and archaea were not recognised and archaea were classified with bacteria as part of the kingdom Monera. However, in

1990 the microbiologist Carl Woese proposed the three-domain system that divided living things into bacteria, archaea and eukaryotes. Archaea differ from bacteria in both their genetics and biochemistry. For example, while bacterial cell membranes are made from phosphoglycerides with ester bonds, archaean membranes are made of ether lipids.

Archaea were originally described in extreme environments, such as hot springs, but have since been found in all types of habitats. Only now are scientists beginning to appreciate how common archaea are in the environment, with crenarchaeota being the most common form of life in the ocean, dominating ecosystems below 150 m in depth. These organisms are also common in soil and play a vital role in ammonia oxidation.

Eukaryotes

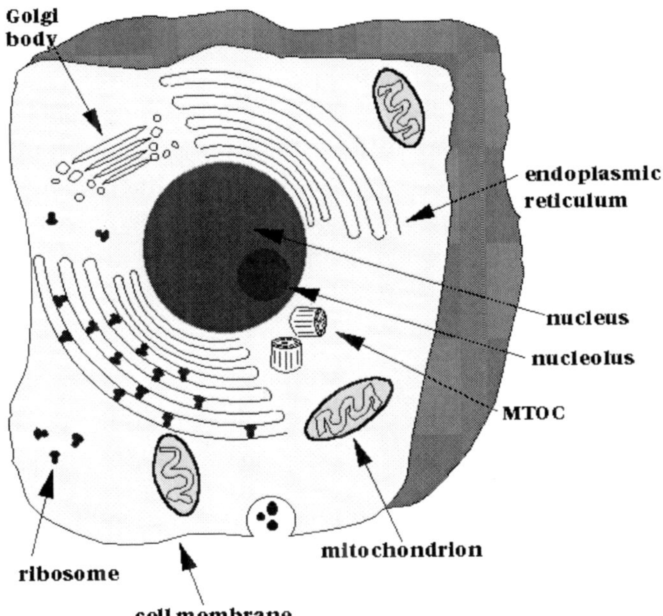

All living things which are *individually* visible to the naked eye are eukaryotes (with few exceptions, such as *Thiomargarita namibiensis*), including humans. However, a large number of eukaryotes are also microorganisms. Unlike bacteria and archaea, eukaryotes contain organelles such as the cell nucleus, the Golgi apparatus and mitochondria in their cells. The nucleus is an organelle which houses the DNA that makes up a cell's genome. DNA itself is arranged in complex chromosomes. Mitochondria are organelles vital in metabolism as they are the site of the citric acid cycle and oxidative phosphorylation. They evolved from symbiotic bacteria and retain a remnant genome. Like bacteria, plant cells have cell walls, and contain organelles such as chloroplasts in addition to the organelles in other eukaryotes. Chloroplasts produce energy from light by photosynthesis, and were also originally symbiotic bacteria.

Unicellular eukaryotes are those eukaryotic organisms that consist of a single cell throughout their life cycle. This qualification is significant since most multicellular eukaryotes consist of a single cell called a zygote at the beginning of their life cycles. Microbial eukaryotes can be either haploid or diploid, and some organisms have multiple cell nuclei (see coenocyte). However, not all microorganisms are unicellular as some microscopic eukaryotes are made from multiple cells.

Protists

Of eukaryotic groups, the protists are most commonly unicellular and microscopic. This is a highly diverse group of organisms that are not easy to classify. Several algae species are multicellular protists, and slime molds have unique life cycles that involve switching between unicellular, colonial, and multicellular forms. The number of species of protozoa is uncertain, since we may have identified only a small proportion of the diversity in this group of organisms.

Animals

All animals are multicellular, but some are too small to be seen by the naked eye. Microscopic arthropods include dust mites and spider mites. Microscopic crustaceans include copepods and the cladocera, while many nematodes are too small to be seen with the naked eye. Another particularly common group of microscopic animals are the rotifers, which are filter feeders that are usually found in fresh water. Micro-animals reproduce both sexually and asexually and may reach new habitats as eggs that survive harsh environments that would kill the adult animal. However, some simple animals, such as rotifers and nematodes, can dry out completely and remain dormant for long periods of time.

Fungi

The fungi have several unicellular species, such as baker's yeast (*Saccharomyces cerevisiae*) and fission yeast (*Schizosaccharomyces pombe*).

Some fungi, such as the pathogenic yeast *Candida albicans*, can undergo phenotypic switching and grow as single cells in some environments, and filamentous hyphae in others. Fungi reproduce both asexually, by budding or binary fission, as well by producing spores, which are called conidia when produced asexually, or basidiospores when produced sexually.

Plants

The green algae are a large group of photosynthetic eukaryotes that include many microscopic organisms. Although some green algae are classified as protists, others such as charophyta are classified with embryophyte plants, which are the most familiar group of land plants. Algae can grow as single cells, or in long chains of cells. The green algae include unicellular and colonial flagellates, usually but not always with two flagella per cell, as well as various colonial, coccoid, and filamentous forms. In the Charales, which are the algae most closely related to higher plants, cells differentiate into several distinct tissues within the organism. There are about 6000 species of green algae.

Habitats and Ecology

Microorganisms are found in almost every habitat present in nature. Even in hostile environments such as the poles, deserts, geysers, rocks, and the deep sea, some types of microorganisms have adapted to the extreme conditions and sustained colonies; these organisms are known as extremophiles. Extremophiles have been isolated from rocks as much as 7 kilometres below the earth's surface, and it has been suggested that the amount of living organisms below the earth's surface may be comparable with the amount of life on or above the surface. Extremophiles have been known to survive for a prolonged time in a vacuum, and can be highly resistant to radiation, which may even allow them to survive in space. Many types of microorganisms have intimate symbiotic relationships with other larger organisms; some of which are mutually beneficial (mutualism), while others can be damaging to the host organism (parasitism). If microorganisms can cause disease in a host they are known as pathogens.

Extremophiles

Extremophiles are microorganisms which have adapted so that they can survive and even thrive in conditions that are normally fatal to most lifeforms. For example, some species have been found in the following extreme environments:

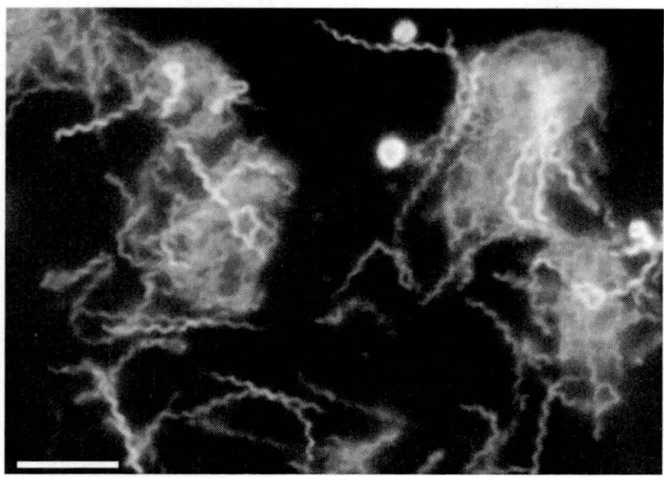

 ❖ *Temperature*: as high as 130 °C (266 °F), as low as -17 °C (1.4 °F)
 ❖ *Acidity/alkalinity*: less than pH 0, up to pH 11.5
 ❖ *Salinity*: up to saturation

❖ *Pressure*: up to 1,000-2,000 atm, down to 0 atm (e.g. of space)
❖ *Radiation*: up to 5kGy

Extremophiles are significant in different ways. They extend terrestrial life into much of the Earth's hydrosphere, crust and atmosphere, their specific evolutionary adaptation mechanisms to their extreme environment can be exploited in bio-technology, and their very existence under such extreme conditions increases the potential for extraterrestrial life.

Soil Microbes

The nitrogen cycle in soils depends on the fixation of atmospheric nitrogen. One way this can occur is in the nodules in the roots of legumes that contain symbiotic bacteria of the genera *Rhizobium, Mesorhizobium, Sinorhizobium, Bradyrhizobium,* and *Azorhizobium.*

Importance

Microorganisms are vital to humans and the environment, as they participate in the Earth's element cycles such as the carbon cycle and nitrogen cycle, as well as fulfilling other vital roles in virtually all ecosystems, such as recycling other organisms' dead remains and waste products through decomposition. Microbes also have an important place in most higher-order multicellular organisms as symbionts. Many blame the failure of Biosphere 2 on an improper balance of microbes.

Use in Food

Microorganisms are used in brewing, winemaking, baking, pickling and other food-making processes.

They are also used to control the fermentation process in the production of cultured dairy products such as yogurt and cheese. The cultures also provide flavour and aroma, and inhibit undesirable organisms.

Use in Water Treatment

Microbes are used in the biological treatment of sewage and industrial waste effluents.

Use in Energy

Microbes are used in fermentation to produce ethanol.

Use in Science

Microbes are also essential tools in biotechnology, biochemistry, genetics, and molecular biology. The yeasts (*Saccharomyces cerevisiae*) and fission yeast (*Schizosaccharomyces pombe*) are important model organisms in science, since they are simple eukaryotes that can be grown rapidly in large numbers and are easily manipulated. They are particularly valuable in genetics, genomics and proteomics. Microbes can be harnessed for uses such as creating steroids and treating skin diseases. Scientists are also considering using microbes for living fuel cells, and as a solution for pollution.

Use in Warfare

In the Middle Ages, dead corpses were thrown over walls during sieges, this meant that any bacteria carrying the disease that killed the person/creature would multiply in the vicinity of the opposing side.

Importance in Human Health

Human Digestion

Microorganisms can form an endosymbiotic relationship with other, larger organisms. For example, the bacteria that live within the human digestive system contribute to gut immunity, synthesise vitamins such as folic acid and biotin, and ferment complex indigestible carbohydrates.

Diseases and Immunology

Microorganisms are the cause of many infectious diseases. The organisms involved include pathogenic bacteria, causing diseases such as plague, tuberculosis and anthrax; protozoa, causing diseases such as malaria, sleeping sickness and toxoplasmosis; and also fungi causing diseases such as ringworm, candidiasis or histoplasmosis. However, other diseases such as influenza, yellow fever or AIDS are caused by pathogenic viruses, which are not living organisms and are not therefore microorganisms. As of 2007, no clear examples of archaean pathogens are known, although a relationship has been proposed between the presence of some methanogens and human periodontal disease.

Hygiene

Hygiene is the avoidance of infection or food spoiling by eliminating microorganisms from the surroundings. As microorganisms, particularly bacteria, are found practically everywhere, this means in most cases the reduction of harmful microorganisms to acceptable levels. However, in some cases it is required that an object or substance be completely sterile, i.e. devoid of all living entities and viruses. A good example of this is a hypodermic needle.

In food preparation microorganisms are reduced by preservation methods (such as the addition of vinegar), clean utensils used in preparation, short storage periods or by cool temperatures. If complete sterility is needed, the two most common methods are irradiation and the use of an autoclave, which resembles a pressure cooker.

There are several methods for investigating the level of hygiene in a sample of food, drinking water, equipment etc. Water samples can be filtrated through an extremely fine filter. This filter is then placed in a nutrient medium. Microorganisms on the filter then grow to form a visible colony. Harmful microorganisms can be detected in food by placing a sample in a nutrient broth designed to enrich the organisms in question. Various methods, such as selective media or PCR, can then be used for detection. The hygiene of hard surfaces, such as cooking pots, can be tested by touching them with a solid piece of nutrient medium and then allowing the microorganisms to grow on it.

There are no conditions where all microorganisms would grow, and therefore often several different methods are needed. For example, a food sample might be analyzed on three different nutrient mediums designed to indicate the presence of "total" bacteria (conditions where many, but not all, bacteria grow), molds (conditions where the growth of bacteria is prevented by e.g. antibiotics) and coliform bacteria (these indicate a sewage contamination).

In Fiction

Microorganisms have frequently played an important part in science fiction, both as agents of disease, and as entities in their own right.

Some notable uses of microorganisms in fiction include:

- *The War of the Worlds*, where microorganisms play important thematic and plot-related roles.
- *Fantastic Voyage*, in which some scientists are miniaturised to microscopic size and observe micro-organisms from a new perspective
- *Blood Music*, in which a colony of microorganisms is given intelligence
- *The Andromeda Strain*, in which extraterrestrial microorganisms kill several people.

❖ *The White Plague*, is created and released in vengeance by John Roe O'Neill for the death of his wife and children, it is designed to kill only women.

❖ *Twelve Monkeys*, James Cole (Bruce Willis) searches for a pure germ in the past, which creates a deadly plague in the future. Also, Brad Pitt (as Jeffery Goines) discusses his germaphobia.

6.3 Microbial Reproduction

One of the reasons microbes have survived and flourished for billions of years is the different ways they reproduce and how fast they do it. Bacteria usually reproduce by simply dividing in two. Each new bacterium is a clone of the original—they each contain a copy of the same DNA. This is called binary fission (*bye-nair-ee fish-un*). If conditions are just right, one bacterium could become a BILLION (1,000,000,000) bacteria in just 10 hours through binary fission! Sometimes bacteria do have a form of sex called conjugation (*con-ju-gay-shun*). One bacterium reaches out to another using a thread-like structure called a pilus. The first bacterium transfers part of its DNA to the other bacterium through the pilus. By this exchange, bacteria can quickly create or pass along new traits that help them adapt to different environments. Some yeasts, which are a type of fungus, also reproduce through binary fission. Other yeasts reproduce by budding. A parent yeast cell forms a growth, or bud, on its surface. As this bud gets bigger, the parent cell's nucleus divides into two by a process called mitosis (*my-toe-sis*). One of these nuclei transfers into the growing bud, which eventually pinches off. One yeast cell can produce up to 24 daughter cells.

Other fungi reproduce through spore formation. Fungi can produce spores either without sex or when two fungal cells fuse and share DNA. Spores are kind of like seeds in that they can be carried on the wind or rain, spreading the fungus to new places. Some species can do both sexual and asexual reproduction, depending on environmental conditions.

Microbes in the protist category reproduce in a number of ways. Algae, for example, can reproduce without sex. The nucleus of the algal cell divides by mitosis and moves to opposite ends of the cell. Then the cell divides into two new cells—this is called cytokinesis (*sigh-toe-kin-ee-sis*). Some species can reproduce sexually. Some do both. Protozoa reproduce by binary fission, budding or a process called schizogony (*skiz-og-uh-knee*). Schizogony is multiple fission: the cell's nucleus divides several times before the cell itself divides into multiple new cells, each with one of those new nuclei.

Viruses cannot reproduce on their own, which is one reason people debate whether they should be counted as living creatures. But when they're inside a suitable host cell, all they do is reproduce and they do it well. They take over the host cell's machinery and use it to make copies of their genes and proteins. These genes and proteins come together to make new viruses, which eventually escape the host cell and drift off to infect new cells.

So why aren't we up to our elbows in bacteria, fungi and protists if microbes can reproduce so quickly? It's because conditions are almost never perfect in the real world. Once there are too many microbes in one place, food runs out, their wastes pile up, they crowd each other and eventually they start dying off.

6.4 Mutations Make a Hardy Microbe

Mutations (*mew-tay-shuns*) are misspellings or changes in the genetic code. They're pretty rare; only about one mutation occurs for every 1-10 million DNA bases.

But microbes reproduce very quickly. For example some bacteria can divide every nine minutes. And the entire DNA string or genome (*gee-nome*) of microbes is relatively small. For example, the average bacterium has only about five million DNA bases. So in each new generation of bacteria, there may be one or two mistakes in the genetic code.

In other words, these random mutations occur often enough in microbes to play a significant role in their ability to survive and adapt. Some mutations are random "mistakes." They can occur when a microbe is exposed to radiation or chemicals that cause changes in DNA. Microbes, like us, have DNA repair proteins that, like a mini construction crew, work hard to repair any such mistakes. They don't always catch every mistake, however.

Other mutations result from a change in a creature's circumstances. For example, if the temperature starts getting hotter and hotter, "jumping genes"— pieces of DNA that can move around—may pick up and move to a new place in the microbe's DNA. This switch may result in giving the microbe a new genetic trait—like greater heat tolerance—that helps it adapt and survive better. It's like a homebuilder who, upon realizing that he's building in a hurricane-prone area, decides that the heavy oak wood he planned to use for hardwood floors is more urgently needed to build the beams that support the house, so he changes it on the blueprint. Jumping genes cause genetic changes in all kinds of microbes.

Another way bacteria can undergo mutation is to take up bits of free-floating DNA in the environment. This is called transformation. Bits of DNA litter the environment when it's released from dying bacterial cells or bacteria burst open by a virus infection. DNA in the form of plasmids—circular bits of DNA that contain genes but exist outside the main chromosome—are also released by bacteria into the environment where they can be taken up by other bacteria.

Sometimes mutations are harmful and the microbes die off before they can reproduce and pass the mutations on to offspring. But other times a mutation gives rise to a new trait that helps the microbe better survive. Take antibiotic resistance among bacteria.

Antibiotics (*an-tea-bye-ah-ticks*) are life-saving drugs that kill bacteria. We discovered and began using them about 60 years ago. In that relatively short amount of time, bacteria have developed new traits through mutations that help protect them against antibiotics—sort of like a helmet protects you against injury.

Other mutations have helped microbes adapt to all sorts of environments from salty to icy to extremely hot and live off everything from decaying leaves to sunlight to bubbling sulfur.

6.5 Microbial Intelligence

Microbial Intelligence (popularly known as *bacterial intelligence*) is the intelligence shown by microorganisms. The concept encompasses complex adaptive behaviour shown by single cells, and altruistic and/or cooperative behaviour in populations of like or unlike cells mediated by chemical signalling that induces physiological or behavioural changes in cells and influences colony structures.

Complex cells, like protozoa or algae, show remarkable abilities to organise themselves in changing circumstances . Shell-building by amoebae, reveals complex discrimination and manipulative skills that are ordinarily thought to occur only in multicellular organisms. Even bacteria, which show primitive behaviour as isolated cells, can display more sophisticated behaviour as a population. These behaviours occur in single species populations, or mixed species populations. Examples are colonies of Myxobacteria, quorum sensing, and biofilms.

It has been suggested that a bacterial colony loosely mimics a biological neural network. The bacteria can take inputs in form of chemical signals, process them and then produce output chemicals to signal other bacteria in the colony. The mechanisms that enable single celled organisms to coordinate in populations presumably carried over in those lines that evolved multicellularity, and were co-opted as mechanisms to coordinate multicellular organisms.

Examples of Microbial Intelligence

- ❖ The formation of biofilms requires joint decision by the whole colony.
- ❖ Under nutritional stress bacterial colonies can organise themselves in such a way so as to maximise nutrient availability.
- ❖ Bacteria reorganise themselves under antibiotic stress.
- ❖ Bacteria can swap genes (such as genes coding antibiotic ressistance) between members of mixed species colonies.
- ❖ Individual cells of myxobacteria and cellular slime moulds coordinate to produce complex structures or move as multicellular entities.
- ❖ Populations of bacteria use Quorum sensing to judge their own densities and change their behaviours accordingly. This occurs in the formation of biofilms, infectious disease processes, and the light organs of bobtail squid.
- ❖ It's known that for any bacterium to enter a host's cell, the cell must display receptors to which bacteria can adhere and be able to enter the cell. Some strains of *E. coli* are able to internalize themselves into a host's cell even without the presence of specific receptors as they bring their own receptor to which they then attach and enter the cell.
- ❖ Under rough circumstances, some bacteria transform into endospores to resist heat and dehydration.
- ❖ A huge array of microorganisms have the ability to overcome being recognized by the immune system as they change their surface antigens so that any defense mechanisms directed against previously present antigens are now useless with the newly expressed ones.

6.6 Molecular Microbial Ecology (MME)

Our research group uses molecular approaches for the study of a variety of natural microbial communities. This includes the development of new techniques to dissect the microbial diversity of complex ecosystems. The long-term goal of this research is to integrate different fields of biology (i.e., genomics, ecology, molecular biology, proteomics, bioinformatics and metagenomics) to provide insight into the survival of environmental microorganisms under stressful conditions. Current research in the lab focuses on the microbial ecology of a number of environments including the air, water, soil, and deep subsurface with an emphasis on the natural distribution of pathogens in the environment. By understanding processes that allow specific bacteria to exploit unique environments we hope to better understand critical mechanisms for survival.

Our laboratory is a center for microarray research at LBNL. We are working on advanced technologies to measure microbial species composition, whole-genome transcriptional analysis, and SNP discovery for epidemiology studies. We use Affymetrix high-density microarrays for much of our work. The combinatorial power of at least 500,000 oligonucleotide probes has greatly increased precision and reproducibility in our measurements. We have designed arrays for the broad characterization of species in a microbial community using the 16S rRNA gene, pathogen specific diagnostics using unique signatures, bacterial gene expression using whole genome sequence, and SNP discovery for genotyping of bacterial strains.

Current projects in our lab include understanding mechanisms of bacterial diversity using 16S rRNA gene sequence to measure relative abundance of individual members of microbial communities. We have developed novel microarray systems to measure dynamic changes and we are working with the Joint Genome Institute in Walnut Creek, CA to develop a rapid system for classifying the thousands of individual sequences from clone libraries that we a constructing. We are also measuring whole genome transcriptional expression of bacteria subjected to different environmental stresses. With collaborators at Stanford University and the University of California at Berkeley, we are developing systems for *Caulobacter crescentus, Dehalococcoides ethenogenes,* and *Thiobacillus ferrooxidans.*

We are also interested in the survival and spread of pathogens in the environment. We are studying two related pathogens with very distinctive disease lifecycles, *Salmonella enterica* serovar Enteritidis and *Yersinia pestis.* We have completed a 10X draft genome sequence of a phage type 4 strain of *Salmonella enteritidis* and have recently finished two strains of *Yersinia pestis,* Nepal516, and Antiqua. To better understand the epidemiology of these pathogens, we are identifying novel SNPs for high accuracy genotyping in a partnership with Perlegen Sciences from Mountain View, CA. We are also assisting Callida Genomics in a project to develop a diagnostic system for the identification of plague (*Yersinia pestis*) and anthrax (*Bacillus anthracis*) using a revolutionary SBH system. Details of individual projects may be seen by selecting the Research Projects link on the left.

6.7 Ecology Department

Scientific Focus Areas

The Ecology Department (ED) intends to maintain the highest quality and highest visibility for its research and development in five areas:

❖ Bioenergy,
❖ Hydroecological engineering advanced decision support,
❖ Molecular microbial ecology,
❖ Real-time assessment of bioavailability and biokinetics, and
❖ Bioremediation and natural attenuation.

These five R&D areas are largely integrated, but contain some domains that are not inclusive. These five areas are considered ED's core competencies.

Bioenergy

Concerns about global warming, dependence on foreign oil, and rising gasoline prices have spurred a strong interest in bioenergy in the United States. ED scientists will be contributing their expertise in microbiology, molecular microbial ecology, and bacterial metabolism to a scientific challenge of great national importance: the efficient production of biofuels. LBNL has been a key strategic partner in large research initiatives for developing bioenergy, including the Joint BioEnergy Institute (JBEI), a Department of Energy-funded research center that will receive $125 million over five years, and the Energy Biosciences Institute (EBI), which is being funded by the energy company BP at a level of $500 million over ten years; both JBEI and EBI began operations in late 2007. The Ecology Department will play an important role in both JBEI and EBI. For example, two ED personnel will serve as Directors in JBEI: the Director of Microbial Communities (responsible for studying the structural and functional dynamics of microbial communities that efficiently degrade lignocellulosic biomass) and the Director of Biofuels Pathways (responsible for discovering genes that can be engineered into metabolic pathways resulting in efficient production of various transportation fuels from cellulose degradation products). In EBI, ED scientists will be involved in the study of Microbially Enhanced Hydrocarbon Recovery and will co-lead a Systems Biology programme, which will entail the intensive, genome-enabled study of certain bioenergy-related microorganisms ("omics" studies, including transcriptomics, proteomics, and metabolomics).

Hydroecological Engineering Advanced Decision Support (HEADS)

The HEADS research focus area has established a strong track record in the rapidly growing, new subject area of Ecological Engineering. The term "hydroecological engineering" signifies the group's concentration on water resources and wastewater engineering. Recognizing the growth in the field of environmental informatics and the application of computer-based models in the development of decision support systems, the group is active in this niche area. Interest in decision support interfaces well with the group's expertise in the deployment of real-time flow and water quality sensors, rapid laboratory assessment techniques, and mathematical models to develop an early warning system for contaminant management and containment.

Molecular Microbial Ecology (MME)

Understanding microbial interactions is key to the study of global warming, biodegradation of harmful compounds, and the exploration of complex microbial communities in their natural environment. The DOE has placed an increased emphasis on the role microbes play in modifying their environment and their impact on energy security. The MME group has responded to these needs by aggressively seeking out new projects and expanding its staff to develop new core capabilities. One of the key challenges has been to harness the explosion of microbial DNA sequence information to accurately measure the microbial dynamics in extreme environments. Since less than one percent of the microbial species can be cultured from these environments, our knowledge of what these organisms may be doing is limited to where they are observed and the similarity of their genomes to studied organisms. By understanding the ecological structure of microbial communities and the fine-scale dynamics resulting from subtle perturbations, it may be possible to identify novel functional pathways and use the diverse microbial capabilities to assist in key DOE missions. The molecular tools being developed in the ESD's Center for Environmental Biotechnology will position us to be leaders in this area.

Real-Time Assessment of Bioavailability and Biokinetics (RABB)

Interactions between environmental pollutants and ecological receptors begin when the pollutants become available to the target sites of a live ecological receptor. The ability to characterize the dynamics of the bioavailability of pollutants, their transformation kinetics, and the subsequent ecological response is a keystone to advancing the science in relevant DOE areas, including biogeochemistry, bioremediation, and exposure and risk assessment. Since 1999, the RABB research group at ED has pioneered the development and application of several cutting-edge technologies, such as synchrotron radiation-based spectromicroscopy, in vitro human gastrointestinal mimetic reactors, and in vivo mouse protocols that allow for the real-time assessment of bioavailability and biokinetics of environmental pollutants. The RABB group intends to establish this capability further by seeking out new projects and new collaborators, as well as expanding our staff, to position us as leaders in the areas of biological and environmental sciences.

Bioremediation and Natural Attenuation

Bioremediation and natural attenuation have been rapidly growing areas of scientific study over the past decade. The acceptance of natural attenuation as a solution for cleaning up contaminated sites, and DOE's recognition that they will have long-term stewardship concerns to address at the most contaminated sites, has greatly increased the urgency for research related to microbial ecology, biogeochemistry, biochemistry, and molecular detection techniques. This type of research is truly enabling for natural attenuation, since characterization, transport-and-fate prediction, and verification monitoring require a strong scientific basis. Natural attenuation is viewed as the best solution for cleaning up many waste sites and will save billions of dollars in cleanup costs. ED scientists and engineers are recognized leaders in the field of bioremediation and natural attenuation. The Center for Environmental Biotechnology provides the primary facilities used by ED, including state-of-the-art equipment for microbiology and environmental engineering. ED investigators have extensive experience in both water treatment and aquifer bioremediation. In addition to basic research, ED investigators have been involved in various aspects of more than 100 field demonstrations and deployments, and have five patents in this area that are licensed to more than 30 companies. The types of contaminants in which ED investigators have expertise include chlorinated solvents, petroleum hydrocarbons (including monoaromatic and polynuclear aromatic hydrocarbons), methyl tert-butyl ether (MTBE), high explosives (especially TNT and RDX), nitrate, plutonium, neptunium, chromium, and uranium. The Bioremediation and Natural Attenuation area has both basic research and field application foci for the ED. The basic research foci are metabolism and physiology (including the genetics and biochemistry underlying pollutant metabolism), biotransformation kinetics, and modeling of biogeochemical processes. Field-application foci are in situ monitoring techniques (including molecular, mass spectrometric, and biogeochemical approaches), co-metabolic techniques, and modeling of attenuation and environmental fate.

Funding

ED personnel are funded by DOE Programmes in (1) the Office of Science, Office of Biological and Environmental Research (OBER) (Natural and

Accelerated Bioremediation Research Programme, Genomics: GTL, and Medical Sciences); (2) the Office of Environmental Management, Environmental Restoration Programmes; (3) the Office of Fossil Research, the Petroleum Environmental Research Forum; and (4) the National Nuclear Security Administration, Office of Nonproliferation Research and Engineering (NN20). In addition, support is provided by the U.S. Department of Homeland Security, the Department of Agriculture; the Department of the Interior, Bureau of Land Management, and Bureau of Reclamation under the CALFED programme, NASA Astrobiology Instititutes, as well as several projects with remediation companies using DOE-patented technologies for in situ bioremediation. ED personnel are also funded by Berkeley Lab's Laboratory Directed Research and Development (LDRD) Programme in the area of microbial fuel cells, fungal rDNA arrays, and FTIR biokinetic analaysis.

7

Environment Biotechnology and Engineering

7.1 Environmental Biotechnology

Biotechnology is "the integration of natural sciences and engineering in order to achieve the application of organisms, cells, parts thereof and molecular analogues for products and services" (EFB General Assembly, 1989). Environmental biotechnology as discussed in this briefing paper is the application of these processes for the protection and restoration of the quality of our environment.

This briefing paper reviews the various areas of environmental biotechnology together with their related issues and implications. The overall aim is to provide balanced information and advance public debate. This paper results from the combined contributions of scientists, industrialists, and governmental and environmental organisations across Europe. It is intended to supply information and does not represent the views or policy of the European Federation of Biotechnology or any other body.

Introduction

Biotechnological processes to protect the environment have been used for almost a century now, even longer than the term 'biotechnology' exists. Municipal sewage treatment plants and filters to purify town gas were developed around the turn of the century. They proved very effective although at the time, little was known about the biological principles underlying their function. Since that time our knowledge base has increased enormously. This briefing paper describes the state-of-the-art and possibilities of environmental biotechnology. It also deals with the societal aspects of environmental biotechnology.

Biotechnological techniques to treat waste before or after it has been brought into the environment are described and exemplified in the section on bioremediation. Biotechnology can also be used to develop products and processes that generate less waste and use less non-renewable resources and energy. In this respect biotechnology is well positioned to contribute to the development of a more sustainable society, a principle which was advocated in the Brundtland Report in 1987 and in Agenda 21 of the second Earth Summit in Rio de Janeiro in 1992 and which has been widely accepted in the mean time. The section on prevention deals with this subject. Biotechnological techniques to monitor the quality of the environment are presented in the section on detection and monitoring. Recombinant DNA technology has improved the possibilities for the prevention of pollution and holds a promise for a further development of bioremediation. These topics will be discussed in the section on genetic modification. The development of modern biotechnology has been accompanied by the establishment or adaptation of regulations to deal with genetically modified organisms. What this means for environmental biotechnology is embodied in the section on legislation. The section on public opinion, dialogue and debate highlights how people feel about environmental biotechnology and ways in which their opinion is influenced.

Bioremediation

Bioremediation is the use of biological systems for the reduction of pollution from air or from aquatic or terrestrial systems. Micro-organisms and plants are the biological systems which are generally used. Biodegradation with micro-organisms is the most frequently occurring bioremediation option. Micro-organisms can break down most compounds for their growth and/or energy needs. These biodegradation processes may or may not need air. In some cases, metabolic pathways which organisms normally use for growth and energy supply may also be used to break down pollutant molecules. In these cases, known as co-metabolism, the micro-organism does not benefit directly. Researchers have taken advantage of this phenomenon and use it for bioremediation purposes. A complete biodegradation results in detoxification by mineralising pollutants to carbon dioxide, water and harmless inorganic salts. Incomplete biodegradation will yield breakdown products which may or may not be less toxic than the original pollutant. Incomplete biodegradation of tri-or tetrachloroethylene for instance can yield vinylchloride, which is more toxic and carcinogenic than the original compounds.

Biodegradation may occur spontaneously, in which case the expressions "intrinsic bioremediation" or "natural attenuation" are often used. In many cases however the natural circumstances are not favourable enough for this to happen due to the lack of enough nutrients, oxygen or suitable bacteria. Such situations may be improved by supplementing one or more of these prerequisites. Extra nutrients were for instance disseminated to speed up the breakdown of the oil spilled on 1000 miles of Alaskan shoreline by the super tanker Exxon Valdez in 1989. The future trend in bioremediation increasingly is to look at the speed of unaided biodegradation first and act only if there is insufficient activity to remove the pollutant quickly enough to prevent any expected risks of the pollutant.

Bioremediation techniques can be used to reduce or to remove hazardous waste which has already polluted the environment. They can also be used to treat waste streams before they leave production facilities: end-of-pipe-processes. Some applications of bioremediation are discussed below.

❖ *Waste water and industrial effluents:* Micro-organisms in sewage treatment plants remove the more common pollutants from waste water before it is discharged into rivers or the sea. Increasing industrial and agricultural pollution has led to a greater need for processes that remove specific pollutants such as nitrogen and phosphorus compounds, heavy metals and chlorinated compounds. New methods include aerobic, anaerobic and physico-chemical processes in fixed-bed filters and in bioreactors in which the materials and microbes are held in suspension. The costs of waste water treatment can be reduced by the conversion of wastes into useful products. For example, heavy metals and sulphur compounds can be removed from waste streams of the galvanisation industry by the aid of sulphur metabolising bacteria and reused. Another example is the production of animal feed from the fungal biomass which remains after the production of penicillin. Most anaerobic waste water treatment systems produce useful biogas.

❖ *Drinking and Process Water:* Abundant supplies of water are vital for modern urban and industrial development. By the turn of this century, it is estimated that two-thirds of the world's nations will be water stressed—using clean water faster than it is replenished in aquifers or rivers. A very important aspect of biotechnology is therefore its potential for the reclamation and purification of waste waters for re-use. Public concern has also increased over the quality of drinking water. Not only does water need to be recycled in the development of sustainable use of resources, overall quality must also be improved to satisfy consumers. In many agricultural regions of the world, animal wastes and excess fertilisers result in high levels of nitrates in drinking water. Biotechnology has provided successful methods by which these compounds can be removed from processed water before it is delivered to customers.

❖ *Air and waste gases:* Originally, industrial waste gas treatment systems were based on cheap compost-filled filters that removed odours. Such systems still exist. However, slow processing rates and the short life of such filters drove research into better methods such as bioscrubbers, in which the pollutants are washed out using a cell suspension and biotrickling filters, in which the pollutant is degraded by micro-organisms immobilised on an inert matrix and provided with an aqueous nutrient film trickling through the device. The selection of micro-organisms that are more efficient at metabolising pollutants has also led to better air and gas purifying biofilters. Examples are a bioscrubber based system for the simultaneous removal of nitrogen and sulphur oxides from the flue gas of blast-furnaces which has been developed as an alternative to the classical limestone gypsum process, and the elimination of styrene from the waste gas of polystyrene processing industries by a biofilter containing fungi.

❖ *Soil and land treatment:* Both in situ (in its original place) and ex situ (somewhere else) methods are commercially exploited for the cleanup of soil and the associated groundwater. In situ treatments may include the introduction of micro-organisms (bio-augmentation), ventilation and/or adding nutrient solutions (biostimulation). Ex situ treatment involves removing the soil and groundwater and treating it above ground. The soil may be treated as compost, in soil banks, or in specialised slurry bioreactors. Groundwater is treated in bioreactors and either pumped back into the ground or drained into surface water. Bioremediation of land (biorestoration) is often cheaper than physical methods and its products are harmless if complete mineralisation takes place. Its action can however, be time-consuming, tying up capital and land. The in situ bioremediation of the ground under petrol stations has already become common practice but also for chlorinated solvents like tri-and tetrachloroethylene in situ bioremediation is possible. The applicability of in situ bioremediation is and probably will remain dependent on the physical parameters of the soil, mainly its transport properties. Bioremediation using plants is called phyto-remediation. This technique is presently already used to remove metals from contaminated soils and groundwater and is being further explored for the remediation of other pollutants. The combined use of plants and bacteria may also be possible. Certain bacteria live closely associated with the roots of plants and depend on substances excreted by the roots. Such rhizobacteria, whose numbers are much higher than those of other soil bacteria, may be genetically modified to break down pollutants. Research is being conducted to test this hypothesis.

❖ *Solid waste:* Domestic solid wastes are a major problem in our consumption society. Their elimination is both costly and warrants constant surveillance in terms of groundwater and air pollution. Yet, for a major part they are composed of readily biodegradable organics. In this respect, source separated bio-wastes can be converted to a valuable resource by composting or anaerobic digestion. In recent years, both processes have seen remarkable developments in terms of process design and control. Particularly, anaerobic digestion of solid wastes in high-rate anaerobic digesters has gained increasing public acceptance because it permits the recovery of substantial amounts of high-value biogas together with a high quality stable organic residue and this without giving rise to environmental nuisance. Moreover, anaerobic digestion of mixed solid wastes is under intensive development because in the near future it may be an important step in recycling of solid wastes and constitute an alternative to incineration.

Prevention

Progressively more industrial companies are developing processes with reduced environmental impact responding to the international call for the development of a sustainable society. There is a pervading trend towards less harmful products and processes; away from "end-of-pipe" treatment of waste streams. Biotechnology is pre-eminently suitable to contribute to this trend and it has already done so in many cases, both by the improvement of existing processes and by the development of new ones.

❖ *Process improvement:* Many industrial processes have been made more environmentally friendly by the use of enzymes. Enzymes are biological catalysts that are highly efficient and have numerous advantages over non-biological catalysts. They are non-toxic and biodegradable, work best at moderate temperatures and in mild conditions, and have fewer side reactions than traditional methods because they are highly specific. Production methods that employ enzymes are generally not only cleaner and safer compared with other methods, but mostly also more economic in energy and resource consumption (see box). Their specificity does however mean that it is not always easy to find the appropriate enzyme for a given application. Enzymes are already widely employed in industry and have been for many years. New techniques and approaches to protein design and molecular modelling are enabling researchers to develop novel enzymes active at high temperatures, in non-aqueous solvents and as solids.

❖ *Product innovation:* Biotechnology also can help to produce new products which have less impact on the environment than their predecessors. The production of new biomaterials like bioplastics avoids the use of non-renewable resources like fossil fuels. Potatoes normally contain 80 per cent amylopectin but also 20 per cent amylose which is unwanted in many applications. For the isolation of pure amylopectin large amounts of water and energy are consumed. A Dutch company has developed a genetically modified potato variety which no longer contains amylose and hence can be processed with less impact on the environment. The use of genetically modified plant varieties that are resistant against insects and/or diseases may considerably diminish the use of pesticides which not only prevents the use of the mostly non-renewable-raw materials, energy and labor necessary for their production but also will reduce the negative impacts of their residues. Many more of these biotechnological solutions for pollution have been developed (see box). Future developments may involve things which at this moment seem science-fiction to most people such as the replacement of chemically produced super-fibres by microbially produced spider web silk. One thing that should not be forgotten however is that the increased use of biological systems in industry should be accompanied by adequate training and protection of workers handling these systems, just like in other sections of industry.

Detection and Monitoring

Detection and monitoring of pollutants: A wide range of biological methods are already in use to detect pollution incidents and for the continuous monitoring of pollutants. Long established measures include: counting the number of plant, animal and microbial species, counting the numbers of individuals in those species or analysing the levels of oxygen, methane or other compounds in water. More recently, biological detection methods using biosensors and immunoassays have been developed and are now being commercialised.

Most biosensors are a combination of biological and electronic devices— often built onto a microchip. The biological component might be simply an enzyme or antibody, or even a colony of bacteria, a membrane, neural receptor, or an entire organism. Immobilised on a substrate, their properties change in response to some environmental effect in a way that is electronically or optically detectable. It is then possible to make quantitative measurements of pollutants with extreme precision or to very high sensitivities. The sensors can be designed to be very selective, or sensitive to a broad range of compounds. For example, a wide range of herbicides can be detected in river water using algal-based biosensors; the stresses inflicted on the organisms being measured as changes in the optical properties of the plant's chlorophyll.

Microbial biosensors are micro-organisms which produce a reaction upon contact with the substance to be sensed. Usually they produce light but cease to do so upon contact with substances which are toxic to them. Both naturally occurring light emitting micro-organisms as well as specially developed ones are used. Positively acting bacterial biosensors have been constructed which start emitting light upon contact (and subsequent reaction) with a specific pollutant. In the USA such a light emitting bacterium has been approved for the detection of polyhalogenated aromatic hydrocarbons in field tests.

Immunoassays use labelled antibodies (complex proteins produced in biological response to specific agents) and enzymes to measure pollutant levels. If a pollutant is present, the antibody attaches itself to it; the label making it detectable either through colour change, fluorescence or radioactivity. Immunoassays of various types have been developed for the continuous, automated and inexpensive monitoring of pesticides such as dieldrin and parathion. The nature of these techniques, the results of which can be as simple as a colour change, make them particularly suitable for highly sensitive field testing where the time and large equipment needed for more traditional testing is

impractical. Their use is however limited to pollutants which can trigger biological antibodies. If the pollutants are too reactive, they will either destroy the antibody or suppress its activity and so also the effectiveness of the test.

Detection and monitoring of micro-organisms used for bioremediation: When laboratory grown micro-organisms are inoculated into a bioremediation site (bio-augmentation) it often becomes necessary to monitor their presence and/or multiplication to check the progress of the process. This is especially true and even required when genetically modified micro-organisms are involved. The traditional technique to detect the presence of micro-organisms in soil is direct plating on selective media. This is greatly facilitated if the organism contains a marker which can be selected for. Newer techniques include the abovementioned immunological and light-based bioreporter techniques. The spatial distribution of specific micro-organisms in a sample can be determined microscopically and non-invasively by using fluorescent in situ hybridisation (FISH) of micro-organisms. The most sensitive and specific technique is the direct isolation and amplification of DNA from soil, which is increasingly being used.

Detection and monitoring of ecological effects: Bioremediation is aimed at improving the quality of the environment by removing pollutants. However, the disappearance of the original pollutant is not the only criterion by which the success of a bioremediation operation is determined. (Even more) toxic metabolites may be produced from the pollutant or the biodegrading bacterium may cause diseases or produce substances that are harmful to useful micro-organisms, plants, animals or humans. All these negative effects, are of course, excluded as much as possible in advance by getting as familiar as possible with the organism through extensive literature searches and microcosm studies in which the bioremediation process is simulated in the laboratory. To avoid unexpected effects, especially after the release of new member of the eco-system like a genetically modified organism, the monitoring of the ecological effects of a bioremediation operation may be required. The problem with monitoring ecological effects is what to monitor. Numerous ecological effects are possible but not all of them may be relevant or permanent or even the result of the bioremediation operation. The parameters to be monitored are usually determined case-by-case. Monitoring techniques may include all of those mentioned in the two previous subsections on detection and monitoring.

Genetic Engineering

Recombinant DNA technology has had amazing repercussions in the last few years. Molecular biologists have mapped entire genomes, many new medicines have been developed and introduced and agriculturists are producing plants with novel types of disease resistance that could not be achieved

More Sustainable Industrial Processes through the use of Enzymes

The leather processing industry has introduced enzymes to replace harsh chemicals traditionally used for cleaning the hide. In textile production, enzymes have superseded chemicals for bleaching, including the "stone washing" of jeans. Chlorine consumption by the pulp and paper industry may soon also be reduced considerably by the use of enzymes. The grease and protein digesting enzymes in washing powders significantly reduce the quantity of detergents needed for a given washing effect. They also mean that the washing temperature can be reduced. Lowering the temperature 20°C saves more than a third of the energy used by the machine. Since in many Western European countries up to 5 per cent of household energy consumption was used for washing, these molecules have made a significant contribution to energy conservation.

Biotechnological Solutions for Pollution

Pigs and chickens cannot utilize phosphate from phytate in their feed, which therefore ends up in their manure. By adding the enzyme phytase to their feed the amount of phosphate which is excreted by these animals can be reduced by more than 30 per cent. In South Africa bacteria are used for the isolation of gold from gold-ore. This so-called biomining saves an enormous amount of smelting energy and generates much less waste.

The chemical production of indigo, the dye which is used for blue jeans, takes eight steps, the use of very toxic chemicals and special protection measures for the process operators and the environment. The biotechnological production of indigo, which uses a genetically modified bacterium containing the right enzymes, takes only three steps, proceeds in water, uses simple raw materials like sugar and salts and generates only indigo, carbon dioxide and biomass which is biodegradable.

Through conventional breeding. Several of the previously mentioned examples like the amylose-free potato and the indigo-producing bacterium also involve the use of organisms genetically modified by recombinant DNA technology. Many enzymes are routinely produced by genetically modified organisms too.

Given the overwhelming diversity of species, biomolecules and metabolic pathways on this planet, genetic engineering can in principle be a very powerful tool in creating environmentally friendlier alternatives for products and processes that presently pollute the environment or exhaust its non-renewable resources. Politics, economics and society will ultimately determine which scientific possibilities will become reality.

Nowadays organisms can also be supplemented with additional genetic properties for the biodegradation of specific pollutants if naturally occurring organisms are not able to do that job properly or not quickly enough. By combining different metabolic abilities in the same micro-organism bottlenecks in environmental cleanup may be circumvented. Until now this has not been done on any significant scale. The main reason being the fact that in most cases naturally occurring organisms can be found or selected for, which are able to clean up a polluted site. Examples have been found where soil bacteria have developed new properties in response to the introduction of xenobiotics (that is, man-made chemicals that are normally not found in nature). In some cases they even appear to have acquired properties from other species. In the USA some genetically modified bacteria have been approved for bioremediation purposes but large scale applications have not yet been reported. In Europe only controlled field tests have been authorized.

Because new organisms can be created by genetic engineering that may never be produced by spontaneous or selection driven evolution, concerns exist about the unpredictability of their possible interactions with the eco-system. Genetically modified organisms which are properly kept within the confines of their approved production facilities are much less a concern than genetically modified organisms which are meant to be released into the environment like disease-resistant plants or soil bacteria for bioremediation.

The possible ecological effects of the latter are even more difficult to evaluate due to the fact that it is well known that soil bacteria frequently exchange genetic material (also between species). This together with the fact that we know little about the great majority of soil inhabiting bacterial species, makes it almost impossible to predict the fate of every DNA copy of a newly introduced genetic property in a soil bacterium. If the extra DNA is derived from another soil bacterium, it may on the other hand be reasonable to argue that the genetically modified bacterium might also have evolved spontaneously some day due to the frequent exchange of genetic material in the soil.

Legislation

Regulation to ensure safe application of novel or modified organisms in the environment is important, not least to maintain public confidence. The European Union has two Directives on the contained use of genetically modified micro-organisms, and on the deliberate release of genetically modified organisms into the environment. These have been implemented in the national legislation of most EU Member States. They require that a detailed experimental protocol, including assessment of potential risks, is approved by competent authorities before a genetically modified organism is released into the environment. The nature and sometimes even the site of the release has to be published in the local press in some countries. After several years' experience using the legislation, the procedures involved are now being revised. Ammendments to clarify and revise Directive 90/219/EEC were published in December 1998. The aim of the European Commission is to maintain the EU's competitiveness globally—both in research and commercial applications-without compromising safety.

Public Opinion, Dialogue and Debate

In spite of the fact that traditional biotechnology already is of great value to bioremediation and modern biotechnology may enhance this even further, there are no recent data on what Europeans specifically think about environmental biotechnology. Generally speaking, Europeans tend to take an "optimistic" view of the developments they expect from modern biotechnology, according to the most recent European Commission public opinion survey which was published in 1997. Unfortunately this survey did not investigate the attitude of the public towards environmental biotechnology. The only environmentally related question in this survey was whether people believed that modern biotechnology would substantially reduce environmental pollution, which 47 per cent did. Whether or not this is only wishful thinking remains to be determined. Ultimately, hard proof expessed in the form of improved environmental parameters will be needed for full acceptance of environmental biotechnology.

Conferences, public debates, seminars and round table meetings have been held to bring people from the public, government, environmental organisations, science and industry together to discuss critical issues. These lively debates do not always lead to consensus, but they can provide a fuller appreciation of all the aspects in a particular issue, facilitating a better understanding of the problems involved. A recent example is the workshop 'How can biotechnology benefit the environment'.

Public information aimed at advancing dialogue and debate is provided by many organisations. A compilation of these can be found in the handbook of information sources which has been published by and can be ordered from the EFB Taskgroup on Public perceptions of Biotechnology.

Conclusion

Environmental biotechnology has a career extending back into the last century.

As the need is better appreciated to move towards less destructive patterns of economic activity, while maintaining improvement of social conditions in spite of increasing population, the role of biotechnology grows as a tool for remediation and environmentally sensitive industry. Already, the technology has been proven in a number of areas and future developments promise to widen its scope. Some of the new techniques now under consideration make use of genetically modified organisms designed to deal efficiently with specific tasks. As with all situations where there is to be a release of new technology into the environment, concerns exist. There is a potential for biotechnology to make a further major contribution to protection and remediation of the environment. Hence biotechnology is well positioned to contribute to the development of a more sustainable society. As we move into the next millennium this will become even more vitally important as populations, urbanisation and industrialisation will continue to climb.

7.2 Environmental Biotechnology: Progess & Constraints

Environmental Biotechnology involves application of biotechnology i.e use of living organism to the management of environmental problems.

All environmental problems (pollution) are off anthropogenic origin and have deep socioeconomic and health effects on man and his crops. Indisouminate exploitation of the nature since the down of human civilization has resulted in large scale deterioration of the environment and poses threat to the existence of life itself. The scientists have been called upon and they have responded with diverse techniques of removal of the pollutants of the environment by the application of the general and biotechnological approaches. This has development into a new branch of biotechnology called bioremediation which includes cleanup operation by microorganisms or higher plants (Phytoremediation). Most environmental scientist are more interested in this aspect of environmental biotechnology.

But environmental biotechnology should be viewed in much wider perspective. Environmental pollution is consisting of two components: 1) Generation of pollution due to human developmental programme, and 2) Accumulation of the pollutant in the environment. Environmental biotechnology aims at both: 1) stoppage of further pollution, and 2) Cleanup of accumulated pollutants. Many scientist are ameliorating the polluted environment by bioremediation, but what about stoppage of generation of pollution? Can biotechnology be of any help in this regards?

Analysis of the attitudes of the politicians and scientist of advanced countries like USA, European Communities and Japan indicates glaring differences. Three different approaches of R & D programmes are 6 by these three advanced nations:

1. *Site-specific cleanup*: This is the focus of USA. USA is the highest GH gas emitting country of the world because of luxurious often wasteful life-style of the population. The country as a whole is reluctant to stop or reduced GH gas or other pollutant generation. During Johanesburg summit in 2002, President Bush was asked about reduction of GH gases especially CO2 and NXO but he declared that "business would be as usal"

   In USA federal and state governments however, enforced legally the cleanup of several highly contaminated sites that pose threats to human health, particularly where the pollutants are gradually seepling through soil into aquifers used for drinking water. EPA has listed more than 1200 locations called superfund sites. Classical examples are: 1) bioremediation of oil spells in ALSKA coast 2) Cleanup of chlorinated compounds especially TCE from the aquifers below the New York city and 3) removal of heavy metals from polluted sites by microbs or by plants like Brassica juncea (Phytoremediation).

2. *European community countries like Germany*: The Netherlands, Belgium and others foster advanced technology for waste and water treatment systems. The Netherlands is the most developed country in this regards, the biotechnology approaches of European countries are mostly concerned with:

   (i) Technology upgrading for waste recycling, waste sludge degradation and biogas (CH4) production are the main foci of R & D programmes. Development of anaerobic sludge digestion using a consortium of anaerobic bacteria that includes methanogenic archae bacteria is of significant help to biogas production from municipal and agro-waste in many developing countries.

   (ii) Removal of organic and inorganic compounds from municipal waste by aerobic bioreactors for preventing utrophication of the water bodies receiving the treated water. Before release of water the microbes are removed by passing through trickling filters, bioscrubbers, biofilms etc.

3. *Japanese approaches is more holistic*: The Japanese Ministry of International Trade and Industry (MITI) advocates long term

approach to environmental problems. Japan is more interested in broader issues like stoppage of GH emission at global scale and the aims are:

(i)   Replacement of drastic reduction of fossil fuels as energy source. R & D programmes are aimed at using microbes producing alternate energy source. The safest and most ecofriendly is hydrogen. The alternate energy sources like solar, wind and tidal forces are also being attempted. The stoppage of GH gas emission especially CO_2 will avert global climate change.

Hydrogen producing cyanobacteria have been patented by Japanese workers which can efficiently produced during biological nitrogen and is of interest because this is used as fuel of space rockets. It is liquid under high pressure and cryogenic conditions, but release hydrogen in normal pressure and temperature.

(ii) Removal of CO_2 emanated. Marine biotech lab of Kamiashi has isolated and patented algae which can convert CO_2 to carbohydrate 10 times faster than any green plants at 400C. Such algae could be grown in bioreactors through which the chimney discharged gas of the industries are passed and the CO_2 is removal. The carbohydrates thus formed is biodegradable and the CO_2 would be cycled back to air. Japan has developed microbes which would remove CO_2 from air and form polymeric polysaccharadies of diverse use.

(iii) Reversal of desert formation: MITI sponsored research has developed a super bioabsorbent from *Alcaligenes latis*, the polysaccharadie composed of glucose and glucuronic acid that can absorb and hold more than thousand times its own weight of water. This super absorbent as soil treatment can reverse desertification when applied to the edge of advancing deserts.

In India DBT and Department of Environment and Forest are sponsoring R & D programmes for development of eco-friendly and sustainable environment. the emphasis on is:

1.   Environmental monitoring through development of biosensors and membrane bioprobes for heavy metals, insecticides pesticides, radionucludes and other contaminants in soil and water bodies. Development of biosensors for BOD determination of water samples and detection of enteric bacteria and viruses in water samples using immunoprecipitations techniques and DNA probes are also encouraged.

2. Cleanup of contaminated soil and sub surface water by screening and development of GM microbes is an urgent need. We need heavy metal, pesticides organic and inorganic xenotrictic degradable microbes and plants for largescale bioremediation and phytoremediation.

3. Recycle and reuse of argo-waste or residues as secondary resources through use of selected, mutant or GM microbes for methane biosynthes and methylotrophs in coal seams paddy soils, wet lands. The cultivation of mushroom using huges rice straw residue of India is encouraged both by central and state governments.

4. Development of alternate agriculture practices to regulate the production of green house gasses and reduction of use of toxic agrochemicals.

5. Development of Geographical Information System (GIS) and appropriate remote sensing techniques for the study of biological diversity.

The Indian government and scientist are in the same line with the Japanese. The approaches are for stoppage of generation of pollution and remediation of accumulated pollutants. Innovative research by Indian Scientist is the need.

Biotransformation: An Useful Tool for Design and Detoxification of Organic Molecules by Microbes

Biotransformation essentially means modification of organic molecules into product of defined structure. Such modifications are possible with microbe, plant and animal cells or their enzyme preparation. However, microbes are considered to be most suitable because it furnishes i) both regio-and stereospecific product ii) the reaction can be run under comfortably controlled condition and iii) the product can be prepared by fermentation in quantity necessary for characterization and clinical trial. Biotransformation by microbes has been tried with a variety of substrates ranging from simple aliphatic hydrocarbon to macrocyclic compounds. However, one of the major applications of biotransformation has been in the area of steroid and antibiotics, in particular, penicillins, the magic molecules used to alleviate the suffering of mankind.

Among various microbial transformations tried with steroid used as substrate, hydroxylation seems to be most important. However, the remarkable potential of microbes in transformation of steroid had to await the development of antiarthritic steroids. In fact, understanding of therapeutic possibilities of cortisone did not begin to develop until 1949 when Hench and his associates announced successful use of cortisone 21-acetate as a palliative in rheumatoid arthritis, the crippling disease of mankind. The chemical synthesis of steroid, cortisone in patricular being costly, search for better alternative led to the development of microbial method. The discovery of Peterson and Murray for 11-oxygenation of steroid by *Rhizopus nigricans* in 1952 is still considered a milestone in the history of microbial transformation of steroid. Since then 11-oxygenation of steroid had been tried by microbes particularly by fungi and in all the cases, the 11-oxygenation was accompanied by 6α-hydroxylation and as a result the yield of 11-oxygenated steroid was affected considerably. It seems pertinent to mention that the presence of an oxygen function at C11 is obligatory for biological activity of steroids. With this background attempts were made to develop microbe for regio-and stereospecific hydroxylation exclusively at C11 in progesterone. Eventually a strain of *Aspergillus ochreceus* TS was isolated from local soil in our laboratory. The isolated strain, which later received the ATCC accession number MYA *Aspergillus ochraceus* TS-1, transformed progesterone exclusively to its 11α-hydroxy derivative both under *in vivo* and *in vitro* conditions and there was no side product formed under the conditions used. The enzyme responsible for the transformation was characterized as a cytochrome P450 linked monooxygenase which was resolved into a reductase, phosphotidyl choline and cytochrome P450, the terminal oxidase. The spores of the test organism were immobilized in different matrices by entrapment technique. The calcium alginate entrapped spores proved to be the best matrix and the insolubilized preparation retained considerable activity of free spores and the product pattern remained unaltered. On the other hand, there was complete change in product pattern when the transformation was tried with the immobilized spores under reduced water activity. Progesterone was converted to $\alpha1$-testosterone and androsta-1,4-dien-3,17-dione (ADD). Evidently there was a cleavage in C17-20 bond furnishing the above products. Thus water activity seems to be an important parameter in steroid transformation by microbes.

Since 1952 various reports are available on hydroxylation of steroid although very little is known about the role of substituents at a particular centre on hydroxylation of steroid by microbe. Attempts to direct hydroxylation to certain sites or to eliminate unwanted side product(s) are receiving attention in respect of design of newer steroids by biological means. It was Schneider, who reported for the first time about the influence of oxygen substituents at C-11, 17, 20 and 21 positions on the 6β-hydroxylation in steroid by *Rhizopus arrhizus*. Later it was reported from our laboratory that substituents at C11 have got good influence on 6β-hydroxylation of a group of closely related C21 steroids by *Syncephalastrum racemosum*. It was demonstrated that chirality at C11 inhibits the hydroxylation at C6 i.e. both the epimeric alcohols at C11 of progesterone remained unaltered while progesterone and 11-ketoprogesterone got hydroxylated at C6 under the conditions used. Moreover, it has been argued that the steric strain introduced due to the presence of a carbonyl function in ring C is transmiited to the B-ring and as a result 11-ketoprogesterone became a better substrate than progesterone itself. Similar distinction was established in transformation of progesterone analogs by a *Bacillus* sp. Again chirality at C11 has been proved by the us to be the key factor behind 14α-hydroxylation of steroids.

During biochemical studies with the P450 linked monooxygenase produced by *Aspergillus ochraceus* TS, the hydroxylating enzyme was found to be inducible in character like other microbial monooxygenases. Among the inducers tried for induction of the enzyme metabolizing progesterone, benzo[a]pyrene (BP), an environmetal pollutant was found to be a substrate-cum-inducer and it got metabolized by the test organism. Resolution of BP-metabolites by high pressure liquid chromatography (HPLC) exhibited the presence of phenols, quinones and dihydrodiols, a profile, which was exactly similar to that produced by mammalian enzyme system. It was interesting to note that the monooxygenase, produced by *Aspergillus ochraceus* TS which metabolizes progesterones distributed between microsomes and post microsomal supernatant. In contrast, in case of benzo[a]pyrene, the entire monooxygenase was found to be located in microsomes only. Moreover, the phase II enzyme, glutathione-S-transferase (GST) was also shown to be present in microsomes of *Aspergillus ochraceus* TS to reestablish the fact that the product of the phase I enzyme, P450 is the substrate of phase II enzyme, the GST. The GST produced by *Aspergillus ochraceus* TShough located in microsomes, shares the properties of both microsomal and cytosolic GSTs.

Apart from steroids, transformation was tried with penicillins. Indiscriminate use of penicillins by the doctors and users to combat infection all over the world during past several decades had led to the development of penicillin resistant strains. The penicillin resistance has been ascribed to three unrelated enzymes which includes ?-lactamase. It has been proved to be the key enzyme, which cleaves the lactam ring in penicillin and thereby makes the drug ineffective. Subsequently there was a two prong attack to handle penicillin resistance. Attempts were made by all concerned to develop good anti-beta-lactamase and produce newer penicllins, which can stand against ?-lactamase and combat infection. The second approach still proves to be good and as a result a number of new semi-synthetic penicillins were produced and used in chemotherapy. 6-Aminopenicillanic acid (6-APA) is the key intermediate used for the synthesis of newer penicillins. Chemical synthesis of 6-APA had been proved to be costly and its biological synthesis is 20 per cent cheaper and its 80 per cent requirement in industry is served by biological means. Penicillin amidase (PA) is the enzyme which can furnish 6-APA from penicillin G or penicillin V and penicillin G/ V from 6-APA and corresponding acyl donors by condensation. Considering high demand of newer penicillins as well as acute penicillin resistance, attempts were made in our laboratory to develop a β-lactamase-free penicillin amidase. In fact, a strain of *Alcaligenes* sp. was isolated adopting a novel strategy. The isolated strain has been shown to produce PA which is free from β-lactamase. The enzyme was found to be located in periplasmic space of the bacterium. Test of homogeneity of the purified enzyme on sodium dodecyl sulphate polyacrylamide gel electrophoresis (SDS-PAGE) showed it to be a heterodimer of Mr. 63 and 22 kDa respectively. The biochemical and biophysical studies of the enzyme showed that it is thermally stable and got tryptophan at its active site. In fact quenching of fluorescence with both acrylamide and potassium iodide was observed. Its N-terminal aminoacid analysis and allignment with the sequence of other PAs produced by microbes showed good homology pattern. Synthesis of newer β-lactams was achieved by immobilized periplasmic fraction using 6-APA and aromatic acyl donors by condensation in organic solvents. Bioassay of the enzymatically synthesized products against *S. aurius* and *S. typhi* showed positive activity. It was interesting to note the synthetic products were more active against gram negative than gram positive organisms.

Thus, biotransformation is a powerful tool which can be used for design and detoxification of both endogenous and exogenous substrates. The check and balance on mechanism of detoxification and activation will be discussed.

7.3 Environmental Engineering

Environmental engineering is the application of science and engineering principles to improve the environment (air, water, and/or land resources), to provide healthy water, air, and land for human habitation and for other organisms, and to remediate polluted sites. Environmental engineering involves water and air pollution control, recycling, waste disposal, and public health issues as well as a knowledge of environmental engineering law. It also includes studies on the environmental impact of proposed construction projects. Environmental engineers conduct hazardous-waste management studies to evaluate the significance of such hazards, advise on treatment and containment, and develop regulations to prevent mishaps. Environmental engineers also design municipal water supply and industrial wastewater treatment systems as well as being concerned with local and worldwide environmental issues such as the effects of acid rain, ozone depletion, water pollution and air pollution from automobile exhausts and industrial sources.

Educational Licensing Requirements

To become an environmental engineer, at least a Bachelor's degree in engineering (usually civil or chemical, and more frequently environmental engineering) is required, usually followed by specialized training at the Master's or Doctoral level. Unlike traditional Engineering disciplines like Civil Engineering, Electrical Engineering or Chemical Engineering, Environmental Engineering evolved as a branch of Chemical and Civil Engineering over the past several decades to deal with increasingly complex environmental challenges. At many universities, Environmental Engineering programmes follow either the Department of Civil Engineering or The Department of Chemical Engineering at Engineering faculties. Environmental "civil" engineers focus on hydrology, water resources management and water treatment plant design. Environmental "chemical" engineers, on the other hand, focus on environmental chemistry, advanced air and water treatment technologies and separation processes. Additionally, engineers are more frequently obtaining specialized training in law (J.D.) and are utilizing their technical expertise in the practices of Environmental engineering law. Most jurisdictions also impose licensing and registration requirements.

Development of Environmental Engineering

Ever since people first recognized that their health and well-being were related to the quality of their environment, they have applied thoughtful principles to attempt to improve the quality of their environment. The ancient Harappan civilization utilized early sewers in some cities. The Romans constructed aqueducts to prevent drought and to create a clean, healthful water supply for the metropolis of Rome. In the 15th century, Bavaria created laws restricting the development and degradation of alpine country that constituted the region's water supply.

Modern environmental engineering began in London in the mid-19th century when it was realized that proper sewerage could reduce the incidence of waterborne diseases such as cholera. The introduction of drinking water treatment and sewage treatment in industrialized countries reduced waterborne diseases from leading causes of death to rarities.

In many cases, as societies grew, actions that were intended to achieve benefits for those societies had longer-term impacts which reduced other environmental qualities. One example is the widespread application of DDT to control agricultural pests in the years following World War II. While the agricultural benefits were outstanding and crop yields increased dramatically, thus reducing world hunger substantially, and malaria was controlled better than it ever had been, numerous species were brought to the verge of extinction due to the impact of the DDT on their reproductive cycles. The story of DDT as vividly told in Rachel Carson's "Silent Spring" is considered to be the birth of the modern environmental movement and the development of the modern field of "environmental engineering."

Conservation movements and laws restricting public actions that would harm the environment have been developed by various societies for millennia. Notable examples are the laws decreeing the construction of sewers in London and Paris in the 19th century and the creation of the U.S. national park system in the early 20th century.

Briefly speaking, the main task of environmental engineering is to protect public health by protecting (from further degradation), preserving (the present condition of), and enhancing the environment.

Scope of Environmental Engineering

Pollutants may be chemical, biological, thermal, radioactive, or even mechanical. Environmental engineering emphasizes several areas: process engineering, environmental chemistry, water and sewage treatment (sanitary engineering), waste reduction/management, and pollution prevention/cleanup. Environmental engineering is a synthesis of various disciplines, incorporating elements from the following:

- ❖ Agricultural engineering
- ❖ Civil engineering
- ❖ Chemical engineering
- ❖ Public health
- ❖ Mechanical engineering
- ❖ Chemistry
- ❖ Biology
- ❖ Geology
- ❖ Ecology

Environmental engineering is the application of science and engineering principles to the environment. Some consider environmental engineering to include the development of sustainable processes. There are several divisions of the field of environmental engineering.

Environmental Impact Assessment and Mitigation

In this division, engineers and scientists assess the impacts of a proposed project on environmental conditions. They apply scientific and engineering principles to evaluate if there are likely to be any adverse impacts to water quality, air quality, habitat quality, flora and fauna, agricultural capacity, traffic impacts, social impacts, ecological impacts, noise impacts, visual (landscape) impacts, etc. If impacts are expected, they then develop mitigation measures to limit or prevent such impacts. An example of a mitigation measure would be the creation of wetlands in a nearby location to mitigate the filling in of wetlands necessary for a road development if it is not possible to reroute the road.

Water Supply and Treatment

Engineers and scientists work to secure water supplies for potable and agricultural use. They evaluate the water balance within a watershed and determine the available water supply, the water needed for various needs in that watershed, the seasonal cycles of water movement through the watershed and they develop systems to store, treat, and convey water for various uses. Water is treated to achieve water quality objectives for the end uses. In the case of potable water supply, water is treated to minimize risk of infectious disease transmittal, risk of non-infectious illness, and create a palatable water flavor. Water distribution systems are designed and built to provide adequate water pressure and flow rates to meet various end-user needs such as domestic use, fire suppression, and irrigation.

Wastewater Conveyance and Treatment

Most urban and many rural areas no longer discharge human waste directly to the land through outhouse, septic, and/or honey bucket systems, but rather deposit such waste into water and convey it from households via sewer systems. Engineers and scientists develop collection and treatment systems to carry this waste material away from where people live and produce the waste and discharge it into the environment. In developed countries, substantial resources are applied to the treatment and detoxification of this waste before it is discharged into a river, lake, or ocean system. Developing nations are striving to obtain the resources to develop such systems so that they can improve water quality in their surface waters and reduce the risk of water-borne infectious disease.

There are numerous wastewater treatment technologies. A wastewater treatment train can consist of a primary clarifier system to remove solid and floating materials, a secondary treatment system consisting of an aeration basin followed by flocculation and sedimentation or an activated sludge system and a secondary clarifier, a tertiary biological nitrogen removal system, and a final disinfection process. The aeration basin/activated sludge system removes organic material by growing bacteria (activated sludge). The secondary clarifier removes the activated sludge from the water. The tertiary system, although not always included due to costs, is becoming more prevalent to remove nitrogen and phosphorus and to disinfect the water before discharge to a surface water stream or ocean outfall.

Air Quality Management

Engineers apply scientific and engineering principles to the design of manufacturing and combustion processes to reduce air pollutant emissions to acceptable levels. Scrubbers, electrostatic precipitators, catalytic converters, and various other processes are utilized to remove particulate matter, nitrogen oxides, sulfur oxides, volatile organic compounds (VOC), reactive organic gases (ROG) and other air pollutants from flue gases and other sources prior to allowing their emission to the atmosphere. Scientists have developed air pollution dispersion models to evaluate the concentration of a pollutant at a receptor or the impact on overall air quality from vehicle exhausts and industrial flue gas stack emissions. To some extent, this field overlaps the desire to decrease carbon dioxide and other greenhouse gas emissions from combustion processes.

Other Applications

- ❖ Contaminated land management and site remediation,
- ❖ Risk assessment,
- ❖ Environmental policy and regulation development,
- ❖ Solid waste management,
- ❖ Hazardous waste management,
- ❖ Environmental health and safety,
- ❖ Natural resource management, and
- ❖ Noise pollution.

7.4 Society of Environmental Engineers

The Society of Environmental Engineers (SEE) is a British professional engineering institution founded in 1959. It is licenced by the Engineering Council UK to assess candidates for inclusion on ECUK's Register of Professional Engineers and Technicians. The Society publishes the quarterly journal "Environmental Engineering", as well as a regular newsletter. Its address is The Manor House, High Street, Buntingford, Herts, SG9 9PL, United Kingdom.

7.5 Environmental Engineering Law

Environmental engineering law is a profession that requires an expertise in both environmental engineering and law. This field includes professionals with both a legal and environmental engineering education. This dual educational requirement is typically satisfied through an ABET accredited degree in environmental engineering and an ABA accredited law degree. Likewise, this profession requires both licensure in professional environmental engineering and admittance to one bar.

Environmental engineering law is the professional application of law, science and engineering principles to improve the environment (air, water, and/or land resources), to provide healthy water, air, and land for human habitation and for other organisms, and to remediate polluted sites. Environmental engineering lawyers seek to promote the advancement of technical engineering knowledge in the legal profession and to enhance informed legal analysis of complex environmental matters.

Practice Areas

Environmental engineering law professionals offer a sound knowledge base in the fields of both environmental engineering and law to address complex environmental problems which demand both professional technical practice and legal expertise. Areas of practice are continually expanding, but frequently include complex land transactions, such as:

- ❖ Brownfields redevelopment,
- ❖ Asbestos Baseline Survey and Building Revaluation due to forthcoming Asbestos Abatements,
- ❖ Soil Contamination Assessment & Remediation, the development of a Remedial Action WorkPlan (RAWP) and Engineering Controls, including an Environmental Land Use Restriction (ELUR), and
- ❖ TMDL Nutrient Loading Studies (ex. for NPDES Wastewater Discharges) and regulatory negotiation of Nutrients Discharge Limits from waste treatment plants, such as phosphorus and nitrogen.

Air Pollution and Its Control

8.1 Air Quality Index

The *Air Quality Index* (AQI) is a standardized indicator of the air quality in a given location. It measures mainly ground-level ozone and particulates (except the pollen count), but may also include sulfur dioxide, and nitrogen dioxide. Various agencies around the world measure such indices, though definitions may change between places. The United States Environmental Protection Agency (EPA) and the Meteorological Service of Canada (MSC) differ on what AQI structure and health classification is used:

Health classifications used by the EPA:

- 0-50 Good is usually green
- 51-100 Moderate is usually yellow
- 101-150 Unhealthy for sensitive groups is usually orange
- 151-200 Unhealthy is usually red
- 201-300 Very unhealthy is usually purple
- 301-500 Hazardous is usually maroon

The EPA's AQI 100 corresponds to 0.08 ppm ozone, and to other levels for other pollutants. Source: EPA

The AQI standards in Canada are relatively more stringent. The current health classifications used by the Meteorological Service of Canada (MSC) are as follows:

- 0-25*: Good (green)
- 26*-50: Fair (yellow)
- 51-100: Poor (orange)
- 101+: Very poor (red)

In Ontario, 31 is the upper limit for good and 32 the lower limit for moderate. Zero to 15 is classified as very good, and is given the colour blue. In June 2007, the EPA proposed a slight possible tightening of the pollution standards associated with smog after an independent EPA scientific board said that the standard "needs to be substantially reduced" and that there is "no scientific justification" for retaining the current, weaker standard. In light of the new scientific findings, one should expect adjustments in the AQI such that pollution currently denoted as "moderate" will in the future be recognized as "unhealthy." The AQI can worsen (go up) due to lack of dilution with fresh air. Stagnant air, often caused by an or temperature inversion, or other lack of winds lets air pollution remain in a local area. On these days, the news media may ask the public to carpool or use public transport, or take other air pollution prevention measures such as teleworking.

Other Indices

Hong Kong

The Air Pollution Index (API) levels for Hong Kong are related to the measured concentrations of ambient respirable suspended particulate (RSP), sulfur dioxide (SO_2), carbon monoxide (CO), ozone (O_3) and nitrogen dioxide (NO_2) over a 24-hour period based on the potential health effects of air pollutants. An API level at or below 100 means that the pollutant levels are in the satisfactory range over 24 hour period and pose no acute or immediate health effects. However, air pollution consistently at "High" levels (API of 51 to 100) in a year may mean that the annual Hong Kong "Air Quality Objectives" for protecting long-term health effects could be violated. Therefore, chronic health effects may be observed if one is persistently exposed to an API of 51 to 100 for a long time.

API	Air Pollution Level	Health Implications
0-25	Low	Not expected.
26-50	Medium	Not expected for the general population.
51-100	High	Acute health effects are not expected but chronic effects may be observed if one is persistently exposed to such levels.
100-200	Very High	People with existing heart or respiratory illnesses may notice mild aggravation of their health conditions. Generally healthy individuals may also notice some discomfort.
201-500	Severe	People with existing heart or respiratory illnesses may experience significant aggravation of their symptoms. There may also be widespread symptoms in the healthy population (e.g. eye irritation, wheezing, coughing, phlegm and sore throats).

"Very High" levels (API in excess of 100) means that levels of one or more pollutant(s) is/are in the unhealthy range. The Hong Kong Environmental Protection Department provides advice to the public regarding precautionary actions to take for such levels.

Malaysia

The air quality in Malaysia is reported as the API or Air Pollution Index. Four of the index's pollutant components (i.e., carbon monoxide, ozone, nitrogen dioxide and sulfur dioxide) are reported in ppmv but particulate matter is reported in ìg/m^3. Unlike the American AQI, the index number can exceed 500. Above 500, a state of emergency is declared in the reporting area. Usually, this means that non-essential government services are suspended, and all ports in the affected area closed. There may also be a prohibition on private sector commercial and industrial activities in the reporting area excluding the food sector.

Singapore

Singapore uses the Pollutant Standards Index to report on its air quality.

United Kingdom

The Met Office of the United Kingdom (UK) issues air quality forecasts wherein the level of pollution is described either as an index (ranging from 1 to 10) or as a banding (low, moderate, high or very high). These levels are based on the health effects of each pollutant.

Index	Banding	Health Effect
1-3	Low	Effects are unlikely to be noticed even by individuals who know they are sensitive to air pollutants.
4-6	Moderate	Mild effects, unlikely to require action, may be noticed amongst sensitive individuals.
7-9	High	Significant effects may be noticed by sensitive individuals and action to avoid or reduce these effects may be needed (e.g. reducing exposure by spending less time in polluted areas outdoors). Asthmatics will find that their 'reliever' inhaler is likely to reverse the effects on the lung.
10	Very High	The effects on sensitive individuals described for 'High' levels of pollution may worsen.

The forecast is produced for a number of different pollutants and their typical health effects are shown in the following table.

Pollutant	Health Effects at High Level
Nitrogen dioxide Ozone	These gases irritate the airways of the lungs, increasing the symptoms of those suffering from lung diseases.
Sulphur dioxide Particulates	Fine particles can be carried deep into the lungs where they can cause inflammation and a worsening of heart and lung diseases

United States

The United States Environment Protection Agency (USEPA) developed the Pollutant Standards Index (PSI) to provide accurate, timely and easily understandable information about daily levels of air pollution. It is no longer in use, having been replaced by the AQI, which is more sensitive. For example, particulate matter with an aerodynamic diameter less than 2.5 micrometres (PM2.5) is a sub index, replacing the less sensitive PM10 component of the PSI.

Notes:

In the context of this article about air quality:
- ❖ ppmv = parts per million by volume = volume of pollutant gas per million volumes of ambient air
- ❖ PM_{10} = particulate matter smaller than 10 ìm in diameter
- ❖ µg/m³ = micrograms per cubic metre of ambient air
- ❖ µm = micrometre

The air quality in the United States has improved dramatically over 23 years.

Air quality by Country or Region

- ❖ British Columbia
- ❖ Hong Kong

8.2 National Ambient Air Quality Standards

The National Ambient Air Quality Standards (NAAQS) are standards established by the United States Environmental Protection Agency that apply for outdoor air throughout the country. Primary standards are designed to protect human health, with an adequate margin of safety, including sensitive populations such as children, the elderly, and individuals suffering from respiratory disease. Secondary standards are designed to protect public welfare from any known or anticipated adverse effects of a pollutant (e.g. building facades, visibility, crops, and domestic animals).

NAAQS requires the EPA to set standards on six criteria air contaminants:

1. Ozone (O_3),
2. Particulate Matter:

   ❖ PM_{10}, coarse particles: 2.5 micrometers (ìm) to 10 ìm in size (although current implementation includes all particles 10 ìg or less in the standard),
   ❖ $PM_{2.5}$, fine particles: 2.5 ìm in size or less,

3. Carbon monoxide (CO),
4. Sulfur dioxide (SO_2),
5. Nitrogen oxides (NO_x),
6. Lead (Pb).

Standards

The standards are listed in Title 40 of the Code of Federal Regulations Part 50.

Pollutant	Type	Standard	Averaging Timea	Regulatory Citation
SO_2	Primary	0.14 ppm (365 ig/m3)	24-hour	40 CFR 50.4(b)
SO_2	Primary	0.030 ppm (80 ig/m³)	annual	40 CFR 50.4(a)
SO_2	Secondary	0.5 ppm (1,300 ig/m³)	3-hour	40 CFR 50.5(a)
PM_{10}	Primary and Secondary	150 ig/m³	24-hour	40 CFR 50.6(a)
$PM_{2.5}$	Primary and Secondary	35 ig/m³	24-hour	40 CFR 50.7(a)
$PM_{2.5}$	Primary and Secondary	15 ig/m³	annual	40 CFR 50.7(a)
CO	Primary	35 ppm (40 mg/m³)	1-hour	40 CFR 50.8(a)(2)
CO	Primary	9 ppm (10 mg/m³)	8-hour	40 CFR 50.8(a)(1)
O_3	Primary and Secondary	0.12 ppm (235 ig/m³)	1-hourb	40 CFR 50.9(a)
O_3	Primary and Secondary	0.075 ppm (235 ig/m³)	8-hour	40 CFR 50.10(a)
NO^2	Primary and Secondary	0.053 ppm (100 ig/m³)	annual	40 CFR 50.11(a) and (b)
Pb	Primary and Secondary	and 1.5 ig/m³	quarterly	40 CFR 50.12

Note: (*a*) Each standard has its own criteria for how many times it may be exceeded, in some cases using a three year average.

(*b*) As of June 15, 2005, the 1-hour ozone standard no longer applies to areas designated with respect to the 8-hour ozone standard (which includes most of the United States, except for portions of 10 states).

8.3 Scrubber

Scrubber systems are a diverse group of air pollution control devices that can be used to remove particulates and/or gases from industrial exhaust streams. Traditionally, the term "scrubber" has referred to pollution control devices that used liquid to "scrub" unwanted pollutants from a gas stream. Recently, the term is also used to describe systems that inject a dry reagent or slurry into a dirty exhaust stream to "scrub out" acid gases. Scrubbers are one of the primary devices that control gaseous emissions, especially acid gases.

Removal and Neutralization

The exhaust gases of combustion may at times contain substances considered harmful to the environment, and it is the job of the scrubber to either remove those substances from the exhaust gas stream, or to neutralize those substances so that they cannot do any harm once emitted into the environment as part of the exhaust gas stream.

Wet Scrubbing

A is used to clean air or other gases of various pollutants and dust particles. Wet scrubbing works via the contact of target compounds or particulate matter with the scrubbing solution. Solutions may simply be water (for dust) or complex solutions of reagents that specifically target certain compounds.

Removal efficiency of pollutants is improved by increasing residence time in the scrubber or by the increase of surface area of the scrubber solution by the use of a spray nozzle, packed towers or an aspirator. Wet scrubbers will often significantly increase the proportion of water in waste gases of industrial processes which can be seen in a stack plume. Typical wet scrubber Compliance agencies typically place minimum DP thresholds on wet scrubber.

Dry Scrubbing

A dry or semi-dry scrubbing system, unlike the wet scrubber, does not saturate the flue gas stream that is being treated with moisture. In some cases no moisture is added; while in other designs only the amount of moisture that can be evaporated in the flue gas without condensing is added. Therefore, dry scrubbers do not have a stack steam plume or wastewater handling/ disposal requirements. Dry scrubbing systems are used to remove acid gases (such as SO_2 and HCl) primarily from combustion sources. There are a number of dry type scrubbing system designs. However, all consist of two main sections or devices: a device to introduce the acid gas sorbent material into the gas stream and a particulate matter control device to remove reaction products, excess sorbent material as well as any particulate matter already in the flue gas.

Dry scrubbing systems can be categorized as dry sorbent injectors (DSIs) or as spray dryer absorbers (SDAs). Spray dryer absorbers are also called semi-dry scrubbers or spray dryers.

Dry sorbent injection involves the addition of an alkaline material (usually hydrated lime or soda ash) into the gas stream to react with the acid gases. The sorbent can be injected directly into several different locations: the combustion process, the flue gas duct (ahead of the particulate control device), or an open reaction chamber (if one exists). The acid gases react with the alkaline sorbents to form solid salts which are removed in the particulate control device. These simple systems can achieve only limited acid gas (SO_2 and HCl) removal efficiencies. Higher collection efficiencies can be achieved by increasing the flue gas humidity (i.e., cooling using water spray). These devices have been used on medical waste incinerators and a few municipal waste combustors.

In *spray dryer absorbers*, the flue gases are introduced into an absorbing tower (dryer) where the gases are contacted with a finely atomized alkaline slurry. Acid gases are absorbed by the slurry mixture and react to form solid salts which are removed by the particulate control device. The heat of the flue gas is used to evaporate all the water droplets, leaving a non-saturated flue gas to exit the absorber tower. Spray dryers are capable of achieving high (80+%) acid gas removal efficiencies. These devices have been used on industrial and utility boilers and municipal waste combustors.

Mercury Removal

Mercury has no known beneficial uses in nature, but it is a common substance found in coal that must also be removed. Wet scrubbers are only effective for mercury removal under certain conditions. Mercury vapor in its elemental form, Hg^0, is insoluble in the scrubber slurry and not removed. Oxidized mercury, Hg^{2+}, compounds are more soluble in the scrubber slurry and can be captured. The type of coal burned as well as the presence of a selective catalytic reduction unit both affect the ratio of elemental to oxidized mercury in the flue gas and thus the degree to which the mercury is removed.

Scrubber Waste Products

One side effect of scrubbing is that the process only moves the unwanted substance from the exhaust gases into a solid paste or powder form. If there is no useful purpose for this solid waste, it must be either contained or buried to prevent environmental contamination. Limestone-based scrubbers can produce a synthetic gypsum of sufficient quality that can be used to manufacture drywall and other industrial products. Mercury removal results in a waste product that either needs further processing to extract the raw mercury, or must be buried in a special hazardous wastes landfill that prevents the mercury from seeping out into the environment.

Bacteria Spread

Until recently, scrubbers have not been associated with health risks involving bacteria spread as a result of inadequate cleaning, unlike other devices such as cooling towers. However, a 2005 outbreak of Legionnaires' disease in Norway was proven to emanate from a scrubber, causing ten deaths and more than fifty cases of infection as it spread the bacteria through the air during a period of only two weeks. This particular system had undergone regular cleaning routines every three weeks. This case is believed to be the first documented case of a scrubber being the source of such bacteria spread.

8.4 Air Pollution Control Equipment

Aircon Systems: Case Study

Aircon Systems: India is a manufacturer and exporter in India of air pollution control equipment, centrifugal air blowers, industrial scrubber, waste water treament plant and effluent treatment plant.

Aircon Systems has a technological commitment to sound and air pollution control. We offer a comprehensive range of products that include:

Products

Centrifugal Air Blowers/Exhauster (Single/Multi Stage)

- ❖ Capacity: 500 to 1,00,000CMH
- ❖ Pressure: 50 to 2,000mm wg.
- ❖ Material of construction: MS, SS, PPGL and MS Rubber lined.

Industrial Air Pollution Control Equipment

- ❖ System Offered: Multi Clones, Bag Filter (Pulse Jet), Industrial Scrubber, Venturi Scrubber.

❖ *Applications*: Drugs & Pharmaceuticals, Cement, Steel Plant, Foundries, Chemical Distilleries, Rice Mill and other allied industries.

Treatment Plant

❖ We offer turnkey solutions for waste water/effluent treatment as per the norms specified by Pollution Board.

Acoustic Enclosure

❖ We manufacture acoustic hoods to reduce noise level to meet standards for generator/process.

Pneumatic Conveying Systems

❖ We manufacture vacuum, pressure cum vacuum conveying system to convey powdery material.

Cooling System

❖ We manufacture air washer, fans and complete cooling system.

Other Products

❖ Axial Flow Fan, Industrial Man-cooler, Roof Extractor
❖ Flash Dryer
❖ Briquetting Machine
❖ Cooling Tower

Industiral Scrbbers

Aircon Systems is a manufacturer and exporter of custom built industrial scrubbers and systems for air pollution control.

Aircon Venturi Scrubbers

Aircon Venturi Scrubbers is designed to remove both plus- micron and sub-micron particulate from various effluent gases, where extremely high collection efficiency is required.

Aircon Wet Scrubbers

The Aircon System wet scrubbers utilizes a variety of techniques to achieve high efficiency contaminant removal. Our strength lies in our ability to integrate these products into a solution tailored to meet the most exacting requirements.

AIRCON Packed Bed/Fume Scrubbers

AIRCON Packed Bed/Fume Scrubbers are Primarily designed to absorb gases In cross and counter flow configuration.

FRP Scrubbers

Efficiencies of 99.9 per cent are achievable.

Multi-channel Bed Scrubbers (MCBS)

The Multi-Channel Bed Scrubbers can continuously scrub sticky particulate and toxic gases without clogging. The low maintenance design is adjustable to allow changes in turbulent layer depth and pressure drop for optimal efficiency.

Aircon Voc/Odor Control (Carbon Adsorber)

Industrial Activated Carbon Adsorbers eliminate volatile organic compounds (VOCs) and odors from industrial, municipal and commercial process exhaust streams. Activated carbon attracts and holds organic molecules, resulting in high efficiency removal and elimination of toxic or noxious discharge to the atmosphere.

Bag Filters (Pulse Jet, Shaker, Fine Dust Filter)

Aircon Systems is a manufacturer and exporter of Bag Filters including Pulse Jet Jet Filter, Fine Dust Filter, Shaker Dust Collector, Product Recovery Filter Collector.

Aircon Pulse Jet Filter

Aircon Systems supplies Pulse Air type BAG HOUSE FILTER for Industrial Utility application. Since 1987 Aircon has supplied 60 such units for a variety of Industries.

Aircon Fine Dust Filter

The Fine Dust Collector combines the coarse filteration of a Cyclone with the high efficiency of Pulse Jet Bag Filter.

- ❖ *Aircon Shaker Dust Collector*
- ❖ *Aircon Filter*
- ❖ *Product Recovery Filter Collector*

Dust Collector and Suppressor

Aircon Systems is a manufacturer and exporter in India of Dust Collector and Suppressor Equipment.

Dust Suppression Systems/Dust Collection Systems

Chem-jet Dust Suppression System is a chemical i.e., Compound MST mixed water-based and is the most reliable and cost-effective method of dust control in bulk material handling plants.

Industrial Vacuum Cleaner Fume/Smoke Collection Systems Multi-Clone

Industrial waste/dust/particles/Smoke/fume is created by processes such as welding, scrubber and plastic processing, high speed machining with coolants, tempering and quenching. From these processes, both wet and dry smoke are created.

Dust Collector and Suppressor

Aircon Systems is a manufacturer and exporter in India of Dust Collector and Suppressor Equipment

Dust Suppression Systems/Dust Collection Systems

Chem-jet Dust Suppression System is a chemical i.e., Compound MST mixed water-based and is the most reliable and cost-effective method of dust control in bulk material handling plants.

Industrial Vacuum Cleaner Fume/Smoke Collection Systems Multi-Clone

Industrial waste/dust/particles/Smoke/fume is created by processes such as welding, scrubber and plastic processing, high speed machining with coolants, tempering and quenching.From these processes, both wet and dry smoke are created.

Pneumatic Conveying System

Aircon Systems is a manufacturer and exporter of Pneumatic Conveying System including Lean Phase Pneumatic Conveying System, Dense Pneumatic Conveying System, Air Handling System, Truck Loading System, Central Vacuum System.

Evaporative Cooling System

How an Evaporative Cooling System Works

An evaporative cooling produces effective cooling by combining a natural process—water evaporation—with a simple, reliable air-moving system. Fresh outside air is filtered through the saturated evaporative media, cooled by evaporation, and circulated by a blower.

Cooling Temperatures

An AIR WASHER will nearly always deliver air cooler than 80 degrees F.

Energy Savings

An evaporative cooling System consumes only one-fourth of the electrical energy required to operate a refrigerated air conditioning unit.

Health Benefits

With evaporative cooling, a complete air change occurs every one-to-three minute. This offers a greathealth advantage over traditional refrigerated air conditioning, which employs a complicated "closed"system that recirculates the same stale dry air over and over.Constant cool air movement pushes heat out—along with stale air, smoke, odors and pollution.The high volume of fresh, cool air produced by the evaporative cooler helps your body ventilate naturally .Evaporative cooling helps maintain natural humidity level.

Industrial Wastewater Treatment Systems

Physical chemical are techniques to remove the coarse fraction.

Oil, fatty acids and suspended solids could be removed by the use of the following techniques:

- ❖ Screens
- ❖ Coaqulation
- ❖ Flocculation
- ❖ Centrifugation
- ❖ Fluidization
- ❖ Settling
- ❖ Precipittion

Aircon manufactures high efficiency clarifiers for precipitates and other non-biological solids settling applications. PVC structured fill media create a high settling area compared to the floor space occupied by the clarifier by 5.5 mtr^2/m^2 for 500 mm deep module. Aircon clarifiers have no moving parts and virtually no wear. Few solids will adhere to this material which is virtually impervious to chemical attack.

General features include:

- ❖ Flows from 100 to over 25,000 LPH,
- ❖ Standard units only 10' overall height,
- ❖ Enhances The Particle agglomeration growth, and
- ❖ Exclusive surface skimming feature.

Centrifugal Air Blowers

Aircon Systems is a manufacturer and exporter in India of Centrifugal Air Blowers including High efficiency Centrifugal Single, Multi stage Blowers & Exhausters, ID Fan & FD Fan, DIDW Fan Centrifugal Blower, Centrifugal Exhauster, Multi Stage, Axial Flow Fan, Man Cooler, Rotary Air Lock.

High efficiency Centrifugal Single, Multi stage Blowers & Exhausters:

- ❖ ID Fan & FD Fan
- ❖ DIDW Fan Centrifugal Blower
- ❖ Centrifugal Exhauster
- ❖ Multi Stage
- ❖ Axial Flow Fan
- ❖ Man Cooler
- ❖ Rotary Air Lock

Acoustic Hood For Dg Sets, Flash Dryer

Aircon Systems is a manufacturer and exporter in India of ACOUSTIC HOOD FOR DG SETS, FLASH DRYER.

We offer Acoustic Hoods designed for quick assembly and dismantling for meeting critical requirements of noise suppression of DG Sets.

The salient features of our Acoustic Canopy are as under:

- Enough space to house panel and Fuel-Tank inside canopy (optional)
- Heavy-duty sturdy steel base frame for D.G. Set.
- Provision of Air-Intake and Air-Exhaust silencer(s) for best results
- Open able doors for easy access to virtually every part of D.G. Set for comfortable maintenance.
- We provide

  (a) st Layer of Tissue Paper.
  (b) Rock-Wool 64 Kg/m 350 mm Thk.
  (c) Total thickness of panel-100mm High Performance most comfortable Sound Reduction level to 70-75 db at a distance of 3 meters.

- Axial Fan(s) to supply air.
- Separate Duct for ventilating hot air coming out of engine (In case of Air Cooled Engines)/Radiator duct for water cooled engines.
- Weather proof structure that can also be erected in open space, totally water proof.
- Anti-Corrosion treatment to the full structure and body for long-life under adverse weather conditions.
- 2 coats of Epoxy primer followed by 2 coats of Enamel Paint.

Flash Dryer

Flash Dryer is a pneumatic system primarily used to dry products requiring the removal of free moisture. Drying takes place in a matter of seconds. Wet material is dispersed into a stream of heated air (or gas) which conveys it through a drying duct.

Using the heat from the air stream, the material dries as it is conveyed. Product is separated using cyclones, and/or bag filters. Typically, cyclones are followed by scrubbers or bag filters for final cleaning of the exhaust gases to meet current emission requirements.

Elevated drying temperatures can be used with many products since the flashing off of surface moisture instantly cools the drying gas without appreciably increasing the product temperature.

For even greater thermal efficiency and where inertisation is required, recycling of exhaust gases can be used. This can be implemented on all our air stream drying systems and retrofitted on customer's existing drying operations.

Water Pollution and Its Control

9.1 Strategic Alliance for Freshwater Information, Resources and Education (SAFFIRE)

Water pollution is a complex issue—from the source of poluution to its impacts, both short-term and long-term on human and other species that depend on water. Point and non-point sources of pollution have a profound impact on the degree of pollution, and many times the actual causes are hidden behind more 'visible' causes. Understanding these causes-behind-causes is critical in developing appropriate responses to reduce pollution. Broad-based awareness of the sources and impacts of pollution—involving a number of stakeholders on the water continuum—is also important to effect lasting solutions.

Resources specific to river pollution are aslo included—River pollution is a result of a complex combination of processes that reduce overall river water quality. Acid rain, industrial pollution, agricultural pollution contribute to river pollution, but so do everyday activities that drain untreated pollutants and leachate into rivers and streams. Transportation has a role to play too—where carbon and one-drop-at-a-time 'oil spills' can also cause pollution through storm run-off. A holistic and integrative understanding of the cause-effect cycles of river pollution is an effective starting point to improve river water quality.

9.2 Water Pollution and Society

Comprising over 70 per cent of the Earth's surface, water is undoubtedly the most precious natural resource that exists on our planet. Without the seemingly invaluable compound comprised of hydrogen and oxygen, life on Earth would be non-existent: it is essential for everything on our planet to grow and prosper. Although we as humans recognize this fact, we disregard it by polluting our rivers, lakes, and oceans. Subsequently, we are slowly but surely harming our planet to the point where organisms are dying at a very alarming rate. In addition to innocent organisms dying off, our drinking water has become greatly affected as is our ability to use water for recreational purposes. In order to combat water pollution, we must understand the problems and become part of the solution.

Point and Nonpoint Sources

According to the American College Dictionary, pollution is defined as: 'to make foul or unclean; dirty.' Water pollution occurs when a body of water is adversely affected due to the addition of large amounts of materials to the water. When it is unfit for its intended use, water is considered polluted. Two types of water pollutants exist; point source and nonpoint source. Point sources of pollution occur when harmful substances are emitted directly into a body of water. The Exxon Valdez oil spill best illustrates a point source water pollution. A nonpoint source delivers pollutants indirectly through environmental changes. An example of this type of water pollution is when fertilizer from a field is carried into a stream by rain, in the form of run-off which in turn effects aquatic life. The technology exists for point sources of pollution to be monitored and regulated, although political factors may complicate matters. Nonpoint sources are much more difficult to control. Pollution arising from non-point sources accounts for a majority of the contaminants in streams and lakes.

Causes of Pollution

Many causes of pollution including sewage and fertilizers contain nutrients such as nitrates and phosphates. In excess levels, nutrients over stimulate the growth of aquatic plants and algae. Excessive growth of these types of organisms consequently clogs our waterways, use up dissolved oxygen as they decompose, and block light to deeper waters. This, in turn, proves very harmful to aquatic organisms as it affects the respiration ability or fish and other invertebrates that reside in water.

Pollution is also caused when silt and other suspended solids, such as soil, washoff plowed fields, construction and logging sites, urban areas, and eroded river banks when it rains. Under natural conditions, lakes, rivers, and other water bodies undergo Eutrophication, an aging process that slowly fills in the water body with sediment and organic matter. When these sediments enter various bodies of water, fish respirationbecomes impaired, plant productivity and water depth become reduced, and aquatic organisms and their environments become suffocated. Pollution in the form of organic material enters waterways in many different forms as sewage, as leaves and grass clippings, or as runoff from livestock feedlots and pastures. When natural bacteria and protozoan in the water break down this organic material, they begin to use up the oxygen dissolved in the water. Many types of fish and bottom-dwelling animals cannot survive when levels of dissolved oxygen drop below two to five parts per million. When this occurs, it kills aquatic organisms in large numbers which leads to disruptions in the food chain.

The pollution of rivers and streams with chemical contaminants has become one of the most crutial environmental problems within the 20th century. Waterborne chemical pollution entering rivers and streams cause tramendous amounts of destruction.

Pathogens are another type of pollution that prove very harmful. They can cause many illnesses that range from typhoid and dysentery to minor respiratory and skin diseases. Pathogens include such organisms as bacteria, viruses, and protozoan. These pollutants enter waterways through untreated sewage, storm drains, septic tanks, runoff from farms, and particularly boats that dump sewage. Though microscopic, these pollutants have a tremendous effect evidenced by their ability to cause sickness.

Additional Forms of Water Pollution

Three last forms of water pollution exist in the forms of petroleum, radioactive substances, and heat. Petroleum often pollutes waterbodies in the form of oil, resulting from oil spills. The previously mentioned Exxon Valdez is an example of this type of water pollution. These large-scale accidental discharges of petroleum are an important cause of pollution along shore lines. Besides the supertankers, off-shore drilling operations contribute a large share of pollution. One estimate is that one ton of oil is spilled for every million tons of oil transported. This is equal to about 0.0001 percent. Radioactive substances are produced in the form of waste from nuclear power plants, and from the industrial, medical, and scientific use of radioactive materials. Specific forms of waste are uranium and thorium mining and refining. The last form of water pollution is heat. Heat is a pollutant because increased temperatures result in the deaths of many aquatic organisms. These decreases in temperatures are caused when a discharge of cooling water by factories and power plants occurs.

Oil pollution is a growing problem, particularly devestating to coastal wildlife. Small quantities of oil spread rapidly across long distances to form deadly oil slicks. In this picture, demonstrators with "oil-covered" plastic animals protest a potential drilling project in Key Largo, Florida. Whether or not accidental spills occur during the project, its impact on the delicate marine ecosystem of the coral reefs could be devastating.

Workers use special nets to clean up a California beach after an oil tanker spill. Tanker spills are an increasing environmental problem because once oil has spilled, it is virtually impossible to completely remove or contain it. Even small amounts spread rapidly across large areas of water. Because oil and water do not mix, the oil floats on the water and then washes up on broad expanses of shoreline. Attempts to chemically treat or sink the oil may further disrupt marine and beach ecosystems.

Classifying Water Pollution

The major sources of water pollution can be classified as municipal, industrial, and agricultural. Municipal water pollution consists of waste water from homes and commercial establishments. For many years, the main goal of treating municipal wastewater was simply to reduce its content of suspended solids, oxygen-demanding materials, dissolved inorganic compounds, and harmful bacteria. In recent years, however, more stress has been placed on improving means of disposal of the solid residues from the municipal treatment processes. The basic methods of treating municipal wastewater fall into three stages: primary treatment, including grit removal, screening, grinding, and sedimentation; secondary treatment, which entails oxidation of dissolved organic matter by means of using biologically active sludge, which is then filtered off; and tertiary treatment, in which advanced biological methods of nitrogen removal and chemical and physical methods such as granular filtration and activated carbon absorption are employed. The handling and disposal of solid residues can account for 25 to 50 percent of the capital and operational costs of a treatment plant. The characteristics of industrial waste waters can differ considerably both within and among industries. The impact of industrial discharges depends not only on their collective characteristics, such as biochemical oxygen demand and the amount of suspended solids, but also on their content of specific inorganic and organic substances. Three options are available in controlling industrial wastewater. Control can take place at the point of generation in the plant; wastewater can be pretreated for discharge to municipal treatment sources; or wastewater can be treated completely at the plant and either reused or discharged directly into receiving waters.

Raw sewage includes waste from sinks, toilets, and industrial processes. Treatment of the sewage is required before it can be safely buried, used, or released back into local water systems. In a treatment plant, the waste is passed through a series of screens, chambers, and chemical processes to reduce its bulk and toxicity. The three general phases of treatment are primary, secondary, and tertiary. During primary treatment, a large percentage of the suspended solids and inorganic material is removed from the sewage. The focus of secondary treatment is reducing organic material by accelerating natural biological processes. Tertiary treatment is necessary when the water will be reused; 99 percent of solids are removed and various chemical processes are used to ensure the water is as free from impurity as possible.

Agriculture, including commercial livestock and poultry farming, is the source of many organic and inorganic pollutants in surface waters and groundwater. These contaminants include both sediment from erosion cropland and compounds of phosphorus and nitrogen that partly originate in animal wastes and commercial fertilizers. Animal wastes are high in oxygen demanding material, nitrogen and phosphorus, and they often harbor pathogenic organisms. Wastes from commercial feeders are contained and disposed of on land; their main threat to natural waters, therefore, is from runoff and leaching. Control may involve settling basins for liquids, limited biological treatment in aerobic or anaerobic lagoons, and a variety of other methods.

Ground Water

Ninety-five percent of all fresh water on earth is ground water. Ground water is found in natural rock formations. These formations, called aquifers, are a vital natural resource with many uses. Nationally, 53 per cent of the population relies on ground water as a source of drinking water. In rural areas this figure is even higher. Eighty one percent of community water is dependent on ground water. Although the 1992 Section 305(b) State Water Quality Reports indicate that, overall, the National's ground water quality is good to excellent, many local areas have experienced significant ground water contamination.

Legislation

Several forms of legislation have been passed in recent decades to try to control water pollution. In 1970, the Clean Water Act provided 50 billion dollars to cities and states to build wastewater facilities. This has helped control surface water pollution from industrial and municipal sources throughout the United States. When congress passed the Clean Water Act in 1972, states were given primary authority to set their own standards for their water. In addition to these standards, the act required that all state beneficial uses and their criteria must comply with the 'fishable and swimmable' goals of the act. This essentially means that state beneficial uses must be able to support aquatic life and recreational use. Because it is impossible to test water for every type of disease-causing organism, states usually look to identify indicator bacteria. One for a example is a bacteria known as fecal coliforms. These indicator bacteria suggest that a certain selection of water may be contaminated with untreated sewage and that other, more dangerous, organisms are present. These legislations are an important part in the fight against water pollution. They are useful in preventing Envioronmental catastrophes. The graph shows reported pollution incidents since 1989-1994. If stronger legislations existed, perhaps these events would never have occurred.

Global Water Pollution

Estimates suggest that nearly 1.5 billion people lack safe drinking water and that at least 5 million deaths per year can be attributed to waterborne diseases. With over 70 percent of the planet covered by oceans, people have long acted as if these very bodies of water could serve as a limitless dumping ground for wastes. Raw sewage, garbage, and oil spills have begun to overwhelm the diluting capabilities of the oceans, and most coastal waters are now polluted. Beaches around the world are closed regularly, often because of high amounts of bacteria from sewage disposal, and marine wildlife is beginning to suffer.

Perhaps the biggest reason for developing a worldwide effort to monitor and restrict global pollution is the fact that most forms of pollution do not respect national boundaries. The first major international conference on environmental issues was held in Stockholm, Sweden, in 1972 and was sponsored by the United Nations (UN). This meeting, at which the United States took a leading role, was controversial because many developing countries were fearful that a focus on environmental protection was a means for the developed world to keep the undeveloped world in an economically subservient position. The most important outcome of the conference was the creation of the United Nations Environmental Programme (UNEP).

UNEP was designed to be 'the environmental conscience of the United Nations¿½ and, in an attempt to allay fears of the developing world, it became the first UN agency to be headquartered in a developing country, with offices in Nairobi, Kenya. In addition to attempting to achieve scientific consensus about major environmental issues, a major focus for UNEP has been the study of ways to encourage sustainable development increasing standards of living without destroying the environment. At the time of UNEP's creation in 1972, only 11 countries had environmental agencies. Ten years later that number had grown to 106, of which 70 were in developing countries.

Water Quality

Water quality is closely linked to water use and to the state of economic development. In industrialized countries, bacterial contamination of surface water caused serious health problems in major cities throughout the mid 1800's. By the turn of the century, cities in Europe and North America began building sewer networks to route domestic wastes downstream of water intakes. Development of these sewage networks and waste treatment facilities in urban areas has expanded tremendously in the past two decades. However, the rapid growth of the urban population (especially in Latin America and Asia) has outpaced the ability of governments to expand sewage and water infrastructure. While waterborne diseases have been eliminated in the developed world, outbreaks of cholera and other similar diseases still occur with alarming frequency in the developing countries. Since World War II and the birth of the 'chemical age', water quality has been heavily impacted worldwide by industrial and agricultural chemicals. Eutrophication of surface waters from human and agricultural wastes and nitrification of groundwater from agricultural practices has greatly affected large parts of the world. Acidification of surface waters by air pollution is a recent phenomenon and threatens aquatic life in many area of the world. In developed countries, these general types of pollution have occurred sequentially with the result that most developed countries have successfully dealt with major surface water pollution. In contrast, however, newly industrialized countries such as China, India, Thailand, Brazil, and Mexico are now facing all these issues simultaneously.

Clearly, the problems associated with water pollution have the capabilities to disrupt life on our planet to a great extent. Congress has passed laws to try to combat water pollution thus acknowledging the fact that water pollution is, indeed, a seriousissue. But the government alone cannot solve the entire problem. It is ultimately up to us, to be informed, responsible and involved when it comes to the problems we face with our water. We must become familiar with our local water resources and learn about ways for disposing harmful household wastes so they don't end up in sewage treatment plants that can't handle them or landfills not designed to receive hazardous materials. In our yards, we must determine whether additional nutrients are needed before fertilizers are applied, and look for alternatives where fertilizers might run off into surface waters. We have to preserve existing trees and plant new trees and shrubs to help prevent soil erosion and promote infiltration of water into the soil. Around our houses, we must keep litter, pet waste, leaves, and grass clippings out of gutters and storm drains. These are just a few of the many ways in which we, as humans, have the ability to combat water pollution. As we head into the 21st century, awareness and education will most assuredly continue to be the two most important ways to prevent water pollution. If these measures are not taken and water pollution continues, life on earth will suffer severely.

Global environmental collapse is not inevitable. But the developed world must work with the developing world to ensure that new industrialized economies do not add to the world's environmental problems. Politicians must think of sustainable development rather than economic expansion. Conservation strategies have to become more widely accepted, and people must learn that energy use can be dramatically diminished without sacrificing comfort. In short, with the technology that currently exists, the years of global environmental mistreatment can begin to be reversed.

9.3 Protect Water Quality

Water quality depends upon well-defined water quality standards and criteria, consistent and well-constructed monitoring and evaluation programmes, and a legal framework that produces effective incentives to control public and private activities that may result in water pollution. ELI's Water Quality Protection activities focus on the laws and implementation strategies that influence:

- ❖ The effectiveness of the federal Clean Water Act in improving water quality. ELI's work includes projects to preserve broad national authority, improve water quality standards, identify impaired waters and address Total Maximum Daily Loads, strengthen permitting programmes, control nonpoint sources and stormwater, and resolve wet weather issues.

- ❖ Innovation in state and local programmes that can improve water quality. ELI examines state laws and local ordinances that address sectors such as agriculture, forestry, and land development; the connection between groundwater and surface water and its uses; and activities in coastal and riparian areas. We identify opportunities for state laws dealing with water quality, taxation, land use, funding, and other subjects to support water quality objectives.

Solid Waste and Soil Pollution Management

10.1 Solid Waste Treatment Technologies

The following page contains a list of different forms of solid waste treatment technologies and facilities employed in waste management infrastructure.
Waste handling facilities

- Civic amenity site (CA Site)
- Transfer station

Established Waste Treatment Technologies

- Composting
- Incineration
- Landfill
- Windrow composting

A dvanced Waste Treatment Technologies

Also called alternative waste treatment technologies

- Anaerobic digestion
- Alcohol/ethanol production

  - Bioconversion of biomass to mixed alcohol fuels (pilot scale)

- Automotive shredder residue (ASR)
- Biodrying
- Gasification

❖ In-vessel composting
❖ Mechanical biological treatment
❖ Mechanical heat treatment
❖ Plasma arc waste disposal (pilot scale)
❖ Pyrolysis
❖ Recycling
❖ Sewage treatment
❖ Tunnel composting
❖ UASB (applied to solid wastes)
❖ Waste autoclave
❖ Waste-to-energy

10.2 Composting

Composting is the aerobic decomposition of biodegradable organic matter, producing compost. (Or in a simpler form: Composting is the decaying of food, mostly vegetables or manure.) The decomposition is performed primarily by facultative and obligate aerobic bacteria, yeasts and fungi, helped in the cooler initial and ending phases by a number of larger organisms, such as ils, and other families representing ants, nematodes and oligochaete worms.

Composting can be divided into home composting and industrial composting. Essentially the same biological processes are involved in both scales of composting, however techniques and different factors must be taken into account.

Importance

Composting recycles or "downcycles" organic household and yard waste and manures into an extremely useful humus-like, soil end-product called compost. Examples are fruits, vegetables and yard clippings. Ultimately this permits the return of needed organic matter and nutrients into the foodchain and reduces the amount of "green" waste going into landfills. Composting is widely believed to speed up the natural process of decomposition appreciably as a result of the raised temperatures that often accompany it. The elevated heat results from exothermic processes, and the heat in turn reduces the generational time of microorganisms and thereby speeds the energy and nutrient exchanges taking place. It is a popular misconception that composting is a "controlled" process; if the right environmental circumstances are present the process virtually runs itself. Hence a popular expression, "compost happens". It is nonetheless very necessary to provide as optimal circumstances as possible for large amounts of organic waste to break down properly. This is especially so when it is accompanied by heating, since at elevated temperatures oxygen within the piles is consumed more rapidly, and if not controlled, will lead to malodor.

Decomposition similar to composting occurs throughout nature as garbage dissolves in the absence of all the conditions that modern composters talk about; however, the process can be slow. For example, in the forest bark, wood and leaves break down into humus over 3-7 . In restricted environments, for example, vegetables in a plastic trash container, decomposition with a lack of air encourages growth of anaerobic microbes, which produce disagreeable odors. Another form of degradation practiced deliberately in absence of oxygen is called anaerobic digestion-an increasingly popular companion to composting as it enables capture of residual energy in the form of biogas, whereas composting releases the majority of bound carbon-energy as excess heat (which helps sanitize the material) as well as copious amounts of biogenic CO_2 to the atmosphere.

It is important to distinguish between terms such as "biodegradable", "compostable", and "compost-compatible".

- ❖ A *biodegradable* material is capable of being broken down completely under the action of microorganisms into carbon dioxide, water and biomass. It may take a very long time for a material to biodegrade depending on its environment (e.g. hardwood in an arid area), but it ultimately breaks down completely.

- ❖ A *compostable* material biodegrades substantially under composting conditions, into carbon dioxide, methane, water and compost biomass. Compost biomass refers to the portion of the material that is metabolized by the microorganisms and which is incorporated into the cellular structure of the organisms or converted into humic acids etc. Compost biomass residues from a compostable material are fully biodegradable. "Compostable" is thus a subset of "biodegradable". The size of the material is a factor in determining compostability because it affects the rate of degradation. Large pieces of hardwood may not be compostable under a specific set of composting conditions, whereas sawdust of the same type of wood may be.

- ❖ A *compost-compatible* material does not have to be compostable or even biodegradable. It may biodegrade too slowly to be compostable itself, or it may not biodegrade at all. However, it is not readily distinguishable from the compost on a macroscopic scale and does not have a deleterious effect on the compost (e.g. it is not a biocide). Compost-compatible materials are generally inert and are present in compost at relatively low levels. Examples of compost-compatible materials include sand particles and inert particles of plastic.

Reducing Waste

Although composting has historically focussed on creating garden-ready soil, it is becoming more important as a tool is reducing solid waste. More than 60 percent of household waste in the United States is recyclable or compostable. But Americans only compost 8 percent of their waste. Surveys have shown that the #1 reason Americans don't compost their waste is because they feel the process is complicated, time-consuming or requires special equipment. However, especially in rural areas, much of the solid waste could be removed from the waste stream by promoting "extremely passive composting" where consumers simply discard their yard waste and kitchen scraps on their own land, regardless of whether the material is ever re-used as "compost". (http://www.nrdc.org/cities/recycling/fover.asp)

Materials

Many different materials are suitable for composting organisms. Composters often refer to "C:N" requirements; some materials contain high amounts of carbon in the form of cellulose which the bacteria need for their energy. Other materials contain nitrogen in the form of protein, which provide nutrients for the energy exchanges. It would however be an over-simplification to describe composting as about carbon and nitrogen, as is often portrayed in popular literature. Elemental carbon—such as charcoal—is not compostable nor is a pure form of nitrogen, even in combination with carbon. Not only this, but a great variety of man-made, carbon-containing products, including many textiles and polyethylene, are not compostable—hence the push for biodegradable plastics.

Suitable ingredients with relatively high carbon content include:

- ❖ Dry, straw-type material, such as cereal straws,
- ❖ Autumn leaves,
- ❖ Sawdust and wood chips,
- ❖ paper and cardboard (such as corrugated cardboard or newsprint with soy-based inks).

Ingredients with relatively high nitrogen content include:

- ❖ Green plant material (fresh or wilted) such as crop residues, hay, grass clippings, weeds,
- ❖ Manure of poultry and herbivorous animals such as horses, cows and llamas,
- ❖ Fruit and vegetable trimmings,

The most efficient composting occurs by seeking to obtain an initial C:N mix of 25/30 by dry chemical weight. Grass clippings have an average ratio of 10-19 to 1 and dry autumn leaves from 55-100 to 1. Mixing equal parts by volume approximates the ideal range.

Poultry manure provides much nitrogen but with a ratio to carbon that is imbalanced. If composted alone, this results in excessive N-loss in the form of ammonia—and some odor. Horse manure provides a good mix of both, although in modern stables, so much bedding may be used as to make the mix too carbonaceous.

For home-scale composting, mixing the materials as they are added increases the rate of decomposition, but it can be easier to place the materials in alternating layers, approximately 15 cm (6 in) thick, to help estimate the quantities. Keeping carbon and nitrogen sources separated in the pile can slow down the process, but decomposition will still occur.

Some people put special materials and activators into their compost. A light dusting of agricultural lime (not on animal manure layers) can curb excessive acidity, especially with food waste. Seaweed meal provides a ready source of trace elements. Finely pulverized rock (rock flour or rock dust) can also provide minerals, while clay and leached rock dust are poor in trace minerals.

Composting in the form of bioremediation can break down petroleum hydrocarbons, TNT and a variety of toxic compounds. This is the bacterial and in some cases fungal content of the compost, that possess the enzymatic properties to de-polymerize the complex man-made molecules. In other words, there is nothing about the composting process *per se* that adds or detracts from this, unless as noted above, by warming, to increase the metabolic rate of the constituent organisms.

Some materials are best left to high-rate, a thermophilic composting system, as they decompose slower, attract vermin and require higher temperatures to kill pathogens than backyard composting provides. These materials include meat, dairy products, eggs, restaurant grease, cooking oil, manure and bedding of non-herbivores, and residuals from the treatment of wastewater and drinking water. Meat and dairy products can be recycled using bokashi, a fermentation method.

Human waste can be composted by industrial, high-heat methods and also composting toilets, even though most composting toilets do not allow for the thermophilic decomposition believed to be necessary rapid kill of pathogens, such as Salmonella This is not a problem, however, since composting toilets also incorporate the essential element of time required to reduce available substrate on which pathogens can feed, while increasing the growth of competing microbes. If these high temperatures are reached, the resulting compost can be safely used as a fertilizer for food crops and even directly edible crops provided it is not illegal in the regions where the sludge is applied. Careful filtration of the compost also prevents contamination.

Approaches

There are two major approaches to composting: active and passive. These terms are somewhat of a misnomer since both active and passive composts can attain high heating, which increases the rate of biochemical processes. But the terms active and passive are appropriate descriptions for the nature of human intervention used.

Active

Active (hot) composting is composting at close to ideal conditions, allowing aerobic bacteria to thrive. Aerobic bacteria break down material faster and produce less odor and fewer pathogens and destructive greenhouse gases than anaerobic bacteria. Commercial-grade composting operations actively control the composting conditions such as the carbon-to-nitrogen ratio. For backyard composters, the charts of carbon and nitrogen ratios in various ingredients and the calculations required to get the ideal mixture can be intimidating, so many rules of thumb exist for approximating it.

Pasteurisation is a misnomer in composting, as no compost will become truly sterilized by high temperatures alone. Rather, in a very hot compost where the temperature exceeds 55 °C (130 °F) for several days, the ability of organisms to survive is greatly compromised. Nevertheless, there are many organisms in nature that can survive extreme temperatures, including the group of pathogenic Clostridium, and so no compost is completely safe. To achieve the elevated temperatures, the compost bin must be kept warm, insulated and damp.

Aerated Composting is an efficient form of composting from the chemical point of view as it produces ultimately only energy in the form of waste heat and CO_2 and H_2O. With aerated composting, fresh air (i.e. oxygen) is introduced throughout the mix of materials using any appropriate mechanism. The air stimulates the microorganisms that are already in the mix, and their by-product is heat. In a properly operated compost system, pile temperatures are sufficient to stabilize the raw material, and the oxygen-rich conditions within the core of the pile eliminate offensive odors. High temperatures also destroy fly larvae and weed seeds, yielding a safe, high-quality finished product.

Finally, aeration expedites the composting process through the mechanism of heating insofar as the elevated heat will drive biochemical processes faster, so that a finished product can be rendered in 60 to 120 days. Aerated compost is an excellent source of macro-and micro-nutrients as well as stable organic matter, all of which support healthy plant growth. In addition, the micro-organisms in compost aid in the suppression of plant pathogens. Finally, compost retains water extremely well resulting in improved drought resistance, a longer growing season, and reduced soil erosion.

In Thailand, May 2008, this system has been used by farmer groups for more than 445 sites. The process needs only 30 days to be finished and 10 metric tons of compost is obtained each time. A blower (15 inch squirrel-caged blower with 3 hp motor) is needed to force the air through 10 piles of compost twice a day and 15 minutes each. The raw materials consist of agricultural wastes and animal manure in the ratio of 3: 1 by volume. For more information please visit www.compost.mju.ac.th/eng

Passive

Passive composting is composting in which the level of physical intervention is kept to a minimum, and often as a result the temperatures never reach much above 30°C (86 °F). It is slower but is the more common type of composting in most domestic garden compost bins. Such composting systems may be either enclosed (home container composting, industrial in-vessel composting) or in exposed piles (industrial windrow composting). Kitchen scraps are put in the garden compost bin and left untended. This scrap bin can have a very high water content which reduces aeration, and so becomes odorous. To improve drainage and airflow, a gardener can mix in wood chips, small pieces of bark, leaves or twigs, or make physical holes through the pile.

Natural

An unusual form of natural composting in nature is seen in the case of the mound-builders (megapodes) of eastern Indonesia, New Guinea, and Australia as well as in the case of bowerbirds of New Guinea and Australia. These Megapodes are fowl-sized birds famous for building nests in the form of huge compost heaps containing leaf litter, in which they incubate their eggs. The birds work constantly to maintain the correct, almost exact, incubation temperatures, by adding and removing leaves from the compost pile. In effect, this teaches us that thermophilic high-temperature composting is not man-made.

Home Composting

Home composters use a range of techniques, varying from extremely passive (throw everything in a pile and leave it for a year or two) to extremely active (monitor the temperature, turn the pile regularly, and adjust the ingredients over time). Some composters use mineral powders to absorb smells, although a well-maintained pile seldom has bad odors. It is usually located in the back garden.

Moisture and Heat

An effective compost pile is about as damp as a well wrung-out sponge. This provides the moisture that all life requires. Microorganisms vary by their ideal temperature and the heat they generate as they digest. Mesophilic bacteria survive best at temperatures of 20 to 44 °C (70 to 120 °F). thermophilic (heat-surviving) bacteria grow optimally at around 55°C (130 °F), and can attain the fastest decomposition, since metabolic processes proceed more rapidly under higher temperatures. Elevated temperature is also preferred since it causes the most rapid pathogen reduction, and is more destructive of weed seeds. To minimally achieve it, the heap should be about 1 m (3 ft) wide, 1 m (3 ft) tall, and as long as is practicable. This provides enough insulating mass to build up heat but also allows aeration. The center of the pile heats up the most.

If the pile does not heat up, common reasons include that:

❖ The heap is too wet, limiting the oxygen which bacteria require,
❖ The heap is too dry for the bacteria to survive and reproduce, and
❖ There is insufficient protein (nitrogen-rich material).

The necessary material should be added, or the pile should be turned to aerate it and bring the outer layers inside and vice versa. You should add water at this time to help keep the pile damp. One guideline is to turn the pile when the high temperature has begun to drop, indicating that the food source for the fastest-acting bacteria (in the center of the pile) has been largely consumed. When turning the pile does not cause a temperature rise, it brings no further advantage. When all the material has turned into dark brown crumbly matter, it is ready to use.

Worm Composting

Worm composting or vermicomposting is a method of composting using Red Wiggler worms in a container. Food waste and moistened bedding are added, and the worms and micro-organisms eventually convert them to rich compost. The worms excrete a soil-nutrient material called worm castings. Worm composting can be done indoors, allowing year-round composting, and providing apartment dwellers with a means of composting.

Worms are low in the food chain, and so are critical to healthy soil. This is why farmers have historically wanted healthy worm populations to live in their fields.

The nutrients and micro-organisms can be concentrated in liquid form called worm tea, made by running distilled water through worm castings. When it is poured into the soil, the microorganisms multiply, creating a healthy growing environment for plants.

Industrial Composting

Industrial composting systems are increasingly being installed as a waste management alternative to landfills, along with other advanced waste processing systems. Industrial composting or anaerobic digestion combined with mechanical sorting of mixed waste streams is called mechanical biological treatment increasingly used in Europe due to stringent new regulations controlling the amount of organic matter allowed in landfills. Treating biodegradable waste before it enters a landfill reduces global warming from fugitive methane; untreated waste breaks down anaerobically in a landfill, producing landfill gas that contains methane, a greenhouse gas even more potent than carbon dioxide.

Most commercial and industrial composting operations use active composting techniques. These ensure that the process does not get out of control especially with the high through-put demand imposed by contracted, incoming waste. This means that as short as possible a processing time must be maintained to keep the facility properly functioning (see compost windrow turner). Partly for this reason composters have declined to support compost maturity standards if it would increase the required holding time. The greatest amount of technological control of composting is seen in systems using an enclosed vessel and controlling its temperature, air flow, moisture and other parameters.

Large-scale composting systems are used by many urban centers around the world. Co-composting is a technique which combines solid waste with de-watered biosolids, which originated in the 1960s and has fallen somewhat out of favour due to difficulties controlling inert and plastic contamination from MSW. In Europe, mixed waste composting is virtually illegal. The world's largest MSW co-composter is the Edmonton Composting Facility in Edmonton in Alberta, Canada, which turns 220,000 tonnes of residential solid waste and 22,500 dry tonnes of biosolids per year into 80,000 tonnes of compost. The facility is 38,690 square metres (416,500 ft²) large (equivalent to 4½ Canadian football fields), and the aeration building alone is the largest stainless steel building in North America, the size of 14 NHL rinks.

10.3 Incineration

Incineration is a waste treatment technology that involves the combustion of organic materials and/or substances. Incineration and other high temperature waste treatment systems are described as "thermal treatment". Incineration of waste materials converts the waste into incinerator bottom ash, flue gases, particulates, and heat, which can in turn be used to generate electric power. The flue gases are cleaned of pollutants before they are dispersed in the atmosphere.

Incineration with energy recovery is one of several waste-to-energy (WtE) technologies such as gasification, pyrolysis and anaerobic digestion. Incineration may also be implemented without energy and materials recovery.

In several countries there are still expert and local community concerns about the environmental impact of incinerators.

In some countries, incinerators built just a few decades ago often did not include a materials separation to remove hazardous, bulky or recyclable materials before combustion. These facilities tended to risk the health of the plant workers and the local environment due to inadequate levels of gas cleaning and combustion process control. Most of these facilities did not generate electricity.

Incinerators reduce the volume of the original waste by 95-96 per cent, depending upon composition and degree of recovery of materials such as metals from the ash for recycling. This means that while incineration does not completely replace landfilling, it reduces the necessary volume for disposal significantly.

Incineration has particularly strong benefits for the treatment of certain waste types in niche areas such as clinical wastes and certain hazardous wastes where pathogens and toxins can be destroyed by high temperatures. Examples include chemical multi-product plants with diverse toxic or very toxic wastewater streams, which cannot be routed to a conventional wastewater treatment plant.

Waste combustion is particularly popular in countries such as Japan where land is a scarce resource. Denmark and Sweden have been leaders in using the energy generated from incineration for more than a century, in localised combined heat and power facilities supporting district heating schemes. In 2005, waste incineration produced 4.8 per cent of the electricity consumption and 13.7 per cent of the total domestic heat consumption in Denmark. A number of other European Countries rely heavily on incineration for handling municipal waste, in particular Luxemburg, The Netherlands, Germany and France.

Technology

Types of Incinerators

An incinerator is a furnace for burning waste. Modern incinerators include pollution mitigation equipment such as flue gas cleaning. There are various types of incinerator plant design: moving grate, fixed grate, rotary-kiln, fluidised bed.

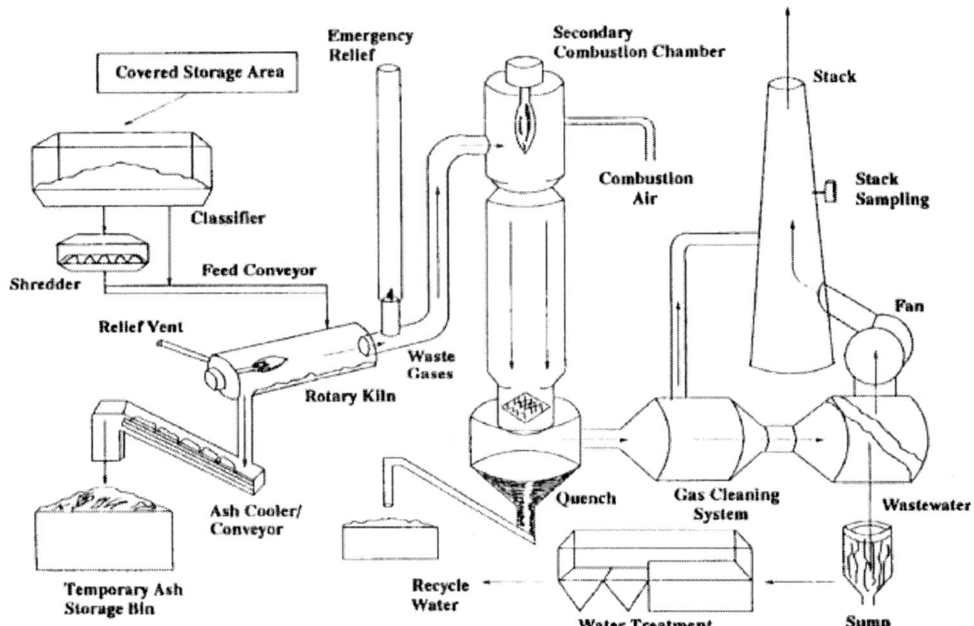

Moving Grate

The typical incineration plant for municipal solid waste is a moving grate incinerator. The moving grate enables the movement of waste through the combustion chamber to be optimised to allow a more efficient and complete combustion. A single moving grate boiler can handle up to 35 tonnes of waste per hour, and can operate 8,000 hours per year with only one scheduled stop for inspection and maintenance of about one months duration . Moving grate incinerators are sometimes referred to as Municipal Solid Waste Incinerators (MSWIs).

The waste is introduced by a waste crane through the "throat" at one end of the grate, from where it moves down over the descending grate to the ash pit in the other end. Here the ash is removed through a water lock.

Part of the combustion air (primary combustion air) is supplied through the grate from below. This air flow also has the purpose of cooling the grate itself. Cooling is important for the mechanical strength of the grate, and many moving grates are also water cooled internally.

Secondary combustion air is supplied into the boiler at high speed through nozzles over the grate. It facilitates complete combustion of the flue gases by introducing turbulence for better mixing and by ensuring a surplus of oxygen. In multiple/stepped hearth incinerators, the secondary combustion air is introduced in a separate chamber downstream the primary combustion chamber.

According to the European Waste Incineration Directive, incineration plants must be designed to ensure that the flue gases reach a temperature of at least 850 °C for 2 seconds in order to ensure proper breakdown of organic toxins. In order to comply with this at all times, it is required to install backup auxiliary burners (often fueled by oil), which are fired into the boiler in case the heating value of the waste becomes too low to reach this temperature alone.

The flue gases are then cooled in the superheaters, where the heat is transferred to steam, heating the steam to typically 400 °C at a pressure of 40 bar for the electricity generation in the turbine. At this point, the flue gas has a temperature of around 200 °C, and is passed to the flue gas cleaning system.

At least in Scandinavia scheduled maintenance is always performed during summer, where the demand for district heating is low. Often incineration plants consist of several separate 'boiler lines' (boilers and flue gas treatment plants), so that waste receival can continue at one boiler line while the others are subject to revision.

Fixed Grate

The older and simpler kind of incinerator was a brick-lined cell with a fixed metal grate over a lower ash pit, with one opening in the top or side for loading and another opening in the side for removing incombustible solids called clinkers. Many small incinerators formerly found in apartment houses have now been replaced by waste compactors.

Rotary-kiln

The rotary-kiln incinerator used by municipalities and by large industrial plants. This design of incinerators have 2 chambers a primary chamber and secondary chamber. The primary chamber in a rotary kiln incinerator consist of an inclined refractory lined cylindrical tube. Movement of the cylinder on its axis facilitates movement of waste. In the primary chamber, there is conversion of solid fraction to gases, through volatilization, destructive distillation and partial combustion reactions. The secondary chamber is necessary to complete gas phase combustion reactions

The clinkers spill out at the end of the cylinder. A tall flue gas stack, fan, or steam jet supplies the needed draft. Ash drops through the grate, but many particles are carried along with the hot gases. The particles and any combustible gases may be combusted in an "afterburner". A diagram of a rotary-kiln incinerator can be found here.

Fluidized Bed

A strong airflow is forced through a sandbed. The air seeps through the sand until a point is reached where the sand particles separate to let the air through and mixing and churning occurs, thus a fluidised bed is created and fuel and waste can now be introduced.

The sand with the pre-treated waste and/or fuel is kept suspended on pumped air currents and takes on a fluid-like character. The bed is thereby violently mixed and agitated keeping small inert particles and air in a fluid-like state. This allows all of the mass of waste, fuel and sand to be fully circulated through the furnace.

Specialized Incineration

Furniture factory sawdust incinerators need much attention as these have to handle resin powder and many flammable substances. Controlled combustion, burn back prevention systems are very essential as dust when suspended resembles the fire catch phenomenon of any liquid petroleum gas.

Use of Heat

The heat produced by an incinerator can be used to generate steam which may then be used to drive a turbine in order to produce electricity. The typical amount of net energy that can be produced per ton municipal waste is about 0.67 MWh of electricity and 2 MJ of district heating . Thus, incinerating about 600 tonnes per day of waste will produce about 17 MW of electrical power and 1200 MJ district heating each day.

Pollution

Incineration has a number of outputs such as the ash and the emission to the atmosphere of flue gas. Before the flue gas cleaning system, the flue gases may contain significant amounts of particulate matter, heavy metals, dioxins, furans, sulfur dioxide, and hydrochloric acid.

In a study from 1994, Delaware Solid Waste Authority found that, for same amount of produced energy, incineration plants emitted fewer particles, hydrocarbons and less SO_2, HCl, CO and NO_x than coal-fired power plants, but more than natural gas fired power plants. According to Germany's Ministry of the Environment, waste incinerators reduce the amount of some atmospheric pollutants by substituting power produced by coal-fired plants with power from waste-fired plants.

Gaseous Emissions

Dioxin and Furans

The most publicized concerns from environmentalists about the incineration of municipal solid wastes (MSW) involve the fear that it produces significant amounts of dioxin and furan emissions. Dioxins and furans are considered by many to be serious health hazards. Older generation incinerators that were not equipped with adequate gas cleaning technologies were indeed significant sources of dioxin emissions. Today, however, due to advances in emission control designs and stringent new governmental regulations, incinerators emit virtually no dioxins. In 2005, The Ministry of the Environment of Germany, where there were 66 incinerators at that time, estimated that "whereas in 1990 one third of all dioxin emissions in Germany came from incineration plants, for the year 2000 the figure was less than 1 per cent. Chimneys and tiled stoves in private households alone discharge approximately twenty times more dioxin into the environment than incineration plants." According to the U.S. EPA, incineration plants are no longer significant sources of dioxins and furans. In 1987, before the governmental regulations required the use of emission controls, there was a total of 10,000 grams of dioxin emissions from U.S. incinerators. Today, the total emissions from the 87 plants are only 10 grams yearly, a reduction of 99.9 per cent. Backyard barrel burning of household and garden wastes, still allowed in some rural areas, generates 580 grams of dioxins yearly. Studies conducted by EPA demonstrate that the emissions from just one family using a burn barrel produces more emissions than an incineration plant disposing of 200 tonnes of waste per day.

Generally the breakdown of dioxin requires exposure of the molecule to a sufficiently high temperature so as to trigger thermal breakdown of the molecular bonds holding it together. When burning of plastics outdoors in a burn barrel or garbage pit such temperatures are not reached, causing high dioxin emissions as mentioned above. While the plastic does burn in an open-air fire, the dioxins remain after combustion and float off into the atmosphere.

Modern municipal incinerator designs include a high temperature zone, where the flue gas is ensured to sustain a temperature above 850 C for at least 2 seconds befores it is cooled down. They are equipped with auxiliary heaters to ensure this at all times. These are often fueled by oil, and normally only active for a very small fraction of the time. A side effect controlling dioxin is the potential for generation of reactive oxides (NO_x) in the flue gas, which must be removed with SCR or SNCR (see below).

For very small municipal incinerators, the required temperature for thermal breakdown of dioxin may be reached using a high-temperature electrical heating element, plus an SCR stage.

CO_2

As for other complete combustion processes, nearly all of the carbon content in the waste is emitted as CO_2 to the atmosphere. MSW contain approximately the same mass fraction of carbon as CO_2 itself (27%), so incineration of one tonne of MSW produce approximately 1 tonne of CO_2.

In the event that the waste was landfilled, one tonne of MSW would produce approximately 62 m³ methane via the anaerobic decomposition of the biodegradable part of the waste. This amount of methane has more than twice the global warming potential than the one tonne of CO_2, which would have been produced by incineration. In some countries, large amounts of landfill gas are collected, but still the global warming potential of the landfill gas emitted to atmosphere in the US in 1999 was approximately 32 per cent higher than the amount of CO_2 that would have been emitted by incineration.

In addition, nearly all biodegradable waste has biological origin. This material has been formed by plants using atmospheric CO_2 typically within the last growing season. If these plants are regrown the CO_2 emitted from their combustion will be taken out from the atmosphere once more.

Such considerations are the main reason why several countries administrate incineration of the biodegradable part of waste as renewable energy. The rest—mainly plastics and other oil and gas derived products—is generally treated as non-renewables.

Different results for the CO_2 footprint of incineration can be reached with different assumptions. Local conditions (such as limited local district heating demand, no fossil fuel generated electricity to replace or high levels of aluminum in the waste stream) can decrease the CO_2 benefits of incineration. The methology and other assumptions may also influence the results significantly. For example the methane emissions from landfills occurring at a later date may be neglected or given less weight, or biodegradable waste may not be considered CO_2 neutral. A recent study by Eunomia Research and Consulting on potential waste treatment technologies in London demonstrated that by applying several of these (according to the authors) unusual assumptions the average existing incineration plants performed poorly for CO_2 balance compared to the theoretical potential of other emerging waste treatment technologies. .

Other Emissions

Other gaseous toxins in the flue gas from incinerator furnaces include sulfur dioxide, hydrochloric acid, heavy metals and fine particles.

The steam content in the flue may produce visible fume from the stack, which can be perceived as a visual pollution. It may be avoided by decreasing the steam content by flue gas condensation, or by increasing the flue gas exit temperature well above its dew point. Flue gas condensation allows the latent heat of vaporization of the water to be recovered, subsequently increasing the thermal efficiency of the plant.

Flue Gas Cleaning

The quantity of pollutants in the flue gas from incineration plants is reduced by several processes.

Particulate is collected by particle filtration, most often electrostatic precipitators (ESP) and/or baghouse filters. The latter are generally very efficient for collecting fine particles. In an investigation by the Ministry of the Environment of Denmark in 2006, the average particulate emissions per energy content of incinerated waste from 16 Danish incinerators were below 2.02 g/GJ (grams per energy content of the incinerated waste). Detailed measurements of fine particles with sizes below 2.5 micrometres ($PM_{2.5}$) were performed on three of the incinerators: One incinerator equipped with an ESP for particle filtration emitted 5.3 g/GJ fine particles, while two incinerators equipped with baghouse filters emitted 0.002 and 0.013 g/GJ $PM_{2.5}$.

Acid gas scrubbers are used to remove hydrochloric acid, nitric acid, hydrofluoric acid, mercury, lead and other heavy metals. Basic scrubbers remove, forming gypsum by reaction with lime.

Waste water from scrubbers must subsequently pass through a waste water treatment plant.

Sulfur dioxide may also be removed by dry desulfurisation by injection limestone slurry into the flue gas before the particle filtration.

NO_x is either reduced by catalytic reduction with ammonia in a catalytic converter (selective catalytic reduction, SCR) or by a high temperature reaction with ammonia in the furnace (selective non-catalytic reduction, SNCR).

Heavy metals are often adsorbed on injected active carbon powder, which is collected by the particle filtration.

Solid Outputs

Incineration produces fly ash and bottom ash just as is the case when coal is combusted. The total amount of ash produced by municipal solid waste incineration ranges from 4-10 per cent by volume and 15-20 per cent by weight of the original quantity of waste and the fly ash amounts to about 10-20 per cent of the total ash. The fly ash, by far, constitutes more of a potential health hazard than does the bottom ash because the fly ash often contain high concentrations of heavy metals such as lead, cadmium, copper and zinc as well as small amounts of dioxins and furans. The bottom ash seldom contain significant levels of heavy metals. While fly ash is always regarded as hazardous waste, bottom ash is generally considered safe for regular landfill after a certain level of testing defined by the local legislation. Ash, which is considered hazardous, may generally only be disposed of in landfills which are carefully designed to prevent pollutants in the ash from leaching into underground aquifers—or after chemical treatment to reduce its leaching characteristics. In testing over the past decade, no ash from an incineration plant in the USA has ever been determined to be a hazardous waste. At present although some historic samples tested by the incinerator operators' group would meet the being ecotoxic criteria at present the EA say "we have agreed" to regard incinerator bottom ash as "non-hazardous" until the testing programme is complete.

Other Pollution Issues

Odour pollution can be a problem with old-style incinerators, but odours and dust are extremely well controlled in newer incineration plants. They receive and store the waste in an enclosed area with a negative pressure with the airflow being routed through the boiler which prevents unpleasant odours from escaping into the atmosphere. However, not all plants are implemented this way, resulting in inconveniences in the locality.

An issue that affects community relationships is the increased road traffic of waste collection vehicles to transport municipal waste to the incinerator. Due to this reason, most incinerators are located in industrial areas.

The Debate Over Incineration

Use of incinerators for waste management is controversial. The debate over incinerators typically involves business interests (representing both waste generators and incinerator firms), government regulators, environmental activists and local citizens who must weigh the economic appeal of local industrial activity with their concerns over health and environmental risk.

People and organizations professionally involved in this issue include the U.S. Environmental Protection Agency and a great many local and national air quality regulatory agencies worldwide.

The Argument for Incineration

❖ The concerns over the health effects of dioxin and furan emissions have been significantly lessened by advances in emission control designs and very stringent new governmental regulations that have resulted in large reductions in the amount of dioxins and furans emissions.

❖ Incineration plants generate electricity and heat that can substitute power plants powered by other fuels at the regional electric and district heating grid, and steam supply for industrial customers.

❖ The bottom ash residue remaining after combustion has been shown to be a non-hazardous solid waste that can be safely landfilled or recycled as construction aggregate.

- ❖ In densely populated areas, finding space for additional landfills is becoming increasingly difficult.
- ❖ Fine particles can be efficiently removed from the flue gases with baghouse filters. Even though approximately 40 per cent of the incinerated waste in Denmark was incinerated at plants with no baghouse filters, estimates based on measurements by the Danish Environmental Research Institute showed that incinerators were only responsible for approximately 0.3 per cent of the total domestic emissions of particulate smaller than 2.5 micrometres ($PM_{2.5}$) to the atmosphere in 2006.
- ❖ Incineration of municipal solid waste avoids the release of methane. Every ton of MSW incinerated, prevents about one ton of carbon dioxide equivalents from being released to the atmosphere.
- ❖ Incineration of medical waste and sewage sludge produces an end product ash that is sterile and non-hazardous.

The Argument Against Incineration

- ❖ The highly toxic fly ash must be safely disposed of. This usually involves additional waste miles and the need for specialist toxic waste landfill elsewhere, sometimes with concerns for local residents. This has been the case in Bishops Cleeve, Gloucestershire, UK. Dr Dick van Steenis MBBSndustrial toxicologist.
- ❖ There are still concerns by many about the health effects of dioxin and furan emissions into the atmosphere from old incinerators; especially during start up and shut down events, or where filter bypass events are required.
- ❖ Incinerators emit varying levels of heavy metals such as vanadium, manganese, chromium, nickel, arsenic, mercury, lead, and cadmium, which can be toxic at very minute levels.
- ❖ Incinerator Bottom Ash (IBA) has high levels of heavy metals with ecotoxicity concerns if not reused properly. Some people have the opinion that IBA reuse is still in its infancy and is still not considered to be a mature or desirable product, despite additional engineering treatments.
- ❖ Alternative technologies are available or in development such as Mechanical Biological Treatment, Anaerobic Digestion (MBT/AD), Autoclaving or Mechanical Heat Treatment (MHT) using steam or Plasma arc gasification PGP, or combinations.
- ❖ Building and operating an incinerator requires long contract periods

to recover initial investment costs, causing a long term lock-in. Incinerator lifetimes normally range from 25-30 years.

❖ Incinerators produce fine particles in the furnace. Even with modern particle filtering of the flue gases, a fraction of these are emitted to the atmosphere. As an example, the baghouse filters in an incineration plant planned for erection in the UK, are only specified to capture 65-70 per cent (in weight) particulate smaller than 2.5 micrometres ($PM_{2.5}$), if the significant filtration in the filter cake is not accounted for. $PM_{2.5}$ is not separately regulated in the European Waste Incineration Directive, even though they are suspected to be linked to infant mortality in the UKr Dick van Steenis MBBSndustrial toxicologist and $PM_{2.5}$ emissions from local incinerators to be a significant $PM_{2.5}$ source here.

❖ Local communities are often opposed to the idea of locating incinerators in their vicinity. (The Not In My Back Yard phenomenon). Studies in Andover, Massachusetts linked 10 per cent property devaluations with close incinerator proximity .

❖ Prevention, waste minimisation, reuse and recycling of waste should all be preferred to incineration according to the waste hierarchy. Supporters of zero waste consider incinerators and other waste treatment technologies as barriers to recycling and separation beyond particular levels, and that waste resources are sacrificed for energy producion.

❖ A recent Eunomia report found that incinerators under some circumstances and assumptions can be seen as contributing to climate change, and are less energy efficient than emerging technologies for treating residual mixed waste.

❖ Some incinerators are architecturally monstrous and ugly. In many countries they require a visually intrusive chimney stack.

Trends in Incinerator Use

The history of municipal solid waste (MSW) incineration is linked intimately to the history of landfills and other waste treatment technology. The merits of incineration are inevitably judged in relation to the alternatives available. Since the 1970s, recycling and other prevention measures have changed the context for such judgements. Since the 1990s alternative waste treatment technologies have been maturing and becoming viable.

Incineration is a key process in the treatment of hazardous wastes and clinical wastes. It is often imperative that medical waste be subjected to the high temperatures of incineration to destroy pathogens and toxic contamination it contains.

Incineration in North America

The first full-scale waste-to-energy facility in the U.S. was the Arnold O. Chantland Resource Recovery Plant, built in 1975 located in Ames, Iowa. This plant is still in operation and produces refuse-derived fuel that is sent to local power plants for fuel. The first commercially-successful incineration plant in the U.S. was built in Saugus, Massachusetts in October 1975 by Wheelabrator Technologies, and is still in operation today.

Several older generation incinerators have been closed; of the 186 MSW incinerators in 1990, only 89 remained by 2007, and of the 6200 medical waste incinerators in 1988, only 115 remained in 2003. Between 1996 and 2007, no new incinerators were built. The main reasons for lack of activity have been:

❖ *Economics*: With the increase in the number of large inexpensive regional landfills and, up until recently, the relatively low price of electricity, incinerators were not able to compete for the 'fuel', i.e., waste. By contrast, a number of Canadian cities are working toward installation of incinerators.

❖ *Tax Policies*: Tax credits for plants producing electricity from waste were rescinded in the 1990s. In Europe, some of the electricity generated from waste is deemed to be from a 'Renewable Energy Source (RES)'. A new law granting tax credits for such plants was implemented in the U.S. in 2004.

Despite these problems, there has been renewed interest in waste-to-energy in the U.S., Canada, and the UK. Projects to add capacity to existing plants are underway, and municipalities are once again evaluating the option of building incinerators rather than continue landfilling municipal wastes.

Incineration in Europe

In Europe, with the ban on landfilling untreated waste, scores of incinerators have been built in the last decade, with more under construction. Recently, a number of municipal governments have begun the process of contracting for the construction and operation of incinerators. In Europe, some of the electricity generated from waste is deemed to be from a 'Renewable Energy Source (RES)' and is thus eligible for tax credits if privately operated.

Incineration in the United Kingdom

The technology employed in the UK waste management industry has been greatly lagging behind that of Europe due to the wide availablility of landfills. The Landfill Directive set down by the European Union led to the Government of the United Kingdom imposing waste legislation including the landfill tax and Landfill Allowance Trading Scheme. This legislation is designed to reduce the release of greenhouse gases produced by landfills through the use of alternative methods of waste treatment. It is the UK Government's position that incineration will play an increasingly large role in the treatment of municipal waste and supply of energy in the UK.

Small Incinerator Units

Small scale incinerators exist for special purposes. For example, the small scale incinerators are aimed for hygienically safe destruction of medical waste in developing countries. Simple, mobile incinerators are becoming more widely used in developing countries where the threat of avian influenza is high. Small incinerators can be quickly deployed to remote areas where an outbreak has occurred to dispose of infected animals quickly and without the risk of cross contamination.

Incinerators

- Allington Incinerator
- Isle of Man Incinerator
- Kirklees Incinerator
- List of incinerators in the UK
- SELCHP
- Sheffield Incinerator

10.4 Landfill

A *landfill*, also known as a *dump* or *tip* (and historically as a *midden*), is a site for the disposal of waste materials by burial and is the oldest form of waste treatment. Historically, landfills have been the most common methods of organized waste disposal and remain so in many places around the world.

Landfills may include internal waste disposal sites (where a producer of waste carries out their own waste disposal at the place of production) as well as sites used by many producers. Many landfills are also used for other waste management purposes, such as the temporary storage, consolidation and transfer, or processing of waste material (sorting, treatment, or recycling).

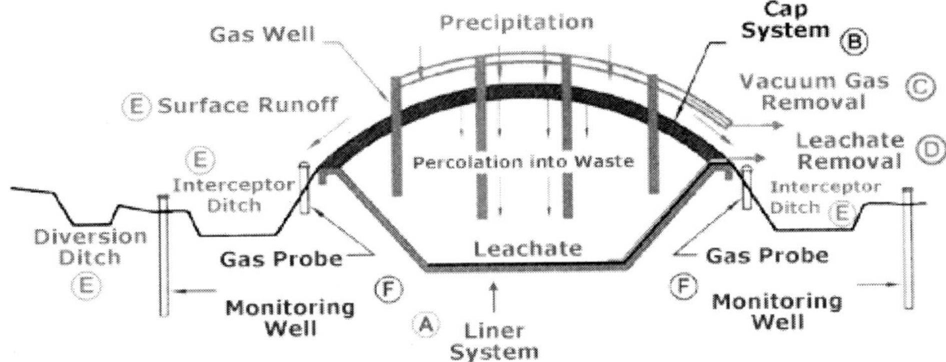

A landfill also may refer to ground that has been filled in with soil and rocks instead of waste materials, so that it can be used for a specific purpose, such as for building houses. Unless they are stabilized, these areas may experience severe shaking or liquefaction of the ground in a large earthquake.

Site Construction Requirements

The construction of a landfill requires a staged approach. Landfill designers are primarily concerned with the viability of a site. To be commercially and environmentally viable a landfill must be constructed in accord with specific requirements, which are related to:

Location

- ❖ Easy access to transport by road
- ❖ Transfer stations if rail network is preferred
- ❖ Land value
- ❖ Cost of meeting government requirements, such as the Environment Agency in England and Wales
- ❖ Location of community served
- ❖ Type of construction (more than one may be used at single site)

  - *Pit*: filling existing holes in the ground, typically left behind by mining,
  - *Canyon*: filling in naturally occurring valleys or canyons,
  - *Mound*: piling the waste up above the ground.

Stability

❖ Underlying geology,
❖ Nearby earthquake faults,
❖ Water table,
❖ Location of nearby rivers, streams, and flood plains,
❖ *Capacity:* The available voidspace must be calculated by comparison of the landform with a proposed restoration profile. This calulation of capacity is based on:

● Density of the wastes,
● Amount of intermediate and daily cover,
● Amount of settlement that the waste will undergo following tipping,
● Thickness of capping, and
● Construction of lining and drainage layers.

Protection of Soil and Water Through

❖ Installation of liner and collection systems.
❖ Storm water control.
❖ Leachate management.
❖ Landfill gas management.

Nuisances and Hazards Management

Costs

❖ Feasibility studies,
❖ Site after care,
❖ Site investigations (costs involved may make small sites uneconomical).

Operations

Typically, in non hazardous waste landfills, in order to meet predefined specifications, techniques are applied by which the wastes are:

1. Confined to as small an area as possible.
2. Compacted to reduce their volume.
3. Covered (usually daily) with layers of soil.

During landfill operations the waste collection vehicles are weighed at a weigh-bridge on arrival and their load is inspected for wastes that do not accord with the landfill's waste acceptance criteria. Afterwards, the waste collection vehicles use the existing road network on their way to the tipping face or working front where they unload their load. After loads are deposited, compactors or dozers are used to spread and compact the waste on the working face. Before leaving the landfill boundaries, the waste collection vehicles pass through the wheel cleaning facility. If necessary, they return to weighbridge in order to be weighed without their load. Through the weighing process, the daily incoming waste tonnage can be calculated and listed in databases. In addition to trucks, some landfills may be equipped to handle railroad containers. The use of 'rail-haul' permits landfills to be located at more remote sites, without the problems associated with many truck trips.

Typically, in the working face, the compacted waste is covered with soil daily. Alternative waste cover materials are several sprayed on foam products and temporary blankets. Blankets can be lifted into place with tracked excavators and then removed the following day prior to waste placement. Chipped wood and chemically 'fixed' bio-solids, may also be used as an alternate daily cover. The space that is occupied daily by the compacted waste and the cover material is called daily cell. Waste compaction is critical to extending the landfill life. Factors such as waste compressibility, waste layer thickness and the number of passes of the compactor over the waste affect the waste densities.

Land Reclamation

As human overcrowding of developed areas intensified during the 20th century, it has become important to develop land re-use strategies for completed landfills. Some of the most common usages are for parks, golf courses and other sports fields. Increasingly, however, office buildings and industrial uses are made on a completed landfill. In these latter uses, methane capture is customarily carried out to minimize explosive hazard within the building.

An example of a Class A office building constructed over a landfill is the Dakin Building at Sierra Point, Brisbane, California. The underlying fill was deposited from 1965 to 1985, mostly consisting of construction debris from San Francisco and some municipal wastes. Aerial photographs prior to 1965 show this area to be tidelands of the San Francisco Bay. A clay cap was constructed over the debris prior to building approval.

Another strategy for landfill reclamation is the incineration of landfill trash at high temperature via the plasma-arc gasification process, which is currently used at two facilities in Japan, and will be used at a planned facility in St. Lucie County, Florida.

Impacts

A number of adverse impacts occur from landfill operations. These impacts can vary: fatal accidents (e.g., scavengers buried under waste piles); infrastructure damage (e.g., damage to access roads by heavy vehicles); pollution of the local environment (such as contamination of groundwater and/or aquifers by leakage and residual soil contamination during landfill usage, as well as after landfill closure); offgassing of methane generated by decaying organic wastes (methane is a greenhouse gas many times more potent than carbon dioxide, and can itself be a danger to inhabitants of an area;) harbouring of disease vectors such as rats and flies, particularly from improperly operated landfills, which are common in Third-world countries; injuries to wildlife; and simple nuisance problems (e.g., dust, odour, vermin, or noise pollution).

Environmental noise and dust are generated from vehicles accessing a landfill as well as from working face operations. These impacts are best to intercept at the planning stage where access routes and landfill geometrics can be used to mitigate such issues. Vector control is also important, but can be managed reasonably well with the daily cover protocols.

Most modern landfills in industrialized countries are operated with controls to attempt manage problems such as these. Analysis of common landfill operational problems are available in .

Some local authorities have found it difficult to locate new landfills. Communities may charge a fee or levy in order to discourage waste and/or recover the costs of site operations. Some landfills are operated for profit as commercial businesses. Many landfills, however, are publicly operated and funded.

Impacts to People Near Landfills in the U.S.

Communities near landfills are increasingly facing health consequences from air and water contamination, particularly from landfills that are poorly constructed and operated. Environmental contamination from landfills is entering waterways and underground aquifer at alarming rates. Liner breaches are not uncommon. Liners can delay contamination but they do not prevent it. With large amounts of toxic solid waste entering landfills today, ground and air contamination pose a significant threat to public health for those living within three to five miles of a landfill, and will eventually degrade the environment far beyond those limits.

Poorly constructed and operated landfills persist with leachate breaks, uncovered trash, and unchecked banned hazardous compounds. Federal laws to protect the public in Sec. 4001, Subtitle D of the Resource Conservation and Recovery Act (RCRA) can be unenforceable to citizens without adequate legal funding. The U.S. Environmental Protection Agency generally relies on the states to enforce their own operating permits and federal laws. If state agencies are not aggressive, violations can worsen, multiplying negative environmental impacts exponentially. There are some notable recorded violations in the U.S., such as for a landfill in Hawaii that was fined $2.8 million in 2006 for operating violations, but this is not common.

Regional Practice

United Kingdom

Landfilling practices in the UK have had to change in recent years to meet the challenges of the European Landfill Directive. The UK now imposes landfill tax upon biodegradable waste which is landfilled. In addition to this the Landfill Allowance Trading Scheme has been established for local authorities to trade landfill quotas.

United States

In the U.S., landfills are regulated by the state's environmental agency that establishes minimum guidelines; however, none of these standards may fall below those set by the United States Environmental Protection Agency (EPA); such as was the case with the Fresh Kills Landfill in Staten Island, which is claimed by many to not only be the world's largest landfill, but the world's largest manmade structure.

The Fresno Municipal Sanitary Landfill, opened in Fresno, California in 1937, is considered to have been the first modern, sanitary landfill in the United States, innovating the techniques of trenching, compacting, and the daily covering of waste with soil. It has been designated a National Historic Landmark, underlining the significance of waste disposal in urban society.

Before the advent of modern landfills in America, most Americans lived in sparsely populated rural farming communities and most burned their garbage. Due to environmental and safety concerns, burning garbage by civilians has been outlawed by most municipalities and can only be performed by landfill managers or people who have obtained permits from the municipality. More information on landfill history in the United States can be found at.

The Resource Conservation and Recovery Act is a US federal law that is designed to protect the public from harm caused by waste disposal. The EPA runs a Landfill Methane Outreach Programme (LMOP), a voluntary assistance programme that helps to reduce methane emissions from landfills by encouraging the recovery and use of landfill gas as an energy resource. U.S. landfills consist of 40 per cent to 50 per cent paper waste, 20 per cent to 30 per cent construction debris, and 1.2 per cent disposable diapers.

Alternatives

The obvious alternative to landfills are waste reduction and recycling strategies. Secondary to not creating waste, there are various alternatives to landfills. In the late 20th century, alternative methods to waste disposal to landfill and incineration have begun to gain acceptance. Anaerobic digestion, composting, mechanical biological treatment, pyrolysis and plasma arc gasification have all began to establish themselves in the market.

In recent years, some countries, such as Germany, Austria, and Switzerland, have banned the disposal of untreated waste in landfills. In these countries, only the ashes from incineration or the stabilised output of mechanical biological treatment plants may still be deposited.

Radiation Pollution

11.1 Radiation Pollution

Radiation pollution is the increase in natural background radiation. There are many sources of radiation pollution such as research laboratories, nuclear power plants, etc. The worst case of the radiation pollution was the chernobyl disaster, which occurred over a centaury ago, but the effects still linger on today.

11.2 Radioactivity

The Industrial Pollution and Radiochemical Inspectorate (IPRI) is responsible for controlling the keeping and use of radioactive material and the disposal of radioactive waste under the Radioactive Substances Act 1993. Certificates of Registration are issued to permit the keeping and use of radioactive material and Certificates of Authorisation are used to permit the accumulation and disposal of radioactive waste. The Inspectorate also enforces the statutory control of the transport of radioactive material.

As well as regulatory responsibilities, IPRI also carries out environmental monitoring of the impact of nuclear discharges into the Irish Sea on the Northern Ireland coastline.

The Inspectorate has an important role to play in the Northern Ireland response should there be an overseas peacetime nuclear incident. During radioactive emergencies IPRI would work closely with other Northern Ireland Departments likely to be involved. We are also responsible for mapping levels of radon, a naturally occurring radioactive gas, in dwellings throughout Northern Ireland.

We are also responsible for mapping levels of Radon, a naturally occurring Radioactive Gas in dwellings throughout Northern Ireland.

11.3 Radioactive Pollution

Surface waters are a powerful factor that causes migration of radionuclides across the territory of Belarus. For this reason it is essential to take into due account the transit role of rivers in the transportation of radionuclides, including transboundary transfer. In watercourses and flowing water bodies concentration of radionuclides are reducing every year, but they tend to accumulate in sttatic water bodies (lakes, ponds, reservoirs, especially in bottom sediments).

Due to the accident at the Chernobyl nuclear power station (CNPS) one quarter of the country's territory was contaminated by caesium-137 (23%), strontium-90 (10%) and plutonium (about 2 %). Concentrations of caesium-137 by the year 2001.

According to the monitoring data the radiation situation in the Dnieper-Sozh and Pripyat basins is stable. The annual average concentration of caesium-137 has decreased significantly in large and small rivers for the observed 1987-2000 period. Exceeding of the National permissible levels (NPL-96, 99) for caesium-137 and strontium-90 were not observed. However, caesium-137 concentration in surface waters are closely connected with the annual volume of river flow, as can be seen from the increase in concentrations of caesium-137 in some rivers, where water supply was lower than the perennial average values.

The data analysis of concentration of caesium-137 during the spring flood in the Pripyat basin in 1999 shows that concentration of caesium-137 in a dissolved form in the Pripyat basin (the city of Mozyr) remains at the level of the average indices for the prior period (1996-1998). Yet concentration of this radionuclide has considerably increased in dredges. This means that caesium-137 is washed out and transported by flood flow with sediments.

According to the measurements of the total radioactivity of caesium-137, transferred by rivers during 12 years (1987-1999), 85 per cent were transferred by the Ipyt during 2 years, 81 per cent by the Sozh during 3 years, the Besed during 4 years, the Pripyat during 6 years, and 71 per cent by the Dnieper during 9 years. The comparison between the indices of caesium-137 transfer for the period of 1987-1993 at the end-point monitoring checkpoints on the Dnieper and the Pripyat in the Ukraine and at the monitoring checkpoints in Belarus shows that only 26 per cent of the total transfer of radioactive caesium-137 were generated within the 30-kilometer Chernobyl zone. Thus, in the river runoff formation large watersheds had greater impact on secondary river pollution within the first few years (3-4 years) after the Chernobyl accident than surface river runoff from highly contaminated territories with relatively small watersheds.

Since the radiation situation has stabilized, transboundary transport of radioactive elements through river flow has significantly decreased. Mainly it is the Pripyat that transports radionuclides, in particular strontium-90, as they are washed out from the 30-kilometer Chernobyl zone.

Due attention is devoted to studies of the radiation state of small rivers, which are tributaries of the Pripyat (the Braginka, Nesvich and Slovechna Rivers) and the Sozh (the Lipa and the Senna) in the most contaminated areas of the Gomel and Mogilev regions. Over the years radioactivity in water tends to decrease. The exception is the dissolved strontium -90, which is a specific feature for the CNPS zone.

Annual data on the content of caesium-137 and strontium-90 (in soluble and suspended forms) in the bottom sediments and water biota suggest that bottom sediments and water biota are significant contributors to the total radioactivity of surface water systems. A tendency towards a reduction in radioactivity of bottom sediments and water biota is minor.

During spring high water and summer-autumn flood, migration of radionuclides into open water systems occurs both in soluble and in absorbed forms on organic and mineral carrieres.

The ratio between concentrations of caesium-137 and strontium-90 at the end checkpoints of the Braginka and Senna Rivers suggests that starting from1992-1993, the concentration of strontium-90 has begun to exceed that of caesium-137. That phenomenon is characteristic for surface watercourses close to the CNPS zone and is explained by an increase in migration ability of strontium-90 due to its release from active particles.

11.4 Radiation Pollution Prevention: A Cash Study of Environment and Heritage Service

The Industrial Pollution and Radiochemical Inspectorate (IPRI) is responsible for controlling the keeping and use of radioactive material and the disposal of radioactive waste under the Radioactive Substances Act 1993. Certificates of Registration are issued to permit the keeping and use of radioactive material and Certificates of Authorisation are used to permit the accumulation and disposal of radioactive waste. The Inspectorate also enforces the statutory control of the transport of radioactive material.

As well as regulatory responsibilities, IPRI also carries out environmental monitoring of the impact of nuclear discharges into the Irish Sea on the Northern Ireland coastline.

The Inspectorate has an important role to play in the Northern Ireland response should there be an overseas peacetime nuclear incident. During radioactive emergencies IPRI would work closely with other Northern Ireland Departments likely to be involved. We are also responsible for mapping levels of radon, a naturally occurring radioactive gas, in dwellings throughout Northern Ireland.

We are also responsible for mapping levels of Radon, a naturally occurring Radioactive Gas in dwellings throughout Northern Ireland.

Radon

Radon is a naturally occurring radioactive gas formed as a result of the radioactive decay of uranium, which is present to some extent in all rocks and soils, but amounts vary from place to place. Radon rises from the soil into the air, outdoors, radon is diluted into the air and the risk it poses is negligible. Where it enters buildings the levels can vary from property to property and even between neighboring dwellings. Radon can build up in homes and, at high concentrations, leads to an increased risk of lung cancer. Smokers who are in a radon affected house are at an increased risk. Radon dominates the radioactive dose to the population contributing about 50 per cent of dose at the average level in houses.

Radioactivity is measured in Becquerels. The Government has established an Action Level for radon in homes of 200 Becquerels per cubic metre (Bq/m3), based on advice from the Health Protection Agency (HPA).

The Radon in Dwellings Report produced in 1999 by the Environment and Heritage Service includes a radon map of Northern Ireland and the risk and information on radon tests carried out in the Province. To date there have been 15700 houses tested in Northern Ireland. The average level is low at 19 Bq/m3, similar to the rest of the United Kingdom. The radon map shows that there are areas in the West of the Province and the Southeast where radon risk is elevated and it is estimated that 4000 homes in Northern Ireland are likely to exceed the action level. Free booklets can be obtained from the Industrial Pollution and Radiochemical Inspectorate (IPRI) in the Klondyke Building, Cromac Avenue, Gasworks Business Park, Lower Ormeau Road Belfast. Telephone: (028) 9056 9299. The booklets are

1. Radon: A Householder's Guide (pdf-313 KB)
2. Radon: A guide to reducing levels in your home (pdf-603 KB)
3. Radon: A guide to homebuyers and sellers (pdf-243 KB)
4. Radon Factsheet (pdf-195 KB)

If you wish to discuss any technical matters about radon please contact (028) 9056 9305. If anyone feels that they would like a test done on their house they could contact HPA (Health Protection Agency), Chilton, Didcot, Oxfordshire, Telephone freephone 0800 614529 and they will issue the test kits which cost in the region of £38.00, HPA will also confirm the cost.

The control of exposure to radon in commercial and public buildings, including schools, is the responsibility of the Health and Safety Executive for Northern Ireland (Ladas Drive, Belfast) of the Department of Enterprise, Trade and Investment or the appropriate District Council.

Anyone wishing to view maps showing levels of Radon in dwellings in the Republic of Ireland should visit the website of the Radiological Protection Institute of Ireland.

Radiation Monitoring

Background levels of radiation are monitored at various locations in the City by Pollution Team officers. Additionally, a radiation monitor continuously registers background levels at Walbrook Wharf. The data from the Walbrook Wharf monitor can be seen on the internet at www.weatherprobe.com. Officers are immediately alerted by the equipment if there are any significant changes in the background data.

Environmental Monitoring

The Environment Agency (EA) is responsible under the Radioactive Substances Act 1993 for authorising discharges from Sellafield. The Centre for Environment, Fisheries and Aquaculture Science (CEFAS), an Executive Agency within the Ministry of Agriculture, Fisheries and Food (MAFF), has been sampling seawater, fish, shellfish and sediments in the Irish Sea since the early 1950's. In Northern Ireland the Environment and Heritage Service's Industrial Pollution and Radiochemical Inspectorate monitors the impact of Sellafield discharges on the Northern Ireland coastal environment.

The Industrial Pollution & Radiochemical Inspectorate arranges for samples of seaweed, sediment, fish, nephros, and winkles to be collected quarterly and forwarded to the CEFAS Research Laboratory at Lowerstoft. The seaweeds are collected in the Ards Peninsula and at Portrush; the marine life samples are obtained as far as possible from commercial landings at Kilkeel and Portavogie. Sediment samples are collected from the five marine loughs.

The Northern Ireland results are published annually in a report entitled "Radioactivity in Food and the Environment" (pdf-3.32 MB) issued by the Environment Agency, Environment and Heritage Service, Food Standards Agency and the Scottish Environment Protection Agency. The results are also published in the Northern Ireland Abstract of Statistics. The levels of radioactivity measured indicate that they are of negligible radiological significance.

Radioactive Emergency Response

Following the Chernobyl accident in April 1996 the Northern Ireland Emergency Committee set up a specialised group, the Northern Ireland Technical Advice Group (NITAG). Civil Contingencies Policy Branch (CCPB) ensure that the most efficient and effective response can be made to assist the public during, and in the aftermath of a civil emergency.

NITAG operates under the Chairmanship of the Chief Radiochemical Inspector with professional membership of the group drawn from the Inspectorate and those Northern Departments likely to be involved in an overseas peacetime nuclear emergency. Under the National Response Plan, the Radiation Incident Monitoring Network (RIMNET) is in place throughout the UK. There are 92 monitoring stations continuously measuring gamma radiation dose rates throughout the United Kingdom with 5 stations in Northern Ireland. The 5 stations are at Ballypatrick, Aldergrove, Glenanne, St Angelo and Carnmoney.

The Inspectorate play a major role in NITAG activities locally, participates in the Radiation Incident Monitoring Co-ordinating Committee (RIMCC) and also take part in various exercises using the RIMNET system to test its reaction capability and effectiveness.

Near-Surface Disposal Facilities for Radioactive Waste

Guidance on Requirements for Authorisation

In 1997, we published guidance on regulating the disposal of low and intermediate level radioactive waste to specialised facilities on land. We have now decided to update this guidance because of developments in UK law and policy, revised international advice and recommendations and also recognition that specific guidance for near-surface facilities was required. We are jointly publishing this guidance with the Environment Agency and the Scottish Environment Protection Agency.

The guidance sets out the principles and requirements that we would expect a developer or operator of a radioactive waste disposal facility to meet to ensure protection of people and the environment, both now and in the future.

Deep Geological Facilities for Radioactive Wastes

Guidance on Requirements for Authorisation

In 1997, we published guidance on regulating the disposal of low and intermediate level radioactive waste to specialised facilities on land. We have now decided to update this guidance because of developments in UK law and policy, revised international advice and recommendations and the need to address disposal of higher activity wastes, which were not covered by the 1997 guidance. We are jointly publishing this guidance with the Environment Agency.

The guidance sets out the principles and requirements that we would expect a developer or operator of a radioactive waste disposal facility to meet to ensure protection of people and the environment, both now and in the future.

Control of Major Accident Hazards

The Seveso II Directive on the control of major accident hazards involving dangerous substances requires that operators of prescribed major hazard sites 'take all measures necessary to prevent major accidents and to limit their consequences for man and the environment'. The Directive is implemented in Northern Ireland by the Control of Major Accident Hazards Regulations (Northern Ireland) 2000.

Threshold quantities above which the Regulations apply are listed for a range of named dangerous substances and also for some classes of substances including 'toxic', 'explosive', 'flammable' and 'dangerous for the environment'. The Regulations are enforced by a Competent Authority comprising the Health and Safety Executive for Northern Ireland (HSENI) and the Environment and Heritage Service. EHS is responsible for assessing measures to protect the environment under the Regulations. This includes examining operators' safety reports and carrying out inspections on establishments covered by the regulations.

Biopollution or Bioinvasion

12.1 Concept and Use

Biological Response to Stress

Organisms are subjected to a number and variety of stressors in the environment, therefore multiple measures of health are needed to help identify and separate anthropogenic-induced effects of stress from those effects caused by natural stressors.

Multiple measures of health include responses that represent different levels of biological organization, response time to stressors, and sensitivity and specificity to stressors.

Environmental stressors can have direct (through metabolic pathways) and indirect (through food and habitat availability) effects on organisms.

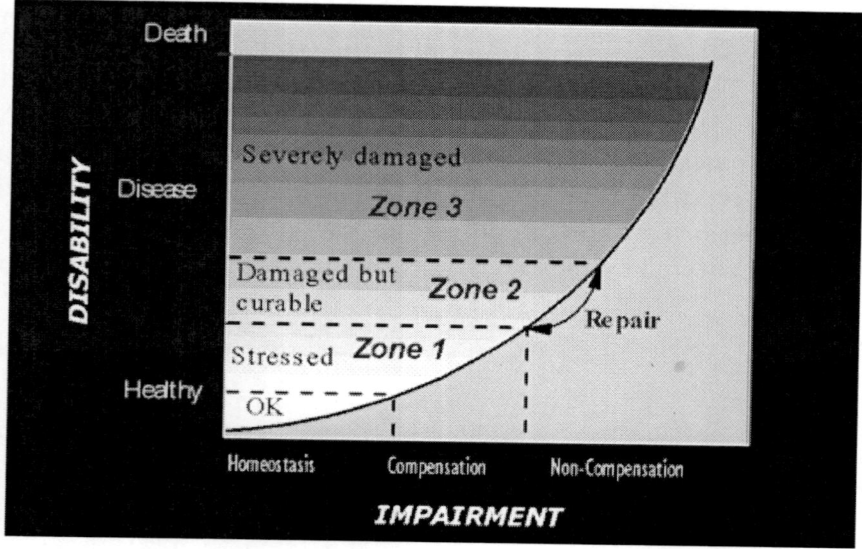

Bioindicators that can be measured to characterize the physiological condition of organisms in each stress zone (see above figure). Indicators in zone 1 are more sensitive to environmental stressors, while those indicators in zone 3 are less sensitive, but have higher ecological relevance.

The physiological condition of organisms can be characterized by three major stress zone based on the level of disability caused by environmental stressors.

If properly calibrated to higher level responses, the lower-level responses can be effectively used in environmental management and in the ecological risk assessment process.

12.2 Mechanical Biological Treatment

A mechanical biological treatment system is a form of waste processing facility that combines a sorting facility with a form of biological treatment such as composting or anaerobic digestion. MBT plants are designed to process mixed household waste as well as commercial and industrial wastes.

Process

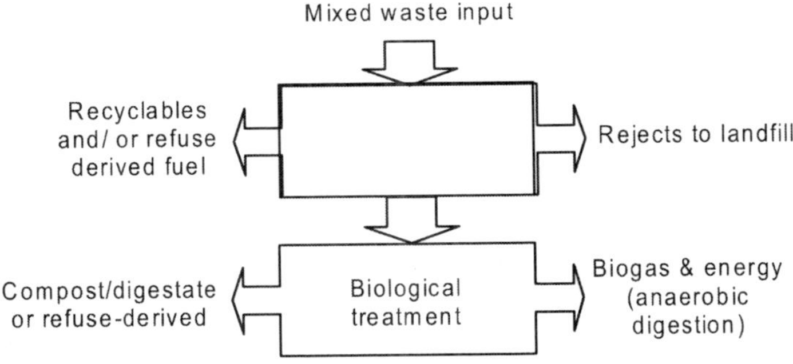

Figure 12.1: Process Flow Chart

The terms 'mechanical biological treatment' or 'mechanical biological pre-treatment' related to a group of solid waste treatment systems. These systems enable the recovery of materials contained within the and stabilisation of the biodegradable component of the material.

The sorting component of the plants resemble a materials recovery facility. This component is either configured to recover the individual elements of the waste or produce a refuse-derived fuel that can be used for the generation of power.

The components of the mixed waste stream that can be recovered include:

❖ Metals
❖ Plastics
❖ Glass

Terminology

MBT is also sometimes termed BMT—biological mechanical treatment—however this simply refers to the order of processing, i.e. the biological phase of the system precedes the mechanical sorting. MBT should not be confused with MHT—*mechanical heat treatment*—which does not include any form of biological degradation or stabilisation.

Mechanical Sorting

The "mechanical" element is usually an automated mechanical sorting stage. This either removes recyclable elements from a mixed waste stream (such as metals, plastics, glass and paper) or processes them. It typically involves factory style conveyors, industrial magnets, eddy current separators, trommels, shredders and other tailor made systems, or the sorting is made by hand. The mechanical element has a number of similarities to a materials recovery facility (MRF).

Some systems integrate a wet MRF to recover & wash the recyclable elements of the waste in a form that can be sent for recycling. MBT can alternatively process the waste to produce a high calorific fuel given the term refuse derived fuel (RDF). RDF can be used in cement kilns or power plants and is generally made up from plastics and biodegradable organic waste. Systems which are configured to produce RDF include the Herhof and Ecodeco Processes. It is a common misconception that all MBT processes produce RDF. This is not the case and depends strictly on system configuration and suitable local markets for MBT outputs.

Biological Processing

The "biological" element refers to either:

❖ Anaerobic digestion
❖ Composting
❖ Biodrying

Anaerobic digestion breaks down the biodegradable component of the waste to produce biogas and soil improver. The biogas can be used to generate electricity and heat.

Biological can also refer to a composting stage. Here the organic component is treated with aerobic microorganisms. They break down the waste into carbon dioxide and compost. There is no green energy produced by systems employing only composting treatment for the biodegradable waste.

In the case of biodrying, the waste material undergoes a period of rapid heating through the action of aerobic microbes. During this partial composting stage the heat generated by the microbes result in rapid drying of the waste. These systems are often configured to produce a refuse-derived fuel where a dry, light material is advantageous for later transport combustion.

Some systems incorporate both anaerobic digestion and composting. This may either take the form of a full anaerobic digestion phase, followed by the maturation (composting) of the digestate. Alternatively a partial anaerobic digestion phase can be induced on water that is percolated through the raw waste, dissolving the readily available sugars, with the remaining material being sent to a windrow composting facility.

By processing the biodegradable waste either by anaerobic digestion or by composting MBT technologies help to reduce the contribution of greenhouse gases to global warming.

Usable wastes for this system:

- ❖ Municipal solid waste
- ❖ Sewage sludge

Products of this system:

- ❖ Recycable materials such as metals, paper, plastics, glass etc.
- ❖ Organic fertilizer (separate collection of organic waste).
- ❖ Unusable materials prepared for their unharmful final deposit (compaction > 1.3 t/m^3).
- ❖ Carbon credits—additional revenues.
- ❖ High calorific fraction (refuse derived fuel—RDF)—additional revenues.

Further advantages:

- ❖ The finally deposited waste is inert.
- ❖ Reduction of the waste volume to be deposited to at least a half (density > 1.3 t/m^3), thus the lifetime of the landfill is at least twice as long as usually.
- ❖ Utilization of the leachate in the process.

❖ No unbidden guests such as birds, dogs, vermin, rats on site
❖ No additional facilities for the collection and combustion of biogas
  as there is no biogas
❖ Daily covering not necessary
❖ Aftercare 3 to 5 years

Consideration of Applications

MBT systems can form an integral part of a region's waste treatment infrastructure. These systems are typically integrated with kerbside collection shemes. In the event that a refuse-derived fuel is produced as a by-product then a combustion facility would be required.

Alternatively MBT solutions can diminish the need for home separation and kerbside collection of recyclable elements of waste. This gives the ability of local authorities and councils to reduce the use of waste vehicles on the roads and keep recycling rates high.

Position of Environmental Groups

Friends of the Earth suggests that the best environmental route for residual waste is to firstly remove the remaining recyclable elements from the waste stream (such as metals, plastics and some paper). The small amount of waste remaining should be composted or anaerobically digested and unless sufficiently clean to be used as a compost should be disposed of by incineration or—as the least favourable option—in landfills. MBT plants that fit in with these statements could therefore play an increasing role in the environmental management of mixed streams.

13

Hazardous Materials/Hazardous Wastes

13.1 Dangerous Goods

A dangerous good is any solid, liquid, or gas that can harm people, other living organisms, property, or the environment. An equivalent term, used almost exclusively in the United States, is *hazardous material* (hazmat). Dangerous goods may be radioactive, flammable, explosive, toxic, corrosive, biohazardous, an oxidizer, an asphyxiant, a pathogen, an allergen, or may have other characteristics that render it hazardous in specific circumstances.

Mitigating the risks associated with hazardous materials may require the application of safety precautions during their transport, use, storage and disposal. Most countries regulate hazardous materials by law, and they are subject to several international treaties as well.

Persons who handle dangerous goods will often wear protective equipment, and metropolitan fire departments often have a response team specifically trained to deal with accidents and spills. These teams train with different organizations at a variety of specialized locations. Some of the most well-known in the U.S. and Canada include the California Specialized Training Institute, the Texas A&M TEEX Academy, Signet North America, the Justice Institute of British Columbia, and the U.S. National Fire Academy. Persons who handle or potentially come into contact with dangerous goods as part of their work are also often subject to monitoring or health surveillance to ensure that their exposure does not exceed occupational exposure limits.

Laws and regulations on the use and handling of hazardous materials may differ depending on the activity and status of the material. For example one set of requirements may apply to their use in the workplace while a different requirements may apply to spill response, sale for consumer use, or transportation. Most countries regulate some aspect of hazardous materials.

The most widely applied regulatory scheme is that for the transportation of dangerous goods. The Committee of Experts on the Transport of Dangerous

Goods of the United Nations Economic and Social Council issues Model Regulations on the Transportation of Dangerous Goods. Most regional and national regulatory schemes for hazardous materials are harmonized to a greater or lesser degree with the UN Model Regulation. For instance, the International Civil Aviation Organization has developed regulations for air transport of hazardous materials that are based upon the UN Model but modified to accommodate unique aspects of air transport. Individual airline and governmental requirements are incorporated with this by the International Air Transport Association to produce the widely used IATA Dangerous Goods Regulations. Similarly, the International Maritime Organization has developed the IMO Dangerous Goods Regulations for transportation on the high seas. Many individual nations have also structured their dangerous goods transportation regulations to harmonize with the UN Model in organization as well as in specific requirements.

Dangerous goods are divided into classes on the basis of the specific chemical characteristics producing the risk.

Note: The graphics and text in this article representing the dangerous goods safety marks are derived from the United Nations-based system of identifying dangerous goods. Not all countries use precisely the same graphics (label, placard and/or text information) in their national regulations. Some use graphic symbols, but without English wording or with similar wording in their national language. Refer to the Dangerous Goods Transportation Regulations of the country of interest.

For example, see the Dangerous Goods Safety Marks in the Canadian Transportation of Dangerous Goods Regulations.

The statement above applies equally to all the Dangerous Goods classes discussed in this article.

Classification and Labeling Summary Tables

Class 1: Explosives

Information on this graphic changes depending on which, "Division" of explosive is shipped. Explosive Dangerous Goods have compatibility group letters assigned to facilitate segregation during transport. The letters used range from A to S excluding the letters I, M, O, P, Q and R. The example above shows an explosive with a compatibility group "A" (shown as 1.1A). The actual letter shown would depend on the specific properties of the substance being transported.
For example, the Canadian Transportation of Dangerous Goods Regulations provides a description of compatibility groups.

1.1 Explosives with a mass explosion hazard
- Ex: TNT, dynamite, nitroglycerine.

1.2 Explosives with a severe projection hazard.

1.3 Explosives with a fire, blast or projection hazard but not a mass explosion hazard.

1.4 Minor fire or projection hazard (includes ammunition and most consumer fireworks).

1.5 An insensitive substance with a mass explosion hazard (explosion similar to 1.1).

1.6 Extremely insensitive articles.

The United States Department of Transportation (DOT) regulates hazmat transportation within the territory of the US.

1.1—Explosives with a mass explosion hazard. (nitroglycerin/dynamite)

1.2—Explosives with a blast/projection hazard.

1.3—Explosives with a minor blast hazard. (rocket propellant, display fireworks)

1.4—Explosives with a major fire hazard. (consumer fireworks, ammunition)

1.5—Blasting agents.

1.6—Extremely insensitive explosives.

Hazardous Materials

Class 1: Explosives

Hazardous Materials

Class 1.1: Explosives

Mass Explosion Hazards

Hazardous Materials

Class 1.2: Explosives

Blast/Projection Hazards

Hazardous Materials

Class 1.3: Explosives

Minor Blast Hazard

Hazardous Materials

Class 1.4: Explosives

Major Fire Hazards

Hazardous Materials

Class 1.5: Explosives

Blasting Agents

Hazardous Materials

Class 1.6: Explosives

Class 2: Gases

Gases which are compressed, liquefied or dissolved under pressure as detailed below. Some gases have subsidiary risk classes; poisonous or corrosive.

2.1 *Flammable Gas:* Gases which ignite on contact with an ignition source, such as acetylene and hydrogen.

2.2 *Non-Flammable Gases:* Gases which are neither flammable nor poisonous. Includes the cryogenic gases/liquids (temperatures of below −100°C) used for cryopreservation and rocket fuels, such as nitrogen and neon.

2.3 *Poisonous Gases:* Gases liable to cause death or serious injury to human health if inhaled; examples are fluorine, chlorine, and hydrogen cyanide.

Class 3: Flammable Liquids

Flammable liquids included in Class 3 are included in one of the following packing groups:

❖ Packing Group I, if they have an initial boiling point of 35°C or less at an absolute pressure of 101.3 kPa and any flash point, such as diethyl ether or carbon disulfide;

❖ Packing Group II, if they have an initial boiling point greater than 35°C at an absolute pressure of 101.3 kPa and a flash point less than 23°C, such as gasoline (petrol) and acetone; or

❖ Packing Group III, if the criteria for inclusion in Packing Group I or II are not met, such as kerosene and diesel.

Note: For further details, check the Dangerous Goods Transportation Regulations of the country of interest.

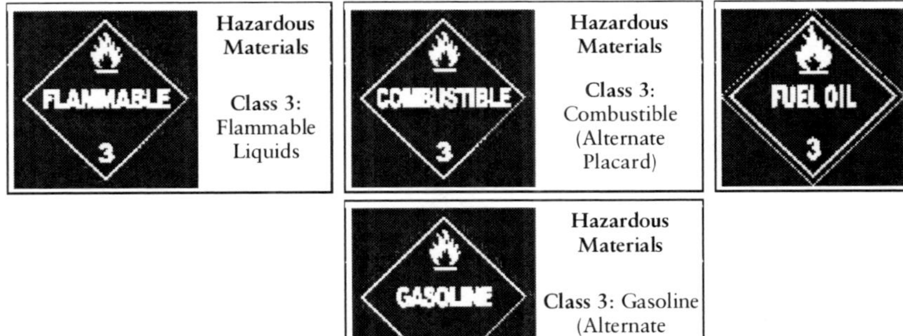

Class 4: Flammable Solids

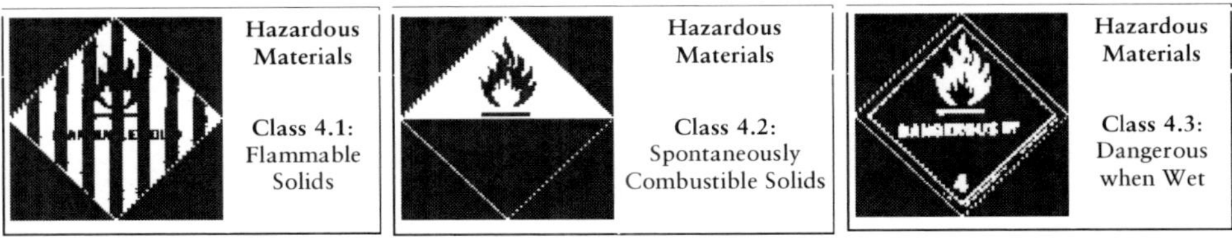

4.1 *Flammable Solids:* Solid substances that are easily ignited and readily combustible (nitrocellulose, magnesium, safety or strike-anywhere matches).

4.2 *Spontaneously Combustible:* Solid substances that ignite spontaneously (aluminium alkyls, white phosphorus).

4.3 *Dangerous when Wet:* Solid substances that emit a flammable gas when wet or react violently with water (sodium, calcium, potassium, calcium carbide).

Class 5: Oxidizing Agents and Organic Peroxides

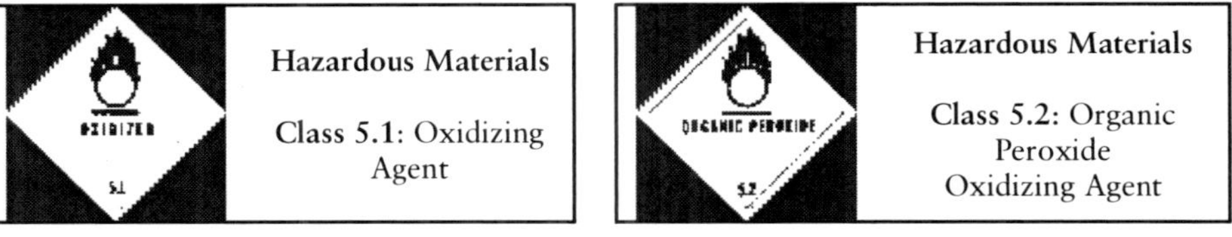

5.1 Oxidizing agents other than organic peroxides (calcium hypochlorite, ammonium nitrate, hydrogen peroxide, potassium permanganate).

5.2 Organic peroxides, either in liquid or solid form (benzoyl peroxides, cumene hydroperoxide).

Class 6: Toxic and Infectious Substances

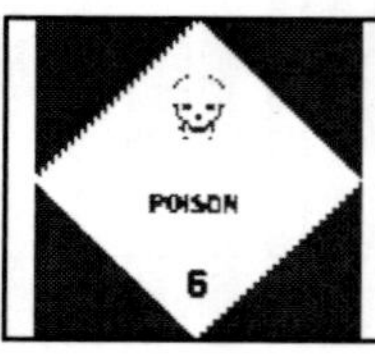

Hazardous
Materials
Class 6.1: Poison

Hazardous
Materials
Class 6.2:
Biohazard

6.1a Toxic substances which are liable to cause death or serious injury to human health if inhaled, swallowed or by skin absorption (potassium cyanide, mercuric chloride).

6.1b (Now PGIII) Toxic substances which are harmful to human health (N.B this symbol is no longer authorized by the United Nations) (pesticides, methylene chloride).

Class 7: Radioactive Substances   Class 8: Corrosive Substances

Hazardous Materials

Class 7: Radioactive

Hazardous Materials

Class 8: Corrosive

Class 9: Miscellaneous

Hazardous
Materials
Hazard Symbol:
C/Corrosive

Hazardous
Materials
Hazard
Symbol:
E/Explosive

Hazardous
Materials
Hazard
Symbol:
F/Flammable

Hazardous
Materials
Hazard Symbol:
N/Environmental
Hazard

Hazardous
Materials
Hazard
Symbol:
O/Oxidizing

Hazardous
Materials
Hazard
Symbol:
T/Toxic

Hazardous
Materials
Hazard
Symbol:
Xn/Harmful;
Xi/Irritant

Australia

Australia uses the standard international UN numbers with a few slightly different signs on the back, front and sides of vehicles carrying hazardous substances. The country uses the same "HAZCHEM" as the UK HAZCHEM Code to provide advisory information to emergency services personnel in the event of an emergency situation.

Canada

Transportation of dangerous goods (hazardous materials) in Canada by road is normally a provincial jurisdiction. The federal government has jurisdiction over air, most marine, and most rail transport. The federal government acting centrally created the federal transportation of dangerous goods act and regulations, which provinces adopted in whole or in part via provincial transportation of dangerous goods legislation. The result is that all provinces use the federal regulations as their standard within their province; some small variances can exist because of provincial legislation. Creation of the federal regulations was coordinated by Transport Canada. Hazard classifications are based upon the UN Model.

The province of Nova Scotia's dangerous goods transportation act.

The province of Nova Scotia's dangerous goods transportation regulations.

Europe

The European Union has passed numerous directives and regulations to avoid the dissemination and restrict the usage of hazardous substances, the most famous being the Restriction of Hazardous Substances Directive and the REACH regulation. There are also long standing European treaties such as ADR and RID that regulate the transportation of hazardous materials by road, rail, river and inland waterways, following the guide of the UN Model Regulation.

United States

Due to the increased threat of terrorism in the early 21st century, funding for greater HAZMAT-handling capabilities was increased throughout the United States, in recognition of the fact that flammable, poisonous, explosive, or radioactive substances in particular could make attractive weapons for terrorist attacks.

The United States Department of Transportation (DOT) regulates hazmat transportation within the territory of the US. The regulations are found in 49 CFR (Title 49 of the Code of Federal Regulations).

The U.S. Occupational Safety and Health Administration (OSHA) regulates the handling of hazardous materials in the workplace as well as response to hazardous materials-related incidents, most notably through HAZWOPER (HAZ-ardous W-aste OP-erations and E-mergency R-esponse) regulations found at 29 CFR 1910.120.

The Environmental Protection Agency regulates hazardous materials as they may impact the community and environment, including specific regulations for environmental cleanup and for handling and disposal of waste hazardous materials.

The Consumer Product Safety Commission regulates hazardous materials that may be used in products sold for household and other consumer uses.

Hazard Classes for Materials in Transport

Following the UN Model, the DOT divides regulated hazardous materials into nine classes, some of which are further divided into divisions. Hazardous materials in transportation must be placarded and have specified packaging and labelling. Some materials must always be placarded, others may only require placarding in certain circumstances.

Trailers of goods in transport are usually marked with a four digit UN (United Nations) number. This number can be referenced by first responders (Firefighters, Police Officers, and ambulance personnel) who can find information about the material in the *Emergency Response Guidebook*.

Fixed Facilities

Different standards usually apply for handling and marking HAZMATs at fixed facilities, including NFPA 704 diamond markings (a consensus standard often adopted by local governmental jurisdictions), OSHA regulations requiring chemical safety information for employees, and CPSC requirements requiring informative labeling for the public, as well as wearing Hazmat suits when handling hazardous materials.

13.2 Hazardous Materials

Chemicals are found everywhere. They purify drinking water, increase crop production, and simplify household chores. But chemicals also can be hazardous to humans or the environment if used or released improperly. Hazards can occur during production, storage, transportation, use, or disposal. You and your community are at risk if a chemical is used unsafely or released in harmful amounts into the environment where you live, work, or play.

Hazardous materials in various forms can cause death, serious injury, long-lasting health effects, and damage to buildings, homes, and other property. Many products containing hazardous chemicals are used and stored in homes routinely. These products are also shipped daily on the nation's highways, railroads, waterways, and pipelines.

Chemical manufacturers are one source of hazardous materials, but there are many others, including service stations, hospitals, and hazardous materials waste sites.

Varying quantities of hazardous materials are manufactured, used, or stored at an estimated 4.5 million facilities in the United States—from major industrial plants to local dry cleaning establishments or gardening supply stores.

Hazardous materials come in the form of explosives, flammable and combustible substances, poisons, and radioactive materials. These substances are most often released as a result of transportation accidents or because of chemical accidents in plants.

13.3 Hazardous Waste

The term hazardous waste comprises all toxic chemicals, radioactive materials, and biologic or infectious waste. These materials threaten workers through occupational exposure and the general public in their homes, communities, and general environment. Exposure to these materials can occur near the site of generation, along the path of its transportation, and near their ultimate disposal sites. Most hazardous waste results from industrial processes that yield unwanted byproducts, defective products, and spilled materials. The generation and disposal of hazardous wastes is controlled through a variety of international and national regulations.

Hazardous waste was formerly known as 'special' waste.

In addition to releasing gases and particles into the atmosphere, humans produce waste that is dumped on the environment. Often, this waste is hazardous and dangerous to both nature and human life.

The levels of dangerous wastes continue to grow. Industries and individuals continue to be largely unaware of this major environmental problem.

As a result, many people and industries are failing to prevent the creation of hazardous waste or to limit the negative effects it produces.

Individuals often throw out goods without realizing that they are headed for a landfill and could be dangerous for the environment. No matter where people put these hazardous waste materials, there is always a chance that they could find their way into the ground, and eventually into our bodies.

Corporations usually want to avoid the costs associated with having to limit creation of hazardous waste. Consequently, they build landfills on site and fill them with waste, or sometimes pay to have their waste removed. Often, hazardous materials are transported to areas that accept money to take the waste.

It may prove very difficult to reduce hazardous waste in the future. Unlike many other environmental problems, waste creation is something people do not often think about.

In the future, people may have to reduce not only their generation of hazardous waste, but also their consumption of many products that end up in landfills.

Environmental Pollution: Bioindicators and Biological markers

14.1 Bioindicators

A bioindicator is an anthropogenically-induced response in biomolecular, biochemical, or physiological parameters that has been causally linked to biological effects at one or more of the organism, population, community, or ecosystem levels of biological organization.

Bioindicators are organisms, such as lichens, birds, and bacteria, that are used to monitor the health of the environment. The organisms are monitored for changes that may indicate a problem within their ecosystem. The changes can be chemical, physiological, or behavioural.

Uses and Types of Bioindicators

Each organism within an ecosystem has the ability to report on the health of its environment. Bioindicators are used to:

- detect changes in the natural environment;
- monitor for the presence of pollution and its effect on the ecosystem in which the organism lives;
- monitor the progress of environmental cleanup; and
- test substances, like drinking water, for the presence of contaminants.

Types of bioindicators and their uses include the following:

- *Plant Indicators*—The presence or absence of certain plant or other vegetative life in an ecosystem can provide important clues about the health of the environment. Lichens, often found on rocks and tree trunks, are organisms consisting of both fungi and algae. They respond to environmental changes in forests, including changes in forest structure, air quality, and climate. The disappearance of lichens in a forest may indicate environmental stresses, such as high levels of sulfur dioxide, sulfur-based pollutants, and nitrogen.

❖ *Animal Indicators*—An increase or decrease in an animal population may indicate damage to the ecosystem caused by pollution. For example, if pollution causes the depletion of important food sources, animal species dependent upon these food sources will also be reduced in number. In addition to monitoring the size and number of certain species, other mechanisms of animal indication include monitoring the concentration of toxins in animal tissues, or monitoring the rate at which deformities arise in animal populations.

❖ *Microbial Indicators*—Microorganisms can be used as indicators of aquatic or terrestrial ecosystem health. Found in large quantities, microorganisms are easier to sample than other organisms. Some microorganisms will produce new proteins, called stress proteins, when exposed to contaminants like cadmium and benzene. These stress proteins can be used as an early warning system to detect low levels of pollution.

The Science

Specific physiological and behavioural changes in bioindicators are used to detect changes in environmental health. The specific changes differ from organism to organism. The use of organisms as bioindicators encompasses many areas of science. Wildlife conservation genetics is an example of how traditional approaches can be combined with emerging biotechnologies to improve accuracy, and to collect information not available through conventional methods. Wildlife conservation genetics combines traditional monitoring of wildlife populations, like racoons, with the scientific discipline of genetics, to gain information about the health of ecosystems.

Behavioural and population changes in a species can be observed by scientists, but physiological changes must be detected using special tests. Bioassays require samples from organisms to detect changes in the environment. These tests may be used to ensure drinking water safety or to measure river health. In the future, as research identifies new ways to use microbes, these uses will expand to include testing of soil and air.

Bioassays can be carried out in traditional ways and with new biotechnology-derived methods. These methods are outlined in more detail below.

Case Study: Testing Water

Bioluminescent bacteria are being used to test water for environmental toxins. If there are toxins present in the water, the cellular metabolism of the bacteria is inhibited or disrupted. This affects the quality or amount of light emitted by the bacteria. Unlike traditional tests, this one is very quick—taking from five to 30 minutes to complete. However, it only indicates the presence of a toxin and cannot identify the specific toxin causing the change in the organism.

Traditional Bioassays

In traditional bioassays, a bioindicator organism is introduced to environmental samples, such as soil or water, and researchers observe any changes that occur as a result of exposure. These methods are based primarily on observation to detect changes. Examples of traditional bioassay methods include the following:

- ❖ Measurement of plant root growth in suspected polluted environments and comparison of the measured growth rate against normal root growth rates.
- ❖ Exposure of microorganisms to an environment and observation of any changes in the organism related to toxin exposure, such as the presence of stress proteins produced when cells are exposed to harmful environmental conditions.

Biotechnology-Based Bioassays

Several biotechnology-based methods use microorganisms to test environmental health. Unlike traditional methods, biotechnology-based bioassays do not rely on observation alone but set out to create specific reactions that indicate the presence of a specific pollutant or an unwanted microorganism. In this way they are similar to traditional chemical analysis of environmental samples.

- ❖ *DNA Microarray Technology*—Using DNA microarray technology, environmental samples, such as water, are tested for the actual genetic material of an organism. This form of testing is used to detect dangerous microorganisms in the environment, such as *E. Coli* bacteria in water. DNA microarrays are stamp-sized glass or silicon microchips that are embedded with thousands of single-stranded DNA or RNA. In this case, the DNA is that of the microorganisms being tested for. The microarrays are manufactured using samples of microorganisms. If the same type of microorganisms are present in the water sample, the DNA or RNA on the array will react with the complementary DNA or RNA of the microorganism in the sample. This identifies its presence in the sample. When these tests are fully developed, it will take as little as four hours to test for microbial presence in environmental samples, such as drinking water and soil. Traditional chemical-based tests take an average of 48 hours.

❖ *Fluorescence in-situ hybridisation (FISH)*—This is a method for detecting the presence of particular genes in a sample. As a bioindicator, FISH can determine whether or not specific microorganisms are polluting certain areas. It does this by testing environmental samples for the presence of microbial genes. A fluorescent marker is attached to the DNA of the type of microorganism being testing. This marked DNA is now called a 'probe.' Environmental samples are fixed onto a slide, and the slide is exposed to the flourescent DNA probe. If the polluting microorganism is present on the slide, its DNA will bind to the fluorescent probe, causing the slide to glow with ultraviolet light. Detection of this ultraviolet light by a special fluorescent microscope demonstrates the presence of polluting microorganisms in the sample.

Biotechnology and Bioindicators

Currently, biotechnology-based tests are being used to identify changes in indicator species to gauge the presence of pollutants in the environment. Many of the tests being developed are designed to detect pollutants in rivers and drinking water sources. These tests will be faster and more accurate than conventional tests in detecting metabolic changes within microorganisms. Microbial bioindicators are also being researched to detect pollution in other substances, such as soil.

Current Research Areas in Bioindicators

Bioindicator research is currently focussed on developing more rapid and reliable tests for the presence of microorganisms in water and soil. Tests for drinking water are a special area of concern for both developed and developing countries. Although biotechnology-based tests currently exist for drinking water, there are still many pollutants that are not detectable. Scientists are busy trying to replace time consuming, traditional methods with newer, faster, and more reliable tests based on biotechnology.

Sustainable Development and Bioindicators

Bioindicators are a method of monitoring or detecting the negative impacts that industrial activity has on the environment. This information helps develop strategies that will prevent or lower such effects and make industry more sustainable. The role of bioindicators in sustainable development will help ensure that industry leaves the smallest footprint possible on the environment.

14.2 Bioindicator

Bioindicators are species or chemicals used to monitor the health of an environment or ecosystem. They are any biological species or group of species whose function, population, or status can be used to determine ecosystem or environmental integrity. An example of such a group are the copepods and other small water crustaceans present in many water bodies. Such organisms are monitored for changes (chemical, physiological, or behavioural) that may indicate a problem within their ecosystem.

Depending on the organism selected and their use, there are three types of bioindicators:

1.   Plant indicators
2.   Animal indicators
3.   Microbial and chemical indicators

Plant Indicators

Plant Indicators—The presence or absence of certain plant or other vegetative life in an ecosystem can provide important clues about the health of the environment-(environmental preservation).

Lichens-(not a *plant*), found on rocks and tree trunks, are organisms comprising both fungi and algae. They respond to environmental changes in forests, including changes in forest structure-(conservation biology), air quality, and climate. The disappearance of lichens in a forest may indicate environmental stresses, such as high levels of sulfur dioxide, sulfur-based pollutants, and nitrogen-oxides. The composition and total biomass of algal species in aquatic systems serves as an important metric for organic pollution and nutrient loading such as nitrogen and phosphorus.

Animal Indicators, and Toxins

Animal Indicators—An increase or decrease in an animal population may indicate damage to the ecosystem caused by pollution. For example, if pollution causes the depletion of important food sources, animal species dependent upon these food sources will also be reduced in number-population decline. Overpopulation, can be the result of opportunistic species growth. In addition to monitoring the size and number of certain species, other mechanisms of animal indication include monitoring the concentration of toxins in animal tissues, or monitoring the rate at which deformities arise in animal populations.

Microbial Indicators and Chemicals

Microbial Indicators—Microorganisms can be used as indicators of aquatic or terrestrial ecosystem health. Found in large quantities, microorganisms are easier to sample than other organisms. Some microorganisms will produce new proteins, called stress proteins, when exposed to contaminants like cadmium and benzene. These stress proteins can be used as an early warning system to detect low levels of pollution.

14.3 Overview

The term bioindicator is used for organisms or organism associations which respond to pollutant load with changes in vital functions, or which accumulate pollutants (Arndt *et al.* 1987). Information about specific biological effects supplements data on air pollutions generated by technical analysis methods. The most important reasons for using bioindicators are:

- ❖ The direct determination of biological effects,
- ❖ The determination of synergetic and antagonistic effects of multiple pollutants on an organism,
- ❖ The early recognition of pollutant damage to plants as well as toxic dangers to humans and
- ❖ Relatively low cost compared to technical measuring methods.

The great potential of bioindicators for environmental monitoring is often confronted with difficult questions of methodology resulting from the use of "living measuring instruments". The effects of environmental load cannot always be clearly differentiated from natural stress factors. Lack of practical experience with certain bioindicators sometimes makes clear interpretion of findings more difficult, especially if no comparable pollutant measurements are available.

Intensive research over the last decades has resulted in the availability of numerous bioindicators which satisfy the requirements of convenience, standardization, cost, and evaluative capability (cf. Arndt 1987, Zimmermann u. Umlauff-Zimmermann 1994).

Bioindicators are commonly grouped into accumulation indicators and response indicators. Accumulation indicators store pollutants without any evident changes in their metabolisms. Response indicators react with cell changes or visible symptoms of damage when taking up even small amounts of harmful substances.

Biomonitoring is divided into passive and active. Passive biomonitoring is the use of organisms, organism associations, and parts of organisms which are a natural component of the ecosystem and appear there spontaneously. Active biomonitoring includes all methods which insert organisms under controlled conditions into the site to be monitored.

Cadastre of Ecological Pollution Effects

The Berlin Department of Urban Development, Environmental Protection, and Technology has conducted comprehensive studies for the determination of air pollution effects since 1991. These studies emphasized the complex of pollutants typical for metropolitan areas (SenStadtUm 1993a). All bioindicator methods employed have advantages and disadvantages. This means in practice that not only individual bioindicators, but a whole series of bioindicators was employed, according to the particular goal. This series is as finely adjusted to each other as possible. The systematic use of bioindicators was used for the Cadastre of Ecological Pollution Effects, conducted for the Ecological Monitoring Berlin and Hinterland. The findings are important supplements to and extensions of air quality monitoring data, especially for the year-long random sample measurement programme of the Berlin Air Quality Monitoring System—BLUME.

A great amount of data and knowledge has been generated by the Cadastre of Ecological Pollution Effects. The size of this data requires presentation in the Environmental Atlas to be limited to findings which have been sufficiently secured. These findings allow a differentiation of pollutant load ranges and could serve as a particularly significant basis for planning.

The Environmental Atlas took active and passive bioindicator methods, as well as accumulation and reaction indicators (cf. Fig. 14.1) almost equally into consideration. The selected results allow an overview of general air pollution in Berlin, manifested by effects upon naturally occuring lichen and exposed lichen transplants, as well as an overview of regional pollutant patterns, indexed by accumulation in pine needles, rye grass, and green kale.

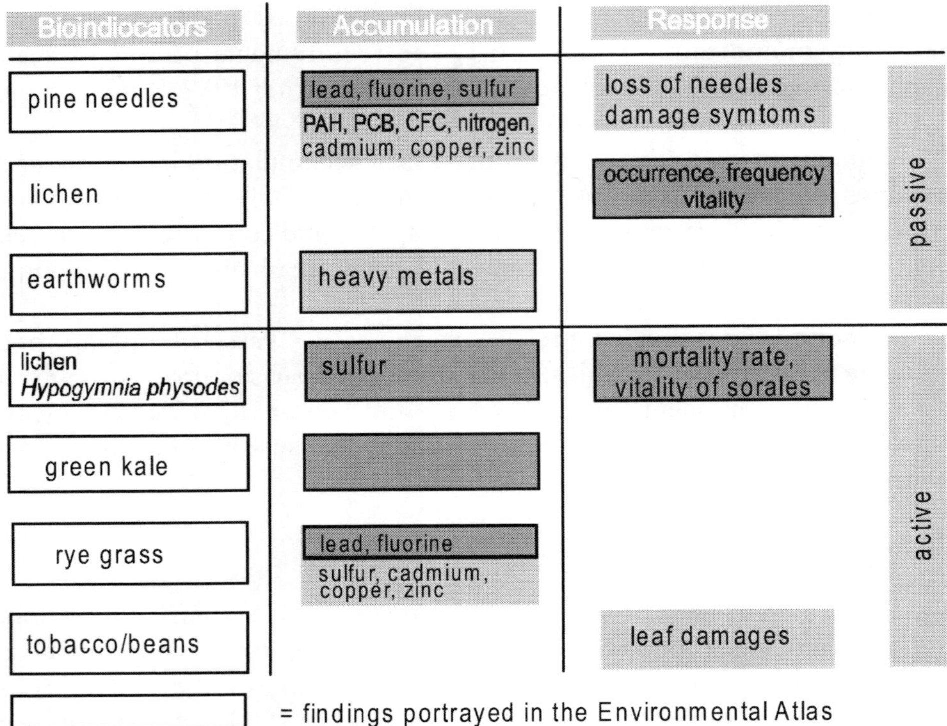

Figure 14.1: Bioindicator Methods Used for the Cadastre of Ecological Pollution Effects

Findings of the Cadastre of Ecological Pollution Effects not presented here can be found in various studies and publications (cf. Literature).

The evaluation of study findings can, in some cases, be supplemented by earlier research. This enables a chronological comparison of general pollutant effects and certain pollutant materials (cf. Cornelius *et al.* 1984). These kinds of reference data are valuable yardsticks for evaluation of the current pollutant situation in Berlin, for changes in air quality over recent decades can thus be documented.

Investigated Pollutants

The selection of analyzed pollutants took into consideration human toxicity, phytotoxicity, and ecosystem aspects. The information can thus be used for an overview of environmental loads in the Berlin area. The Cadastre of Ecological Pollution Effects focused mainly on the pollutants described below.

Until the end of the 80's, Berlin air had relatively high SO_2 and dust pollutions, especially during low-exchange winter weather periods. This was mainly due to the large amounts of sulfur-rich lignite brown coal used for heating in older residential areas (cf. Map 03.01, SenStadtUm 1994).

SO_2 in the air enters plants as a gas or in water solution and can cause damage. Chemical reactions of SO_2 create acids. Precipitation can transport these acids into the soil and trigger indirect damages, such as lack of nutrients and acid stress. Value limits for direct damaging effects on plants are clearly lower than for animals and humans. High sulfur dioxide concentrations lead to air pollution damages on coniferous trees, including non-specific chloroses, necroses, growth inhibition, and impairment of reproduction.

Lead is a heavy metal and has special environmental relevance because it is toxic. Leaded gasoline is still the main source. Most lead particles remain on the surface of leaves. Only a small amount becomes physiologically active and damages plants. This process can lead to accumulation in the food chain. Toxic effects for grazing animals or humans cannot be ruled out.

Fluorine is released from various industrial processes, waste burning, and by fossil fuel energy production. Hydrogen fluoride damages in plants appear primarily as necroses on leaf edges (marginal) and tips (terminal). Fluoride is accumulated in plants, damages leaves and inhibits growth.

Polycyclic aromatic hydrocarbons (PAH) are emitted primarily as incomplete combustion products, such as from internal combustion motors and heating plants. They also enter the environment as wash-outs from surfaces covered by tar, asphalt and bitumen. These substances are taken up by the human organism mainly by breathing, but also in food and by skin contact. The considerable carcinogenic and mutagenic potential of this group makes it more important for humans than for plants. Benzo(a)pyrene (BaP) is a leading component of polycyclic aromatic hydrocarbons.

Polychlorinated biphenyls (PCB) are ubiquitous in the environment today. Production of these chemicals was stopped in Germany in 1983, but considerable diffuse emissions continue from waste disposal sites, burning of waste-oil and waste, and leakage. PCBs enter the human organism primarily in the form of animal products, through accumulation along the food chain. Accumulation in plants seems to be of little phytotoxic significance. Great attention must be given to the exposure of the population to these substances, in view of their damaging effects on embryos and a well-founded suspicion of carcinogenic effects.

Polychlorinated dibenzodioxin (PCDD) and polychlorinated dibenzofuran (PCDF) are undesired side products of chlorine created in chemical and thermic processes. The main sources of emissions in Berlin are industrial (chemical cleaners, metallurgical processes) waste combustion plants, power plants, metal recycling plants, traffic, and domestic burning. Secondary sources are contaminated areas. The group known as "dioxins" accumulate in the food chain. They primarily enter the human organism in foods containing animal fats. The behaviour of this material in the environment as well as its toxicity for human is not yet sufficiently known. Studies on these materials conducted for the Cadastre of Ecological Pollution Effects thus represent an important contribution to the environmental relevance of these materials in Berlin.

Bioindicators were also used to study the effects of **highly volatile chlorinated hydrocarbons, photooxidants** (ozone), and **nitrogen oxides**. These results are not presented, but can be found in the studies.

Bioindicator Methods

Maps 03.07.1-4 contain findings on pollution effects obtained by bioindicator use in Berlin since 1991. They are differentiated between 1) the determination of biological effects caused by a general complex of air pollutant factors working as a sum and 2) the analytical detection of specific pollutants.

The response of lichen to pollutants in the course of a "screening" programme was the first indication for the presence of a pollutant load (Map 03.07.1). The *Hypogymnia physodes* lichen was employed for differentiation of small-area pollutant patterns in areas where a rich variety of natural lichen species can no longer exist (Map 03.07.2). Additionally, relevant pollutants were isolated with suitable accumulation indicators and evaluated for human toxicity, phytotoxicity, and ecosystem effects (Map 03.07.3/4).

14.4 Biological Markers: Their Use in Quantitative Assessments

Biological markers can be conceptualized in terms of categories of markers that form a continuum representing a sequence of events from exposure to disease. These categories include markers of internal dose, biologically effective dose, early response, and disease. Outside of this sequence are susceptibility factors that can act at any point along the way to modify the effects of external exposures on disease outcomes. Examples of the use of these different types of markers in epidemiologic research are provided. There are many factors that one must consider when selecting a biological marker for use in an epidemiologic study. These factors include: the objectives of the study, the availability and specificity of potential markers, the feasibility of measuring the markers in various biological media, the invasiveness of the techniques necessary to measure the markers, the amount of biological specimen needed for analysis, the time to appearance of the markers in the biological media, the persistence of the markers in biological media, the variability of marker levels within and between individuals, the stability of markers in storage, as well as the cost, sensitivity, specificity, and reliability of the assays used to measure the markers.

14.5 Biomarker (medicine)

In medicine, a biomarker is an indicator of a particular disease state or a particular state of an organism.

An NIH study group committed to the following definition in 1998: "a characteristic that is objectively measured and evaluated as an indicator of normal biologic processes, pathogenic processes, or pharmacologic responses to a therapeutic intervention."

In the past, biomarkers were primarily physiological indicators such as blood pressure or heart rate. More recently, biomarker is becoming a synonym for molecular biomarker, such as elevated prostate specific antigen as a molecular biomarker for prostate cancer, or using enzyme assays as liver function tests. Biomarkers also cover the use of molecular indicators of environmental exposure in epidemiologic studies such as human papilloma virus or certain markers of tobacco exposure such as 4-(methylnitrosamino)-1-(3-pyridyl)-1-butanone (NNK).

15

Biosorption of Metals

15.1 Biosorption

Biosorption is a property of certain types of inactive, dead, microbial biomass to bind and concentrate heavy metals from even very dilute aqueous solutions. Biomass exhibits this property, acting just as a chemical substance, as an ion exchanger of biological origin. It is particularly the cell wall structure of certain algae, fungi and bacteria which was found responsible for this phenomenon. Opposite to biosorption is metabolically driven active bioaccumulation by living cells. That is an altogether different phenomenon requiring a different approach for its exploration.

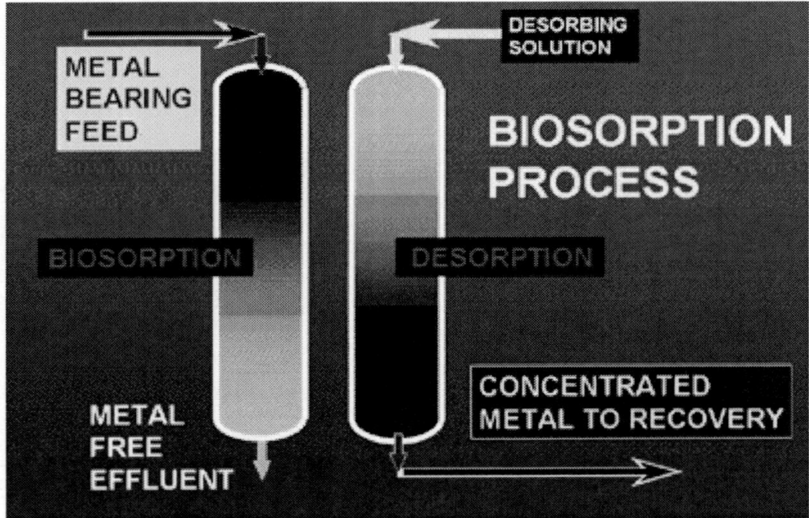

Pioneering research on biosorption of heavy metals at McGill University in Montreal has led to identification of a number of microbial biomass types which are extremely effective in concentrating metals. Some of the biomass types come as a waste by-product of large-scale industrial fermentations (the mold *Rhizopus* or the bacterium *Bacillus subtilis*). Other metal-binding biomass types, certain abundant seaweeds (particularly brown algae e.g. *Sargassum, Ecklonia*), can be readily collected from the oceans. These biomass types, serving as a basis for metal biosorption processes, can accumulate in excess of 25 per cent of their dry weight in deposited heavy metals: Pb, Cd, U, Cu, Zn, even Cr and others. Research on biosorption is revealing that it is sometimes a complex phenomenon where the metallic species could be deposited in the solid biosorbent through different sorption processes of ion exchange, complexation, chelation, microprecipitation, etc.

A whole new family of suitably "formulated" biosorbents can be used in the process of metal removal and detoxification of industrial metal-bearing effluents. The sorption packed-column configuration is the most effective mode of application for the purpose. Recovery of the deposited metals from saturated biosorbent can be accomplished because they can often be easily released from the biosorbent in a concentrated wash solution which also regenerates the biosorbent for subsequent multiple reuse. This and extremely low cost of biosorbents makes the process highly economical and competitive particularly for environmental applications in detoxifying effluents of e.g.

- ❖ metal-plating and metal-finishing operations,
- ❖ mining and ore processing operations,
- ❖ metal processing, battery and accumulator manufacturing operations,
- ❖ thermal power generation (coal-fired plants in particular), and
- ❖ nuclear power generation, (etc.).

Different types of science background, from engineering to biochemistry, can make a significant contribution in elucidating the biosorption phenomenon. Interdisciplinary efforts are mandatory and represent quite a challenge. Preparing biosorption for application as a process requires mainly chemical engineering background. Good understanding of the sorption operation is mandatory.

While ongoing research is essential for improving and optimizing metal biosorption effectiveness, wastewater purification applications of the biosorption process are readied for pilot testing of this alternative new technology. Optimization of specific biosorption process applications has to be done in conjunction with industrial users/clients and requires specific process engineering expertise and a serious development capital commitment.

15.2 Introduction

Biosorption is the binding and concentration of heavy metals from aqueous solutions (even very dilute ones) by certain types of inactive, dead, microbial biomass.

Pioneering research on biosorption of heavy metals has led to the identification of a number of microbial biomass types that are extremely effective in concentrating metals. Some types of biomass are wastebyproducts of large-scale industrial fermentations (*e.g.*, the mold *Rhizopus* or the bacterium *Bacillus subtilis*). Other metal-binding biomass types, such as certain abundant seaweeds (particularly brown algae, *e.g.*, *Sargassum, Ecklonia*), can be readily harvested from the oceans. These biomass types can accumulate in excess of 25 per cent of their dry weight in deposited heavy metals: Pb, Cd, U, Cu, Zn, Cr and others. Research on biosorption is revealing that it is sometimes a complex phenomenon where the metallic species could be deposited in the solid biosorbent through various sorption processes, such as ion exchange, complexation, chelation, microprecipitation, etc.

Individuals with different backgrounds, from engineering to biochemistry, can make significant contributions to the understanding of biosorption. Interdisciplinary efforts are essential to exploit this technology commercially. A chemical engineering background is particularly useful for expanding the application of this technology in large-scale process industries.

Threat from the Environment

The greatest demand for metal sequestration today comes from the need to immobilize the metals released to the environment (or mobilized) by and partially lost through human technological activities. It has been established that dissolved metals (particularly heavy metals) escaping into the environment pose a serious health hazard. They accumulate in living tissues throughout the food chain, which has humans at its top, multiplying the danger. Thus, it is necessary to control emissions of heavy metals into the environment.

Am example of one method for prioritizing the recovery of ten metals is presented in Table 1. This may be simplistic, but it provides a useful direction by ranking metals into three general priority categories:

1. Environmental Risk (ER)
2. Reserve Depletion Rate (RDR)
3. Combination of ER and RDR

Environmental risk assessment could be based on a number of different factors, which could also be weighted.

Table 15.1: Ranking of Risks Associated with Various Metals.

Relative Priority	Environmental Risk	Reserve Depletion	Combined Factors
High	Cd	Cd	Cd
	Pb	Pb	Pb
	Hg	Hg	Hg
	—	Zn	Zn
Medium	Cr	—	—
	Co	Co	Co
	Cu	Cu	Cu
	Ni	Ni	Ni
	Zn	—	—
Low	Al	—	Al
	—	Cr	Cr
	Fe	Fe	Fe

The Need for Novel Technology

Conventional techniques to remove toxic metals and radionuclides, such as ion exchange and precipitation, lack specificity and are ineffective at low metal ion concentrations. The need for effective and economically viable technologies is driven by environmental pressures such a:

❖ Stricter regulations with regard to the metal discharges are being enforced, particularly in industrialized countries.

❖ Toxicology studies confirm the dangerous impacts of heavy metals.

❖ Current technologies for the removal of heavy metals from industrial effluents often create secondary problems with metal-bearing sludge.

Biosorption Mechanisms

Various metal-binding mechanisms have been postulated to be active in biosorption, such as:

❖ Chemisorption by ion exchange, complexation, coordination and/or chelation,

❖ Physical Adsorption,

❖ Microprecipitation,

❖ Oxidation/Reduction.

Due to the complexity of the biomaterials used, it is possible that at least some of these mechanisms are acting simultaneously to varying degrees, depending on the biosorbent and the solution environment.

Ion Exchange

Ion exchange is a reversible chemical reaction wherein an ion in a solution is exchanged for a similarly charged ion attached to an immobile solid particle. These solid ion-exchange particles are either naturally occurring inorganic zeolites or synthetically produced organic resins. Synthetic organic resins are the predominant type used today because their characteristics can be tailored to specific applications.

Ion exchange reactions are stoichiometric and reversible, and as such they are similar to other solution-phase reactions. For example, in the reaction

$$NiSO_4 + Ca(OH)_2 \rightarrow Ni(OH)_2 + CaSO_4$$

the nickel ions of the nickel sulfate ($NiSO_4$) are exchanged for the calcium ions of the calcium hydroxide $Ca(OH)_2$ molecule.

Chelation

The word *chelation* is derived from the Greek word *chele,* which means *claw,* and is defined as the firm binding of a metal ion with an organic molecule (ligand) to form a ring structure. The resulting ring structure protects the mineral from entering into unwanted chemical reactions. Examples include the carbonate (CO_3^{2-}) and oxalate ($C_2O_4^{2-}$) ions:

Coordination (Complex Formation)

A coordination complex is any combination of cations with molecules or anions containing free pairs of electrons. Bonding may be electrostatic, covalent or a combination of both; the metal ion is coordinately bonded to organic molecules. Example of the formation of a coordination compound are:

$$Cu^{2+} + 4H_2O \rightarrow [Cu(H_2O)]_4^{2+}$$
$$Cu^{2+} + 4Cl^- \rightarrow [CuCl_4]^{2-}$$

where coordinate covalent bonds are formed by donation of a pair of electrons from H_2O and Cl^- (Lewis bases) to Cu^{2+} (Lewis acid).

In general, biosorption of toxic metals and radionuclides is based on non-enzymatic processes such as adsorption. Adsorption is due to the non-specific binding of ionic species to polysaccharides and proteins on the cell surface or outside the cell. Bacterial cell walls and envelopes, and the walls of fungi, yeasts and algae, are efficient metal biosorbents that bind charged groups. The cell walls of gram-positive bacteria bind larger quantities of toxic metals and radionuclides than the envelopes of gram-negative bacteria.

Bacterial sorption of some metals can be described by the linearized Freundlich adsorption equation:

$$\log S = \log K + n \log C$$

where:

S is the amount of metal absorbed in $\mu mol/g$.
C is the equilibrium solution concentration in $\mu mol/L$.
K and n are the Freundlich constants.

Biomass deriving from several industrial fermentations may provide an economical source of biosorptive materials. Many species have cell walls with high concentrations of chitin, a polymer of N-acetyl-glucosamine that is an effective biosorbent.

Biosorption uses biomass raw materials that are either abundant (*e.g.*, seaweeds) or wastes from other industrial operations (*e.g.*, fermentation wastes). The metal-sorbing performance of certain types of biomass can be more or less selective for heavy metals, depending on the type of biomass, the mixture in the solution, the type of biomass preparation, and the chemical-physical environment.

It is important to note that the concentration of a specific metal in solution can be reduced either during the sorption uptake by manipulating the properties of the biosorbent or upon desorption during the regeneration cycle of the biosorbent.

Sources of Biomass for Biosorption

Sources of biomass include:

- ❖ Seaweeds,
- ❖ Microorganisms (bacteria, fungi, yeast, molds),
- ❖ Activated sludge,
- ❖ Fermentation waste, and
- ❖ Other specially propagated biomasses.

Biosorbents must be hard enough to withstand the application pressures, porous and/or "transparent" to metal ion sorbate species, and have high and fast sorption uptake even after repeated regeneration cycles.

Granulation of biomass materials into suitable cost-effective biosorbents is a crucial step for the successful application of biosorption processes.

The objectives of granulation are to:

- ❖ Establish the behaviour of native biomass in a packed-bed reactor,
- ❖ Establish the effectiveness of biomass granulation and reinforcement,
- ❖ Determine the effect of size reduction on sorption capacity, and
- ❖ Determine the feasibility of biomass processing.

Conventional granulation technologies are rather advanced, and their adaptation will likely yield desirable biosorbent granules. Because of the wide variety of biomass types, extensive experimentation will undoubtedly be required.

The need to transport raw biomass may also present some logistical problems. Microbial biomass has a high water content and is prone to decay, so drying may be required if it cannot be processed and/or granulated directly on location in the wet state.

Equilibrium Modeling

Biosorption has been studied as simplified sorption systems, usually containing one heavy metal. This is an appropriate simplification for effective experimentation.

Some of the simple sorption isotherm models that are most frequently applied. A particular model may not apply to a particular situation, and in some cases more than one model may explain the biosorption mechanism. There is no critical reason to use a more-complex model if a two-parameter model (such as the Langmuir and Freundlich isotherm models) can fit the data reasonably well.

The sorption uptake, q, can be expressed in different units depending on the purpose of the exercise:

- ❖ For practical and engineering process evaluation purposes eventually concerned with process mass balances, it is customary to use weight per (dry) weight (*e.g.*, mg of metal sorbed per gram of the (dry) sorbent material).
- ❖ Ultimately, mainly because of reactor volume considerations (*e.g.*, a packed-bed column), the uptake may also be expressed on a per volume basis (*e.g.*, mg/L). However, the porosity may complicate the quantitative comparison of biosorption performance.

❖ Only when working on the stoichiometry of the process and when studying the functional groups and metal-binding mechanisms might it be useful to express q on a molar or charge equivalent basis—again, per unit weight or volume of the sorbent (*e.g.,* mmol/g or mequiv/g).

It is relatively easy to convert among these units; the only problem may arise with the sorbent weight-volume conversions. For scientific interpretations, the sorbent material dry-weight basis is thus preferred.

The use of "wet biomass weight" should be discouraged, unless the wet-weight-to-dry-weight conversion is well specified. Different biomass types are likely to retain different moisture contents, intracellular as well as that trapped in the interstitial space between the cells or tissue particles (*e.g.,* seaweed particles). Different types of biomass obviously compact in a different ways. When centrifuging biomass, the g-force and time need to be specified, and even then it is difficult to make any comparisons. All this makes the "wet biomass weight" citation very approximate at best and generally undesirable.

Biomass Types

The assessment of the metal-binding capacity of some types of biomass has gained momentum since 1985. Indeed, some biomass types are very effective in accumulating heavy metals.

Availability is a major factor to be taken into account to select biomass for clean-up purposes. The economics of environmental remediation dictate that the biomass must come from nature, or even be a waste material. Seaweeds, molds, yeasts, bacteria, and crab shells, among other kinds of biomass have been tested for metal biosorption with very encouraging results.

Biosorbents

Some biosorbents can bind and collect a wide range of heavy metals with no specific priority, whereas others are specific for certain types of metals. When choosing the biomass for metal biosorption experiments, its origin is a major factor to be considered.

Biomass can come from:

❖ industrial wastes which should be obtained free of charge,
❖ organisms that can be obtained easily in large amounts in nature (*e.g.,* bacteria, yeast, algae),

❖   fast-growing organisms that are specifically cultivated or propagated for biosorption purposes (crab shells, seaweeds).

Organisms for Biosorption

There is a wide variety of microorganisms (Table 4), including bacteria, fungi, yeast, and algae, that can interact with metals and radionuclides and transform them through several mechanisms.

Table 15.4: Examples of toxic heavy metals accumulating microorganisms.

Organism	Element
Citrobacter sp.	Lead, Cadmium
Thiobacillus ferrooxidans	Silver
Bacillus cereus	Cadmium
Bacillus subtilis	Chromium
Pseudomonas aeruginosa	Uranium
Micrococcus luteus	Strontium
Rhisopus arrhizus	Mercury
Aspergillus niger	Thorium
Saccharomyces cerevisiae	Uranium

Cost-effectiveness is the main attraction of metal biosorption. This cost-effectiveness can be maintained by using the microbial biomass directly where possible. In addition, biosorbents derived from microbial biomass through a simple process are expected to be the lowest-priced and most-economical for metal removal.

It has been suggested that numerous chemical groups contribute to biosorption metal binding, by either whole organisms such as algae and bacteria or by molecules such as biopolymers. These include hydroxyl, carbonyl, carboxyl, sulfhydryl, thioether, sulfonate, amine, imine, amide, imidazole, phosphonate, and phosphodiester groups. The importance of any given group for biosorption of a certain metal by a certain biomass depends on such factors as the number of sites in the biosorbent material, the accessibility of the sites, the chemical state of the sites (*i.e.,* availability), and the affinity between the site and the metal (*i.e.,* binding strength). For covalent metal binding, even an occupied site is theoretically available; the extent to which the site can be used by a given metal depends on its binding strength and concentration compared to the metal already occupying the site.

Some types of industrial fermentation waste biomass are excellent metal sorbers. It is necessary to realize that some "waste" biomass is actually a commodity, not a waste. This applies particularly to the ubiquitous brewer's yeasts sold on the open market, usually as animal fodder. Activated sludge from wastewater treatment plants has not demonstrated high enough metal-sorbing capacities. Some types of seaweed biomass offer excellent metal-sorbing properties, and sometimes a local economy can benefit from turning seaweeds into a resource.

As a fallback, biomass with a high metal-sorbing capacity can be specifically grown relatively cheaply in fermenters using low-cost or even waste carbohydrate-containing growth media such as molasses or cheese whey.

Experimental Sorption Isotherms

It is relatively easy to obtain equilibrium sorption data for a single sorbate in the laboratory. A small amount of the sorbent is brought into contact with a solution containing the sorbate of interest. The conditions of the sorption system, particularly pH, must be carefully controlled at the required values over the entire period of contact until the sorption equilibrium is reached. This may take a few hours or much longer, depending on the size of the sorbent particles and the time it takes until they attain sorption equilibrium.

A simple preliminary sorption kinetics test will establish the exposure time necessary for the given sorbent particles to reach the equilibrium state. The following procedure provides an example for obtaining the experimental sorption equilibrium data points for the isotherm:

1. Prepare the sorbate in solution at the highest concentration of interest.
2. Prepare dilutions covering the entire concentration range, from 0 (blank) to the maximum.
3. Adjust the conditions, *e.g.*, pH, ionic strength, etc.
4. Determine the sorbate initial concentrations (C_i) in all the liquid samples.
5. Distribute the samples into containers of appropriate volumes (30-150 mL of liquid) such as flasks or test tubes; prepare samples in duplicate, triplicate or as required.
6. Accurately weigh each quantity of the biosorbent solids to be used in the tests and record the weights (S, mg). It may help to be able to roughly estimate the anticipated sorption uptake so that there is an easily detectable final sorbate concentration in each sample solution at equilibrium. If too much sorbent is added, there may be virtually

no sorbate left in the solution, precluding a reliable analysis. Varying the initial concentration could cause the sorbent weight to fluctuate, which has to be precisely known for each sample. Metal depletion in the solution must be avoided because it renders such samples useless.

7. Add the sorbent solids into each sample solution and provide rather gentle mixing over the contact period.

8. Make sure the conditions (especially pH) are controlled at constant values during the contact period. Use an appropriate acid or base for this; do not dilute the sorption system by adding excessive volume.

9. At the end of the contact period, separate the solids from the liquid by decantation, filtration, centrifugation, etc.

10. Analyze the liquid portion to determine the residual final sorbate concentration (C_f).

11. Calculate the sorbate uptake: $q = V$ [L] $\times (C_i - C_f)$ [mg/L]/S [g]. Note that q could also be determined directly by analyzing the separated solids and thus closing the material balance on the sorbate in the system. However, this usually presents analytical difficulties (digestion-liquefaction of solids, and/or very sophisticated analytical methods may be required).

12. Plot the sorption isotherm q vs. (C_f).

Comparison of Sorption Performance

The performance of sorbing materials needs to be evaluated and often compared. The simplest situation is when there is only one sorbate species in the system, in which case it is best to base the single-sorbate sorption performance on a complete single-sorbate sorption isotherm curve.

To fairly compare two or more sorbents, the comparison must be done under uniform conditions. These may be restricted by the environmental factors under which sorption may have to take place (pH, temperature, ionic strength, etc.), which may not necessarily be easily or widely adjustable. In particular, it is important to compare sorption performance under the same pH conditions, since isotherms can vary with pH.

The performance of the sorbent is usually gauged by its uptake (q). Sorbents can be compared based on their respective maximum uptake values (q_{max}), which can be calculated by fitting the Langmuir isotherm model to the actual experimental data (if it fits). This approach is feasible if q_{max} reaches a plateau. Some isotherms might not exhibit the asymptotic plateau represented by the Langmuir equation.

In general, one is looking for a "good" sorbent with a high sorption uptake capacity (q_{max}). Surface area in biosorption is not particularly important.

Types of Biosorption

Biosortpion can be carried out as a batch process, a continuous process, or a two-stage process with continuous metal recovery. Biomass should be defrosted and washed with deionized water. To ensure equal quality of the biomass during all experiments, different kinds of biomass should be mixed together to obtain a uniform mixture.

Batch Process

Batch biosorption experiments can be done in a stirred with a working volume of approximately 100 mL. A minimal amount of concentrated solution of $Pb(NO_3)_2$ (metal) can be added into a suspension of fungal pellets in water of various concentrations (25, 50, 100, 150 and 200 g of wet biomass per L of biomass suspension) to produce the desired initial metal concentrations of 10, 20, 50, 100 and 300 mg/L Pb (metal). The decreasing metal concentration can be recorded as a function of the initial metal concentration and the biomass loading.

Continuous Process

Continuous process experiments can be carried out in a glass column having an inner diameter of 5-8 cm and filled with a packed bed of biomass pellets of varying heights (20, 40 and 55 cm), set with an adjustable plug. The effluent solution of metal ions can be fed from the top of the column with the help of a pump using varying flowrates. An inert bed of glass spheres can be placed at the bottom of the column below the active biomass bed to ensure homogenous distribution of the feed. The remaining metal concentration can be measured online in the effluent at the top of the column. The breakthrough curves can be recorded as a function of the flowrate and bed height.

Measurements of metal ion concentrations in the solution can be made online with metal-detecting electrodes or ion-selective electrodes, and may be verified with an atomic absorption spectrometer.

Two-Stage Process with Continuous Metal Recovery

Two-stage continuous biosorption and metal recovery can be carried. This process is similar to continuous biosorption, although the metal solution is adsorbed in two stages. After initial adsorption and filtration in stage one, the effluent is fed with fresh biosorbent into stage two, where further biosorption of the metal ion takes place. The effluent from the second stage is filtered to recover the metal ions and biosorbents. The effluent sample can be analyzed using an ion meter or by adsorption spectroscopy.

Desorption

Regeneration of loaded biosorbent is critical to keeping costs down and to recovering the metal(s) extracted from the liquid phase. The deposited metals are washed out (desorbed) and the biosorbent is regenerated for another cycle. The desorption process should result in:

- ❖ high-concentration metal effluent,
- ❖ undiminished metal uptake upon re-use, and
- ❖ no physico-chemical damage to the biosorbent.

The desorption and sorbent regeneration studies might require somewhat different methodologies, beginning with screening for the most effective regenerating solution.

Because different metal ions have different affinities for the biosorbent, the uptake has some degree of metal selectivity. The selectivity of the elution-desorption operation may be different, which may serve as another means of eventually separating metals from one another if desirable.

The concentration ratio (*CR*) is used to evaluate the overall concentration effectiveness of the whole sorption-desorption process:

Obviously, the higher the *CR*, the better the overall performance of the sorption process, making the eventual recovery of the metal more feasible with higher eluate concentrations.

Recovery of the metal from these concentrated desorption solutions is carried out in a different plant by electrowinning. Following desorption of the metal(s), the column may be further pre-treated (*e.g.*, pre-saturated with protons such as Ca, K, etc.) for optimum operation in the next metal uptake cycle. The specific type of pre-treatment used to optimize the column performance may vary.

Feasibility of Biosorption

For successful application on a large scale, any operation needs to be economically viable. The feasibility of a biosorption process depends on such factors as:

❖ biosorbent uptake performance,
❖ the source of the raw biomass,
❖ biomass granulation and treatment, and
❖ the desorption and regeneration processes used.

Often, the source of the biosorbent has a major impact on the feasibility of the operation. Biosorbents (biomass) should always be obtained from the least-expensive source, such as from the effluent of a fermenter, seaweeds from nearby bodies of water, algae, etc. The spent biosorbents can be regenerated at very low cost using water, so the material can be reused many times. Hence, considering the overall unit operations involved in biosorption, we can conclude that the process is generally economically viable.

Advantages of Biosorption

Biosorption is highly competitive with the presently available technologies like ion exchange, electrodialysis, reverse osmosis, etc. Some of the key features of biosorption compared to conventional processes include:

❖ competitive performance
❖ heavy metal selectivity
❖ cost-effectiveness
❖ regenerative
❖ no sludge generation

Biosorption is particularly economical and competitive for environmental applications in detoxifying effluents from, for example:

❖ metal plating and metal finishing operations
❖ mining and ore processing operations
❖ metal processing
❖ battery and accumulator manufacturing operations
❖ thermal power generation (coal-fired plants in particular)
❖ nuclear power generation

Conclusion

There appear to be many modes of non-active metal uptake by microbial biomass. Any one or a combination of them can be functional in immobilizing metallic species on biosorbents. A number of anionic ligands participate: phosphoryl, carbonyl, sulfhydryl and hydroxyl groups can all be active to various degrees in binding the metal.

Many scientific studies are currently underway to provide a deeper understanding of biosorption and to support its effective application. Some pollution seems inevitable, and one might wonder what should be done to minimize it. Human populations need methods and technologies to clean waters and diminish the environmental dangers related to technological progress. Biosorption can be one such solution to clean up heavy metal contamination.

The Challenge

- ❖ Heavy metals such as lead, mercury, cadmium, chromium, copper, to name a few, are very **toxic.**
- ❖ Toxic heavy metals released by industries poison the environment and threaten **water** supplies.
- ❖ Stricter environmental disposal **regulations** are being enforced to avert the health hazards.

Biosorption Technology

- ❖ A new technology alternative has been pioneered to remove toxic heavy metals.
- ❖ from industrial effluents. It is called biosorption. It is effective, simple and cheap.
- ❖ It is very similar to the use of ion exchange resins which can also do the clean-up job.
- ❖ The difference is in the price of the substances used:

  - ● Man-made synthetic ion exchange resins are marketed for US$ 30-50/kg.
  - ● With the same performance, natural new biosorbents could cost less than $(3-5)/kg.

Cost-effective biosorption technology has a special edge for opening up huge environmental markets.

Biosorption

Biosorption uses the extraordinary capacity of certain types of microbial and seaweed biomass to bind and concentrate metals. These bio-materials, are used dead—just like "magic granules" which remove and concentrate heavy-metals from industrial effluents.

Suitable biomass comes as a waste material: from fermentation industries or it is renewable, growing in the oceans (seaweeds).

In either case, the costs of biomass raw materials are extremely low.

New biosorbent materials can thus be extremely competitive and cost effective.

A McGill University based biosorption research group has been a recognized leader discovering and pioneering the novel process of biosorption for removal of toxic metals.

Biosorption Process

- ❖ It uses sorption columns.
- ❖ Process principles are well established and well understood.
- ❖ The difference is in the "magic granules" inside the column—newly discovered biosorbents.
- ❖ Performing like much costlier ion exchange resins, biosorbents can be also regenerated for multiple re-use: better cost-effectiveness.

Sorption Columns

A contemporary high-performance sorption process uses the most effective column configuration. Sorbent granules inside the column bind metals from flowing solution until they get saturated. Then the column is taken out of operation and the desorption-washing procedure follows directly in the same column releasing the metal in the concentrated form in the wash solution. Standard process equipment is used: columns, pumps, valves and pipes. Collected metal can be economically routinely **recovered** for re-sale from pre-concentrated column wash solutions. Any size of operation can be attained by adding columns. Purifying large volumes of contaminated wastewater can be economically feasible only with very cheap new biosorbents. Examples of large-scale installations operating on the same principle of continuous-flow columns.

15.4 Biosorption of Heavy Metals

The commonly used procedures for removing metal ions from aqueous streams include chemical precipitation, lime coagulation, ion exchange, reverse osmosis and solvent extraction (Rich and Cherry, 1987). The process description of each method is presented below:

- ❖ *Reverse osmosis*: It is a process in which heavy metals are separated by a semi-permeable membrane at a pressure greater than osmotic pressure caused by the dissolved solids in wastewater. The disadvantage of this method is that it is expensive.

- ❖ *Electrodialysis:* In this process, the ionic components (heavy metals) are separated through the use of semi-permeable ionselective membranes. Application of an electrical potential between the two electrodes causes a migration of cations and anions towards respective electrodes. Because of the alternate spacing of cation and anion permeable membranes, cells of concentrated and dilute salts are formed. The disadvantage is the formation of metal hydroxides, which clog the membrane.

- ❖ *Ultrafiltration*: They are pressure driven membrane operations that use porous membranes for the removal of heavy metals. The main disadvantage of this process is the generation of sludge.

- ❖ *Ion-exchange:* In this process, metal ions from dilute solutions are exchanged with ions held by electrostatic forces on the exchange resin. The disadvantages include: high cost and partial removal of certain ions.

- ❖ *Chemical precipitation:* Precipitation of metals is achieved by the addition of coagulants such as alum, lime, iron salts and other organic polymers. The large amount of sludge containing toxic compounds produced during the process is the main disadvantage.

- ❖ *Phytoremediation*: Phytoremediation is the use of certain plants to clean up soil, sediment, and water contaminated with metals. The disadvantages include that it takes a long time for removal of metals and the regeneration of the plant for further biosorption is difficult.

Hence the disadvantages like incomplete metal removal, high reagent and energy requirements, generation of toxic sludge or other waste products that require careful disposal has made it imperative for a cost-effective treatment method that is capable of removing heavy metals from aqueous effluents.

The search for new technologies involving the removal of toxic metals from wastewaters has directed attention to biosorption, based on metal binding capacities of various biological materials. Biosorption can be defined as the ability of biological materials to accumulate heavy metals from wastewater through metabolically mediated or physico-chemical pathways of uptake (Fourest and Roux, 1992). Algae, bacteria and fungi and yeasts have proved to be potential metal biosorbents (Volesky, 1986). The major advantages of biosorption over conventional treatment methods include (Kratochvil and Volesky, 1998 a):

- ❖ Low cost;
- ❖ High efficiency;
- ❖ Minimisation of chemical and lor biological sludge;
- ❖ No additional nutrient requirement;
- ❖ Regeneration of biosorbent; and
- ❖ Possibility of metal recovery.

The biosorption process involves a solid phase (sorbent or biosorbent; biological material) and a liquid phase (solvent, normally water) containing a dissolved species to be sorbed (sorbate, metal ions). Due to higher affinity of the sorbent for the sorbate species, the latter is attracted and bound there by different mechanisms. The process continues till equilibrium is established between the amount of solid-bound sorbate species and its portion remaining in the solution. The degree of sorbent affinity for the sorbate determines its distribution between the solid and liquid phases.

Biosorbent Material

Strong biosorbent behaviour of certain micro-organisms towards metallic ions is a function of the chemical make-up of the microbial cells. This type of biosorbent consists of dead and metabolically inactive cells.

Some types of biosorbents would be broad range, binding and collecting the majority of heavy metals with no specific activity, while others are specific for certain metals. Some laboratories have used easily available biomass whereas others have isolated specific strains of microorganisms and some have also processed the existing raw biomass to a certain degree to improve their biosorption properties;

Recent biosorption experiments have focused attention on waste materials, which are by-products or the waste materials from large-scale industrial operations. For e.g. the waste mycelia available from fermentation processes, olive mill solid residues (Pagnanelli, *et al.* 2002), activated sludge from sewage treatment plants (Hammaini et aI. 2003), biosolids (Norton *et al.* 2003), aquatic macrophytes (Keskinkan et aI. 2003), etc.

Norton *et aI.* 2003, used dewatered waste activated sludge from a sewage treatment plant for the biosorption of zinc from aqueous solutions. The adsorption capacity was determined to be 0.564 mM/g of biosolids. The use of biosolids for zinc adsorption was favourable compared to the bioadsorption rate of 0.299 mM/g by the seaweed *Durvillea potatorum* (Aderhold *et aI*. 1996). Keskinkan *et al*. 2003 studied the adsorption characteristics of copper, zinc and lead on submerged aquatic plant *Myriophyllum spicatum*. The adsorption capacities were 46.69 mg/g for lead, 15.59 mg/g for zinc and 10.37 mg/g for copper. Table 1 gives a comparison of heavy metal uptakes of various macrophytes.

Pagnanelli, *et al*. 2002 have carried out a preliminary study on the 'Use of oli ve mill residues as hea vy metal sorbent material The results revealed that copper was maximally adsorbed in the range of 5.0 to 13.5 mg/g under different operating conditions.

The simultaneous biosorption capacity of copper, cadmium and zinc on dried activated sludge (Hammaini *et al*. 2003) were 0.32 mmoI/g for metal system such as CuCd; 0.29 mmoI/g for Cu-Zn and 0.32 mmoI/g for Cd-Zn. The results showed that the biomass had a net preference for copper followed by cadmium and zinc.

Another inexpensive source of biomass where it is available in copious quantities is in oceans as seaweeds, representing many different types of marine macro-algae. However most of the contributions studying the uptake of toxic metals by live marine and to a lesser extent freshwater algae focused on the toxicological aspects, metal accumulation, and pollution indicators by live, metabolically active biomass. Focus on the technological aspects of metal removal by algal biomass has been rare.

Although abundant natural materials of cellulosic nature have been suggested as biosorbents, very less work has been actually done in that respect.

The mechanism of biosorption is complex, mainly ion exchange, chelation, adsorption by physical forces, entrapment in inter and intrafibrilliar capillaries and spaces of the structural polysaccharide network as a result of the concentration gradient and diffusion through cell walls and membranes.

There are several chemical groups that would attract and sequester the metals in biomass: acetamido groups of chitin, structural polysaccharides of fungi, amino and phosphate groups in nucleic acids, amido, amino, sulphhydryl and carboxyl groups in proteins, hydroxyls in polysaccharide and mainly carboxyls and sulphates in polysaccharides of marine algae that belong to the divisions Phaeophyta, Rhodophyta and Chlorophyta. However, it does not necessarily mean that the presence of some functional group guarantees biosorption, perhaps due to steric, conformational or other barriers.

Choice of Metal for Biosorption Process

The appropriate selection of metals for biosorption studies is dependent on the angle of interest and the impact of different metals, on the basis of which they would be divided into four major categories: (i) toxic heavy metals (ii) strategic metals (iii) precious metals and (iv) radio nuclides. In terms of environmental threats, it is mainly categories (i) and (iv) that are of interest for removal from the environment and/or from point source effluent discharges.

Apart from toxicological criteria, the interest in specific metals may also be based on how representative their behaviour may be in terms of eventual generalization of results of studying their biosorbent uptake. The toxicity and interesting solution chemistry of elements such as chromium, arsenic and selenium make them interesting to study. Strategic and precious metals though not environmentally threatening are important from their recovery point of view.

Mechanisms

Biosorption Mechanisms

The complex structure of microorganisms implies that there are many ways for the metal to be taken up by the microbial cell. The biosorption mechanisms are various and are not fully understood. They may be classified according to various criteria.

According to the dependence on the cell's metabolism, biosorption mechanisms can be divided into:

1. Metabolism dependent, and
2. Non-metabolism dependent.

According to the location where the metal removed from solution is found, biosorption can be classified as:

1. Extra cellular accumulation/ precipitation,
2. Cell surface sorption/ precipitation, and
3. Intracellular accumulation.

Transport of the metal across the cell membrane yields intracellular accumulation, which is dependent on the cell's metabolism. This means that this kind of biosorption may take place only with viable cells. It is often associated with an active defense system of the microorganism, which reacts in the presence of toxic metal.

During non-metabolism dependent biosorption, metal uptake is by physico-chemical interaction between the metal and the functional groups present on the microbial cell surface. This is based on physical adsorption, ion exchange and chemical sorption, which is not dependent on the cells' metabolism. Cell walls of microbial biomass, mainly composed of polysaccharides, proteins and lipids have abundant metal binding groups such as carboxyl, sulphate, phosphate and amino groups. This type of biosorption, i.e., non-metabolism dependent is relatively rapid and can be reversible (Kuyucak and Volesky, 1988).

In the case of precipitation, the metal uptake may take place both in the solution and on the cell surface (Ercole, *et al.* 1994). Further, it may be dependent on the cell's' metabolism if, in the presence of toxic metals, the microorganism produces compounds that favour the precipitation process. Precipitation may not be dependent on the cells' metabolism, if it occurs after a chemical interaction between the metal and cell surface.

- ❖ *Transport across cell membrane*: Heavy metal transport across microbial cell membranes may be mediated by the same mechanism used to convey metabolically important ions such as potassium, magnesium and sodium. The metal transport systems may become confused by the presence of heavy metal ions of the same charge and ionic radius associated with essential ions. This kind of mechanism is not associated with metabolic activity. Basically biosorption by living organisms comprises of two steps. First, a metabolism independent binding where the metals are bound to the cell walls and second, metabolism dependent intracellular uptake, whereby metal ions are transported across the cell membrane. (Costa, et.al., 1990, Gadd et.al., 1988, Ghourdon et.al., 1990, Huang et.al., 1990., Nourbaksh et.al., 1994)
- ❖ *Physical adsorption*: In this category, physical adsorption takes place with the help of van der Waals' forces. Kuyucak and Volesky 1988, hypothesized that uranium, cadmium, zinc, copper and cobalt biosorption by dead biomasses of algae, fungi and yeasts takes place through electrostatic interactions between the metal ions in solutions and cell walls of microbial cells. Electrostatic interactions have been demonstrated to be responsible for copper biosorption by bacterium *Zoogloea ramigera* and alga *Chiarella vulgaris* (Aksu *et al.* 1992), for chromium biosorption by fungi *Ganoderma lucidum* and *Aspergillus niger*.

* *Ion Exchange:* Cell walls of microorganisms contain polysaccharides and bivalent metal ions exchange with the counter ions of the polysaccharides. For example, the alginates of marine algae occur as salts of K+, Na+, Ca^2+, and Mg^2+. These ions can exchange with counter ions such as CO^2+, Cu^2+, Cd^2+ and Zn^2+ resulting in the biosorptive uptake of heavy metals (Kuyucak and Volesky 1988). The biosorption of copper by fungi *Ganoderma lucidium* (Muraleedharan and Venkobachr, 1990) and *Aspergillus niger* was also up taken by ion exchange mechanism.

* *Complexation:* The metal removal from solution may also take place by complex formation on the cell surface after the interaction between the metal and the active groups. Aksu *et al.* 1992 hypothesized that biosorption of copper by C. *vulgaris* and Z. *ramigera* takes place through both adsorption and formation of coordination bonds between metals and amino and carboxyl groups of cell wall polysaccharides. Complexation was found to be the only mechanism responsible for calcium, magnesium, cadmium, zinc, copper and mercury accumulation by *Pseudomonas syringae.* Microorganisms may also produce organic acids (e.g., citric, oxalic, gluonic, fumaric, lactic and malic acids), which may chelate toxic metals resulting in the formation of metallo-organic molecules. These organic acids help in the solubilisation of metal compounds and their leaching from their surfaces. Metals may be biosorbed or complexed by carboxyl groups found in microbial polysaccharides and other polymers.

* *Precipitation:* Precipitation may be either dependent on the cellular metabolism or independent of it. In the former case, the metal removal from solution is often associated with active defense system of the microorganisms. They react in the presence of a toxic metal producing compounds, which favour the precipitation process. In the case of precipitation not dependent on the cellular metabolism, it may be a consequence of the chemical interaction between the metal and the cell surface. The various biosorption mechanisms mentioned above can take place simultaneously.

* *Use of Recombinant bacteria for metal removal:* Metal removal by adsorbents from water and wastewater is strongly influenced by physico-chemical parameters such as ionic strength, pH and the concentration of competing organic and inorganic compounds. Recombinant bacteria are being investigated for removing specific metals from contaminated water. For example a genetically engineered *E.coli,* which expresses Hg^{2+} transport system and metallothionin (a

metal binding protein) was able to selectively accumulate 8??mole Hg^{2+}/g cell dry weight. The presence of chelating agents Na^+, Mg^{2+} and Ca^{2+} did not affect bioaccumulation.

❖ *Factors affecting Biosorption:* The investigation of the efficacy of the metal uptake by the microbial biomass is essential for the industrial application of biosorption, as it gives information about the equilibrium of the process which is necessary for the design of the equipment.

The metal uptake is usually measured by the parameter 'q' which indicates the milligrams of metal accumulated per gram of biosorbent material and 'qH' is reported as a function of metal accumulated, sorbent material used and operating conditions.

The following factors affect the biosorption process:

1. Temperature seems not to influence the biosorption performances in the range of 20-35 °C (Aksu *et al.* 1992)

2. pH seems to be the most important parameter in the biosorptive process: it affects the solution chemistry of the metals, the activity of the functional groups in the biomass and the competition of metallic ions (Friis and Myers-Keith, 1986, Galun *et al.* 1987)

3. Biomass concentration in solution seems to influence the specific uptake: for lower values of biomass concentrations there is an increase in the specific uptake (Fourest and Roux, 1992; Gadd *et al.* 1988). Gadd *et al.* 1988 suggested that an increase in biomass concentration leads to interference between the binding sites. Fourest and Roux, 1992 invalidated this hypothesis attributing the responsibility of the specific uptake decrease to metal concentration shortage in solution. Hence this factor needs to be taken into consideration in any application of microbial biomass as biosorbent.

4. Biosorption is mainly used to treat wastewater where more than one type of metal ions would be present; the removal of one metal ion may be influenced by the presence of other metal ions. For example: Uranium uptake by biomass of bacteria, fungi and yeasts was not affected by the presence of manganese, cobalt, copper, cadmium, mercury and lead in solution (Sakaguchi and Nakajima, 1991). In contrast, the presence of Fe^{2+} and Zn^{2+} was found to influence uranium uptake by *Rhizopus arrhizus* (Tsezos and Volesky, 1982) and cobalt uptake by different microorganisms seemed to be completely inhibited by the presence of uranium, lead, mercury and copper (Sakaguchi and Nakajima, 1991).

Biosorption Equilibrium Models—Assessment of Sorption Performance

Examination and preliminary testing of solidliquid sorption system are based on two types of investigations: (a) equilibrium batch sorption tests and (b) dynamic continuous flow sorption studies.

The equilibrium of the biosorption process is often described by fitting the experimental points with models (Gadd, *et al.* 1988) usually used for the representation of isotherm adsorption equilibrium.

But the above said adsorption isotherms may exhibit an irregular pattern due to the complex nature of both the sorbent material and its varied multiple active sites, as well as the complex solution chemistry of some metallic compounds (Volesky and Holan, 1995). Evaluation of equilibrium sorption performance needs to be supplemented by process-oriented studies of its kinetics and eventually by dynamic continuous flow tests.

Biosorption by Immobilized Cells

Microbial biomass consists of small particles with low density, poor mechanical strength and little rigidity. The immobilization of the biomass in solid structures Qeates a material with the right size, mechanical strength and rigidity and porosity necessary for metal accumulation. Immobilisation can also yield beads and granules that can be stripped of metals, reactivated and reused in a manner similar to ion exchange resins and activated carbon.

Various applications are available for biomass immobilization. The principal techniques that are available in literature for the application of biosorption are based on adsorption on inert supports, on entrapment in polymeric matrix, on covalent bonds in vector compounds, or on cell cross-linking.

Adsorption on Inert Supports

Support materials are introduced prior to sterilization and inoculation with starter culture and are left inside the continuous culture for a period oftime, after which a film of microorganisms is apparent on the support surfaces. This technique has been used by Zhou and Kiff, 1991 for the immobilization of *Rhizopus arrhizus* fungal biomass in reticulated foam biomass support particles; Macaskie *et al.* 1987, immobilised the bacterium *Citrobacter sp.* by this technique. Scott and Karanjakar 1992, used activated carbon as a support for *Enterobacter aerogens* biofilm. Bai and Abraham, 2003 immobilized *Rhizopus nigricans* on polyurethane foam cubes and coconut fibres.

Entrapment in Polymeric Matrices

The polymers used are calcium alginate (Babu *et al.* 1993, Costa and Leite, 1991, Peng and Koon, 1993, Gulay Bayramoglu *et al.* 2002), polyacrylamide (Macaskie et aI., 1987, Michel *et al.* 1986, Sakaguchi and Nakajima *et al.* 1991, Wong and Kwok, 1992), polysulfone (Jeffers *et al.* 1991, Bai and Abraham, 2003) and polyethylenimine (Brierley and Brierley, 1993). The materials obtained from immobilization in calcium alginate and polyacrylamide are in the form of gel particles. Those obtained from immobilization in polysulfone and polyethyleneimine are the strongest.

Covalent Bonds to Vector Compounds

The most common vector compound (carrier) is silica gel. The material obtained is in the form of gel particles. This technique is mainly used for algal immobilization (Holan *et al.* 1993, Mah:!mn and Holocombe, 1992).

Cross-linking

The addition of the cross-linker leads to the formation of stable cellular aggregates. This technique was found useful for the immobilization of algae (Holan *et al.* 1993). The most common cross linkers are: formaldehyde, glutaric dialdehyde, divinylsulfone and formaldehyde—urea mixtures.

Desorption

If the biosorption process were to be used as an alternative to the wastewater treatment scheme, the regeneration of the biosorbent may be crucially important for keeping the process costs down and in opening the possibility of recovering the metals extracted from the liquid phase. For this purpose it is desirable to desorb the sorbed metals and to regenerate the biosorbent material for another cycle of application.

Bioremediation

16.1 Bioremediation

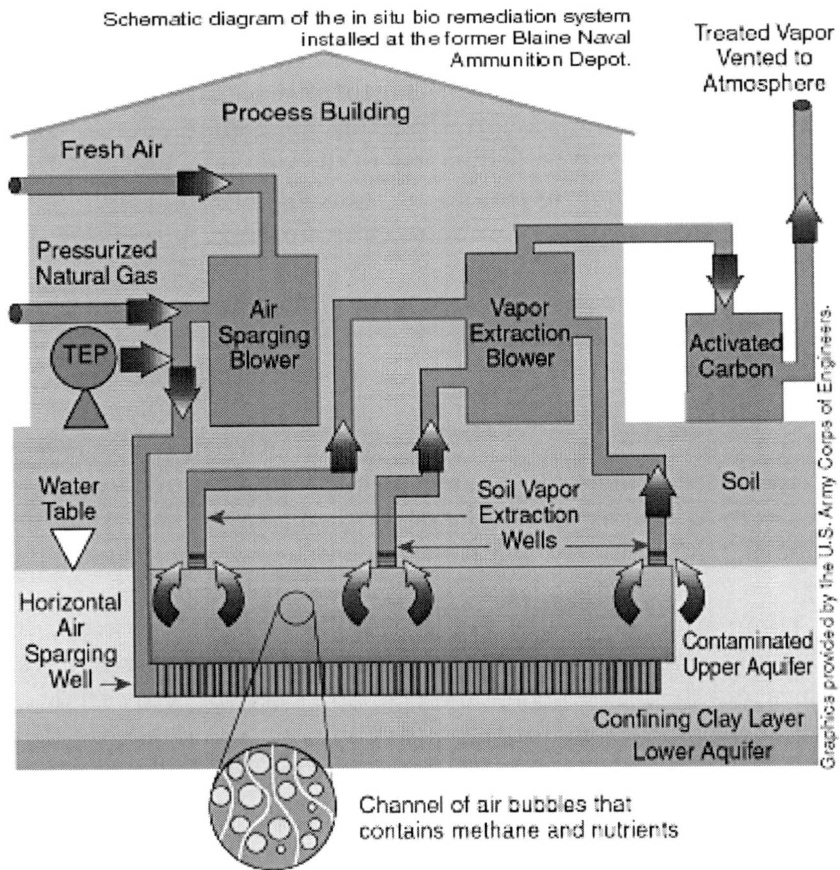

Bioremediation can be defined as any process that uses microorganisms, fungi, green plants or their enzymes to return the natural environment altered by contaminants to its original condition. Bioremediation may be employed to attack specific soil contaminants, such as degradation of chlorinated hydrocarbons by bacteria. An example of a more general approach is the cleanup of oil spills by the addition of nitrate and/or sulfate fertilisers to facilitate the decomposition of crude oil by indigenous or exogenous bacteria.

Overview and Applications

Naturally occurring bioremediation and phytoremediation have been used for centuries. For example, desalination of agricultural land by phytoextraction has a long tradition. Bioremediation technology using microorganisms was reportedly invented by George M. Robinson. He was the assistant county petroleum engineer for Santa Maria, California. During the 1960's, he spent his spare time experimenting with dirty jars and various mixes of microbes.

Bioremediation technologies can be generally classified as *in situ* or *ex situ*. *In situ* bioremediation involves treating the contaminated material at the site while *ex situ* involves the removal of the contaminated material to be treated elsewhere. Some examples of bioremediation technologies are bioventing, landfarming, bioreactor, composting, bioaugmentation, rhizofiltration, and biostimulation.

Not all contaminants, however, are easily treated by bioremediation using microorganisms. For example, heavy metals such as cadmium and lead are not readily absorbed or captured by organisms. The assimilation of metals such as mercury into the food chain may worsen matters. Phytoremediation is useful in these circumstances, because natural plants or transgenic plants are able to bioaccumulate these toxins in their above-ground parts, which are then harvested for removal. The heavy metals in the harvested biomass may be further concentrated by incineration or even recycled for industrial use.

The elimination of a wide range of pollutants and wastes from the environment is an absolute require increasing our understanding of the relative importance of different pathways and regulatory networks to carbon flux in particular environments and for particular compounds and they will certainly accelerate the development of bioremediation technologies and biotransformation processes.

Genetic Engineering Approaches

The use of genetic engineering to create organisms specifically designed for bioremediation has great potential. The bacterium Deinococcus radiodurans (the most radioresistant organism known) has been modified to consume and digest toluene and ionic mercury from highly radioactive nuclear waste.

Advantages

There are a number of cost/efficiency advantages to bioremediation, which can be employed in areas that are inaccessible without excavation. For example, hydrocarbon spills (specifically, petrol spills) or certain chlorinated solvents may contaminate groundwater, and introducing the appropriate electron acceptor or electron donor amendment, as appropriate, may significantly reduce contaminant concentrations after a lag time allowing for acclimation. This is typically much less expensive than excavation followed by disposal elsewhere, incineration or other *ex situ* treatment strategies, and reduces or eliminates the need for "pump and treat", a common practice at sites where hydrocarbons have clean groundwater.

Monitoring Bioremediation

The process of bioremediation can be monitored indirectly by measuring the *Oxidation Reduction Potential* or redox in soil and groundwater, together with pH, temperature, oxygen content, electron acceptor/donor concentrations, and concentration of breakdown products (e.g. carbon dioxide). This table shows the (decreasing) biological breakdown rate as function of the redox potential.

Process	Reaction	Redox potential (E_h in mV)
Aerobic:	$O_2 + 4e^- + 4H^+ \rightarrow 2H_2O$	600 ~ 400
Anaerobic:		
Denitrification	$2NO_3^- + 10e^- + 12H^+ \rightarrow N_2 + 6H_2O$	500 ~ 200
Manganese IV reduction	$MnO_2 + 2e^- + 4H^+ \rightarrow Mn^{2+} + 2H_2O$	400 ~ 200
Iron III reduction	$Fe(OH)_3 + e^- + 3H^+ \rightarrow Fe^{2+} + 3H_2O$	300 ~ 100
Sulfate reduction	$SO_4^{2-} + 8e^- + 10 H^+ \rightarrow H_2S + 4H_2O$	0 ~ ⁻150
Fermentation	$2CH_2O \rightarrow \sigma\ CO_2 + CH_4$	⁻150 ~ ⁻220

This, by itself and at a single site, gives little information about the process of remediation.

1. It is necessary to sample enough points on and around the contaminated site to be able to determine contours of equal redox potential. Contouring is usually done using specialised software, e.g. using Kriging interpolation.
2. If all the measurements of redox potential show is that electron

acceptors have been used up, it's in effect an indicator for total microbial activity. Chemical analysis is also required to determine when the levels of contaminants and their breakdown products have been reduced to below regulatory limits.

16.2 Remediate

"Remediate" means to solve a problem, and "bio-remediate" means to use biological organisms to solve an environmental problem such as contaminated soil or groundwater.

In a non-polluted environment, bacteria, fungi, protists, and other microorganisms are constantly at work breaking down organic matter. What would occur if an organic pollutant such as oil contaminated this environment? Some of the microorganisms would die, while others capable of eating the organic pollution would survive. Bioremediation works by providing these pollution-eating organisms with fertilizer, oxygen, and other conditions that encourage their rapid growth. These organisms would then be able to break down the organic pollutant at a correspondingly faster rate. In fact, bioremediation is often used to help clean up oil spills.

Bioremediation of a contaminated site typically works in one of two ways. In the case described above, ways are found to enhance the growth of whatever pollution-eating microbes might already be living at the contaminated site. In the second, less common case, specialized microbes are added to degrade the contaminants.

Bioremediation provides a good cleanup strategy for some types of pollution, but as you might expect, it will not work for all. For example, bioremediation may not provide a feasible strategy at sites with high concentrations of chemicals that are toxic to most microorganisms. These chemicals include metals such as cadmium or lead, and salts such as sodium chloride.

Nonetheless, bioremediation provides a technique for cleaning up pollution by enhancing the same biodegradation processes that occur in nature. Depending on the site and its contaminants, bioremediation may be safer and less expensive than alternative solutions such as incineration or landfilling of the contaminated materials. It also has the advantage of treating the contamination in place so that large quantities of soil, sediment or water do not have to be dug up or pumped out of the ground for treatment.

16.3 Types of Bioremediation

There are two main types of bioremediation

In-Situ Remediation

In-situ Bioremediation treats the contaminated soil or groundwater in the location in which it is found. In this technology oxygen and occasionally nutrients are pumped under pressure into the soil through wells. The nutrients are spread on the surface to infiltrate into the contaminated area of material or the saturated zone.

Ex-Situ Remediation

Ex-situ Bioremediation requires pumping of the groundwater or excavation of contaminated soil prior to remediation treatments. Ex-situ Bioremediation can be further broken down into two main components or processes; Slurry-phase and solid-phase treatment.

- ❖ *Slurry-phase*: This treatment involves the initial combination of water with the contaminated soil and later degradation in a bioreactor.
- ❖ *Solid-phase*: This treatment achieves the similar goal of the former treatment yet, in this process, the contaminated soil is placed in a bed and nourished with nutrients, moisture and oxygen in hopes that decomposition will occur.

16.4 Bioremediation of Chromium Pollutants

Chromium(Cr), the transition metal has found wide variety of industrial uses in metallurgical, chemical and refractory industries. These anthropogenic activities generate large volumes of Cr waste in various chemical forms and discharge them into the environment. In the industrial wastes chromium is present chiefly in two stable states, Cr(III) and Cr(VI) which differ markedly in terms of their mobility and toxicity. The US EPA has designated chromium as a priority pollutant and it has been shown to affect the generation time of bacteria, spore germination, mycelial proliferation, microbial respiration, photosynthesis and nitrogenase activity. In animals, including human beings chromium is considered as an essential micronutrient but in the event of exposure to excessive concentration or ingestion above the stipulated limit it is toxic to both animals and human beings, leading to cytotoxicity, mutation and carcinogenesis. These toxic effects on diverse life forms ranging from microorganisms to humans have been attribute to altered genetic material and altered metabolism and physiological reactions.

Why Bioremediation?

Such toxicity problems have evoked general awareness among scientists, industrialists, chemists and microbiologists in particular to develop ways and means for amelioration of chromium toxicity in the environment. While the traditional techniques for remediating chromate-contaminated water involve reduction of Cr(VI) to Cr(III) by chemical or electrochemical means, other physicochemical methods such as adsorption, ion-exchange, membrane separation and solvent extraction are also being practiced for Cr removal. Most of these physicochemical methods, however, require high energy or large quantities of chemicals. More practical and economic methods are therefore, necessary for effective control of Cr-pollutants.

Biological methods for reducing or removing heavy metals like Cr are in continuous competition with chemical and physical technologists. Bioremediation, defined as the spontaneous or spontaneous managed process in which biological agents, specially microorganisms act on pollutants and thereby removes or eliminates environmental contamination caused by humankind's activities. As a process it is nothing but an extension of biological treatment processes used to biodegrade environmental pollutants. It is a "green technology" with very little or no effect on ecological balance. The process involves i) the use of indigenous population of microorganisms to degrade the pollutants more rapidly or ii) addition of effective microorganisms where there is no effective indigenous flora. Either or both could be considered for broad application to the polluted site or for the application in the bioreactors.

Microbes Involved

It is noteworthy to mention that microorganism by virtue of their wide degree of adaptability to contaminated environments could be an effective tool for bioremediation or detoxification of pollutants. The general trend is to isolate resistant microorganisms from polluted environment, determination of their resistance mechanisms and employ them for effective bioremediation purpose. Till date, studies on the isolation of Cr-resistant microorganisms from Cr-contaminated samples have been reported from different parts of the world. We have reported the isolation of over 250 Cr-resistant bacterial isolates from Cr-contaminated sediments of tanning industries in and around Kolkata as well as from other industrial effluents and a significant population of these isolates were able to grow in medium containing 1000 ?g/ml of Cr(VI). Application of these resistant microorganisms for the bioremediation of chromium pollutants are on the process of development.

Biosorption of Chromium

Biosorption is the phenomenon of retention of metal ions from solution by dead microbial cells. It is considered as a collective term for a number of passive accumulation processes and may include physical and/or chemical adsorption, ion exchange, coordination, complexation chelation and microprecipitation. Biosorption of chromium by microbial cells has been recognized as a potential alternative to existing technologies for removing heavy metals from industrial wastewaters including tannery effluents. Microbial biomass of different types such as algae, filamentous fungi, and yeasts have been used for removal of Cr(VI) from aqueous solutions. Bacterial biomass derived from *Pseudomonas mendocina; Desufovibrio desulfuricans; Thibacillus ferroxidans* also serve as an effective tool for removal of Cr(VI) from aqueous solution. The waste biomass of *Streptomyces noursei* from the pharmaceutical industry has also been effectively utilized for uptake of Cr(III) from aqueous solution.

Bioreduction of Chromate

The discovery of microorganisms those preferentially reduce toxic Cr(VI) to less toxic Cr(III) has led to their application in the field of bioremediation of toxic chromate. The chromate reducing bacteria are ubiquitous, occurring both in Crcontaminated and non-contaminated environments and carry out Cr(VI) transformation under *in vivo* and *in vitro* conditions. Both aerobic and anaerobic reduction systems are known to occur with different bacteria. Aerobic reduction of chromate is generally associated with soluble proteins, whereas anaerobic reduction is found to occur with membrane preparations. When reduction takes place at the cell surface, Cr(III)) forms extracellular insoluble chromium hydroxide that subsequently precipitate in the culture medium thus making it less available to biological systems.

The rate and extent of Cr(VI) reduction by aerobic systems reported so far are too poor to assure acceptable efficiency of chromate treatment. Current research is directed toward using anaerobic reduction systems as a possible means of bioremediation. Feasibility studies have been performed exclusively with pure cultures of *Enterobacter cloacae* in bioreactors. The reactor systems have the advantage that the microbial activities could be maintained by optimized environmental conditions that cannot be realized in field conditions. Moreover, bacterial reduction of Cr(VI) as a means of bioremediation has several potential advantages: i) reduction occurs under mild conditions; ii) it requires neither chemical additives nor aeration; iii) anaerobic reduction minimizes excess sludge production in aqueous systems; iv) no toxic byproduct is formed; and v) the activity is reproducible and reusable. Finally it should be emphasized that though the potential of chromate-reducing bacteria to detoxify Cr(VI) is well recognized, no work has been done on the engineering aspects until recently. Engineering studies would demonstrate how bacteria could be efficiently applied for the purpose of chromate treatment. Such studies are essential for assessing the possibility of bioremediation of Cr(VI) pollution in waste water including the tannery effluents.

Conclusion

Bacterial reduction and biosorption of hexavalent chromium are of special interest from microbiological viewpoint and could be very promising in the bioremediation of industrial effluents, including those of tanning industries. Both processes, thus far appeared to be quite successful in cleaning up of dilute waste water under laboratory conditions. Studies are now proceeding to the developmental stage. Further efforts are still necessary for establishing the full-scale processes, which will find wide application in the control of environmental pollution.

An Investigation on the Effects of Some Heavy Metals in Pisum Sativum in Vitro Condition

Heavy metals have for long been known to be important constituent of aerosol causing air pollution. Aerosol contain different toxic heavy metals, the size and setting velocity of which constitute important factors in spreading pollution. Compounds containing aluminium, cadmium, copper, lead, mercury, nickel and tellurium have been observed to induce chromosomal aberration or abnormal cell division in animal and plant cell. The alteration involves both chromosome structure and number as well as genetic component.

In this presentation attempts will be made to understand the role of mercury, cadmium and lead (belonging to the same group of periodic table) using in vitro grown tissues of Pisum sativum with respect to the following parameters:

- ❖ To determine the relative threshold value and degree of toxicity due to the effects of heavy metal compounds;
- ❖ To determine clastogenic and cytotoxic effects on chromosomes;
- ❖ To determine changes in nucleolar number and shape;
- ❖ To determine nucleolar and nuclear area and finally;
- ❖ To compare the effect of the heavy metal compounds at the molecular level with respect to DNA, RNA and protein.

Advantages of Tissue Culture Technique

- ❖ Physiological condition is uniform.
- ❖ Increase in genetic variability may be induced in large and homogenous population and homogenous population of plant cells or in callus tissues by exposing culture to mutagenic agents.
- ❖ Furthermore, tissue culture have certain definite advantages over the whole plants as the mutagenic induced nuclear and biochemical changes in the tissues grown *in vitro* can be easily studied and the inherited changes if any can be detected in *in vitro* system for studying the long range indirect effect of environmental chemical mutagen at the chromosomal, biochemical and also at the developmental level.
- ❖ Plant tissue cultures offer a meaningful tool for isolatinginduced as well as spontaneously occurring variant cell lines and for the increase of genetic diversity. In our investigation comparative assessment of heavy metal compounds such as HgCl2, CdCl2, $Pb(NO2)_2$ was study of cytotoxic effects on *P. sativum* in *in vitro* condition. Aseptic transfer of heavy metal treated embryoids in the test tubes containing MS medium supplemented with 2, 4-D and kinetin compared the germination phenomenon, decreasing plant height with the increase in concentrations. Complete ceasation of plant growth started in 10-2M concentration, from 12 days.
- ❖ There were a significant stimulation of calli initiation and their growth at lower concentrations. Complete ceasation of callus initiation was observed at 10-2M concentration of heavy metal compounds for both the root and shoot explants. Callus growth gradually decreased with the increase in concentration and time. Profuse callusing was found at 10-5M conc. for all the metallic components. Fresh weight and dry weight deceased with the increase in concentration and duration.
- ❖ Regarding the frequency and distribution of chromosomal aberration induced by heavy metal compounds (HgCl2, CdCl2, $Pb(NO3)_2O$ in root and shoot callus culture, it was also found that percentage of aberrent cells were increased with the increase in concentration, chromosomal anomalies found were clumped metaphase, micronuclei, C-metaphase, laggards, anaphasic bridge, chromocentric nuclei, breakage and fragments, binucleate cell polyploid and aneuploid cells. Mitotic inhibition, increased polyploidy and degeneration of cytoplasm at high doses were commonly observed.
- ❖ Regarding the nucleolar, nuclear effects, it was found that nucleolar indices (N.I.) as well as Nu/No showed positive correlation with the

increasing concentration and duration of treatment. Uninucleolate, binucleolate stages were common but tri, tetra and polynucleolated stages were found in less frequent in root and shoot calli. In relation to these, a gradual decrease innuclear area/cell was obtained at increasing concentration and time. Among the different shapes of nuclei, round, elongate, ellipsoidal and dumbbell shaped nuclei were common in the cells of calli for different concentration and duration of the treatments.

❖ Comparative analysis of the effect of the heavy metal at the molecular level indicate the inhibitory effects on DNA, RNA and protein contents of leaf samples in *Pisum sativum* collected from T1 population growing in field. In general, the amounts gave more effective values at higher concentration and duration of treatment. They might be correlated with the chromosomal anomalies produced by the heavy metal and fungicidal compounds.

❖ Present investigation suggest that all the compounds under study possess clastogenic, mitotoxic and cytotoxic properties. Mutagenic potentialities of the compounds have been evidenced by the cytogenetic implication of their effects. At the molecular level heavy metal may interact with protein leading to denaturation allosteric effects or enzyme inhibition or it may bind to nucleic acid resulting irreversible conformational changes as well as alteration in chromosomal configuration and cell and chromosome division.

16.5 Bioremediation-Services

Bioremediation

On-site remediation of soils and media contaminated with petrochemicals, pesticides, explosives or hazardous organic materials can be an expensive and time-consuming process. There are also serious legal and regulatory issues to consider when decontaminating soils on-site—issues that can jeopardize the viability of a construction or redevelopment project. Waste Management, North America's largest solid waste services company, has developed a collection of innovative off-site remediation technologies to help companies deal effectively with contaminated soils. These bioremediation services include:

❖ *Toss^{sm} (two-step static system):* TOSS^{SM} is a two-stage, solid-phase bioremediation technology that involves both anaerobic and aerobic treatment stages. In the first stage, explosives-contaminated soil is combined with a carbon source, an inoculum, vitamins and water to achieve anaerobic conditions. The resulting mixture is formed into a static pile or placed in a bermed construction or box to facilitate the chemical reduction of nitroaromatic and nitramine explosives. In the second stage, the anaerobically treated soil is combined with yard waste compost and built into an aerated biopile. The biopile may be aerated by forced air conveyed through perforated piping buried within the pile or by turning the pile with a compost turner. Previous testing of TOSS[SM] has demonstrated TNT removal efficiencies of greater than 99 per cent.

❖ *The biosite[sm] system:* The BioSite[SM] System is Waste Management's proprietary system for the large-scale bioremediation of soils contaminated with:

- Petrochemicals including, but not limited to:

    ■ acetone, alcohols, benzene, ethylbenzene, methyl ethyl ketone (MEK), methyl isobutyl ketone (MIBK), petroleum hydrocarbons, toluene, two-and three-ring PAHs, xylene

- other contaminants, including:

    ■ aliphatic chlorinated hydrocarbons (e.g., trichloroethylene), spent molecular sieve from packing towers, chemical manufacturing wastes, pesticides

  Regulated compounds including underlying hazardous constituents (UHC) are screened prior to acceptance. Soils co-contaminated with metals may be accepted depending on their concentration.

❖ *Bio-In-A-Box[SM]:* Bio-In-A-Box[SM] is based on the same principles as both BioSite[SM] and TOSS[SM], but is designed to operate indoors on a relatively smaller scale. Instead of being formed into long earthen mounds, the contaminated soil is moistened, mixed with nutrients and custom-grown microorganisms, and then placed in enclosed containers called "solid phase bioreactors" for incubation. These containers may or may not be linked to aeration and vacuum pipes, depending on the contaminants being processed. In just a few weeks, the decontaminated soil will be ready for landfill disposal or reintroduction into the environment.

- Residential
- Business
- City/Town (Municipality)
- Recycling
- Landfills
- Waste to Energy
- WasteByRail
- Renewable Energy
- Bioremediation

16.6 Bioremediation Services: A Case Study

Bioremediate.com uses a range of microbial probiotic products that are suitable for the bioremediation of terrestrial, wetland, marine, farm land and industrial areas and is specifically suited to the restoration of natural ecological environments. The pace of the Bioremediation Industry has moved quickly. From the discovery that natural indigenous microbes had the ability to remediate pollutants, to the identification, isolation and commercial production of viable products, it continues to progress. Manufacturers are now able to isolate and mass-produce standardized bacteria and fungi in concentrated liquid formats and dry endosporic (dormant) state, a tremendous advancement from the 1st generation liquids. Recently, traceable scientific studies for these new, standardized, CFU count microbial products have been conducted and demonstrate the efficacy of microbes for the safe remediation of numerous contaminants. Remediation is achieved by using a combination of microbial products that remove hydrocarbon spills, biodegrades organic pollutants, reduce phosphorous levels and then, by adding microbial supplements and biostimulants, promote natural soil conditions. This process of bio-augmentation enhances the viability of all planting, minimizes transplant shock, encourages germination and regrowth to the treated areas. Our team of scientists will formulate a product line and methodology specific to customer requirements. This assures that you will achieve project closure approvals more quickly and economically than most mechanical or hard chemical technologies.

Water

Pond and Lake Management Programme

Eutro-Clear—Controlled experiments clearly demonstrates that Eutro-Clear has the capacity to remove organic pollutants from the water column through microbial bioremediation. Once the nutrient source has been removed and the water is brought back into ecological equilibrium, algae and cyanobacteria growth and pond scum and eutrophication will be reduced.

Aquaculture Therapy

Using a combination of nitrifying bacteria Nitro-Clear with Eutro-Clear Probiotic formula will address ammonia and reduce the organic sludge within the system. The pro-biotics assist in feed conversion and digestion. Excess food, feces build-up is reduced. Turbidity is reduced, water becomes cleaner. (more information on Aquaculture Treatments)

Wastewater Treatment for Industrial And Municipal Systems

This environmentally responsible alternative to chemical treatments, provides specialized microbial species in high standardized concentrations, which in turn quickly break down the organic waste found in typical systems. (more information on wastewater treatments)

Ammonia Reduction with Nitrifying Bacteria

For high ammonia nitrogen Nitro-Clear—nitrifying bacteria is designed to control ammonia and nitrite. Nitrifying bacteria is designed to provide a biological oxidation of ammonia for shrimp/prawn, fish hatchery pond, ornamental koi fish pond, tropical fish pond, fish farms, lobster pens, prawn farms, natural lakes, ponds and rivers. Nitrification is the biological or biochemical process in which ammonia is oxidized to nitrite, and nitrite oxidized to nitrate.

Hydrocarbons

Hydrocarbon Remediation

Converts harmful petroleum products into beneficial elements—basic carbon and nitrogen. Fast, effective bioremediation for gasoline, oil, hydraulic fluid and pesticide spills around the shop and on the golf course. Stop these spills from creating a hazard to the turf or to irrigation ponds. (more information on Hydrocarbon remediation—oil spills and petrochemical treatments)

Hydrocarbon Remediation In Water

Based on years of research, a series of all-natural microbial systems has been isolated to bio-remediate hydrocarbon contamination in either salt or fresh water environments. These microbial systems contain high concentrations of safe, non-pathogenic, microorganisms.(more information hydrocarbon spills in aqueous environments)

Indoor Air Quality

Mould Inspection and Remediation in Vancouver BC

Mold Remediation and Abatement In Atlanta

- Mold Remediation services for domestic, commercial or industrial buildings.
- Mold Inspection and Testing in the greater Atlanta area.
- Mold Remediation and Abatement.
- Mold cleaning tips for small DIY remediation projects—FAQ.

On-line sales

- *Eutro-Clear*: Microbial pond products for small problematic ponds.
- *Petro-Clear*: Microbial products for small hydrocarbon spills on soil-oil, diesel, gasoline.
- *BOD-Clear*: Septic Tank, Leach Field and Grease Trap treatments.

Bioremediate.com, LLC. Remediation Services and Consultancy

- We provide microbial remediation solutions for a wide range of pollutants.
- We offer personalized service for both large and small projects.
- We solve the problem; recommending the most cost effecitive solutions.
- We use products specifically chosen for your particular environmental problem.

16.7 Bioremediation in Practice: A Case Study

The Problem

The problem in Hanahan, South Carolina, a quiet suburb of Charleston, was not particularly unusual. In 1975, a massive leak from a military fuel storage facility released about 80,000 gallons of kerosine-based jet fuel. Immediate and extensive recovery measures managed to contain the spill, but could not prevent some fuel from soaking into the permeable sandy soil and reaching the underlying water table. Soon, ground water was leaching such toxic chemicals as benzene from the fuel-saturated soils and carrying them toward a nearby residential area.

By 1985, contamination had reached the residential area, and the facility was faced with a serious environmental problem. Removing the contaminated soils was technically impractical, and removing contaminated ground water did not address the source of the contaminants. How could contaminated ground water be kept from seeping toward the residential area in the future?

One possible solution was a new technology called bioremediation. Studies by the U.S. Geological Survey (USGS) had shown that microorganisms naturally present in the soils were actively consuming fuel-derived toxic compounds and transforming them into harmless carbon dioxide. Furthermore, these studies had shown that the rate of these biotransformations could be greatly increased by the addition of nutrients. By "stimulating" the natural microbial community through nutrient addition, it was theoretically possible to increase rates of biodegradation and thereby shield the residential area from further contamination.

In 1992, this theory was put into practice by USGS scientists. Nutrients were delivered to contaminated soils through infiltration galleries, contaminated ground water was removed by a series of extraction wells, and the arduous task of monitoring contamination levels began. By the end of 1993, contamination in the residential area had been reduced by 75 percent. Nearer to the infiltration galleries (the source of the nutrients), the results were even better. Ground water that once had contained more than 5,000 parts per billion toluene now contained no detectable contamination.

Bioremediation had worked!

Why Bioremediation Works?

The success of the Hanahan Bioremediation Project was no accident. It was the result of many years of intensive effort by many USGS scientists.

In the early 1980's, little was known about how toxic wastes interact with the hydrosphere. This lack of knowledge was crippling efforts to remediate environmental contamination under the new Superfund legislation—-the Comprehensive Environmental Response, Compensation, and Liability Act. Faced with this problem, Congress directed the USGS to conduct a programme to provide this critically needed information. By means of this programme, known as the Toxic Substances Hydrology Programme, the most important categories of wastes were systematically investigated at sites throughout the United States. One of the principal findings of this programme was that microorganisms in shallow aquifers affect the fate and transport of virtually all kinds of toxic substances. For example:

Crude Oil Spill, Bemidji, Minnesota

In 1979, a pipeline carrying crude oil burst and contaminated the underlying aquifer. USGS scientists studying the site found that toxic chemicals leaching from the crude oil were rapidly degraded by natural microbial populations. Significantly, it was shown that the plume of contaminated ground water stopped enlarging after a few years as rates of microbial degradation came into balance with rates of contaminant leaching. This was the first and best-documented example of intrinsic bioremediation in which naturally occurring microbial processes remediates contaminated ground water without human intervention.

Sewage Effluent, Cape Cod, Massachusetts

Disposal of sewage effluent in septic drain fields is a common practice throughout the United States. Systematic studies of a sewage effluent plume at Massachusetts Military Reservation (formerly known as Otis Air Force Base) led to the first accurate field and laboratory measurements of how rapidly natural microbial populations degrade nitrate contamination (denitrification) in a shallow aquifer.

Chlorinated Solvents, New Jersey

Chlorinated solvents are a particularly common contaminant in the heavily industrialized Northeast. Because their metabolic processes are so adaptable, microorganisms can use chlorinated compounds as oxidants when other oxidants are not available. Such transformations, which can naturally remediate solvent contamination of ground water, has been extensively documented by USGS scientists at Picatinny Arsenal, New Jersey.

Pesticides, San Francisco Bay Estuary

Pesticide contamination of rivers and streams is a matter of concern throughout the United States. Field and laboratory studies in the Sacramento River and San Francisco Bay have shown the effects of biological and non-biological processes in degrading commonly used pesticides, such as molinate, thiobencarb, carbofuran, and methyl parathion.

Agricultural Chemicals in the Midcontinent

Agricultural chemicals affect the chemical quality of ground water in many Midwestern States. Studies in the midcontinent have traced the fate of nitrogen fertilizers and pesticides in ground and surface waters. These studies have shown that many common contaminants, such as the herbicide atrazine, are degraded by biological (microbial degradation) and non-biological (photolytic degradation) processes.

Gasoline Contamination, Galloway, New Jersey

Gasoline is probably the most common contaminant of ground water in the United States. Studies at this site have demonstrated rapid microbial degradation of gasoline contaminants and have shown the importance of processes in the unsaturated zone (the zone above the water table) in degrading contaminants.

Creosote Contaminants, Pensacola, Florida

Creosote and chlorinated phenols have been used extensively as wood preservatives throughout the United States. Contaminants leaked to the underlying aquifer through several unlined ponds and were transported toward nearby Pensacola Bay. Studies at this site have demonstrated that microorganisms can adapt to extremely harsh chemical conditions and that microbial degradation was restricting migration of the contaminant plume.

Together, these studies laid the technical foundation that enabled bioremediation to be applied at Hanahan.

Technology Transfer

The Hanahan Bioremediation Project is just one of many successful bioremediation experiments that can be traced to basic research carried out by USGS scientists. Methods and technology developed in the Toxic Substances Hydrology Programme are now being used by private contractors, State environmental managers, and other Federal agencies to address contaminant problems throughout the United States.

Stretching Remediation Dollars

Cleaning up existing environmental contamination in the United States could cost as much as $1 *trillion* dollars. Bioremediation can help contain costs as follows:

Treating Contamination in Place

Most of the cost associated with traditional cleanup technologies is associated with physically removing and disposing of contaminated soils. Because engineered bioremediation can be carried out in place by delivering nutrients to contaminated soils, it does not incur removal-disposal costs.

Harnessing Natural Processes

At some sites, natural microbial processes can remove or contain contaminants without human intervention. In these cases where intrinsic bioremediation (natural attenuation) is appropriate, substantial cost savings can be realized.

Reducing Environmental Stress

Because bioremediation methods minimize site disturbance compared with conventional cleanup technologies, post-cleanup costs can be substantially reduced.

Future Challenges

Although bioremediation holds great promise for dealing with intractable environmental problems, it is important to recognize that much of this promise has yet to be realized. Specifically, much needs to be learned about how microorganisms interact with different hydrologic environments. As this under-standing increases, the efficiency and applicability of bioremediation will grow rapidly. Because of its unique interdisciplinary expertise in microbiology, hydrogeology, and geochemistry, the USGS will continue to be at the forefront of this exciting and rapidly evolving technology.

16.8 A Citizen's Guide to Bioremediation

Bioremediation is a treatment process that uses naturally occurring microorganisms (yeast, fungi, or bacteria) to break down, or *degrade*, hazardous substances into less toxic or nontoxic substances. Microorganisms, just like humans, eat and digest organic substances for nutrients and energy. In chemical terms, "organic" compounds are those that contain carbon and hydrogen atoms. Certain microor-ganisms can digest organic substances such as fuels or solvents that are hazardous to humans. The microorganisms break down the organic contaminants into harmless products—mainly carbon dioxide and water. Once the contaminants are degraded, the microorganism population is reduced because they have used all of their food source. Dead microorganisms or small populations in the absence of food pose no contamination risk.

How Does it Work?

Microorganisms must be active and healthy in order for bioremediation to take place. Bioremediation technologies assist microorganisms' growth and increase microbial populations by creating optimum environmental conditions for them to detoxify the maximum amount of contaminants. The specific bioremediation technology used is determined by several factors, for instance, the type of microorganisms present, the site conditions, and the quantity and toxicity of contaminant chemicals. Different microorganisms degrade different types of compounds and survive under different conditions.

Indigenous microorganisms are those microorganisms that are found already living at a given site. To stimulate the growth of these indigenous microorganisms, the proper soil temperature, oxygen, and nutrient content may need to be provided.

If the biological activity needed to degrade a particular contaminant is *not* present in the soil at the site, microorganisms from other locations, whose effectiveness has been tested, can be added to the contaminated soil. These are called *exogenous* microorganisms. The soil conditions at the new site may need to be adjusted to ensure that the exogenous microorganisms will thrive.

Bioremediation can take place under *aerobic* and *anaerobic* conditions. In aerobic conditions, microorganisms use available atmospheric oxygen in order to function. With sufficient oxygen, microorganisms will convert many organic contaminants to carbon dioxide and water. *Anaerobic* conditions support biological activity in which no oxygen is present so the microorganisms break down chemical compounds in the soil to release the energy they need. Sometimes, during aerobic and anaerobic processes of breaking down the original contaminants, intermediate products that are less, equally, or more toxic than the original contaminants are created.

Bioremediation can be used as a cleanup method for contaminated soil and water. Bioremediation applications fall into two broad categories: *in situ* or *ex situ*. In situ bioremediation treats the contaminated soil or groundwater in the location in which it was found. Ex situ bioremediation processes require excavation of contaminated soil or pumping of groundwater before they can be treated.

In Situ Bioremediation of Soil

In situ techniques do not require excavation of the contaminated soils so may be less expensive, create less dust, and cause less release of contaminants than ex situ techniques. Also, it is possible to treat a large volume of soil at once. In situ techniques, however, may be slower than ex situ techniques, may be difficult to manage, and are most effective at sites with *permeable* (sandy or uncompacted) soil.

The goal of aerobic in situ bioremediation is to supply oxygen and nutrients to the microorganisms in the soil. Aerobic in situ techniques can vary in the way they supply oxygen to the organisms that degrade the contaminants. Two such methods are *bioventing* and *injection of hydrogen peroxide*. Oxygen can be provided by pumping air into the soil above the water table (bioventing) or by delivering the oxygen in liquid form as hydrogen peroxide. In situ bioremediation may not work well in clays or in highly layered subsurface environments because oxygen cannot be evenly distributed throughout the treatment area. In situ remediation often requires years to reach cleanup goals, depending mainly on how biodegradable specific contaminants are. Less time may be required with easily degraded contaminants.

Bioventing: Bioventing systems deliver air from the atmosphere into the soil above the water table through injection wells placed in the ground where the contamination exists. The number, location, and depth of the wells depend on many geological factors and engineering considerations.

An air blower may be used to push or pull air into the soil through the injection wells. Air flows through the soil and the oxygen in it is used by the microorganisms. Nutrients may be pumped into the soil through the injection wells. Nitrogen and phosphorous may be added to increase the growth rate of the microorganisms.

Injection of Hydrogen Peroxide: This process delivers oxygen to stimulate the activity of naturally occurring microorganisms by circulating hydrogen peroxide through contaminated soils to speed the bioremediation of organic contaminants. Since it involves putting a chemical (hydrogen peroxide) into the ground (which may eventually seep into the groundwater), this process is used only at sites where the groundwater is already contaminated.

A system of pipes or a sprinkler system is typically used to deliver hydrogen peroxide to shallow contaminated soils. Injection wells are used for deeper contaminated soils.

In Situ Bioremediation of Groundwater

In situ bioremediation of groundwater speeds the natural biodegradation processes that take place in the watersoaked underground region that lies below the water table. For sites at which both the soil and groundwater are contaminated, this single technology is effective at treating both.

Generally, an in situ groundwater bioremediation system consists of an extraction well to remove groundwater from the ground, an above-ground water treatment system where nutrients and an oxygen source may be added to the contaminated groundwater, and injection wells to return the "conditioned" groundwater to the subsurface where the microorganisms degrade the contaminants.

One limitation of this technology is that differences in underground soil layering and density may cause reinjected conditioned groundwater to follow certain preferred flow paths. Consequently, the conditioned water may not reach some areas of contamination.

Another frequently used method of in situ groundwater treatment is *air sparging,* which means pumping air into the groundwater to help flush out contaminants. Air sparging is used in conjunction with a technology called soil vapor extraction and is described in detail in the document entitled *A Citizen's Guide to Soil Vapor Extraction and Air Sparging.*

Ex Situ Bioremediation of Soil

Ex situ techniques can be faster, easier to control, and used to treat a wider range of contaminants and soil types than in situ techniques. However, they require excavation and treatment of the contaminated soil before and, sometimes, after the actual bioremediation step. Ex situ techniques include *slurry-phase bioremediation* and *solid-phase bioremediation.*

❖ *Slurry-phase bioremediation*: Contaminated soil is combined with water and other additives in a large tank called a "bioreactor" and mixed to keep the microorganisms—which are already present in the soil—in contact with the contaminants in the soil. Nutrients and oxygen are added, and conditions in the bioreactor are controlled to create the optimum environment for the microorganisms to degrade the contaminants. Upon completion of the treatment, the water is removed from the solids, which are disposed of or treated further if they still contain pollutants.

  Slurry-phase biological treatment can be a relatively rapid process compared to other biological treatment processes, particularly for contaminated clays. The success of the process is highly dependent on the specific soil and chemical properties of the contaminated material. This technology is particularly useful where rapid remediation is a high priority.

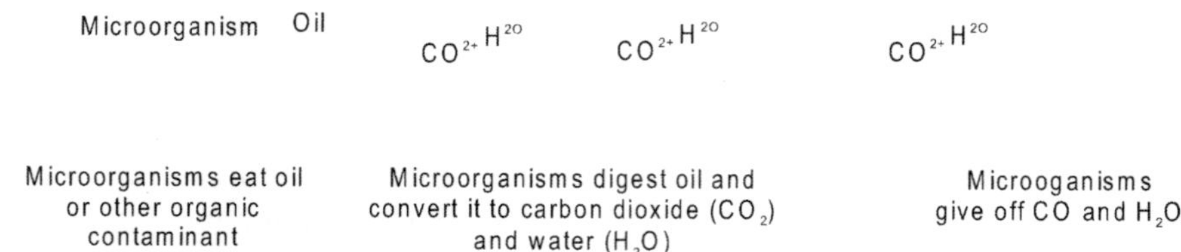

Figure 16.1: Schematic Diagram of Aerobic Biodegradation in Soil

❖ *Solid-phase bioremediation:* Solid-phase bioremediation is a process that treats soils in above-ground treatment areas equipped with collection systems to prevent any contaminant from escaping the treatment. Moisture, heat, nutrients, or oxygen are controlled to enhance biodegradation for the application of this treatment. Solid-phase systems are relatively simple to operate and maintain, require a large amount of space, and cleanups require more time to complete than with slurry-phase processes. Solid-phase soil treatment processes include *landfarming, soil biopiles,* and *composting.*

❖ *Landfarming:* In this relatively simple treatment method, contaminated soils are excavated and spread on a pad with a built-in system to collect any "leachate" or contaminated liquids that seep out of contaminant-soaked soil. The soils are periodically turned over to mix air into the waste. Moisture and nutrients are controlled to enhance bio-remediation. The length of time for bioremediation to occur will be longer if nutrients, oxygen or temperature are not properly controlled. In some cases, reduction of contaminant concentrations actually may be attributed more to volatilization than biodegradation. When the process is conducted in enclosures controlling escaping volatile contaminants, volatilization losses are minimized.

❖ *Soil biopiles:* Contaminated soil is piled in heaps several meters high over an air distribution system. Aeration is provided by pulling air through the heap with a vacuum pump. Moisture and nutrient levels are maintained at levels that maximize bioremediation. The soil heaps can be placed in enclosures. Volatile contaminants are easily controlled since they are usually part of the air stream being pulled through the pile.

❖ *Composting:* Biodegradable waste is mixed with a bulking agent such as straw, hay, or corn cobs to make it easier to deliver the optimum levels of air and water to the micro-organisms. Three common designs are *static pile composting* (compost is formed into piles and aerated with blowers or vacuum pumps), *mechanically agitated in-vessel*

❖ *Composting* (compost is placed in a treatment vessel where it is mixed and aerated), and *windrow composting* (compost is placed in long piles known as windrows and periodically mixed by tractors or similar equipment).

Will it Work at Every Site?

Biodegradation is useful for many types of organic wastes and is a cost-effective, natural process. Many techniques can be conducted on-site, eliminating the need to transport hazardous materials.

The extent of biodegradation is highly dependent on the toxicity and initial concentrations of the contaminants, their biodegradability, the properties of the contaminated soil, and the particular treatment system selected.

Contaminants targeted for biodegradation treatment are non-halogenated volatile and semi-volatile organics and fuels. The effectiveness of bioremediation is limited at sites with high concentrations of metals, highly chlorinated organics, or inorganic salts because these compounds are toxic to the microorganisms.

Where has it been Used?

At the Scott Lumber Company Superfund site in Missouri, 16,000 tons of soils contaminated with polyaromatic hydrocarbons (PAHs) were biologically treated using land treatment application. PAH concentrations were reduced by 70 per cent.

At the French Ltd. Superfund site in Texas, slurry-phase bioremediation was used to treat 300,000 tons of lagoon sediment and tar-like sludge contaminated with volatile organic compounds, semi-volatile organic compounds, metals, and pentachlorophenol. Over a period of 11 months, the treatment system was able to meet the cleanup goals set by EPA.

Some additional examples of Superfund sites where different types of bioremediation have been selected as a treatment method are listed in Table 16.1.

For a listing of Superfund sites at which innovative treatment technologies have been used or selected for use, contact NCEPI at the address in the box below for a copy of the document entitled Innovative Treatment Technologies: Annual Status Report (7th Ed.), EPA 542-R-95-008. Additional information about the sites listed in the Annual Status Report is available in database format. The database can be downloaded free of charge from EPA's Cleanup Information bulletin board (CLU-IN). Call CLU-IN at 301-589-8366 (modem). CLU-IN's help line is 301-589-8368. The database also is available for purchase on diskettes. Contact NCEPI for details.

Table 16.1: Examples of superfund sites using Bioremediation Technologies.

Name of Site	Treatment	Contaminants
Applied Environmental Services, NY	Bioventing	Volatile organic compounds (VOCs), semi-volatile organic compounds (SVOCs)
Onalaska Municipal Landfill, WI	Bioventing	VOCs, polyaromatic hydrocarbons (PAHs)
Eielson Air Force Base, AK	Bioventing	VOCs, SVOCs, PAHs
Brown Wood Preserving, FL	Land treatment	PAHs
Vogel Paint & Wax, IA	Land treatment	VOCs
Broderick Wood Products, CO	Land treatment/Bioventing	SVOCs, PAHs, dioxins
Burlington Northern (Somers), MT	Land treatment/ In Situ Bioremediation	SVOCs, PAHs

What is an Innovative Treatment Technology?

Treatment technologies are processes applied to the treatment of hazardous waste or contaminated materials to permanently alter their condition through chemical, biological, or physical means. Innovative treatment technologies are those that have been tested, selected or used for treatment of hazardous waste or contaminated materials but lack well-documented cost and performance data under a variety of operating conditions.

16.9 Petro-Clear Hydrocarbon and Pesticide Remediation—Microbial Solutions

We have bioremediation solutions for of all weights of petroleum hydrocarbons, sites contaminated with solvents and other volatile organic compounds (such as trichloroethane, trichloroethene, benzene, toluene, ethylbenzene, and xylenes, BTEX) and fuels such as diesel or gasoline. Converts harmful petroleum products into carbon dioxide and water.

Many of the most toxic environmental contaminants are now candidates for bioremediation.

Bioremediation of hydrocarbons is best accomplished with a process called Bio-augmentation. That is, the addition of a large number of selected, laboratory grown, standardized microorganisms to a contaminated matrix. These contaminant specific microbes capable of degrading these compounds breaking them down into carbon dioxide and water. The microbes will survive and consume their contaminant food source until the unwanted pollutant is remediated.

Successful Bioaugmentation with Microbes

Bioaugmentation with microbes requires a few criteria to be successful.

- ❖ The first is the ability for the added microbes to reach the contaminant.
- ❖ The second is oxygen, microbes require oxygen.
- ❖ In the bio-degradation of petroleum based products, microbial activity is most frequently limited by insufficient oxygen due to slow rates of diffusion into the interior of the soil layers or piles and into the center of soil aggregates. Generally, the greater the mass of oxygen that can be distributed the more rapid and complete the cleanup. Much of their requirement will come from the atmosphere. However, if the ground is compacted, the ground should be tilled or raked to allow as much oxygen as possible to permeate the soil. If this is not possible, please contact us for other suggestions on how this can be achieved. We have had success with Oxy-Gen, an organic oxygen generating compound, that releases oxygen so that indigenous and augmented aerobic microbes may remediate the organic containments.
- ❖ The third is temperature. The working range is between 5-45° C (with 28° C being the optimum).
- ❖ The fourth is pH. The microbes a have a working range of pH 6.5 to 8.5.
- ❖ The fifth is the addition of nutrients to the soil and moisture. We will advise you of the necessary nutrients required for each remediation project.

As long as the microbes can reach the contaminants with the necessary oxygen, nutrients, moisture, and the temperature and pH are within the microbes working range, then remediation will be successful.

How Long will a Typical Clean up Take?

This depends on the contaminant involved and it's concentration levels. It can range anywhere from a week for simple gasoline spills to several months for complicated crude oils.

Contamination will either be dissolved in the groundwater, floating on the groundwater as free product, "sorbed" (bound) to the soil. In each case, we offer a specific product whose microbial types and levels have been optimized to cope with the various scenarios.

Hydrocarbons in Water

A series of all-natural microbial systems has been isolated to bioremediate hydrocarbon contamination in either salt or fresh water environments. These microbial systems contain high concentrations of safe, non-pathogenic, microorganisms.

Will reduce and eliminate hydrocarbons from contributing to contamination in:

- ❖ stormwater runoff,
- ❖ at commercial and industrial sites,
- ❖ oil spill cleanup in lakes, rivers, groundwater, and ocean,
- ❖ treatment of bilge and ballast water, and
- ❖ marina spills.

These products are ideally suited for immediate application in treating the typical contamination from boats on a marina's surface water.

Hydrocarbons or Organic Solvents—On Soil, Landfarming or In-Situ

Specifically formulated for use in soil applications, landfarming, underground storage tanks, large industrial clean-ups and spills on turf-soil-grass. Also, formulas for concrete, pavement and parking lots clean-ups.

16.10 Biological Treatment, Bioremediation, and Natural Attenuation: A Case Study of Center for Environmental Biotechnology

Biological treatment, bioremediation and natural attenuation has been a rapidly growing area of science over the past decade. The acceptance of natural attenuation as a solution for cleaning up contaminated sites and DOE's recognition that they will have long-term stewardship issues that they must address at the most contaminated sites has greatly increased the urgency for basic and applied research related to microbial ecology and biogeochemistry. This type of research is truly enabling for natural attenuation since characterization, predictions, and verification monitoring require a strong scientific basis. Natural attenuation is viewed as the best solution for cleaning up many waste sites and will save billions of dollars in cleanup costs. Bioremediation, both *in situ* and *ex situ* have also enjoyed strong scientific growth, in part due to the increased use of natural attenuation, since most natural attenuation is due to biodegradation. Bioremediation and Natural Attenuation are also seen as a solution for emerging contaminant problems, e.g. MTBE, endocrine disrupters, landfill stabilization, mixed waste biotreatment, and biological carbon sequestration. The Center for Environmental Biotechnology is a recognized leader in the field of biological treatment, bioremediation, and natural attenuation. The Center includes a world-class bioengineering laboratory and CEB investigators have extensive experience in both water treatment and bioremediation, especially co-metabolic biodegradation and the treatment of inhibitory compounds. In addition to basic research, CEB investigators have been involved in various aspects of more than 60 field demonstrations and deployments, and have 5 patents in this area that are licensed to more than 30 companies. The types of contaminants that CEB investigators have expertise with include chlorinated solvents, petroleum hydrocarbons, polynuclear aromatic hydrocarbons, ketones, MTBE, TNT, inorganic nitrogen (NO_3, NH_4), tritium, Pu, Np, Cr, and U. The Biotreatment, Bioremediation and Natural Attenuation area has both basic research and field application foci for the CEB. The basic research foci are co-metabolism, biotreatability, biotransformation kinetics, and modeling of biogeochemical processes. The field application foci are co-metabolic techniques, biogeochemical assessment techniques, and modeling of attenuation and environmental fate.

Basic Research Areas

- ❖ *Co-Metabolism*: CEB scientists are recognized leaders in the field of co-metabolic pollutant transformation. Research conducted by scientists at CEB has demonstrated that co-metabolism is a dominant process for the degradation of PAHs, chlorinated solvents, and fuel oxygenates. Co-metabolic processes are difficult to study and require novel experimental approaches. One approach under development is the use of genetic probes to identify the presence of metabolic pathways implicated in co-metabolic processes. In this research, physiological responses are being linked to gene probe signatures to develop methods for identifying co-metabolic potential in environmental samples. The second approach uses a kinetic evaluation of partial transformation reactions to compare and characterize bacterial enzyme systems implicated in co-metabolism. Co-metabolic processes are competitive enzyme reactions and can be tested and modeled as such. Both these approaches are yielding unique and valuable advances in our understanding and application of co-metabolic and partial transformations. *Biotreatability*. CEB does a number of types of treatability tests for contaminants in soil and water, using soil columns, respirometers, bioreactors, and field respiration tests. Real-time direct biogeochemical techniques are being developed to provide the most direct possible methods to measure the rates of biodegradation and the effect stimulants and environmental conditions have on both the functional microbial components and the biogeochemistry of the environment being studied. For example, mixed waste biotreatment includes both the engineering of bioreactors for the treatment of high strength waste streams and understanding how natural attenuation or active bioremediation can be applied to mitigate the impact of the DOE weapons legacy. Past projects have examined how bacterial blooms on solvents may effect the environmental fate of actinides in the subsurface. Current studies are examining the biotreatment of tritiated solvent wastes and the transformation of uranium on bacteria surfaces. Biological processes are increasingly being recognized by DOE as important factors effecting the fate of radionuclides in the environment and are more often being considered as useful tools in the clean-up of mixed wastes and waste sites. *Biotransformation Kinetics.* CEB is linking engineering and microbiology in an integrated programme to examine biotransformation kinetic for pollutant clean-up and microbial product formation. The kinetic programme examines microbial response and activity as a function of substrate concentrations and time. The results of kinetic studies are being used in reactor design and operations analysis. Future research will focus on the application of kinetic approaches to predicting microbial population response to pollutant inputs.

❖ *Modeling:* CEB currently does not have a modeling programme, but is developing collaboration with a modeling groups within ESD. Modeling is targeted as an area for increased collaboration in the next three years. In particular, there is an need to link biodegradation kinetic information with subsurface fate and transport of pollutants. Modeling is also needed to further understand the control and operation of co-metabolizing biological reactors.

Field Applications

❖ *Co-Metabolic Techniques*: Increasingly we are finding that contaminants can be deleterious at extremely low concentrations, eg. endocrine disrupters at parts per trillion concentrations. These concentrations of contaminants are too low to allow the contaminant to be a carbon or energy source for the microflora. Thus, a secondary carbon and energy source is needed to stimulate the microbes that are capable of degrading the contaminants of concern, since the excess enzymes produced will degrade the contaminants co-metabolically or fortuitously. CEB scientists have demonstrated the ability of methane to be injected with air into soil and groundwater contaminated with chlorinated solvents and petroleum compounds and shown how this stimulation of methanotrophs can cleanup sites to non-detect levels of these contaminants. They have also shown how various other gases can be added to supplement nutrient deficiencies in both nitrogen and phosphorus, thus allowing the microbes to achieve higher biomass levels and higher rates of biodegradation of the contaminants. Current studies are focusing on chlorinated solvent contaminants at DOD and DOE sites, MTBE at commercial sites, PAHs at DOD sites and municipal solid water landfills. CEB is also focusing other nutrients and surfactants that might stimulate bioremediation of more recalcitrant and strongly sorbed contaminants.Electrokinetic treatment of soil, ground water and ex-situ treatment systems. This technology being studied in collaboration with EETD is particularly applicable to metal ion contamination, which may be removed without damage to the indigenous microorganisms. The electrokinetic treatment also provides gentle heat to soil, which is also biocompatible, and the current can also supply nutrients and even transport organisms themselves.*Biogeochemical Assessment Techniques.* Our ability to determine the rates of bioremediation or biodegradation in the field is quite often hampered by the need to collect samples and transport them back to the laboratory for analysis. This obviously introduces bottle effect, collection contamination, and changes the in situ conditions so that the final measurements may not reflect accurately what is going on in that environment. In order to get a truer understanding of conditions as they occur we have been studying techniques that allow direct assessment, eg. helium tracer tests, soil gas measurements, isotopic analysis of contaminants and daughter products, and immediate lipid and nucleic acid extraction from sediment and groundwater. These techniques are being applied at a number of field sites to determine their efficiency over conventional baseline laboratory analyses.

❖ *Modeling of Attenuation & Environmental Fate:* Using the powerful codes developed for fate and transport by the Yucca Mt. Programme (TOUGH & ITOUGH) CEB is applying kinetic information relevant to biodegradation but especially aerobic processes that have received little attention. Using the computing facilities at LBNL (NERSC) and collaborations with the Nuclear Waste Programme, we are currently developing contaminant fate and transport codes that will be used to model aerobic injection into municipal solid waste landfills. These models may allow us to better control the composting type processes that are occurring and discover better parameters to monitor and characterize the process.

Collaborations

LBNL (EETD, ALS, CSD, LSD), UC Berkeley, Washington State University, Geokinetics, Earth Tech, North Carolina State, Utah State University, USGS, NSF, Yolo County, INEEL, ORNL, PNNL, SRTC, IconGenetics, Maxygen, Physical Optics Systems, American Home Products, DuPont Agrobusiness, UMBI, VECTOR (Russia)) International Institute for Applied Microbiology (Ukraine); Utah Water Research Laboratory at Utah State University, Sandia National Lab, Department of Genetics at Cambridge University in the Great Britain, Institute of Physics at Georgian Academy of Sciences, Brain Tumor Research Center UCSF, Department of Molecular and Cell Biology UC Davis, Department of Chemical and Biochemical Engineering UC Irvine,; Department of Pathology at Stanford University Medical Center, Department of Physics and School of Medicine at Boston University. EETD/CEB collaborates closely with Geokinetic/EDA in Berkeley for electrochemical/electrokinetic technology. EETD also has several close collaborations with UCB Chemistry, Materials Science and Chemical Engineering Departments with CEB associated investigators.FacilitiesLab Space: 2321 ft². Office Space: 2553 ft². EquipmentSterilGARD II 4-foot vertical laminar-flow, biological safety cabinet (Baker) Avanti J-25 high performance centrifuge (Beckman) GS-6R tabletop centrifuge (Beckman) DU 640 UV/VIS scanning spectrophotometer (Beckman) Ultra-low temperature freezer (Revco) Axioskop RLF for DIC, phase contrast, and epifluorescence with microphotography (Carl Zeiss) Stereo microscope with microphotography (Carl Zeiss) Integrated environmental SpeedVac (Savant) MilliQ PF water purification systems (Millipore) GeneAmp PCR system 9600 (Perkin-Elmer) GeneAmp PCR system 9700 (Perkin-Elmer) RoboCycler Gradient Temperature Cycler 96 (Stratagene) Expedite 8909 DNA synthesizer (PerSeptive Biosystems) Vision Workstation BioCAD (PerSeptive Biosystems) Model 377 ABI Prism automated DNA sequencer (Perkin Elmer) TriCarb Liquid Scintillation Counter (Packard Instruments) TopCount microplate counter (Packard Instruments) CHEF DRII pulsed field electrophoresis equipment (Bio-Rad) MIDI identification system (Hewlett Packard) High sensitivity MSD mainframe for the HP 6890 GC (Hewlett Packard) Accelerated solvent extractor, model ASE 200 (DIONEX) BIOLOG microbial identification system (BIOLOG) Innova 4230 environmental shaker (New Brunswick) Innova 4900 Multi Shaker Environmental Chamber (New Brunswick) Landtec landfill gas analyzers2 20-liter US Filter fluidized bed bioreactorssoil gas sampling kitrespirometer3 55-gal landfill bioreactorsMark 4 Helium detector5-liter New Brunswick fermentorOther necessary support equipment and installations such as autoclaves, DI-water, refrigerators, freezers for low temperature storage of temperature sensitive materials, balances, ice-maker, shakers, incubators, magnetic stirrer, hotplates, microcentrifuges, computers, different electrophoresis boxes and power supplies, chemical fume hoods are also available.

Green Technology Biofertilizers and Biopesticides

17.1 Green Technology

The term "technology" refers to the application of knowledge for practical purposes.

The field of "green technology" encompasses a continuously evolving group of methods and materials, from techniques for generating energy to non-toxic cleaning products.

The present expectation is that this field will bring innovation and changes in daily life of similar magnitude to the "information technology" explosion over the last two decades. In these early stages, it is impossible to predict what "green technology" may eventually encompass.

The goals that inform developments in this rapidly growing field include:

- *Sustainability:* meeting the needs of society in ways that can continue indefinitely into the future without damaging or depleting natural resources. In short, meeting present needs without compromising the ability of future generations to meet their own needs.
- *"Cradle to cradle" design:* ending the "cradle to grave" cycle of manufactured products, by creating products that can be fully reclaimed or re-used.
- *Source reduction:* reducing waste and pollution by changing patterns of production and consumption.
- *Innovation:* developing alternatives to technologies—whether fossil fuel or chemical intensive agriculture—that have been demonstrated to damage health and the environment.
- *Viability:* creating a center of economic activity around technologies and products that benefit the environment, speeding their implementation and creating new careers that truly protect the planet.

Examples of green technology subject areas:

Energy

Perhaps the most urgent issue for green technology, this includes the development of alternative fuels, new means of generating energy and energy efficiency.

Green Building

Green building encompasses everything from the choice of building materials to where a building is located.

Environmentally Preferred Purchasing

This government innovation involves the search for products whose contents and methods of production have the smallest possible impact on the environment, and mandates that these be the preferred products for government purchasing.

Green Chemistry

The invention, design and application of chemical products and processes to reduce or to eliminate the use and generation of hazardous substances.

Green Nanotechnology

Nanotechnology involves the manipulation of materials at the scale of the nanometer, one billionth of a meter. Some scientists believe that mastery of this subject is forthcoming that will transform the way that everything in the world is manufactured. "Green nanotechnology" is the application of green chemistry and green engineering principles to this field.

17.2 Environmental Technology

Environmental technology (abbreviated as EnviroTech) or green technology (abbreviated as GreenTech) or clean technology (abbreviated as CleanTech) is the application of the environmental sciences to conserve the natural environment and resources, and to curb the negative impacts of human involvement. Sustainable development is the core of *environmental technologies*. When applying *sustainable development* as a solution for *environmental issues*, the solutions need to be socially equitable, economically viable, and environmentally sound.

Related Technologies

Some *environmental technologies* that retain *sustainable development* are; recycling, water purification, sewage treatment, remediation, flue gas treatment, solid waste management, and renewable energy. Some technologies assist directly with energy conservation, while other technologies are emerging that help the environment by reducing the amount of waste produced by human activities. Energy sources such as solar power create less problems for the environment than traditional sources of energy like coal and petroleum.

Scientists continue to search for clean energy alternatives to our current power production methods. Some technologies such as anaerobic digestion produce renewable energy from waste materials. The global reduction of greenhouse gases is dependent on the adoption of energy conservation technologies at industrial level as well as this clean energy generation. That includes using unleaded gasoline, solar energy and alternative fuel vehicles, including plug-in hybrid and hybrid electric vehicles.

Since electric motors consume 60 per cent of all electricity generated, advanced energy efficient electric motor (and electric generator) technology that are cost effective to encourage their application, such as the brushless wound-rotor doubly-fed electric machine and energy saving module, can reduce the amount of carbon dioxide (CO_2) and sulfur dioxide (SO_2) that would otherwise be introduced to the atmosphere, if electricity is generated using fossil fuels. Greasestock is an event held yearly in Yorktown Heights, New York which is one of the largest showcases of environmental technology in the United States.

Examples

* Alternative energy,
* Energy Saving Modules,
* Brushless Wound-Rotor Doubly-Fed Electric Machine,
* Green building,
* Green syndicalism,
* Hybrid vehicle, and
* Solar power:

  - Solar cell,

- ● Solar heating.

- ❖ Sustainable engineering.

Criticism

Some groups, including , have criticised the concept of environmental technology. From their viewpoint, technology is seen as a system rather than a specific physical tool. Technology, it is argued, *requires* the exploitation of the environment through the creation and extraction of resources, and the exploitation of people through labour, specialisation and the division of labor. There is no "neutral" form of technology, as things are always created in a certain context with certain aims and functions. Thus, green technology is rejected as an attempt to reform this exploitative system, merely changing it on the surface to make it seem environmentally friendly, despite continued unsustainable levels of human and natural exploitation.

17.3 Case Study: City of Chicago

Chicago Green Tech's building was originally constructed in 1952. Since then a number of different companies have owned the building. When it came to the attention of the Chicago Department of Environment (DOE) in 1995, the building and its 17 acres were owned by Sacramento Crushing, a company which had a permit to collect limited construction and demolition debris. The Department of Environment became involved because Sacramento Crushing had gone far beyond the scope of its permit and had filled all 17-acres with illegally dumped debris. The site was littered with 70-foot high piles of rubble, one of which was so dense it sank 15 feet into the ground.

The Department of Environment successfully fought Sacramento Crushing in court and not only closed down their operation but also became the owner of the site itself. It was then DOE's job to clean up this Brownfield. The clean up took 18 months to complete and cost about $9 million. In this process, the site was cleared of over 600,000 tons of concrete, which took 45,000 truck loads to remove. The city recouped some of the clean up cost by selling the concrete and other materials to recycling firms and to other city departments for use in their projects. For example, some of the crushed concrete was used by the Chicago Department of Transportation to lay the foundation of the parking garage at the new Millennium Park.

In 1999, DOE was the proud owner of a cleaned site and vacant building. Rather than simply renovating the building using traditional methods, DOE seized the opportunity to create an energy efficient building using the highest standards of green technology available. The Chicago Chapter of the American Institute of Architects Committee on the Environment formed a design team for the project. This team of local architects, led by Farr Associates, designed the building using a set of guidelines established by the US Green Building Council called LEED (Leadership in Environmental and Energy Design). To match the vision of the building, DOE selected tenants that are environmentally focused. The three tenants are:

❖ Greencorps Chicago, the city's community landscaping and job training programme.
❖ WRD Environmental, an urban landscape design/build firm.

The building opened to the public in May of 2002.

17.4 Visiting Chicago Green Technology

The Chicago Center for Green Technology is located at 445 North Sacramento Boulevard in Chicago. It is open to the public Monday, Wednesday, and Friday from 9:00 a.m. to 5:00 p.m., Tuesday and Thursday from 9:00 a.m. to 8:00 p.m., and Saturday from 9:00 a.m. to 4:00 p.m. The Center is closed on Sunday. Visitors are welcome to explore the public areas of the building and campus. Brochures for self-guided tours are available at the building. The Chicago Department of Environment offers a variety of free educational programmes at Chicago Green Tech for professionals and the public.

Free guided tours of Chicago Green Tech are available for groups of 10 or more persons and are subject to staff availability. You can email the completed form to greentech@cityofchicago.org or fax to (312) 746-9192.

By Bike

The Chicago Center for Green Technology is located at 445 N. Sacramento Blvd (between Chicago Ave. and Lake St.), 60612. Bike racks are located in front and behind the building. Find your route at http://www.ci.chi.il.us/Transportation/bikemap/keymap.html.

By Car

From the Eisenhower Expressway (I-290) eastbound: Exit at Sacramento Blvd. Turn left (north) onto Sacramento and travel about one mile to the Chicago Center for Green Technology. It is located on the right (east) side of the street, at 445 N. Sacramento, between Lake St. and Chicago Ave. Visitor parking lots are located directly in front and behind the building.

From the Eisenhower Expressway (I-290) westbound: Exit at California Ave. Turn right (north) onto California and travel 8 blocks to Lake St. Turn left (west) onto Lake St. and travel three blocks to Sacramento Blvd. Turn right (north) onto Sacramento and travel four blocks to the Chicago Center for Green Technology. It is on the right (east) side of the street, at 445 N. Sacramento, between Lake St. and Chicago Ave. Visitor parking lots are located directly in front and behind the building.

From the Kennedy Expressway (1-90) eastbound: Exit at California Ave. Turn right (south) onto California and travel four blocks to Fullerton Ave. Turn right (west) onto Fullerton and travel four blocks to Sacramento Blvd. Turn left (south) onto Sacramento and travel about 2.5 miles, through Humboldt Park, to the Chicago Center for Green Technology. The Center is on the left (east) side of the street, at 445 N. Sacramento, about.5 mile past Chicago Ave. Visitor parking lots are located directly in front and behind the building.

From the Kennedy Expressway (1-90) westbound: Exit at North Ave. Turn left (west) onto North Ave. and travel approximately 2 miles to North Humboldt Blvd. Turn left (south) onto North Humboldt Blvd. After driving south through Humboldt Park, the name of the street will change to Sacramento Ave. Continue south on Sacramento for about one mile to the Chicago Center for Green Technology, which is on the left (east) side of the street, at 445 N. Sacramento, about.5 mile past Chicago Ave. Visitor parking lots are located directly in front and behind the building.

17.5 The Chicago Green Tech Design Team

LEAD ARCHITECT
Farr Associates
53 W. Jackson, Suite 650
Chicago, IL 60604
312.408.1661
http://www.farrside.com/

MECHANICAL, ELECTRICAL, AND FIRE PROTECTION ENGINEER
IBC Engineering
N8 W22195 Johnson Dr. Suite 180
Waukesha, WI 53186
262.549.1190
http://www.ibcengineering.com/

ELECTRICAL ENGINEER

Spectrum Engineering
633 Skokie Boulevard, Suite 350
Northbrook, IL 60062
847.753.9640

CIVIL ENGINEER

Terra Engineering
505 N. LaSalle, Suite 250
Chicago, IL 60610
312.467.0123
http://www.terraengineering.com

STRUCTURAL ENGINEER

Tylk, Gustafson, Reckers, Wilson, Andrews
407 S. Dearborn, Suite 900
Chicago, IL 60605
312.341.0055

INDOOR AIR QUALITY

O'Donnell, Wickland, Pagozzi and Peterson
111 West Washington, Suite 2100
Chicago, IL 60602
312.332.9600
http://www.owpp.com/

LANDSCAPE ARCHITECTS

Greencorps Chicago
445 North Sacramento Blvd.
Chicago, IL 60612
312.746.9777

Site design group
836 South Michigan Ave.
Chicago, IL 60605
312.427.7240

WRD Environmental
445 North Sacramento Blvd., Suite 201
Chicago, IL 60612
773.722.9870
http://www.wrdenvironmental.com

CONSTRUCTION WASTE MANAGEMENT
Michael Roy Iversen Architects
144 N. Lombard Ave.
Oak Park, IL 60302
708.383.1189

COMPUTER ENERGY MODELING
Serena Sturm Architects
3351 Commercial Ave.
Northbrook, IL 60062
847.564.0370
http://www.serenasturm.com

LIGHTING DESIGN
Sieben Energy Associates
333 N. Michigan Ave., Suite 2106
Chicago, IL 60601
312.899.1000

17.6 Biofertilizers

Biofertilizers are ready to use live formulates of such beneficial microorganisms which on application to seed, root or soil mobilize the availability of nutrients by their biological activity in particular, and help build up the micro-flora and in turn the soil health in general.

With the introduction of green revolution technologies the modern agriculture is getting more and more dependent upon the steady supply of synthetic inputs (mainly fertilizers), which are products of fossil fuel (coal+ petroleum). Adverse effects are being noticed due to the excessive and imbalanced use of these synthetic inputs. This situation has lead to identifying harmless inputs like biofertilizers. Use of such natural products like biofertilizers in crop cultivation will help in safeguarding the soil health and also the quality of crop products.

Benefits from using Biofertilizers:

❖ Increase crop yield by 20-30 per cent.
❖ Replace chemical nitrogen and phosphorus by 25 per cent.
❖ Stimulate plant growth.
❖ Activate the soil biologically.
❖ Restore natural soil fertility.
❖ Provide protection against drought and some soil borne diseases.

Advantages of Bio-Fertilizers

1. Cost effective.
2. Suppliment to fertilizers.
3. Eco-friendly (Friendly with nature).
4. Reduces the costs towards fertilizers use, especially regarding nitrogen and phosphorus.

Types of Biofertilizers are Available

1. For Nitrogen

    ❖ Rhizobium for legume crops.
    ❖ Azotobacter/Azospirillum for non legume crops.
    ❖ Acetobacter for sugarcane only.
    ❖ Blue-Green Algae (BGA) and Azolla for low land paddy.

2. For Phosphorous

    ❖ Phosphatika for all crops to be applied with Rhizobium, Azotobacter, Azospirillum and Acetobacter

3. For enriched compost

    ❖ Cellulolytic fungal culture
    ❖ Phosphotika and Azotobacter culture

Biofertilizers Recommended for Crops

❖ Rhizobium + Phosphotika at 200 gm each per 10 kg of seed as seed treatment are recommended for pulses such as pigeonpea, green gram, black gram, cowpea etc, groundnut and soybean.
❖ Azotobacter + Phosphotika at 200 gm each per 10 kg of seed as seed treatment are useful for wheat, sorghum, maize, cotton, mustard etc.
❖ For transplanted rice, the recommendation is to dip the roots of seedlings for 8 to 10 hours in a solution of Azospirillum + Phosphotika at 5 kg each per ha.

Biofertilizers Applied to Crops

1. *Seed treatment:* 200 g of nitrogenous biofertilizer and 200 g of Phosphotika are suspended in 300-400 ml of water and mixed

thoroughly. Ten kg seeds are treated with this paste and dried in shade. The treated seeds have to be sown as soon as possible.

2. *Seedling root dip:* For rice crop, a bed is made in the field and filled with water. Recommended biofertilizers are mixed in this water and the roots of seedlings are dipped for 8-10 hrs.

3. *Soil treatment:* 4 kg each of the recommended biofertilizers are mixed in 200 kg of compost and kept overnight. This mixture is incorporated in the soil at the time of sowing or planting.

Response to Biofertilizer Application

- Biofertilizer product must contain good effective strain in appropriate population and should be free from contaminating microorganisms.
- Select right combination of biofertilizers and use before expiry date.
- Use suggested method of application and apply at appropriate time as per the information provided on the label.
- For seed treatment adequate adhesive should be used for better results.
- For problematic soils use corrective methods like lime or gypsum pelleting of seeds or correction of soil pH by use of lime.
- Ensure the supply of phosphorus and other nutrients.

Probable Reasons for not Getting Response from the Application of Biofertilizers

1. On account of quality of product:

   - Use of ineffective strain.
   - Insufficient population of microorganisms.
   - High level of contaminants.

2. On account of inadequate storage facilities:

   - May have been exposed to high temperature.
   - May have been stored in hostile conditions.

3. On account of usage:

   - Not used by recommended method in appropriate doses.
   - Poor quality adhesive.
   - Used with strong doses of plant protection chemicals.

4. On account of soil and environment:

- High soil temperature or low soil moisture.
- Acidity or alkalinity in soil.
- Poor availability of phosphorous and molybdenum.
- Presence of high native population or presence of bacteriophages.

Precautions one should take for using biofertilizers:

- Biofertilizer packets need to be stored in cool and dry place away from direct sunlight and heat.
- Right combinations of biofertilizers have to be used.
- As Rhizobium is crop specific, one should use for the specified crop only.
- Other chemicals should not be mixed with the biofertilizers.
- While purchasing one should ensure that each packet is provided with necessary information like name of the product, name of the crop for which intended, name and address of the manufacturer, date of manufacture, date of expiry, batch number and instructions for use.
- The packet has to be used before its expiry, only for the specified crop and by the recommended method of application.
- Biofertilizers are live product and require care in the storage
- Both nitrogenous and phosphatic biofertilizers are to be used to get the best results.
- It is important to use biofertilizers along with chemical fertilizers and organic manures.
- Biofertilizers are not replacement of fertilizers but can supplement plant nutrient requirements.

One of the major concerns in today's world is the pollution and contamination of soil. The use of chemical fertilizers and pesticides has caused tremendous harm to the environment. An answer to this is the biofertilizer, an environmentally friendly fertilizer now used in most countries. Biofertilizers are organisms that enrich the nutrient quality of soil. The main sources of biofertilizers are bacteria, fungi, and cynobacteria (blue-green algae). The most striking relationship that these have with plants is symbiosis, in which the partners derive benefits from each other.

Plants have a number of relationships with fungi, bacteria, and algae, the most common of which are with mycorrhiza, rhizobium, and cyanophyceae. These are known to deliver a number of benefits including plant nutrition, disease resistance, and tolerance to adverse soil and climatic conditions. These techniques have proved to be successful biofertilizers that form a health relationship with the roots.

Biofertilizers will help solve such problems as increased salinity of the soil and chemical run-offs from the agricultural fields. Thus, biofertilizers are important if we are to ensure a healthy future for the generations to come.

Mycorrhiza

Mycorrhizae are a group of fungi that include a number of types based on the different structures formed inside or outside the root. These are specific fungi that match with a number of favourable parameters of the the host plant on which it grows. This includes soil type, the presence of particular chemicals in the soil types, and other conditions.

These fungi grow on the roots of these plants. In fact, seedlings that have mycorrhizal fungi growing on their roots survive better after transplantation and grow faster. The fungal symbiont gets shelter and food from the plant which, in turn, acquires an array of benefits such as better uptake of phosphorus, salinity and drought tolerance, maintenance of water balance, and overall increase in plant growth and development.

While selecting fungi, the right fungi have to be matched with the plant. There are specific fungi for vegetables, fodder crops, flowers, trees, etc.

Mycorrhizal fungi can increase the yield of a plot of land by 30 per cent-40 per cent. It can absorb phosphorus from the soil and pass it on to the plant. Mycorrhizal plants show higher tolerance to high soil temperatures, various soil-and root-borne pathogens, and heavy metal toxicity.

Legume-rhizobium Relationship

Leguminous plants require high quantities of nitrogen compared to other plants. Nitrogen is an inert gas and its uptake is possible only in fixed form, which is facilitated by the rhizobium bacteria present in the nodules of the root system. The bacterium lives in the soil to form root nodules (i.e. outgrowth on roots) in plants such as beans, gram, groundnut, and soybean.

Blue-green Algae

Blue-green algae are considered the simplest, living autotrophic plants, i.e. organisms capable of building up food materials from inorganic matter. They are microscopic. Blue-green algae are widely distributed in the aquatic environment. Some of them are responsible for water blooms in stagnant water. They adapt to extreme weather conditions and are found in snow and in hot springs, where the water is 85 °C. Certain blue-green algae live intimately with other organisms in a symbiotic relationship. Some are associated with the fungi in form of lichens. The ability of blue-green algae tophotosynthesize food and fix atmospheric nitrogen accounts for their symbiotic associations and also for their presence in paddy fields. Blue-green algae are of immense economic value as they add organic matter to the soil and increase soil fertility. Barren alkaline lands in India have been reclaimed and made productive by inducing the proper growth of certain blue-green algae.

Biofertilizers include microorganisms and their metabolites that are capable of enhancing soil fertility, crop growth, and/or yield. These include both indigenous microbes and microbial inoculants, that is, microorganisms that replace fertilizers or increase a crops fertilizer use efficiency. Soil microorganisms such as bacteria, ectomycorhiza, arbuscular mycorrhizal fungi, and soil algae, especially the N_2-fixing cyanobacteria have potential as biofertilizers. Nitrogen-fixing inoculants based on *Rhizobium* species were among the first biofertilizers introduced into agroecosystems back in the nineteenth century. In the twenty-first century, biofertilizers will become an increasingly important area of research and development. The use of fertilizers and pesticides has increased steadily since the 1970s; consequently, concerns about the impacts of these chemicals on land, air, and water have become significant environmental issues. Biofertilizers provide an alternative to agricultural chemicals as more sustainable and ecologically sound practices to increase crop productivity. Biofertilizer sales forecasts in the United States for the years 2001 and 2006 represent \$690 million and \$1.6 billion, respectively. Examples of some biofertilizers currently in use worldwide are shown in Table 17.1.

Table 17.1: Organisms, mode of action, crops, and producers of biofertilizers currently in use for agriculture.

Type	Mode of action	Crop	Used in
Rhizobium spp.	N_2 fixation	Legumes	Russia; several countries
Cyanobacteria	N_2 fixation	Rice	Japan; several countries
Azospirillum spp.	N_2 fixation	Cereals	Several countries
Mycorrhizae	Nutrient acquisition	Conifers	Several countries
Penicillium bilaii	P solubilization	Cereals, legumes	Western Canada
Directed compost	Soil fertility	All plants	Several countries
Earthworm	Humus formation	Vegetables, flowers	Cottage industry

17.7 Biopesticides

Biopesticides (also known as biological pesticides) are certain types of pesticides derived from such natural materials as animals, plants, bacteria, and certain minerals. For example, garlic, mint, and baking soda all have pesticidal applications and are considered biopesticides. At the end of 1998, there were approximately 175 registered biopesticide active ingredients and 700 products. Biopesticides fall into three major categories:

1. *Microbial pesticides* contain a microorganism (bacterium, fungus, virus, protozoan or alga) as the active ingredient. The most widely known microbial pesticides are varieties of the bacterium *Bacillus thuringiensis*, or Bt, which can control certain insects in cabbage, potatoes, and other crops. Bt produces a protein that is harmful to specific insect pests. Certain other microbial pesticides act by out-competing pest organisms. Microbial pesticides need to be continuously monitored to ensure they do not become capable of harming non-target organisms, including humans.

2. *Plant-pesticides* are pesticidal substances that plants produce from genetic material that has been added to the plant. For example, scientists can take the gene for the Bt pesticidal protein, and introduce the gene into the plants= own genetic material. Then the plantBinstead of the Bt bacterium—manufactures the substance that destroys the pest. Both the protein and its genetic material are regulated by EPA; the plant itself is not regulated.

3. *Biochemical pesticides* are naturally occurring substances that control pests by non-toxic mechanisms. Conventional pesticides, by contrast, are synthetic materials that usually kill or inactivate the pest. Biochemical pesticides include substances that interfere with growth or mating, such as plant growth regulators, or substances that repel or attract pests, such as pheromones. Because it is sometimes difficult to determine whether a natural pesticide controls the pest by a non-toxic mode of action, EPA has established a committee to determine whether a pesticide meets the criteria for a biochemical pesticide.

Microbial and Antimicrobial Pesticides

These are two separate and distinct types of pesticides registered by EPA.

Microbial Pesticides are microbes, including bacteria, that help to control insects and weeds, as well as fungi and bacteria that cause plant diseases. These are one type of biopesticide.

Antimicrobial Pesticides are pesticides that control unwanted microbes on inanimate objects, in water, and on selected foods under certain circumstances. These pesticides are almost always chemicals, and they act by killing or inactivating microbes that are pests. Antimicrobial pesticides include the disinfectants used in swimming pools, drinking water supplies, and in hospitals to control microbes that can cause disease.

What are the Advantages of Using Biopesticides?

- ❖ Biopesticides are inherently less harmful than conventional pesticides.
- ❖ Biopesticides are designed to affect only one specific pest or, in some cases, a few target organisms, in contrast to broad spectrum, conventional pesticides that may affect organisms as different as birds, insects, and mammals.
- ❖ Biopesticides often are effective in very small quantities and often decompose quickly, thereby resulting in lower exposures and largely avoiding the pollution problems caused by conventional pesticides.
- ❖ When used as a component of Integrated Pest Management (IPM) programmes, biopesticides can greatly decrease the use of conventional pesticides, while crop yields remain high.

To use biopesticides effectively, however, users need to know a great deal about managing pests.

How does EPA Encourage the Development and use of Biopesticides?

In 1994, the Biopesticides and Pollution Prevention Division was established in the Office of Pesticide Programmes to facilitate the registration of biopesticides. This Division promotes the use of safer pesticides, including biopesticides, as components of IPM programmes. The Division also coordinates the Pesticide Environmental Stewardship Programme (PESP). (See related fact sheets on IPM and PESP.)

Since biopesticides tend to pose fewer risks than conventional pesticides, EPA generally requires much less data to register a biopesticide than to register a conventional pesticide. In fact, new biopesticides are often registered in less than a year, compared with an average of more than 3 years for conventional pesticides.

While biopesticides require less data and are registered in less time than conventional pesticides, EPA must always conduct rigorous reviews to ensure that pesticides will not have adverse effects on human health or the environment. For EPA to be sure that a pesticide is safe, the Agency requires that registrants submit a variety of data about the composition, toxicity, degradation, and other characteristics of the pesticide.

Biopesticides are certain types of pesticides derived from such natural materials as animals, plants, bacteria, and certain minerals. For example, canola oil and baking soda have pesticidal applications and are considered biopesticides. At the end of 2001, there were approximately 195 registered biopesticide active ingredients and 780 products. Biopesticides fall into three major classes:

1. Microbial pesticides consist of a microorganism (e.g., a bacterium, fungus, virus or protozoan) as the active ingredient. Microbial pesticides can control many different kinds of pests, although each separate active ingredient is relatively specific for its target pest[s]. For example, there are fungi that control certain weeds, and other fungi that kill specific insects.

   The most widely used microbial pesticides are subspecies and strains of Bacillus thuringiensis, or Bt. Each strain of this bacterium produces a different mix of proteins, and specifically kills one or a few related species of insect larvae. While some Bt's control moth larvae found on plants, other Bt's are specific for larvae of flies and mosquitoes. The target insect species are determined by whether the particular Bt produces a protein that can bind to a larval gut receptor, thereby causing the insect larvae to starve.

2. Plant-Incorporated-Protectants (PIPs) are pesticidal substances that plants produce from genetic material that has been added to the plant. For example, scientists can take the gene for the Bt pesticidal protein, and introduce the gene into the plant's own genetic material. Then the plant, instead of the Bt bacterium, manufactures the substance that destroys the pest. The protein and its genetic material, but not the plant itself, are regulated by EPA.

3. Biochemical pesticides are naturally occurring substances that control pests by non-toxic mechanisms. Conventional pesticides, by contrast, are generally synthetic materials that directly kill or inactivate the pest. Biochemical pesticides include substances, such as insect sex pheromones, that interfere with mating, as well as various scented plant extracts that attract insect pests to traps. Because it is sometimes difficult to determine whether a substance meets the criteria for classification as a biochemical pesticide, EPA has established a special committee to make such decisions.

Biopesticides are usually inherently less toxic than conventional pesticides. Biopesticides generally affect only the target pest and closely related organisms, in contrast to broad spectrum, conventional pesticides that may affect organisms as different as birds, insects, and mammals. Biopesticides often are effective in very small quantities and often decompose quickly, thereby resulting in lower exposures and largely avoiding the pollution problems caused by conventional pesticides. When used as a component of Integrated Pest Management (IPM) programmes, biopesticides can greatly decrease the use of conventional pesticides, while crop yields remain high. To use biopesticides effectively, however, users need to know a great deal about managing pests.

How does EPA Encourage the Development and use of Biopesticides?

In 1994, the Biopesticides and Pollution Prevention Division was established in the Office of Pesticide Programmes to facilitate the registration of biopesticides. This Division promotes the use of safer pesticides, including biopesticides, as components of IPM programmes. The Division also coordinates the Pesticide Environmental Stewardship Programme (PESP).

Since biopesticides tend to pose fewer risks than conventional pesticides, EPA generally requires much less data to register a biopesticide than to register a conventional pesticide. In fact, new biopesticides are often registered in less than a year, compared with an average of more than 3 years for conventional pesticides.

While biopesticides require less data and are registered in less time than conventional pesticides, EPA always conducts rigorous reviews to ensure that pesticides will not have adverse effects on human health or the environment. For EPA to be sure that a pesticide is safe, the Agency requires that registrants submit a variety of data about the composition, toxicity, degradation, and other characteristics of the pesticide.

17.8 Regulating Biopesticides

Before a pesticide can be marketed and used in the United States, the Federal Insecticide, Fungicide, and Rodenticide Act (FIFRA) requires that EPA evaluate the proposed pesticide to assure that its use will not pose unreasonable risks of harm to human health and the environment. This regulation involves an extensive review of health and safety information.

Biopesticides include naturally occurring substances that control pests (biochemical pesticides), microorganisms that control pests (microbial pesticides), and pesticidal substances produced by plants containing added genetic material (plant-incorporated protectants) or PIPs.

The Biopesticide Registration Tools provide links to information tools to assist applicants. The Fact Sheets sections provides links to information about each of the biopesticide active ingredients. The Product Lists section provides various lists of individual biopesticide products to assist the public in identifying appropriate biopesticide product for pest problems. Finally, the PIPs section provides extensive information regarding the regulation of genetically engineered plants.

Biopesticide Registration Tools

The federal pre-marketing approval of pesticides—termed registration—is a complex process. The documents linked from this page augment the general registration process as they relate specifically to the registration of biopesticides. The e-mail address bppdconsistency@epa.gov has been created to respond to issues concerning biopesticide registration inconsistency that affect processing of submissions.

Biopesticide Regulatory Action Leaders will meet to discuss these issues and provide an answer usually within two to three weeks. Resolution of these issues will be posted to this site.

Since this e-mail address is only for the submission of generic issues regarding consistency in the regulation of biopesticides. Other questions regarding biopesticides should be directed to the appropriate biopesticide Regulatory Action Leader or the Biopesticide Ombudsman, Brian Steinwand (steinwand.brian@epa.gov). If you have questions about conventional pesticides or antimicrobials, please contact the Ombudsman for the Registration Division or the Antimicrobial Division.

Biopesticide Active Ingredient Fact Sheets

This collection of fact sheets contains chemical specific information about biopesticide active ingredients. Additionally, some of the fact sheets include a detailed technical document, bibliographies, regulatory history, Federal Register notices, and/or registrant and product lists.

Biopesticide Ingredient & Product Lists

Lists of EPA registered (approved) biopesticide ingredients & products are provided, including a listing by active ingredient PDF (43 pp, 378 K, about PDF), and lists of biopesticide active ingredients by year first approved.

Plant Incorporated Protectants (PIPs)

Plant-Incorporated Protectants are pesticidal substances produced by plants and the genetic material necessary for the plant to produce the substance. For example, scientists can take the gene for a specific Bt pesticidal protein, and introduce the gene into the plant's genetic material. Then the plant manufactures the pesticidal protein that controls the pest when it feeds on the plant. Both the protein and its genetic material are regulated by EPA; the plant itself is not regulated.

Cosmposting

18.1 Composting

Composting is the aerobic decomposition of biodegradable organic matter, producing compost. (Or in a simpler form: Composting is the decaying of food, mostly vegetables or manure.) The decomposition is performed primarily by facultative and obligate aerobic bacteria, yeasts and fungi, helped in the cooler initial and ending phases by a number of larger organisms, such as springtails, ants, nematodes and oligochaete worms.

Composting can be divided into home composting and industrial composting. Essentially the same biological processes are involved in both scales of composting, however techniques and different factors must be taken into account.

Importance

Composting recycles or "downcycles" organic household and yard waste and manures into an extremely useful humus-like, soil end-product called compost. Examples are fruits, vegetables and yard clippings. Ultimately this permits the return of needed organic matter and nutrients into the foodchain and reduces the amount of "green" waste going into landfills. Composting is widely believed to speed up the natural process of decomposition appreciably as a result of the raised temperatures that often accompany it. The elevated heat results from exothermic processes, and the heat in turn reduces the generational time of microorganisms and thereby speeds the energy and nutrient exchanges taking place. It is a popular misconception that composting is a "controlled" process; if the right environmental circumstances are present the process virtually runs itself. Hence a popular expression, "compost happens". It is nonetheless very necessary to provide as optimal circumstances as possible for large amounts of organic waste to break down properly. This is especially so when it is accompanied by heating, since at elevated temperatures oxygen within the piles is consumed more rapidly, and if not controlled, will lead to malodor.

Decomposition similar to composting occurs throughout nature as garbage dissolves in the absence of all the conditions that modern composters talk about; however, the process can be slow. For example, in the forest bark, wood and leaves break down into humus over 3-7 years. In restricted environments, for example, vegetables in a plastic trash container, decomposition with a lack of air encourages growth of anaerobic microbes, which produce disagreeable odors. Another form of degradation practiced deliberately in absence of oxygen is called anaerobic digestion—an increasingly popular companion to composting as it enables capture of residual energy in the form of biogas, whereas composting releases the majority of bound carbon-energy as excess heat (which helps sanitize the material) as well as copious amounts of biogenic CO_2 to the atmosphere.

It is important to distinguish between terms such as "biodegradable", "compostable", and "compost-compatible".

❖ A *biodegradable* material is capable of being broken down completely under the action of microorganisms into carbon dioxide, water and biomass. It may take a very long time for a material to biodegrade depending on its environment (e.g. hardwood in an arid area), but it ultimately breaks down completely.

❖ A *compostable* material biodegrades substantially under composting conditions, into carbon dioxide, methane, water and compost biomass. Compost biomass refers to the portion of the material that is metabolized by the microorganisms and which is incorporated into the cellular structure of the organisms or converted into humic acids etc. Compost biomass residues from a compostable material are fully biodegradable. "Compostable" is thus a subset of "biodegradable". The size of the material is a factor in determining compostability because it affects the rate of degradation. Large pieces of hardwood may not be compostable under a specific set of composting conditions, whereas sawdust of the same type of wood may be.

❖ A *compost-compatible* material does not have to be compostable or even biodegradable. It may biodegrade too slowly to be compostable itself, or it may not biodegrade at all. However, it is not readily distinguishable from the compost on a macroscopic scale and does not have a deleterious effect on the compost (e.g. it is not a biocide). Compost-compatible materials are generally inert and are present in compost at relatively low levels. Examples of compost-compatible materials include sand particles and inert particles of plastic.

Reducing Waste

Although composting has historically focussed on creating garden-ready soil, it is becoming more important as a tool for reducing solid waste. More than 60 percent of household waste in the United States is recyclable or compostable. But Americans only compost 8 percent of their waste. Surveys have shown that the #1 reason Americans don't compost their waste is because they feel the process is complicated, time-consuming or requires special equipment. However, especially in rural areas, much of the solid waste could be removed from the waste stream by promoting "extremely passive composting" where consumers simply discard their yard waste and kitchen scraps on their own land, regardless of whether the material is ever re-used as "compost".

Materials

Many different materials are suitable for composting organisms. Composters often refer to "C:N" requirements; some materials contain high amounts of carbon in the form of cellulose which the bacteria need for their energy. Other materials contain nitrogen in the form of protein, which provide nutrients for the energy exchanges. It would however be an over-simplification to describe composting as about carbon and nitrogen, as is often portrayed in popular literature. Elemental carbon—such as charcoal—is not compostable nor is a pure form of nitrogen, even in combination with carbon. Not only this, but a great variety of man-made, carbon-containing products, including many textiles and polyethylene, are not compostable—hence the push for biodegradable plastics.

Suitable ingredients with relatively high carbon content include:

- ❖ Dry, straw-type material, such as cereal straws,
- ❖ Autumn leaves,
- ❖ Sawdust and wood chips, and
- ❖ paper and cardboard (such as corrugated cardboard or newsprint with soy-based inks).

Ingredients with relatively high nitrogen content include:

❖ Green plant material (fresh or wilted) such as crop residues, hay, grass clippings, weeds,

❖ Manure of poultry and herbivorous animals such as horses, cows and llamas, and

❖ Fruit and vegetable trimmings.

The most efficient composting occurs by seeking to obtain an initial C:N mix of 25-30 to 1 by dry chemical weight. Grass clippings have an average ratio of 10-19 to 1 and dry autumn leaves from 55-100 to 1. Mixing equal parts by volume approximates the ideal range.

Poultry manure provides much nitrogen but with a ratio to carbon that is imbalanced. If composted alone, this results in excessive N-loss in the form of ammonia—and some odor. Horse manure provides a good mix of both, although in modern stables, so much bedding may be used as to make the mix too carbonaceous.

For home-scale composting, mixing the materials as they are added increases the rate of decomposition, but it can be easier to place the materials in alternating layers, approximately 15 cm (6 in) thick, to help estimate the quantities. Keeping carbon and nitrogen sources separated in the pile can slow down the process, but decomposition will still occur.

Some people put special materials and activators into their compost. A light dusting of agricultural lime (not on animal manure layers) can curb excessive acidity, especially with food waste. Seaweed meal provides a ready source of trace elements. Finely pulverized rock (rock flour or rock dust) can also provide minerals, while clay and leached rock dust are poor in trace minerals.

Composting in the form of bioremediation can break down petroleum hydrocarbons, TNT and a variety of toxic compounds. It is the bacterial and in some cases fungal content of the compost that possess the enzymatic properties to de-polymerize the complex man-made molecules. In other words, there is nothing about the composting process *per se* that adds or detracts from this, unless as noted above, by warming, to increase the metabolic rate of the constituent organisms.

Some materials are best left to a high-rate thermophilic composting system, as they decompose slower, attract vermin and require higher temperatures to kill pathogens than backyard composting provides. These materials include meat, dairy products, eggs, restaurant grease, cooking oil, manure and bedding of non-herbivores, and residuals from the treatment of wastewater and drinking water. Meat and dairy products can be recycled using bokashi, a fermentation method.

Human waste can be composted by industrial, high-heat methods and also composting toilets, even though most composting toilets do not allow for the thermophilic decomposition believed to be necessary for a rapid kill of pathogens, such as Salmonella. This is not a problem, however, since composting toilets also incorporate the essential element of time required to reduce the available substrate on which pathogens can feed, while increasing the growth of competing microbes. If high temperatures are reached, the resulting compost can be safely used as a fertilizer for food crops and even directly edible crops provided it is not illegal in the regions where the humanure is applied. Careful filtration of the compost also prevents contamination. Humanure fertilizer is, however, used throughout the less developed world and is becoming more accepted as a garden amendment in the developed world (see humanure).

Approaches

There are two major approaches to composting: active and passive. These terms are somewhat of a misnomer since both active and passive composts can attain high heating, which increases the rate of biochemical processes. But the terms active and passive are appropriate descriptions for the nature of human intervention used.

Active

Active (hot) composting is composting at close to ideal conditions, allowing aerobic bacteria to thrive. Aerobic bacteria break down material faster and produce less odor and fewer pathogens and destructive greenhouse gases than anaerobic bacteria. Commercial-grade composting operations actively control the composting conditions such as the carbon-to-nitrogen ratio. For backyard composters, the charts of carbon and nitrogen ratios in various ingredients and the calculations required to get the ideal mixture can be intimidating, so many rules of thumb exist for approximating it.

Pasteurisation is a misnomer in composting, as no compost will become truly sterilized by high temperatures alone. Rather, in a very hot compost where the temperature exceeds 55 °C (130 °F) for several days, the ability of organisms to survive is greatly compromised. Nevertheless, there are many organisms in nature that can survive extreme temperatures, including the group of pathogenic Clostridium, and so no compost is completely safe. To achieve the elevated temperatures, the compost bin must be kept warm, insulated and damp.

Aerated Composting is an efficient form of composting from the chemical point of view as it produces ultimately only energy in the form of waste heat and CO_2 and H_2O. With aerated composting, fresh air (i.e. oxygen) is introduced throughout the mix of materials using any appropriate mechanism. The air stimulates the microorganisms that are already in the mix, and their by-product is heat. In a properly operated compost system, pile temperatures are sufficient to stabilize the raw material, and the oxygen-rich conditions within the core of the pile eliminate offensive odors. High temperatures also destroy fly larvae and weed seeds, yielding a safe, high-quality finished product.

Finally, aeration expedites the composting process through the mechanism of heating insofar as the elevated heat will drive biochemical processes faster, so that a finished product can be rendered in 60 to 120 days. Aerated compost is an excellent source of macro-and micro-nutrients as well as stable organic matter, all of which support healthy plant growth. In addition, the micro-organisms in compost aid in the suppression of plant pathogens. Finally, compost retains water extremely well resulting in improved drought resistance, a longer growing season, and reduced soil erosion.

In Thailand this system has been used by farmer groups for more than 445 sites (May, 2008). The process needs only 30 days to finish without turning need and 10 metric tons of compost is obtained each time. A blower (15 inch squirrel-caged blower with 3 hp motor) is needed to force the air through 10 static piles of compost twice a day and 15 minutes each. The raw materials consist of agricultural wastes and animal manure in the ratio of 3: 1 by volume. For more information please visit www.compost.mju.ac.th/eng

Passive

Passive composting is composting in which the level of physical intervention is kept to a minimum, and often as a result the temperatures never reach much above 30°C (86 °F). It is slower but is the more common type of composting in most domestic garden compost bins. Such composting systems may be either enclosed (home container composting, industrial in-vessel composting) or in exposed piles (industrial windrow composting). Kitchen scraps are put in the garden compost bin and left untended. This scrap bin can have a very high water content which reduces aeration, and so becomes odorous. To improve drainage and airflow, a gardener can mix in wood chips, small pieces of bark, leaves or twigs, or make physical holes through the pile.

Natural

An unusual form of natural composting in nature is seen in the case of the mound-builders (megapodes) of eastern Indonesia, New Guinea, and Australia as well as in the case of bowerbirds of New Guinea and Australia. These Megapodes are fowl-sized birds famous for building nests in the form of huge compost heaps containing leaf litter, in which they incubate their eggs. The birds work constantly to maintain the correct, almost exact, incubation temperatures, by adding and removing leaves from the compost pile. In effect, this teaches us that thermophilic high-temperature composting is not man-made.

Home Composting

Home composters use a range of techniques, varying from extremely passive (throw everything in a pile and leave it for a year or two) to extremely active (monitor the temperature, turn the pile regularly, and adjust the ingredients over time). Some composters use mineral powders to absorb smells, although a well-maintained pile seldom has bad odors. It is usually located in the back garden.

Moisture and Heat

An effective compost pile is about as damp as a well wrung-out sponge. This provides the moisture that all life requires. Microorganisms vary by their ideal temperature and the heat they generate as they digest. Mesophilic bacteria survive best at temperatures of 20 to 44 °C (70 to 120 °F). thermophilic (heat-surviving) bacteria grow optimally at around 55°C (130 °F), and can attain the fastest decomposition, since metabolic processes proceed more rapidly under higher temperatures. Elevated temperature is also preferred since it causes the most rapid pathogen reduction, and is more destructive of weed seeds. To minimally achieve it, the heap should be about 1 m (3 ft) wide, 1 m (3 ft.) tall, and as long as is practicable. This provides enough insulating mass to build up heat but also allows aeration. The center of the pile heats up the most.

If the pile does not heat up, common reasons include that:

❖ The heap is too wet, limiting the oxygen which bacteria require,
❖ The heap is too dry for the bacteria to survive and reproduce, and
❖ There is insufficient protein (nitrogen-rich material).

The necessary material should be added, or the pile should be turned to aerate it and bring the outer layers inside and vice versa. You should add water at this time to help keep the pile damp. One guideline is to turn the pile when the high temperature has begun to drop, indicating that the food source for the fastest-acting bacteria (in the center of the pile) has been largely consumed. When turning the pile does not cause a temperature rise, it brings no further advantage. When all the material has turned into dark brown crumbly matter, it is ready to use.

Worm Composting

 composting or vermicomposting is a method of composting using Red Wiggler worms in a container. Food waste and moistened bedding are added, and the worms and micro-organisms eventually convert them to rich compost. The worms excrete a soil-nutrient material called worm castings. Worm composting can be done indoors, allowing year-round composting, and providing apartment dwellers with a means of composting.

Worms are low in the food chain, and so are critical to healthy soil. This is why farmers have historically wanted healthy worm populations to live in their fields.

The nutrients and micro-organisms can be concentrated in liquid form called worm tea, made by running distilled water through worm castings. When it is poured into the soil, the microorganisms multiply, creating a healthy growing environment for plants.

Industrial Composting

Industrial composting systems are increasingly being installed as a waste management alternative to landfills, along with other advanced waste processing systems. Industrial composting or anaerobic digestion combined with mechanical sorting of mixed waste streams is called mechanical biological treatment increasingly used in Europe due to stringent new regulations controlling the amount of organic matter allowed in landfills. Treating biodegradable waste before it enters a landfill reduces global warming from fugitive methane; untreated waste breaks down anaerobically in a landfill, producing landfill gas that contains methane, a greenhouse gas even more potent than carbon dioxide.

Most commercial and industrial composting operations use active composting techniques. These ensure that the process does not get out of control especially with the high through-put demand imposed by contracted, incoming waste. This means that as short as possible a processing time must be maintained to keep the facility properly functioning (see compost windrow turner). Partly for this reason composters have declined to support compost maturity standards if it would increase the required holding time. The greatest amount of technological control of composting is seen in systems using an enclosed vessel and controlling its temperature, air flow, moisture and other parameters. See In-vessel composting (indoor composting).

Large-scale composting systems are used by many urban centers around the world. Co-composting is a technique which combines solid waste with de-watered biosolids, which originated in the 1960s and has fallen somewhat out of favour due to difficulties controlling inert and plastic contamination from MSW. In Europe, mixed waste composting is virtually illegal. The world's largest MSW co-composter is the Edmonton Composting Facility in Edmonton in Alberta, Canada, which turns 220,000 tonnes of residential solid waste and 22,500 dry tonnes of biosolids per year into 80,000 tonnes of compost. The facility is 38,690 square metres (416,500 ft²) large (equivalent to 4½ Canadian football fields), and the aeration building alone is the largest stainless steel building in North America, the size of 14 NHL rinks.

18.2 Windrow Composting

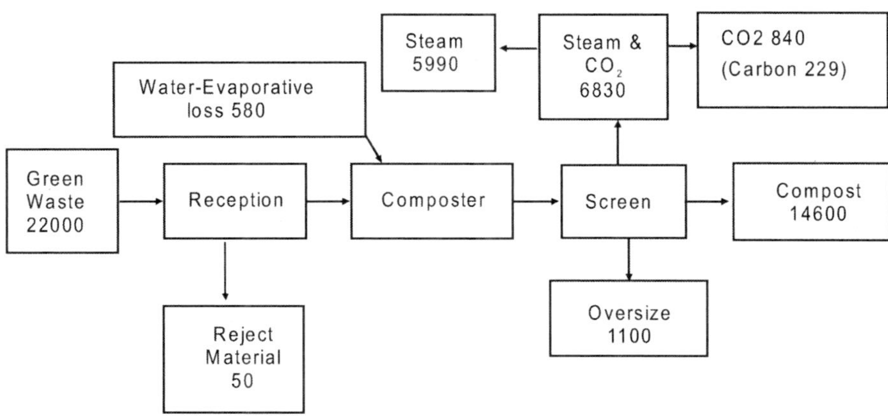

In agriculture, windrow composting is the production of compost by piling organic matter or biodegradable waste, like animal manure and crop residues, in long rows (*windrows*). This method is suited to producing large volumes of compost. These piles are generally turned to improve porosity and oxygen content, mix in or remove moisture, and redistribute cooler and hotter portions of the pile. Windrow composting is the most commonly used of farm scale composting methods. Process control parameters include the initial ratios of carbon and nitrogen rich materials, the amount of bulking agents added to assure air porosity, the pile size, moisture content, and turning frequency.

Compost Windrow Turners

Compost windrow turners were developed to produce compost on a large scale. They are a traditionally a large machine that straddles a windrow of approximately eight feet high by 13 feet across. Although smaller machines exist for smaller windrows, most operations need the larger machines to be efficient. Turners drive through the windrow at a slow rate of forward movement. They have a steel drum with paddles that are rapidly turning. As the turner moves through the windrow, fresh air containing oxygen is injected into the compost by the drum/paddle assembly and waste gases produced by harmful bacteria are eliminated. The oxygen feeds the beneficial composting bacteria and thus speeds the eventual composting process. This process is then extended by windrow dynamics.

Utilization

To properly utilize a compost windrow turner, it is ideal to compost on a concrete pad; haul in the feed stocks, and place them into windrows. Heavy-duty compost windrow turners can be built that allow the user to obtain optimum results. By using 4-wheel-drive or tracks the windrow turner is capable of turning compost in windrows located in remote locations. With a self-trailering option this allows the compost windrow turner to convert itself into a trailer to be pulled by a semi-truck tractor. These two options combined allow the compost windrow turner to be easily hauled anywhere and to turn the compost in muddy and wet locations.

18.3 What is a Composting Toilet System?

Composting toilet systems (sometimes called biological toilets, dry toilets and waterless toilets) contain and control the composting of excrement, toilet paper, carbon additive, and, optionally, food wastes. Unlike a septic system a composting toilet system relies on unsaturated conditions (material cannot be fully immersed in water), where aerobic bacteria and fungi break down wastes, just as they do in a yard waste composter. Sized and operated properly, a composting toilet breaks down waste to 10 to 30 percent of its original volume. The resulting end-product is a stable soil-like material called "humus," which legally must be either buried or removed by a licensed seepage hauler in accordance with state and local regulations in the United States. In other countries, humus is used as a soil conditioner on edible crops.

The primary objective of the composting toilet system is to contain, immobilize or destroy organisms that cause human disease (pathogens), thereby reducing the risk of human infection to acceptable levels without contaminating the immediate or distant environment and harming its inhabitants.

This should be accomplished in a manner that:

❖ is consistent with good sanitation (minimizing both human contact with unprocessed excrement and exposure to disease vectors, such as flies).
❖ produces an inoffensive and reasonably dry end-product that can be handled with minimum risk.
❖ minimizes odor.

A secondary objective is to transform the nutrients in human excrement into fully oxidized, stable plant-available forms that can be used as a soil conditioner for plants and trees.

The main components of a composting toilet are:

❖ a composting reactor connected to one or more dry or micro-flush toilets;
❖ a screened exhaust system (often fan-forced) to remove odors, carbon dioxide, water vapor, and the by-products of aerobic decomposition;
❖ a means of ventilation to provide oxygen (aeration) for the aerobic organisms in the composter;
❖ a means of draining and managing excess liquid and leachate;
❖ process controls, such as mixers, to optimize and manage the process; and
❖ an access door for removal of the end-product.

The composting reactor should be constructed to separate the solids from the liquids and produce a stable, humus material with less than 200 most probable number (MPN) per gram of fecal coliform.

General Types of Composting Toilet Systems

Composting toilet systems can be classified in several ways:

Self-Contained versus Centralized

Composting toilet systems are either *self-contained,* whereby the toilet seat and a small composting reactor are one unit (typically small cottage models), or *centralized* or *remote,* where the toilet connects to a composting reactor that is somewhere else.

Manufactured versus Site-Built

One can either purchase a *manufactured* composting toilet system or have a *site-built* composting toilet system constructed (however, the latter can be difficult to get permitted by local health agents).

Batch (Multiple-Chamber) versus Continuous (Single-Chamber)

Most composting toilet systems use one of two approaches to manage the composting process: either *single-chamber continuous* composting or *multi-chamber batch* composting processes

A continuous composter (including Clivus Minimuses and such brands as CTS, Clivus Multrum, Phoenix, BioLet, Sun-Mar) features a single chamber into which excrement is added to the top, and the end-product is removed from the bottom.

A batch composter (such as the EcoTech Carousel, all Vera systems, BioLet NE, and many site-built composters) utilizes two or more interchangeable composting reactors. One is filled at a time, then allowed to cure while another reactor fills, just as with two-and three-bin yard composters.

Proponents of continuous composting maintain that it is simple (takes place in one fixed reactor), allows urine to constantly moisten the process, and allows the center of the mass to heat up through uninterrupted microbial activity.

Advocates of the batch-composting approach say that by not continually adding fresh excrement and urine to older, more advance material, the material composts more thoroughly, uninterrupted by the added nutrients, pathogens, salts and ammonia in fresh excrement. Also, by dividing the material, it can have more surface area, and thus better aeration. Batch composting also offers an opportunity for unlimited capacity, as one simply adds compost reactors to the process to add capacity.

Batch systems require monitoring the level of the composter to determine when a chamber has filled and a new one must be moved into place. However, because there is more surface area and the material is divided, there is often less or no mixing and raking of the material.

What is true is that complete composting needs time to ecologically cascade all the by-products of the myriad organisms in the composting food web until all of the organic matter is finally transformed into safe, stable humus.

In continuous composters, urine or flush water can leach fresh excrement into the finished compost removal area. Segregating the material into batches reduces the risk of having living disease organisms in the finished products.

(One of the authors [Del Porto] has designed, sold and serviced both types of composters. He prefers batch composting-based systems, because they are more forgiving and users seem to be happier with them. However, he would never say that single-chamber continuous systems do not work. They do!

It is difficult to generalize about which process affords the greatest opportunity for complete processing and minimizes the potential for pathogen survival. In a batch system, a finite supply of nutrients is cycled and recycled through microbe populations until the nutrients, both the free ones and those bound in microbial protoplasm and cell walls, are ultimately converted to stable, fully oxidized forms, and the fungi have performed their work on the remaining lignin and cellulose compounds, releasing antibiotics in the process.

Definitive research is needed in this area.

Active versus Passive

As with solar systems, composting systems are usually either *passive* or *active*. Passive systems are usually simple moldering reactors in which ETPA (excrement, toilet paper and additive) is collected and allowed to decompose in cool environments without active process control (heat, mixing, aeration).

Active systems may feature automatic mixers, pile-leveling devices, tumbling drums, thermostat-controlled heaters, fans, and so forth. The trend in the composting of municipal solid waste (garbage and trash), sewage sludge and yard and agricultural residues is toward active systems. By making the process active, the size of the composter can be reduced, because composting efficiency is speeded up (and the volume of the material reduced faster).

Passive systems are designed to optimize the process by design, not mechanical action, allowing only time, gravity, ambient temperature and the shape of the container to control the process. Passive composters are often referred to as moldering toilets, as the process at work is natural uncontrolled decay at cool in-ground temperatures at or below 68° F. In this cool environment, molds (fungi and actinomycetes) are the primary biological decomposers, because it is a bit too cool for the faster-acting mesophilic and thermophilic bacteria.

18.4 Composting

The Composting Toilet—An ecologically responsible solution

The composting toilet solves with one blow the problems of "black waste water" and loaded nutrient salts in the recipient and overuse of our precious water resources.

Water Consumption

Our daily consumption of water is 180-200 litres which translates to approximately 70,000 litres of drinking water/person/year. Ordinary flush toilets are responsible for 15-20,000 litres of this amount. Composting toilets do away with this water wastage and at the same time keep feces, paper and even urine out of the sewage system.

The Composting Toilet

When one mentions the word "composting toilet" or "mulch latrine", people immediately protest and come up with such comments as, "you don't mean you expect us to go back to the use of stinky latrines ?!" and "doesn't it smell like slurry ?!" But on this matter I must be firm—today we have an odourless and comfortable solution very suitable for installation inside the home. In contrast to many so-called "ingenious" solutions such as incineration toilets, freezer-bag toilets, dry toilets and the like, which often require the use of large amounts of energy, a genuine composting toilet consists of a large container in which a composting process takes place before the contents are removed, and a toilet stool + seat very similar in appearance to that of an ordinary flush toilet. The end product of the composting toilet is an odourless humus material totally unlike manure or the contents of a latrine. (there is no danger in handling it and it can be used as fertilizer for shrubs and decorative plants as long as it has had soil contact for 19 months)

Denmark's Neighboring Countries

During the past few decades, individuals in Norway and Sweden have been continuously working on the development of composting toilets, so that today we have a new generation of composting toilets. In 1982, according to the Nordic Ministercounsil, there were 300,000 composting toilets installed throughout Scandinavia. I suppose that Denmark accounts for only about

100 of these. Development has continued at a fast rate in the neighbouring countries and the latest encouraging news is that Tannum "municipality", near Goteborg, Sweden, has recently forbidden flush toilets from the year 2000 and currently builds 70 per cent of their new housing without flush toilets.

Development in Denmark

Because of the fact that 7 per cent of Danish housing (in rural areas) aren't connected to central sewage systems, some research on composting toilets has been undertaken at The Danish Technical University and the most current news in this field is that a project called "Kompost-toiletter 1994" by Simon Wris-Berg and Peter Friis Moller has been awarded a prize by Denmark's agricultural school. Some Danish firms have experienced an increase in inquiries about composting toilets during the past few years due both to increasing environmental awareness as well as ever increasing water costs.

The Design of a Composting Toilet

Composting toilets vary in appearance depending upon the manufacturer, but common for them all, is that they have a collection container for feces, paper, in some cases kitchen garbage, while some separate urine into a second container. The contents of the feces container are composted by aerobic process, which takes about one year, into a humus like material. In the company "Renewable Energy", we have developed a year-round model called the "Eco-composter" for family homes. The collection container of our toilets is made of fibre reinforced concrete which makes for a reduction in use of fibreglass and epoxy. The toilet stool, which is made of either fibre-concrete or porcelain, is designed so that urine and feces are kept separate. The feces, paper and kitchen garbage drop vertically down into one tank, while the urine is collected at the front—a design appropriate for both female and male use. In this way the urine can, without coming into contact with the feces, be collected in a tank and because there is no danger of disease, be used as liquid fertilizer when mixed with 4-5 times its amount in water. The urine can also, of course, be mixed with household "grey waste water". But this must be regarded as the second solution. Our porcelain stool is designed in such a way that the urine compartment can be flushed with 0.1 litres of water. The tank has ventilation channels to enable the oxidizing process to take place, and all air is drawn up through a vent which reaches up through and over the roof of the house. This means that the composting toilet itself functions as a ventilation system for the entire toiletroom, keeping out unwanted smell. This makes for quite an advantage—there is never any unpleasant smell connected with a "visit" to the toiletroom.

As indicated in table 18.1, a family's yearly production of compost humus is approximately 100-200 litres (the original amount of uncomposted feces having been greatly reduced after composting).

Table 18.1: Yearly production per adult human.

400-500 litres urine which contains:

5 kg	nitrogen
0.4 kg	phosphate
0.9 kg	potassium

50-60 litres feces which contains:

0.1 kg	nitrogen
0.2 kg	phosphate
0.2 kg	potassium

Table 18.2: The amounts of nutrients & heavy metals in manure and soil

Substance	Unit	Sewage Sludge	Human Manure	Animal Manure	Soil
N	kg/t	30	250	25	1-2
P	kg/t	20	35	10	0.4
K	kg/t	2	45	17	0.5
Ca	kg/t	25	30	12	25
Mg	kg/t	4	7	4	7
Zn	g/t	1750	200	100-800	26
Cu	g/t	250	30	20-350	8
Ni	g/t	20	2	1-36	5
Cd	g/t	7	0.4	0.3	0.2
Pb	g/t	300	1	5-15	17
Hg		g/t	5	0.5	0.05

Source: J. Aa. Hansen and J. C. Tjell, 1982 w/Jacob Vester.

The components of our composting toilet take up the same amount of space as a traditional flush toilet in the toiletroom itself, while the collection-container and ante-chamber are located beneath the floor of the toiletroom or from a connected "lightbox" located outdoors.

Infection and the Authorities

The Danish authorities require that the same minimum criteria are fullfilled as with sludge delivery, due to a risk of disease-carrying bacteria and parasites, although the compost can always be spread on the ground because the amount of heavy metals in human feces + urine is extremely low when not mixed with other wastewater (see table 18.1) The compost-humus could of course be pasteurized at 70 degrees for a minimum of 1 hour, but this holds no interest for us, because we want to avoid the use of large amounts of energy and retain the good balance with nutrients which are present in the compost-humus. We don't want to use energy for no reasons. We require that the compost is brought in contact with soil bacteria and kept behind a little fence for yet another year before being spread out on gardens or fields. This requirement is easy to live with in cases where here is a complete plan for the composting toilet and the "grey wastewater", for instance first via a microbiological cleaning in a sandfilter and finally by means of gravitation in an "energy-forest draining system" so that we really can put the compost in the "energy-forest". Today we are no longer confronted with the same resistance to this installation design as 10 years ago, due to greater environmental awareness and stricter wastewater treatment requirements.

The Future

We will experience fast and hopefully creative development in the area of composting toilets combined with more research which initially can find support in the results of our neighbouring countries. Because the lack of fresh water is a global problem, these solutions are quite suitable for export to other countries.

18.5 A Novel Technology of Composting of Rural and Urban Wastes by Two Step Fermentation System

Introduction

After introduction of high-yielding varieties of crops, intensive tillage system with high dose of chemical fertilizers and pesticides for the last three to four decades deteriorated the physical, chemical, biochemical and microbial properties of the soil and also caused imbalanced of plant nutrients in soil (Reganold *et al.*, 1988; Reganold *et al.*, 1995). The high temperature of tropical and subtropical soil ensures greater turnover of soil organic matter by evolution of CO_2, through divergent types of soil microbes (Slott *et al.*, 1986; Santruckova and Straskova, 1991; Ayanablee and Jenkinson, 1990). Present day intensive tillage system with chemical fertilizers helps further greater turnover of soil organic matter (Reicosky and Lindstrom, 1993; Rasmussen *et at.*, 1989; Lynch and Panting 1980). As a result global declining phase of crop yield has set in since last one decade (Hossain, 1995). *The clearly indicates that chemically fertile soil is not a productive soil. So organic farming or new alternative agriculture or precision farming system is essentially needed for sustainable productivity, maintaining soil good health and pollution free environment (Bidwell, 1986; Papendick, 1987; Reganold et al., 1990; Clancy, 1986; Vareijken, 1986). At present the soil is hungry for organic matter specially in tropics and subtropics to give optimum crop yield.*

Inputs of crop residues of annual species, or farm yard manures (FYM), compost or organic manures with relatively "short-time residence substrate" in soil system in the present day intensive tillage system, require high bulk amount (10-20 t ha-1) for their low plant nutritional value but inputs from woody resources having high lignin content organic matter and long lag-phase or "long residence time" in soil system require lower amount (Geijn and van Veen, 1993). As for example, two-third of total dry matter of crop residues incorporation in soil is lost in 6 months, but only onethird of lignin is lost. Consequently, residues initially containing 14.6 per cent lignin content became approximately twice that amount after one and half year later (Bartlett and Norman, 1938). Extensive studies on microbial degradation of lignin were made by several workers (Kirk *et al.*, 1980; Ander ami Eriksson, 1978; Crtaford, 1981). Further, 'because of the increase in relative aboundance of aromatic constituent, lignin, modified lignin molecules or degraded lignins like polyphenols, polyquinones when polymerised and condensed with proteins, peptides of amino acid, have a significant role in the synthesis of humic substances (Waksman, 1938; Stevenson, 1982; Alexander, 1961). However, in the aquatic condition, pathways of humus synthesis is different (Hayes and Wilson, 1997). It was also reported that humic substances having high molecular weight and in highly polymerised and condensed form when applied to the soil system formed a *"Clay-Humus-Mineral Complex Structure"*, which is a rigid structure, resistant to microbial degradation or undergoes degradation at a lesser rate and thus have long residence time in soil system (Stevenson, 1982), best suitable technology in tropical and sub-tropical soil like India for sustainable soil for soil productivity in the tropical soil system research should be clarified.

India has 140 million hectare of cultivable land. ICAR has tried to increase the cultivable and area but so far they are failed. For annual single application of organic manure is the total demand of 1400 million tonne (@ 10t/ha). Total available compostable resource is approximately amounted 2500 million tone per annum. Vershney (1991) reported that 80 per cent of total organic resources are utilized as fuel for cooking purpose, rest 20 per cent from electricity, fuel gas, coal and other sources. So actual available compostable resources is 500 million tone per annum. For decomposition process it will reduce to 250-300 million tone per annum. So practically demand and supply is about 1000-1100 million per annum for manure purpose. In our field survey fuel scarcity in rural area is a great problem as well as compost preparation. So urban waste utilization and entrepreneurs can minimize the demand supply there, Saw dust (2.1 million tone per annum), crop residue and weeds will be used for mechanical composting plant as well as decomposable urban waste.

The review clearly revealed that humus and unrecompensed lignin or partially modified lignin are relatively strongly resistant for decomposition and persist longer time in soil system. Thus they have long residence time and they made be considered as the best natural soil conditioners and most suitable 'for our tropical, subtropical and temperate soil environment. Besides, the purpose of composting is to obtained humus colloid from the microbial decomposition of organic matter (Tenaile, 1974). The value of organic manure depends on the amount of humus it produces and the amount of plant nutrients added to the soil (Daji, 1957). Kononova and Alexandrova (1973) reported that humic substances are formed through polymerisation and condensation reaction of phenolic compounds derived from lignin degradation. Humus is cream of compost as comparable to the cream of milk. On these backgrounds the experiment of the *two-step composting process* was designed. Rapid degradation of lignocellulosic wastes by white-rot fungi into simpler units like polyphenols, polyquinones etc. occurred in the *first-step of* composting and the *second-step of composting* involved high humification by polymerisation and condensation reactions by admixing with N-rich organic dairy wastes.

Phosphocompost Production

The major characteristics of this compost are as follows:

1. Completely free from Human, Animal and Plant Pathogens and completely pollution free of toxic element calculated feeding materials will cause free from heavy metal and other toxic compound and high quality produce.
2. Rather full of beneficial soil microbes as well as full of toxicless fungi, bacterial, virus for disease control of cropping systems. We can claim we are the pioneer in the world or first time in the world-or in this aspect those are very vital points are appropriate, precision and perfect technology for appropriate end products most suitable in our tropical

and subtropical soils of India. This technology will build up present soil organic matter from 0.3-0.5 to 1.5-2 per cent in 3-4 years cultivation due to very lower turn over of soil organic matter in dynamic soils system.

3. Quality point of view the compost or enriched product is "Phospho-Nitro-Sulfo-Humus Compost is the Cream of Compost comparable to most valuable Cream of Milk". For such highly desirable excellent quality compost not reported before in global perceptive. But the precession technology will be pioneer in the world as it is a great challenge to the tropical agricultural scientist for sustainable soil productivity.

4. Calculation made in such a way that feeding materials enrich "specific organic component of compost" which is stable 2-3 years even in our tropical soil system and even in intensive tillage of cropping system. Which form *"Soil-Mineral-Humus Complex Structure"* and control release of plant minerals slowly. This "Complex components of Compost is strong resistant and long persistent in maintenance of soil health for long time in tropical soils. Consecutive application 3-5 year of condensed compost will build up soil carbon from 0.2-0.5 to 2-3 per cent. All soil problem will be solved for sustainable soil productivity maintaining soil good health". This most appropriate, perfect, precession technology is suitable in Indian Agro-climatic condition.

5. Conventional mechanical Solid Waste Management produced so called compost prepared by physicochemical process of Hitachi in 8-10 days or biogas slurry of Volorga (France). WMC process (UK) WBG (Sweden). Dran Co (Belgium) Organic Matter, not full matured Organic Manure rather mixture of organic matter and manure of low grade fertility contents. Very less stable in our tropical soil system (80% loss in nine months).

6. Not suitable in our tropical soil system in modern intensive tillage system. Require 10-15 time more than most precision components and enriched compost in our system which is most suitable in our system which is most suitable in our tropical soil system. Very low turn over of organic carbon in soil system which ultimately build up soil organic matter in long run (low amount as well as but long interval) i.e. 1/1.5 ton/ha.

7. Very small amount requirement of this type of "high quality condensed components of compost" having plant nutrient in soil plant nutritional system.

8. Quality point of view, this type of compost or enriched compost or phospho-nitrog-sulfo-humus compost in the mechanical composting plant it will be the cream of compost comparable to most valuable cream of milk. For such high desirable, excellent, maximum synthesis of Humus (18-34%) which about 3-6 times more than conventional compost, 3-5 times more total N (2-5%), C: N ration 11-15, CEC 111 to 225, Total P 2-3 per cent citral soluble 0.4 per cent (Phospho-Nitro-Sulfo-Humus Compost)

9. It requires only one-tenth i.e. about 1.0/1.5 tonne/ha instead of 10-15 tonne/ha less carrying cost more and more condensed form.

10. First time in the world we have used ligninolytic white-rot fungi cultures for rapid decomposition of lignin and cellulose (lignocellulose of sawdust, cereal straw and weeds etc.).

11. Reducing the composting time into 2 months even sal-hard wood and simul soft-wood dusts, woody shrubs, crop residue, weeds instead of 6-12 months in the conventional compost whereas in mechanical composting plant prepared within 8-21days (Hitachi) which I can say that the product is partially degraded organic matter not organic manures.

12. Conventional mechanical solid wastes management so-called compost prepared by physicochemical process of Hitachi in 8-10 days or Bio-gas slurry of, Volarga (France), WMC process (KU), WBG (Sweden), Dran Co (Belgium) is organic matter not fully matured organic manure or mixture of organic matter and manure of low grade fertility content. Very low stable in our high temperature tropical soil system (80% loss in nine months).

13. Calculation made in such a way that feeding materials enrich with specific organic component of compost which is stable 2-3 years even in our tropical soil system and even in intensive tillage system, which form *"Clay-Mineral-Humus-Complex Structure* and control release of plant minerals slowly. This complex components of compost is relatively strong resistant and long persistent in maintenance of soil health of long time and sustainable soil productivity.

14. Due to very low turnover of soil carbon, consecutive application of 3-5 years of this type of condensed compost will build up our topical carbon from pleasant 0.2-0.5 per cent to 1.5-3 per cent.

15. Most vital point is those will be free from all types of human, animal and plant pathogen and completely free from toxic elements, rather fully enriched with the beneficial soil microbes, specially N2-fixer, P-solubilizer and bio-control agent etc. We can

claim that we are the pioneer in the world or first time in the world in this aspect, those are very vital points and an appropriate technology and appropriate end products most suitable in our tropical soils of Indian.

In India phosphate rock deposits in different parts of the country as estimated by various agencies are about 145 million tones. Most of these rock phosphate are of low grade and suitable for their direct use without chemical treatments with costly foreign exchange or in Indian conditions. However, in both way we are not in position to furnish plant requirement as per cropped area.

During last two decades, the efforts made by the Microbiologist of India, under All India Coordinated Research Project on "Microbiological De composition and Recycling of Organic Wastes" have clearly manifested that the compost prepared by Modern Technology of Composting may be an alternative system to supplement P for crop production through organo-phosphocompost having the potential equivalent at part to synthetically applied P as a cheaper input and quite easy in practice and easily adoptable by the farmers.

It is envisaged that P enriched compost as P supplement or substitute for phosphatic fertilizer not only enhanced the available P concentration but also augment the microbial activity of the composting system. Since it is emphasized that the low grade quality of Indian Rock Phosphate could be utilized for production of phosphate rich compost.

It's a such preparation through composting in which crop residues, leaf falls, cattle shade liters, natural waste/weeds, urine, dung, city wastes etc. as raw material when combined with cattle dung, soil and well matured compost along with 5-10 per cent. Missouri rock phosphate, and 1.5-3 per cent and urea N 1 per cent are composted for a period of 3-4 months adopting modern technology of composting may provide highly available form of phosphate for crop utilization and better production, called phosphocompost.

Technology of Phoshocompost Preparation

1. Prepare a compost pit or earth up soil and make a trench and spread a good quality sheet to avoid leaching (Pucca-to avoid nutrient losses) at high level of ground (no water logging) with a size of 2.0 M width and 1.0 M depth and length should be as per the need or availability of material or make a trench and spread a polythene

sheet and make a shed and cover by colour polythene sheet during rainy season.

2. Collect crop residues material waste/weeds, city wastes (free from polythene, glasses, stones) and if possible cut to a conventional size 0-15 cm. and cattle dung, soil and matured compost.

3. Make a layer of residue/organic wastes on the surface of composting pit.

4. Cattle dung, soil and matured compost (15: 2: 2 ratio) along with rock phosphate (5-10%) should be mixed together and make a slurry of all these taking desired quantity of water and sprinkle uniformly over the organic waste material. Add sufficient quantity of water (60-80%) on total weight basis of used input and cellulolytic or ligninolytic culture (1-2% basis) and Psoluble cultures.

5. Repeat the layering system till the pit is completely filled up.

6. Have a fine layering of mud on the top of composting pit cover by polythene sheet.

7. Turning (manual with the help of handhoe) of composting material (3rd and 5th week) should be preferred at a interval of 21 days upto maturity of composting materials (60-90 days). After 4th week inoculated. Azobacter and Biocontrol agents.

8. Such prepared compost at 3-4 months should be thoroughly mixed and samples should be drawn after making homogenous mixture for estimations of total, water and citric acid soluble P for their standardization in respect of P for supplementation.

9. Spread the whole mature compost in shed so that moisture content not more than 20-30%.

The composition of mixture and to produce one lone phosphocompost is given in Table 18.3:

Table 18.3: Composting of mixtures on dry weight basis is as follows

1.	Organic materials	60%
2.	Cattle dung	20%
3.	Soil	10%
4.	Compost	10%

Ratio Comes = 6: 2: 1: 1

To obtain 1.0 tonne of phosphocompost one can use the following materials in Table 18.4.

Table 18.4: Phosphocompost preparation requirement

1.	Organic wastes	600 kg
2.	Cattle dung	100 kg
3.	Soil	50 kg
4.	Compost	120 kg
5.	MRP	100 kg
6.	Pyrite	20-30 kg

The quantity of such composts is that it is highly rich in citric acid soluble P while being poorly enriched with water soluble P. For instance, compost enriched with 5.0 per cent P_2O_5 equivalent to rock phosphate, had 6.7 per cent total P, and 10.15 mg/g citric acid soluble P. Overall composition of phosphocompost is given in table 18.5.

Table 18.5: Composition of compost

Characteristic	Ordinary Compost	Phosphocompost
Organic carbon (%)	24.3	18.60
Total Nitrogen (%)	1.15	0.82
Total Phosphorus (%)	0.213	6.70
Water soluble P (mg/g)	0.26	0.17
Sodium bicarbonate (mg/g) soluble P	1.38	0.86

Several multilocational trials were carried out at national levels in ICAR project to assess the fertilizer value of phosphocompost when applied to crops on equal P_2O_5 vis-a-vis super phosphate, phosphocompost, by and large, could act as substitute for super phosphate, when evaluated, when evaluated in term of crop yields. P uptake and available P in soil after the crops harvest. The quantities of phosphocompost work out to be handy. For example, for 60 kg, P_2O_5 ha^{-1} the quantity of phosphocompost work out to be 1-1, 5th-1 when the compost is made by charging with 2-3 per cent P_2O_5MRP.

Vermicompost and Phoshocompost

19.1 Vermicompost

Vermicompost (also called worm compost, vermicast, worm castings, worm humus or worm manure) is the end-product of the breakdown of organic matter by some species of earthworm. Vermicompost is a nutrient-rich, natural fertilizer and soil conditioner. The process of producing vermicompost is called *vermicomposting*.

The earthworm species (or composting worms) most often used are Red Wigglers (*Eisenia foetida*) or Red Earthworms (*Lumbricus rubellus*). These species are commonly found in organic rich soils throughout Europe and north America and especially prefer the special conditions in rotting vegetation, compost and manure piles. Composting worms are available from nursery mail-order suppliers or angling (fishing) shops where they are sold as bait. Small-scale vermicomposting is well-suited to turn kitchen waste into high-quality soil, where space is limited.

Together with bacteria, earthworms are the major catalyst for decomposition in a healthy vermicomposting system, although other soil species also play a contributing role: these include insects, other worms and molds.

Bins

Vermicomposting bins (also known as worm bins) vary drastically depending on the desired size of the system. Garden-scale containers can be made out of bricks arranged in the shape of a box. Bins for an apartment or similar dwelling can be anything from reused plastic buckets to purpose-built commercial containers.

Small Scale

Small-scale systems may use a wide variety of bins. Often, small-scale composters build their own bins. Companies also sell such bins. Commonly, bins are made of old plastic containers, wood, Styrofoam containers, or metal containers.

Some materials are less desirable than others in bin construction. Styrofoam is believed to release toxins into the earthworms' environment. Metal containers often conduct heat too readily, are prone to rusting, and may release heavy metals into compost.

Bins should have holes in the sides to allow air to flow, and a spout that can be opened or closed or holes in the bottom to drain into a collection tray. Plastic bins require more drainage than wooden ones because they are non-absorbent. The design of a small bin usually depends on where an individual wishes to store the bin and how they wish to feed the worms. Most small bins can be grouped into three categories:

❖ *Non-continuous:* an undivided container. A layer of bedding materials is placed in the bin, lining the bottom. Worms are added and organic matter for composting is added in a layer above the bedding. Another layer is added on top of the organic matter and the worms will start to compost the organic matter and bedding. This type of bin is often used because it is small and easy to build. But it is relatively difficult to harvest because all the materials and worms must be emptied out when harvesting.

❖ *Continuous vertical flow:* a series of trays stacked vertically. The bottom-most tray is filled first, in a similar fashion to any other bin, but is not harvested when it is full. Instead, a thick layer of bedding is added on top and the tray above is used for adding organic material. Worms finish composting the materials in the bottom tray and then migrate to the one above. When a sufficient number of worms have migrated, the vermicompost in the bottom tray can be collected and should be relatively free of worms. These bins provide an easier method of harvesting.

❖ *Continuous horizontal flow*: a series of trays lined horizontally. This method too relies on the earthworms migrating towards a food source in order to ease the process of harvesting. The bin is usually constructed to be similar to a non-continuous bin but longer horizontally. It is divided in half, usually by a large gauge screen of chicken wire. One half is used until it becomes full, then the other half is filled with bedding and organic matter. In time, the worms migrate to the side with the food and the compost can be collected. These bins are larger than a non-continuous system but still small enough to be convenient.

Large Scale

There are two main methods of large-scale vermiculture. Some systems use a windrow, which consists of bedding materials for the earthworms to live in (see bedding below) and acts as a large bin; organic material is added to it. Although the windrow has no physical barriers to prevent worms from escaping, in theory they should not due to an abundance of organic matter for them to feed on. Often windrows are used on a concrete surface to prevent predators from killing the worm population. When large scale windrows are fed on one side consistently, a wave motion is generated over time.

The second type of large-scale vermicomposting system is the raised bed or flow-through system. Here the worms are fed an inch of "worm chow" across the top of the bed, and an inch of castings are harvested from below by pulling a breaker bar across the large mesh screen which forms the base of the bed. Because red worms are surface dwellers and are constantly moving towards the new food source, the flow-through system eliminates the need to separate worms from the castings before packaging. Flow-through systems are well suited to indoor facilities, making them the preferred choice for operations in colder climates. One major Australian company wormswork has successfully commissioned this continous flow system at their manufacturing plant in Millicent South Australia. with the capacity of producing 12,000 tonnes of finished solid organic fertilizer and 6 million litres of organic liquid fertilizer annually. This can be seen at www.wormswork.com.au

Starting Off

When beginning a vermicomposting bin, moist bedding is put into the bin and the worms are added. In hot climates, the bin is placed away from direct sunlight. Appropriate waste can be added daily or weekly. At first, the worms are fed at most half their body weight per day. After they have established themselves, they can be fed up to their entire body weight. It is best not to add new food on top of old food until the old food has been processed by the worms. However, new food can be added in a different location in the bin.

Bedding

Bedding is the living medium and also a food source for the worms. It is material high in carbon and made to mimic decaying dried leaves on the forest floor, the worms' natural habitat. The bedding should be moist (often similar to the consistency of a wrung-out sponge) and loose to enable the worms to breathe and to facilitate aerobic decomposition of the food that is buried in it.

A wide variety of bedding materials can be used, including shredded newspaper, sawdust, hay, cardboard, coir, burlap coffee sacks, peat moss, pre-composted (aged) manure, and dried leaves. Cat litter, and pet and human waste should not be used.

Most vermicomposters avoid using glossy paper from newspapers and magazines, junk mail, and shredded paper from offices, because they may contain toxins which may disrupt the system. Also, coated cardboard that contains wax or plastic, such as milk boxes, cannot be used. Newspaper and phone books printed on regular, non-glossy paper with non-toxic soy ink are safe for use, and decompose relatively quickly. Some bedding is easier to use and add food scraps to than others.

Temperature

Worms used in composting systems prefer temperatures of 55 to 70 degrees Fahrenheit (12-21 degrees Celsius). The temperature of the bedding should not drop below freezing or above 89.6 °F (32 °C). This temperature range means that indoor vermicomposting is suitable for homes in all but tropical climates.

Food

Worms and other composting organisms have a preferred ratio of carbon to nitrogen (C:N), approximately 30:1. As some waste is richer in carbon and others in nitrogen, waste must be mixed to approximate the ideal ratio. "Brown matter", or wood products such as shredded papers, is rich in carbon. "Green matter", such as food scraps, has more nitrogen, which is related to the amount of protein in the waste. If the waste is mostly vegetable and fruit scraps, and does not regularly include animal products or high-protein vegetable foods like beans, the resulting vermicompost and waste liquid will be low in nitrogen.

Suitable

Kitchen waste suitable for worms includes coffee grounds and filters, tea bags and plate scrapings, as well as rotting fruit (including citrus fruit but NOT citrus peel), vegetable peels, leftovers, moldy bread, etc. These materials can be raw or cooked. They do not have to be ground up, as the micro-organisms in the bin will gradually soften them. However, if a large quantity of dry food (e.g., moldy bread) is added and covered with bedding, pour a little purified water over the bedding to moisten the mixture.

If too much kitchen waste is added, the bin mixture putrifies before the worms can process it and becomes harmful to the worms. High-protein foods like beans are particularly susceptible. Check the bin at least once a week, give the materials a stir to oxygenate, and add bedding if the bin appears too moist.

Soft garden wastes such as carrot tops and tomato leaves are suitable foods. An occasional sprinkling of garden soil in the bin gives the worms the grit they need to digest food. It's not harmful to throw in an entire plant, but the worms will not process the woody parts or large roots and these will have to be hand-removed later from the finished vermicompost.

Compostable plates, cups, etc. are also suitable, but in small bins they should be torn first into smaller pieces so as not to block oxygen flow.

Unsuitable

High-water-content materials like watermelon rinds provide very little food for the worms while disrupting the moisture level of the system. They should be added sparingly. Grass clippings and other products sprayed with pesticides should be avoided. Some banana peels are heavily sprayed, and can kill everything if added to a small bin. Although worms can digest proteins and fats in meat scraps, these materials can attract scavengers. Too much oil or fat can hinder the breathing of the worms, as they breathe through their skin. Worms cannot break down bone and are said to dislike highly spiced foods such as onions, garlic, and salt. If possible, sticky food labels, rubber bands, tea bag staples, and other inedibles should be removed before placing the food in the worm bin, as these items will not decompose. Fruit pits need not be removed from decaying fruit before adding, as the worms will eat all the soft matter. Compostable cutlery takes too long to degrade in vermicompost.

Bin Maintenance

Worms and other composting microorganisms require oxygen, so the bin must "breathe". This can be accomplished by regularly removing the composted material, adding holes to the bin, or using a continuous-flow bin. If insufficient oxygen is available, the decay becomes anaerobic, like that in swamps and bogs, producing a strong odor and creating a toxic environment for the worms. The moisture level and oxygen flow in a home worm bin should be checked at least once a week.

Over the long term, care should be taken to maintain optimum moisture levels. In a non-continuous-flow vermicomposting bin, excess liquid can be drained via a tap and used as plant food. A continuous flow bin does not retain excess liquid and, depending on the foods used, may require sprinklings of water to keep the bedding moist.

The pH should be slightly alkaline. Alkalinity can be increased by occasionally adding a handful of calcium carbonate, sold as "garden lime." Do not confuse calcium carbonate with regular lime (Calcium oxide), which is far too alkaline and will kill worms. Adding many citrus peels can hinder the worms, but probably due not to acidity but to d-limonene, a fragrant chemical present in the rind of citrus fruits. Coffee grounds have sometimes been blamed for acidity, but analysis shows they are only mildly acidic, with a pH of 6.2.

Feeding

There are two methods of adding matter to the bin:

❖ *Top feeding*—organic matter is placed directly on top of the existing layer of bedding in a bin and then covered with another layer of bedding. This is repeated every time the bin is fed.

- ❖ *Pocket feeding*—a top layer of bedding is maintained and food is buried beneath. The location of the food is changed each time, rotating around the bin to give the worms time to decompose the food in the previously fed pockets. The top layer of bedding is replenished as necessary.

Vermicomposters often use a combination of both methods. Sometimes unburied food can attract fruit flies, so food should be buried under at least one inch of bedding material.

Harvesting

Vermicompost is ready for harvest when it contains few to no scraps of uneaten food or bedding. Even a properly composted mixture will contain large items that should be discarded, such as peach or date pits, glassine-like sheets from melon skins, and twigs. Small seeds from composted food such as tomatoes and apples cannot be removed from the vermicompost and may sprout later in the seed-starting pots or garden.

There are several methods of harvesting, depending on the purpose for which the vermicompost will be used, and whether or not the composter wishes to salvage as many worms and worm eggs as possible from the vermicompost.

Vermicompost Properties

Vermicompost, also known as worm castings and vermicast, is richer in many nutrients than compost produced by other composting methods. It is also rich in microbial life which helps break down nutrients already present in the soil into plant-available forms. Unlike other compost, worm castings also contain worm mucus which keeps nutrients from washing away with the first watering and holds moisture better than plain soil. For this reason, some fruit and seed pits are reported to germinate in vermicompost easily. Vermicompost made from ordinary kitchen scraps will contain small seeds, especially those of tomatoes, peppers, and eggplants, that may sprout weeks later.

Vermicompost benefits soil by:

- ❖ improving its physical structure;
- ❖ enriching soil in micro-organisms, adding plant hormones such as auxins and gibberellic acid, and adding enzymes such as phosphatase and cellulase;

- ❖ attracting deep-burrowing earthworms already present in the soil;
- ❖ improving water holding capacity;
- ❖ enhancing germination, plant growth, and crop yield; and
- ❖ improving root growth and structure.

Vermicompost can be used to make compost tea (worm tea), by mixing some vermicompost in water and steeping for a number of hours or days. The resulting liquid is used as a fertilizer.

The dark brown waste liquid that drains into the bottom of some vermicomposting systems, as water-rich foods break down, is also excellent as fertilizer. However, the pH and nutrient contents of these liquids (as well as vermicompost) varies, depending on the food fed to the worms and whether or not lime has been added to the system. pH and nitrogen, phosphorus, and potassium (NPK) measurements should be taken periodically to determine the fertilizer composition before use. Home kits for testing are sold in hardware stores and nurseries.

Problems

Odor, usually due to overabundance of "greens" (wet waste) in the bin, results from too much nitrogen combining with hydrogen to form ammonia. To neutralize the odors, add a fair amount of shredded newspaper or other "browns" to the mix to absorb excess moisture, remove the smelly waste, and stop adding food to the bin until a substantial portion of the uneaten food has been turned into compost. The carbon will balance the nitrogen and form a compound that is not smelly. The higher level of carbon means that decomposition will be slower. Also, always add new material deep in the bin to disallow access to would-be pests. Consistently doing so will greatly reduce any nuisance from odor and undesirable organisms.

Pests such as rodents and flies may be attracted by certain materials and odors, especially lots of kitchen waste and especially meat. This problem is largely avoided if a sealed bin is used where the pests cannot access the material, although many proponents recommend having ample access to air. This promotes natural decomposition, as worms and beneficial bacteria require oxygen. Fruit flies are common in warm weather if the food is not thoroughly covered with bedding.

Commercially Available Vermicomposting Systems

Home vermicompost systems or worm bins are available from many retailers.

Clive Roberts BSc invented the very first 'Wormery' in the late 1980s and in 1991 patented his design for The Original Wormery, in the UK.

In the United States, state Cooperative Extension Services and local governments may provide assistance and training in home vermicomposting.

19.2 Case Study: Vermicompost and Flow Through Vermicompost Bins

Vermicompost is the finished product of processing organic waste with earthworms.

The key benefit to Vermicompost is the diverse and extremely high microbial population. Further benefits with Vermicompost are:

- ❖ Increased microbial activity in the soil,
- ❖ Improved water holding capability in the soil,
- ❖ Higher nutrient levels and availability of those nutrients to the plant,
- ❖ Increased decomposing of residuals,
- ❖ Increased Humus levels,
- ❖ Less compaction leading to better aerated soils, and
- ❖ Decrease in plant and soil susceptibility to pest and disease.

When Vermicompost is used for the starter to Compost Tea the result is a soil amendment that is teaming with beneficial bacteria and fungi. The Compost Tea can then be applied giving all the benefits of compost in a liquid form. When Vermicompost in the liquid form of Compost Tea is applied as a soil drench the microbial population is quickly available to the soil as well as the plant root. Compost Tea also allows for foliar application as well, providing protection for the plant surface from pest and disease.

Traditional methods of vermicomposting have been based on beds or windrows on the ground containing waste up to 18" deep, but such methods have numerous drawbacks. They require large areas of land for large-scale production and are relatively labor intensive, even when machinery is used for adding waste to beds. More importantly, such systems process wastes relatively slowly, taking anywhere from 6 to 18 months for processing to be complete. There is good evidence that a large proportion of the essential plant nutrients, that are in a relatively soluble form, are washed out and also a significant proportion can volatilize during this long processing period. Such nutrient losses are undesirable, particularly in relation to ground water pollution and result in a poor finished product. Our flow through bins require no compost turning and require no separating of worms and partially decomposed waste from the finished compost.

19.3 Growing Plants with Worm Poop: Vermicompost as an Amendment for Soilless Media

Today most floriculture crops are produced in soilless media that often contains an organic component. Certainly peat is one of the most widely employed organic materials contained in commercial soilless media. However, as peat supplies diminish because of over-harvesting, prices will inevitably rise. Therefore, identification of comparably priced substitutes for peat is imperative. One approach is to amend peat-based soilless media with other organic materials such as carbonized bark and compost. While these substances are not intended to substitute for peat, they do reduce the amount of peat used by 10 to 20 per cent, while at the same time alter the physical and chemical properties of the media to provide optimal growing conditions.

At the same time there is growing environmental concern about the disposal of organic wastes such as sewage sludge, yard and garden waste, and animal manure. Therefore, bio-solid wastes must either be contained or converted to a non-hazardous form that will not pollute the environment. This, of course, has led to the skyrocketing of the cost of disposal. Composting such wastes is one environmentally sound way to convert biological waste into a form that can be easily disposed and provide a mechanism to recoup some of the costs incurred.

Incorporation of compost into potting and container media is one potential use for this material. Indeed, the addition of compost has been shown to contribute positively to the chemical and physical properties of potting media. But there are downsides as well; foremost of which is the difficulty in producing a reproducible and stabilized product. The primary reason for this is because the composting process is dependent on the aerobic action of soil microorganisms that break down the complex organic molecules to simple substances that can be taken up and utilized by plants. This bio-oxidative process is highly dependent on variable environmental conditions such as temperature and the availability of oxygen that are hard to control.

Another "composting" process that has not received as much attention is vermicomposting that relies on the action of earthworms to break down complex organic compounds. The worm castings or fecal material produced by the worms is a rich source of a variety of essential plant nutrients. In contrast to conventional composts, vermicomposts are less variable and much more stable. In addition, the time to complete the composting process is much shorter for vermicomposts: one pound of worms can convert one pound of pig manure to compost in 48 hours!

Could vermicompost be an economical and useful amendment to soilless potting media? To answer this question, Gary Bachman, a graduate student of mine, and I have been collaborating with Drs. Clive Edwards and Scott Subler from the Entomology Department at The Ohio State University. We have been examining the effects of vermicompost, derived from pig manure, on the growth of bedding plants and assessing whether or not the use of vermicompost in commercial soilless media is economically viable.

In initial studies we found that incorporation of vermicompost in a commercial potting mix resulted in faster growth, increased height, and higher fresh and dry weights for a variety of bedding plant species. The optimal amount of vermicompost need to produce positive results was relatively low: between 10 and 20 per cent by volume. We found that the incorporation of vermicompost in the media essentially eliminated the need for additional fertilizer in the production tomato plugs. These results suggest that the routine incorporation of vermicompost in soilless media may provide the grower with significant savings in plug production by reducing the costs associated with fertilizing.

We also observed that the positive effects of the vermicompost on plant quality were sustained for a substantial period of time after seedling emergence. In other words, incorporation of vermicompost in media used to grow bedding plants may help to maintain crop quality and salability even after the plants reach the garden center. Additional evidence for this was obtained through a consumer preference study conducted prior to an OSU football game involving nearly 400 adults. An overwhelming majority (around 80%) of the participants said that they would choose to buy tomato plants grown in media amended with 10 per cent vermicompost over those grown using standard fertilization methods.

Although much of the effect of vermicompost on plant growth can be attributed to the addition of important nutrients, there are other unknown factors involved. Plants grown in vermicompost-amended media still grow better than those grown in unamended media with similar nutritional levels. Furthermore, the promotive effect of vermicompost on growth is lost if it is sterilized and can not be restored by adding additional nutrients. We are presently trying to determine the nature of the factor(s) contained in vermicompost that promote plant growth.

Regardless, the use of vermicompost in commercial soilless media holds great promise if the economics hold out. The widespread use of vermicompost as an amendment will ultimately depend on its cost of production. However, with the increased pressure to find alternate ways to safely dispose of animal manure, the techniques for vermicomposting on a large scale are now being developed that can economically process enormous amounts of animal waste.

19.4 Vermiposting

Purpose

To help earthworms transform foodscraps or fish sludge into a rich fertilizer for plants.

Background

Imagine eating your own weight in food everyday. That's what the earthworm does. As soil full of partially decomposed organic matter passes through the earthworm's gut, it is metabolized into rich castings (manure) that nourish plants. When earthworms burrow through soil, they make tunnels that bring air, water and their rich nutrients to plant roots. With earthworms, you can compost anywhere, on your roof, porch or classroom.

The earthworm is nature's plough; producing the finest fertilizer for plants. In every way, the earthworm surpasses any machine that people have yet invented to plough, to cultivate or to fertilize the soil. Earthworm manure, called *vermicompost* or *castings*, is a rich living fertilizer for plants and especially good mixed in potting soil for seedlings. The diversity of microbes in earthworm manure protects against soil disease. The humus in vermicompost holds in and increases the availability of plant nutrients, helps control plant diseases and stimulates healthy plant growth. Vermicompost helps the soil drink in and hold water.

Earthworms can double their population in two to three months. Earthworms are hermaphrodites—each worm is both male and female and can produce fertilized eggs with any other earthworm. Inside the swollen white band on the worm a tiny cocoon is produced. Following fertilization, in about three weeks the cocoon hatches from one to five tiny white baby worms. A mature worm can produce two to three cocoons per week.

Inquiry

Can we find earthworms under leaves in a forest or a moist shady leaf corner of our schoolyard? Under a pile of old grass clippings or mulch? Is there a heap of rotted manure in a nearby farmyard? What are the conditions that earthworms need to thrive?

Observation: When you find earthworms, handle them gently with moist open hands. Place an earthworm on a tray with moist soil. Can you identify the anatomical parts of the earthworm?

Materials

Red earthworms (Eisenia fetida), organic bedding material such as decomposing leaves, old hay, or moist shredded coconut hulls, a home in a wooden or plastic bin with many air and drainage holes covered with nylon screen secured with duct tape.

Activity—Earthworm Terrarium—Purpose—To learn the natural conditions in which earthworms thrive.Gather forest soil teeming with life, bugs and earthworms. Place a layer of leaves and twigs on the bottom of a clear container or aquarium, then add the forest soil. Carefully transplant the delicate forest ferns, moss and a rotting log. Cover with a screen to let air circulate while keeping small creatures inside. Mist the terrarium to keep the soil moist but never soggy. Occasionally add small bits of rotting leaves, grass or vegetable foodscraps for the earthworms.

How to Vermicompost?

1. *Earthworm habitat*: Enlightened zookeepers know that for creatures to thrive in captivity, they need to recreate the wild habitat as closely as possible. Find or build a wooden box and drill holes in the bottom for drainage and on the sides for aeration. Cover bottom holes with nylon screen and tape down the edges with electric or duct tape. Place the box on rocks over plastic for a good air flow. Fill with moist natural bedding (soft rotted trees turning to forest soil, shredded leaves, old plants, old well-rotted farm manure without any smell). Keep bedding light and fluffy. Mix in water as necessary to keep lightly moist, but never soggy. Cover with moss and leaves.

2. *Add red wiggler earthworms*: Bury small amounts of *pre-composted* organic foodwaste—vegetable and fruit scraps, tea and coffee grounds that have decomposed in a compost pile for about one month. Soak old bread so it is moist then bury.

3. *Care*: Don't over feed. Water lightly as needed. Bacteria and other organisms decompose the organic matter. The earthworms eat the semi-decomposed food waste, bedding and bacteria and transform it into humus, a nutrient-rich food for growing healthy plants. An earthworm farm of half a pound of worms needs to be supplied with one-half a pound of organic materials every two days.

Tips

❖ The best quality vermicompost is produced by feeding red earthworms partially *pre-composted* foodscraps, ready after the thermophilic heating up stage.

❖ It's good to add *compost activator-innoculant* to speed up the process.

❖ Keep the Balance—Avoid feeding only acidic scraps such as citrus peels. Worms prefer pH-balanced foods, especially breads, cooked grains and vegetables.

❖ Smell? A bad smell indicates the ecosystem is out of balance. If there is uneaten food going rotten—stop feeding for a while. If the bin is too wet, drain by gently tipping. Dig a trench in the middle and sop up excess water with paper towels or a turkey baster, and increase aeration by adding more holes.

❖ Fruitflies? Avoid fruitflies by burying all fruit peels deep under the bedding. If fruitflies are a problem, vacuum up the flies when you open the bin, bury fruitscraps deep under and trap any remaining flies by putting sweet soft drink in a jar or cup. Tape the corner of a plastic bag over the rim with a hole in the corner that protrudes downward. Avoid adding fruit scraps if the flies continue to be a problem.

Harvesting Vermicompost

When you want to harvest the vermicompost for use, either:

❖ Spread out the vermicompost on a tarp and shine a bright light on the surface. The earthworms will escape away from the light and dryness. Skim off the upper layer of castings. Repeat till the worms are concentrated in the bottom. Mix vermicompost in with potting soil for seedlings or straight into the garden. Or

❖ Place a nylon net on the vermicompost surface that is covered with moist, freshly decomposed foodscraps and old hay, etc. The hungry earthworms will migrate up to this new layer of food. Repeat till the worms are out.

Outdoor Earthworm Composting

Dig a trench of any size in a location where you plan a garden next season. Bury your vegetable foodscraps. Add earthworms if you have them, or wait and they will come. Cover with leaves and soil. Keep watered. Next season, plant right in the trench-bed and continue or start a new trench next to the old one.

Green Technology: Phytoremediation

20.1 Phytoremediation

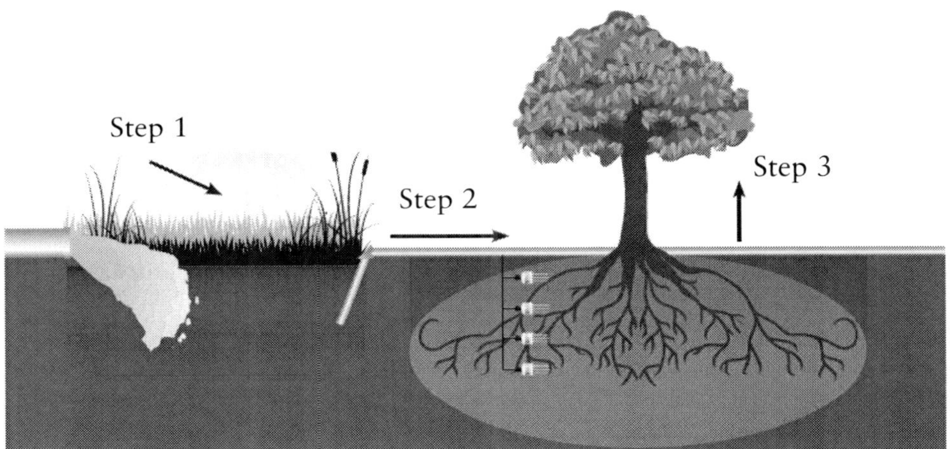

An illustrated guide to Phytoremediation:

Step 1: An artificial wetland is created and contaminated water is pumped in. Anaerobic bacterial break down the TCE and PCE. 1,4-diorama is not broken down by these bacterial organisms. Further steps are needed to break this chemical down.

Step 2: Water is pumped from wetland and irrigated into a grove of deciduous trees. The trees absorb the water along with the 1,4-dioxane.

Step 3: The tree evaporates the 1,4-dioxane contaminated water into the atmosphere, where sunlight rapidly breaks it down into harmless methane and carbon dioxide.

Phytoremediation is the use of certain plants to clean up soil, sediment, and water contaminated with metals and/or organic contaminants such as crude oil, solvents, and polyaromatic hydrocarbons (PAHs). It is a name for the expansion of an old process that occurs naturally in ecosystems as both inorganic and organic constituents cycle through plants. Plant physiology, agronomy, microbiology, hydrogeology, and engineering are combined to select the proper plant and conditions for a specific site. Phytoremediation is an aesthetically pleasing mechanism that can reduce remedial costs, restore habitat, and clean up contamination in place rather than entombing it in place or transporting the problem to another site.

Phytoremediation can be used to clean up contamination in several ways:

❖ *Phytovolatilization*: Plants take up water and organic contaminants through the roots, transport them to the leaves, and release the contaminants as a reduced of detoxified vapor into the atmosphere.

❖ *Microorganism stimulation*: Plants excrete and provide enzymes and organic substances from their roots that stimulate growth of microorganisms such as fungi and bacteria. The microorganisms in the root zone then metabolize the organic contaminants.

Table 201: Media, contaminants and typical plants for the types of phytoremediation

Application	Media	Contaminants	Typical Plants
1.Phytovolatization	Soil, groundwater, Landfill leachate, land application of wastewater	Herbicides (atrazine, alachlor); Aromatics (BTEX); Chlorinated aliphatics(TCE); Nutrients; Ammunition wastes (TNTDX)	Phreatophyte trees (poplarillow, cottonwoodspen); Grasses (rye, Bermuda, sorghum, fescue); Legumes (clover, alfalfa, cowpeas)
2.Microorganism stimulation	Soil, sediments, Land application of waste water	Organic contaminants (pesticides aromatic, and polynuclear aromatic hydrocarbons	Phenolics releasers (mulberry, applesage orange); Grasses with fribous roots (ryeescueermuda); Aquatic plants for sediments
3.Phytostabilization	Soil, sediments	Metals (Pbdnsure), Hydrophobic Organics (PAHCBDTieldrin)	Phreatophyte trees to transpire large amounts of water (hydraulic control); Grasses to stabalize soil erosion; Dense root systems are needed to sorb/bind contaminants
4.Phytoaccumulation/ extraction	Soil, Brownfields, sediments	Metals (Pbdnsure) with EDTA addition for Pb Selenium	Sunflowers; Indian Mustard; Rape seed plants; Barle, Hops; Crucifers; Serpentine plants; Nettles, dandelions
5.Degradation	Soil, groundwater, Landfill leachate, land application of wastewater	Herbicides (atrazine, alachlor); Aromatics (BTEX); Chlorinated aliphatics (TCE); Nutrients; Ammunition wastes (TNTDX)	Phreatophyte trees (poplarillow, cottonwoodspen); Grasses (rye, Bermuda, sorghum, fescue); Legumes (clover, alfalfa, cowpeas)

- ❖ *Phytostabilization*: Plants prevent contaminants from migrating by reducing runoff, surface erosion, and ground-water flow rates. "Hydraulic pumping" can occur when tree roots reach ground water, take up large amounts of water, control the hydraulic gradient, and prevent lateral migration of contaminants within a ground water zone.
- ❖ *Phytoaccumulation/extraction*: Plant roots can remove metals from contaminated sites and transport them to leaves and stems for harvesting and disposal or metal recovery through smelting processes.
- ❖ *Phytodegradation by plants*: Organic contaminants are absorbed inside the plant and metabolized (broken down) to non-toxic molecules by natural chemical processes within the plant.

The Table 20.1 gives the media, contaminants and typical plants for the types of phytoremediation listed above.

Advantages and Disadvantages of Phytoremediation

When using phytoremediation there are many positive and negative aspects to consider. The advantages and disadvantages are listed below:

Advantages	Disadvantages
Works on a variety on organic and inorganic compounds	May take several years to remediate
Can be either *In Situ/ Ex Situ*	May depend on climatic conditions
Easy to implement and maintain	Restricted to sites with shallow contamination within rooting zone
Low-cost compared to other treatment methods	Harvested biomass from phytoextraction may be classified as a RCRA hazardous waste
Environmentally Friendly and aesthetically pleasing to the public	Consumption of contaminated plant tissue is also a concern.
Reduces the amount wastes to be landfilled	Possible effect on the food chain

A major advantage that is listed above is the low cost. For example, the cost of cleaning up one acre of sandy loam soil at a depth of 50cm with plants is estimated at \$60,000-\$100,000 compared to \$400,000 for the conventional excavation and disposal method. One reason for this low cost is phytoremediation may not require expensive equipment or highly specialized personnel, and can be relatively easy to implement.

One major concern with phytoremediation is the possible affects on the food chain. For example vegetation is used that absorbs toxic or heavy metals and moles or voles eat the metal contaminated plants. The predators of the moles or voles then become victims of intoxication. All though the possibilities of such scenarios are being looked at, more fieldwork and analysis is necessary to understand the possible effects phytoremediation can have.

Regulatory Issues

As of now phytoremediation is too new to be approved by regulatory agencies such as the EPA. Eventually the main question that regulators will focus on is will phytoremediation remediate the site to the standards and reduce the risk to human health and the environment. In developing regulations for phytoremediation the following questions will need answering.

- ❖ Can it cleanup the site below action levels? On what scale?
- ❖ Does it create any toxic intermediate or products?
- ❖ Is it cost effective as alternative methods?
- ❖ Does the public accept the technology?

The word's etymology comes from the Greek φυτο (phyto) = plant, and Latin « remedium » = restoring balance, or remediating. Phytoremediation consists in depolluting contaminated soils, water or air with plants able to contain, degrade or eliminate metals, pesticides, solvents, explosives, crude oil and its derivatives, and various other contaminants, from the mediums that contain them.

It is clean, efficient, inexpensive and non-environmentally disruptive, as opposed to processes that require excavation of soil. The definitive textbook on phytoremediation was published in 2003 with contributed, peer reviewed articles from all major research groups involved in phytoremediation research (Phytoremediation: Transformation and Control of Contaminants, edited by Steven C. McCutcheon and Jerald L. Schnoor).

Various Phytoremediation Processes

A range of processes mediated by plants are useful in treating environmental problems:

- ❖ *Phytoextraction*: uptake and concentration of substances from the environment into the plant biomass.
- ❖ *Phytostabilization*: reducing the mobility of substances in the environment, for example by limiting the leaching of substances from the soil.
- ❖ *Phytotransformation*: chemical modification of environmental substances as a direct result of plant metabolism, often resulting in

their inactivation, degradation (phytodegradation) or immobilization (phytostabilization).

❖ *Phytostimulation*: enhancement of soil microbial activity for the degradation of contaminants, typically by organisms that associate with roots. This process is also known as *rhizosphere degradation.*

❖ *Phytovolatilization*: removal of substances from soil or water with release into the air, sometimes as a result of phytotransformation to more volatile and/or less polluting substances.

❖ *Rhizofiltration*: filtering water through a mass of roots to remove toxic substances or excess nutrients. The pollutants remain absorbed in or adsorbed to the roots.

Phytoextraction

Phytoextraction (or *phytoaccumulation*) uses plants to remove contaminants from soils, sediments or water into harvestable plant biomass. Phytoextraction has been growing rapidly in popularity world-wide for the last twenty years or so. Generally this process has been tried more often for extracting heavy metals than for organics. At the time of disposal contaminants are typically concentrated in the much smaller volume of the plant matter than in the initially contaminated soil or sediment. 'Mining with plants', or *phytomining*, is also being experimented with.

The plants absorb contaminants through the root system and store them in the root biomass and/or transport them up into the stems and/or leaves. A living plant may continue to absorb contaminants until it is harvested. After harvest a lower level of the contaminant will remain in the soil, so the growth/ harvest cycle must usually be repeated through several crops to achieve a significant cleanup. After the process, the cleaned soil can support other vegetation.

Two versions of phytoextraction:

❖ *Natural hyper-accumulation*, where plants naturally take up the contaminants in soil unassisted, and

❖ *Induced or assisted hyper-accumulation*, in which a conditioning fluid containing a chelator or another agent is added to soil to increase metal solubility or mobilization so that the plants can absorb them more easily.

Examples of Phytoextraction from Soils

❖ Arsenic, using the Sunflower (*Helianthus annuus*), or the Chinese

      Brake fern ["Pteris spp"], a hyperaccumulator. Chinese Brake fern stores arsenic in its leaves.

❖ Cadmium and zinc, using alpine pennycress (*Thlaspi caerulescens*), a hyperaccumulator of these metals at levels that would be toxic to many plants. On the other hand, the presence of copper seems to impair its growth (see table for reference).

❖ Lead, using Indian Mustard (*Brassica juncea*), Ragweed (*Ambrosia artemisiifolia*), Hemp Dogbane (*Apocynum cannabinum*), or Poplar trees, which sequester lead in its biomass.

❖ Salt-tolerant (moderately halophytic) barley and/or sugar beets are commonly used for the extraction of Sodium chloride (common salt) to reclaim fields that were previously flooded by sea water.

❖ Uranium, using sunflowers, as used after the Chernobyl accident.

❖ Mercury, selenium and organic pollutants such as polychlorinated biphenyls (PCBs) have been removed from soils by transgenic plants containing genes for bacterial enzymes .

Phytostabilization

Phytostabilization focuses on long-term stabilization and containment of the pollutant. For example, the plant's presence can reduce wind erosion, or the plant's roots can prevent water erosion, immobilize the pollutants by adsorption or accumulation, and provide a zone around the roots where the pollutant can precipitate and stabilize. Unlike phytoextraction, phytostabilization mainly focuses on sequestering pollutants in soil near the roots but not in plant tissues. Pollutants become less bioavailable and livestock, wildlife, and human exposure is reduced. An example application of this sort is using a vegetative cap to stabilize and contain mine tailings.

Phytotransformation

In the case of organic pollutants, such as pesticides, explosives, solvents, industrial chemicals, and other xenobiotic substances, certain plants, such as Cannas, render these substances non-toxic by their metabolism. In other cases, microorganisms living in association with plant roots may metabolize these substances in soil or water. These complex and recalcitrant compounds cannot be broken down to basic molecules (water, carbondioxide etc) by plant molecules, and hence the term phytotransformation represents a change in chemical structure without complete breakdown of the compound. The term "Green Liver Model" is used to describe phytotransformation, as plants behave similar to the human liver when dealing with these xenobiotic compounds (foreign compound/pollutant). After uptake of the xenobiotics, plant enzymes increase the polarity of the xenobiotics by adding functional groups such as hydroxyl groups (-OH). This is known as Phase I metabolism, similar to the way the human liver increases the polarity of drugs and foreign compounds (Drug Metabolism. While in the human liver, enzymes like Cytochrome P450s are responsible for the initial reactions, in plants enzymes such as nitroreductases carry out the same role. In the second stage of phytotransformation, known as Phase II metabolism, plant biomolecules such as glucose and amino acids are added to the polarized xenobiotic to further increase the polarity (known as conjugation). This is again similar to the processes occurring in the human liver wherein glucuronidation (addition of glucose molecules by the UGT (e.g. UGT1A1) class of enzymes) and glutathione addition reactions occur on reactive centers of the xenobiotic. Phase I and II reactions serve to increase the polarity and reduce the toxicity of the compounds, although many exceptions to the rule are seen at least in the case of the human liver. The increased polarity also allows for easy transport of the xenobiotic along aqueous channels. In the final stage of phytotransformation (Phase III metabolism), a sequestration of the xenobiotic occurs within the plant. The xenobiotics polymerize in a lignin-like manner and get a complex structure which is sequestered in the plant. This ensures that the xenobiotic is safely stored in the plant, and does not affect the functioning of the plant. However, preliminary studies have shown that these plants can be toxic to small animals (such as snails) and hence plants involved in phytotransformation may need to be maintained in a closed enclosure. The human liver differs from plants in Phase III metabolism, since the liver can transport the xenobiotics into the bile for eventual excretion. Since plants have no excretory mechanisms, they sequester the modified xenobiotics. Hence, the plants reduce toxicity (with exceptions) and sequester the xenobiotics in phytotransformation. Trinitrotoluene phytotransformation has been extensively researched a transformation pathway has been proposed .

The Role of Genetics

Breeding programmes and genetic engineering are powerful methods for enhancing natural phytoremediation capabilities, or for introducing new capabilities into plants. Genes for phytoremediation may originate from a micro-organism or may be transferred from one plant to another variety better adapted to the environmental conditions at the cleanup site. For example, genes encoding a nitroreductase from a bacterium were inserted into tobacco and showed faster removal of TNT and enhanced resistance to the toxic effects of TNT .

Advantages and Limitations

Advantages:

- ❖ the cost of the phytoremediation is lower than that of traditional processes both *in situ* and *ex situ*.
- ❖ the plants can be easily monitored.
- ❖ the possibility of the recovery and re-use of valuable metals (by companies specializing in "phytomining").
- ❖ it is the least harmful method because it uses naturally occurring organisms and preserves the natural state of the environment.

Limitations:

- ❖ phytoremediation is limited to the surface area and depth occupied by the roots.
- ❖ slow growth and low biomass require a long-term commitment.
- ❖ with plant-based systems of remediation, it is not possible to completely to prevent the leaching of contaminants into the groundwater (without the complete removal of the contaminated ground which in itself does not resolve the problem of contamination).
- ❖ the survival of the plants is affected by the toxicity of the contaminated land and the general condition of the soil.
- ❖ possible bio-accumulation of contaminants which then pass into the food chain, from primary level consumers upwards.

Hyperaccumulators and Biotic Interactions

A plant is said to be a hyperaccumulator if it can concentrate the pollutants in a minimum percentage which varies according to the pollutant involved (for example: more than 1000 mg/kg of dry weight for nickel, copper, cobalt, chromium or lead; or more than 10,000 mg/kg for zinc or manganese. Most of the 215 metal-hyperaccumulating species included in their review hyperaccumulate nickel. They listed 145 hyperaccumulators of nickel (around 300 Ni accumulators are known; see Hyperaccumulators table-2: Nickel and its notes), 26 of cobalt, 24 of copper, 14 of zinc, four of Lead, and two of Chromium. This capacity for accumulation is due to *hypertolerance*, or *phytotolerance*: the result of adaptative evolution from the plants to hostile environments along multiple generations. Boyd and Martens list 4 biotic interactions that may be affected by metal hyperaccumulation, to which can be added the biofilm as a particular aspect of micorrhizae:

1. Protection
2. Interferences with neighbour plants of different species
3. Mutualism (Mycorrhizal associations or micorrhizae, and Pollen and seed dispersal)
4. Commensalism
5. The biofilm

Protection

More and more evidence show that the metals in hyperaccumulating plants give them some protection from various bacteria, fungi and/or insects. For instance, with foliar Ni concentrations as low as 93 mg/kg, the larval weight of *Spodoptera exigua* (*Lepidoptera: Noctuidae*) (beet army worm) is reduced and time to pupation extended (Boyd & Moar, subm.).

Information published supporting the defence hypothesis of metal hyperaccumulation.

Paper	Plant species	Metal	Organism(s) affected
Ernst 1987	*Silene vulgaris* (Moench) Garke	Cu (400 mg g⁻¹)	*Hadena cucubalis* Schiff. (*Lepidoptera: Noctuidae*)
Boyd *et al.* 1994	*Streptanthus polygaloides* Gray	Ni	*Xanthomonas campestris* (Gram-negative bacterium)
Boyd *et al.* 1994	*Streptanthus polygaloides* Gray	Ni	*Alternaria brassicicola* (Imperfect fungus)
Boyd *et al.* 1994	*Streptanthus polygaloides* Gray	Ni	*Erisyphe polygoni* (Powdery mildew)
Martens & Boyd 1994	*Streptanthus polygaloides*	Ni	*Pieris rapae* L. (*Lepidoptera: Pieridae*)
Boyd & Martens 1994	*Thlaspi montanum* L. var. *montanum*	Ni	*Pieris rapae*
Pollard & Baker 1997	*Thlaspi caerulescens* J. and C. Presl.	Zn	*Schistocerca gregaria* (Forsk.) (*Orthoptera: Acrididae*)
Pollard & Baker 1997	*Thlaspi caerulescens* J. and C. Presl.	Zn	*Deroceras carvanae* (Pollonera) (*Pulmonata: Limacidae*)
Pollard & Baker 1997	*Thlaspi caerulescens* J. and C. Presl.	Zn	*Pieris brassicae* L. (*Lepidoptera: Pieridae*)

The defense against viruses is not always supported. Davis *et al.* (2001) have compared two close species *S. polygaloides Gray* (Ni hyperaccumulator) and *S. insignis Jepson* (non-accumulator), inoculating them with Turnip mosaic virus. They showed that the presence of nickel weakens the plant's response to the virus.

Circumvention of plants' elemental defences by their predators may occur in three ways : (1) selective feeding on low-metal tissues, (2) use of a varied diet to dilute metal-containing food (likely more efficient in large-sized herbivores), and (3) tolerance of high dietary metal content.

1. *Avoidance of an elemental defence via selective feeding*: Mishra & Kar (1974) reported nickel to be transported through the xylem of crop plants. Similarly, Kramer *et al.* (1996) showed that Ni is transported as a complex with the amino-acid histidine in the xylem. This implies that phloem fluid may contain little nickel; thus phloem fluid may be used by able organisms as a rich source of carbohydrates. Pea aphids (*Acyrthosiphon pisum* [Harris]; *Homoptera*: *Aphididae*) feeding on *Streptanthus polygaloides* Gray (*Brassicaceae*) have equal survival and reproduction rates for plants containing ca. 5000 mg/kg nickel amended with $NiCl_2$, and those containing 40 mg/kg nickel. This means that either the phloem fluid is poor in nickel even for nickel hyperaccumulators, or that the aphids tolerate nickel. Moreover the aphids feeding on high nickel-content plants only show a small increase of nickel content in their bodies, relatively to the nickel content of aphids feeding on low-nickel plants . On the other hand, aphids (*Brachycaudus lychnidis* L.) fed on the zinc-tolerant plant *Silene vulgaris* (Moench) Garcke (*Caryophyllaceae*) —which can contain up to 1400 mg/kg zinc in its leaves—were reported showing elevated (9000 mg/kg) zinc in their bodies.

3. *Metal tolerance*: Hopkin (1989) and Klerks (1990) demonstrated it for animal species; Brown & Hall (1990) for fungal species; and Schlegel & al. (1992) and Stoppel & Schlegel (1995) for bacterial species. Plants of *Streptanthus polygaloides* (*Brassicaceae*, Ni hyperaccumulator) can be parasited by *Cuscuta californica* var. *breviflora* Engelm. (*Cuscutaceae*). Metal contents of *Cuscuta* ranged from 540-1220 mg/kg Ni, 73-fold higher than the metal contents of *Cuscuta* parasitizing a co-occurring non-hyperaccumulator plant species. *Cuscuta* plants are therefore very Ni-tolerant -10 mg Ni/kg is sufficient for growth to start decreasing in unadapted plants . According to Boyd & Martens (subm.) this is "the first well-documented instance of the transfer of elemental defences from a hyperaccumulating host to a seed plant parasite".

Interferences with Neighbour Plants of Different Species

Its likelihood between hyperaccumulators and neighbouring plants was suggested but no mechanism was proposed. Gabrielli *et al.* (1991), and Wilson & Agnew (1992), suggested a decrease in competition experienced by the hyperaccumulators for the litterfall from hyperaccumulators' canopy.

This mechanism mimics allelopathy in its effects, although technically due to redistribution of an element in the soil rather than to the plant manufacturing an organic compound. Boyd et Martens call it "elemental allelopathy"—without the autoxicity problem met in other types of allelopathy (Newman 1978).

Mutualism

Two types of mutualism are considered here, mycorrhizal associations or *mycorrhizae*, and animal-mediated pollen or seed dispersal.

1. *Mycorrhizal associations or mycorrhizae*: There are two types of mycorrhizal fungi: ectomycorrhizae and endomycorrhizae. Ectomycorrhizae form sheaths around plant roots, endomycorrhizae enter cortex cells in the roots.

   Mycorrhizae are the symbiotic relationship between a soil-borne fungus and the roots of a plant. Some hyperaccumulators may form mycorrhizae and, in some cases, the latter may have a role in metal treatment. . In soils with low metal levels, vesicular arbuscular mycorrhizae enhance metal uptake of non-hyperaccumulating species . On the other hand, some mycorrhizae increase metal tolerance by decreasing metal uptake in some low-accumulating species. Mycorrhizae thus assists *Calluna* in avoiding Cu and Zn toxicity . Most roots need about 100 times the amount of carbon than do the hyphae of its associated ectomycorrhizae in order to develop across the same amount of soil . It is therefore easier for hyphae to acquire elements that have a low mobility than it is for plant roots. Caesium-137 and strontium-90 both have low mobilities.

   Mycorrhizal fungi depend on host plants for carbon, while enabling host plants to absorb the soil's nutrients and water with more efficiency. In mycorrhizae, nutrient uptake is enhanced for the plants while they provide energy-rich organic compounds to the fungus. Although certain plant species that are normally symbiotic with mycorrhizal fungi can exist without the fungal association, the fungus greatly enhances the plant's growth. Hosting mycorrhizae is much more energy effective to the plant than producing plant roots. The *Brassicaceae* family reportedly forms few mycorrhizal associations. But Hopkins (1987) notes mycorrhizae associated with *Streptanthus glandulosus* Hook. (*Brassicaceae*), a non-accumulator. Some fungi tolerate easily the generally elevated metal contents of serpentine soils. Some of these fungal species are mycorrhizal . High levels of phosphate in the soil inhibit mycorrhizal growth.

The uptake of radionuclides by fungi depends on its nutritional mechanism (mycorrhizal or saprotrophic) . *Pleurotus eryngii* absorbs Cs ❖best over Sr and Co, while *Hebeloma cylindrosporum* favours Co. But increasing the amount of K increases the uptake of Sr (chemical analogue to Ca) but not that of Cs (chemical analogue to K). Moreover, the uptake of Cs decreases with *Pleurotus eryngii* (mycorrhizal) and *Hebeloma cylindrosporum* (saprotrophic) if the Cs content is increased, but that of Sr increases if its content is increased—this would indicate that the uptake is independent from the nutritional mechanism.

2. *Pollen and seed dispersal*: Some animals obtain food from the plant (nectar, pollen, or fruit pulp—Howe & Westley 1988). Animals feeding from hyperaccumulors high in metal content must either be metal-tolerant or dilute it with a mixed diet. Alternatively hyperaccumulators may rely on abiotic vectors or non-mutualistic animal vectors for pollen or seed transport, but we lack information on seed and pollen dispersal mechanisms for hyperaccumulating plants.

Jaffré & Schmid 1974; Jaffré *et al.* 1976; Reeves *et al.* 1981; have studied metal contents of entire flowers and/or fruits. They have recorded elevated metal levels in these. We find an exception with *Walsura monophylla* Elm. (*Meliaceae*), originating from the Philippines and showing 7000 mg/kg Ni in leaves but only 54 mg/kg in fruits . Some plants may thus have a mechanism by which metal or other contaminants is excluded from their reproductive structures.

Commensalism

This is an interaction benefiting one organism while being of neutral value to another. The most likely one with hyperaccumulators would be epiphytism. But this is most noticeable in humid habitats, whereas only a few detailed field studies of hyperaccumulators have been conducted in such habitats, and those studies (mostly to do with humid tropical forests on serpentine soils) pay little or no attention to that point (e.g., Proctor *et al.* 1989; Baker *et al.* 1992). Proctor *et al.* (1988) studied the tree *Shorea tenuiramulosa*, which can accumulate up to 1000 mg Ni/kg dry weight in leaf material. They estimated covers of epiphytes on the boles of trees in Malaysia, but did not report values for individual species. Boyd *et al.* (1999) studied the occurrence of epiphytes on leaves of the Ni hyperaccumulating tropical shrub *Psychotria douarrei* (Beauvis.). Epiphyte load increased significantly with increasing leaf age, up to 62 per cent for the oldest leaves. An epiphyte sample of leafy liverworts removed from *P. douarrei*, was found to contain 400 mg Ni /kg dry weight (far less than the host plant, whose oldest and most heavily epiphytized leaves contained a mean value of 32,000 mg Ni/kg dry weight). High doses of Ni therefore do not prevent colonization of *Psychotria douarrei* by epiphytes.

Chemicals that mediate host-epiphyte interactions are most likely to be located in the outermost tissues of the host (Gustafsson & Eriksson 1995). Also, most of the metal accumulates in epidermal or subepidermal cell walls or vacuoles (Ernst & Weinert 1972; Vazquez *et al.* 1994; Mesjasz-Rzybylowicz *et al.* 1996; Gabrielli *et al.* 1997). These findings suggest that epiphytes would experience higher metal levels when growing on hyperaccumulator leaves. But Severne (1974) measured the release of metal via leaching of leaves from the Ni hyperaccumulator *Hybanthus floribundus* (Lindl.) F. Muell. (*Violaceae*) from western Australia; he concluded that its leaves do not easily leach Ni.

In theory another commensal interaction could exist, if the high metal content of the soil under hyperaccumulator plants was needed for another plant species to establish itself. No evidence is known showing such effect.

The Biofilm

This section needs be developed. See relevant articles on biofilm and *Pseudomonas aeruginosa*. A biofilm is a layer of organic matter and microorganism formed by the attachment and proliferation of bacteria on the surface of the object. Biofilms are characterised by the presence of bacterial extracellular polymers glycocalyx that create a thin visible slimy layer on solid surface.

Table of Hyperaccumulators

A comprehensive literature survey of hyperaccumulating plants and their uses was started by Stevie Famulari for her students at the University of New Mexico. It is now considerably increased in size and has had to be split into 3 sections:

❖ Hyperaccumulators table 1: Al, Ag, As, Be, Cr, Cu, Mn, Hg, Mo, Naphtalene, Pb, Pd, Pt, Se, Zn.

❖ Hyperaccumulators table 2: Nickel.

❖ Hyperaccumulators table 3: Radionuclides (Cd, Cs, Co, Pu, Ra, Sr, U), Hydrocarbures, Organic Solvents.

20.2 Overview

Phytoremediation is a process that uses plants to remove, transfer, stabilize, or destroy contaminants in soil, sediment, and groundwater. The mechanisms of phytoremediation include enhanced rhizosphere biodegradation (takes place in soil or groundwater immediately surrounding plant roots), phytoextraction (also known as phytoaccumulation, the uptake of contaminants by plant roots and the translocation/accumulation of contaminants into plant shoots and leaves), phytodegradation (metabolism of contaminants within plant tissues), and phytostabilization (production of chemical compounds by plants to immobilize contaminants at the interface of roots and soil). Phytoremediation applies to all biological, chemical, and physical processes that are influenced by plants (including the rhizosphere) and that aid in cleanup of the contaminated substances. Plants can be used in site remediation, both through the mineralization of toxic organic compounds and through the accumulation and concentration of heavy metals and other inorganic compounds from soil into aboveground shoots. Phytoremediation may be applied in situ or ex situ, to soils, sludges, sediments, other solids, or groundwater.

20.3 Phytoremediation—Using Plants to Clean up Polluted Soil

Polluted soil poses a severe problem for both ecosystem health and land development. Because soil lies at the confluence of many natural systems, soil pollution can be spread to other parts of the natural environment. Groundwater, for instance, percolates through the soil and can carry the soil pollutants into streams, rivers, wells and drinking water. Erosion can create the same problem. Plants growing on polluted soil may contain harmful levels of pollutants themselves, and this can be passed on to the animals and people that eat them. Dust blown from polluted soil can be inhaled directly by passersby. Additionally, in an urban setting such as Fairfax County, polluted soil makes valuable open land unusable for parks, recreation or commercial development.

Despite the benefits of cleaning polluted soil, remediation often never takes place because of the cost and effort of the work. Both soil minerals and soil pollutants carry small electric charges that can cause each to bond with each other, thus making polluted soil very hard to clean. Additionally, soil is a dense medium. This causes excavation of polluted soil for off site treatment or disposal to be very expensive because of the time, labor and heavy machinery necessary to do the job. Therefore, cheaper on-site, or in-situ, remediation techniques have been the focus of much attention and research lately. One of the most interesting and promising of these in-situ techniques is phytoremediation.

Phytoremediation is the use of specialized plants to clean up polluted soil. While most plants exposed to high levels of soil toxins will be injured or die, scientists have discovered that certain plants are resistant, and an even smaller group actually thrive. Both groups of plants are of interest to researchers, but the thriving plants show a particular potential for remediation because it has been shown that some of them actually transport and accumulate extremely high levels of soil pollutants within their bodies. They are therefore aptly named hyper-accumulators.

Hyper-accumulators already are being used throughout the country to help clean up heavy metal polluted soil. Heavy metals are some of the most stubborn soil pollutants. They can bond very tightly to soil particles, and they cannot be broken down by microbial processes. Most heavy metals are also essential plant nutrients, so plants have the ability to take up the metals and transport them throughout their bodies. However, on polluted soil, the levels of heavy metals are often hundreds of times greater than normal, and this overexposure is toxic to the vast majority of plants. Hyper-accumulators, on the other hand, actually prefer these high concentrations. Essentially, hyper-accumulators are acting as natural vacuum cleaners, sucking pollutants out of the soil and depositing them in their above ground leaves and shoots. Removing the metals is as simple as pruning or cutting the hyper-accumulators above ground mass, not excavating tons of soil.

Resistant, but not hyper-accumulating, plants also have a role in phytoremediation. Organic toxins, those that contain carbon such as the hydrocarbons found in gasoline and other fuels, can be broken down by microbial processes. Plants play a key role in determining the size and health of soil microbial populations. All plant roots secrete organic materials that can be used as food for microbes, and this creates a healthier, larger, more diverse and active microbial population, which in turn causes a faster breakdown of pollutants. Resistant plants can thrive on sites that are often too toxic for other plants to grow. They in turn give the microbial processes the boost they need to remove organic pollution more quickly from the soil.

Both forms of phytoremediation have the added benefit of not disturbing the soil. While excavation is an effective way to get rid of pollution, it removes the organic matter rich topsoil and, because of the use of heavy machinery, compacts the soil that is left behind. Phytoremediation does not degrade the physical or chemical health of the soil. Actually, it creates a more fertile soil. Soil organic matter is increased as a result of root secretions and falling stems and leaves, and the roots create pores through which water and oxygen can flow. Additionally, few would argue that a dusty excavation site is more aesthetically pleasing than a nicely planted field.

There are, however, many *limitations to phytoremediation*. It is a slow process that may take many growing seasons before an adequate reduction of pollution is seen, whereas soil excavation and treatment cleans up the site quickly. Also, hyper-accumulators can be a pollution hazard themselves. For instance, animals can eat the metal rich hyper-accumulators and cause the toxins to enter the food chain. If the concentration of metals in the plants is thought to be high enough to cause toxicity, there must be a way to segregate the plants from humans and wildlife, which may not be an easy task. Additionally, phytoremediation is in its infancy, and its effectiveness in cleaning up various toxins compared to conventional means of treatment is not always known. However, with more research and practice, the practicality of using phytoremediation should increase.

The growing popularity of phytoremediation can be seen in the number of businesses beginning to appear in urban areas of the United States that are involved in the field. In New Jersey, a company named Phytotech is using mustard greens to remove lead from polluted yards in Boston and hydroponically grown sunflowers to take radioactive metals out of the water surrounding the Chernobyl nuclear power plant. Right here in Fairfax County, Edenspace™ Systems Corp. of Chantilly has more than two dozen current and future contracts to utilize phytoremediation in the field, and they even sell an arsenic accumulating Edenfern® that can be planted in your backyard today.

Description

Phytoremediation is a bioremediation process that uses various types of plants to remove, transfer, stabilize, and/or destroy contaminants in the soil and groundwater. There are several different types of phytoremediation mechanisms. These are:

1. *Rhizosphere biodegradation.* In this process, the plant releases natural

substances through its roots, supplying nutrients to microorganisms in the soil. The microorganisms enhance biological degradation.

2. *Phyto-stabilization.* In this process, chemical compounds produced by the plant immobilize contaminants, rather than degrade them.

3. *Phyto-accumulation (also called phyto-extraction).* In this process, plant roots sorb the contaminants along with other nutrients and water. The contaminant mass is not destroyed but ends up in the plant shoots and leaves. This method is used primarily for wastes containing metals. At one demonstration site, water-soluble metals are taken up by plant species selected for their ability to take up large quantities of lead (Pb). The metals are stored in the plant's aerial shoots, which are harvested and either smelted for potential metal recycling/recovery or are disposed of as a hazardous waste. As a general rule, readily bioavailable metals for plant uptake include cadmium, nickel, zinc, arsenic, selenium, and copper. Moderately bioavailable metals are cobalt, manganese, and iron. Lead, chromium, and uranium are not very bioavailable. Lead can be made much more bioavailable by the addition of chelating agents to soils. Similarly, the availability of uranium and radio-cesium 137 can be enhanced using citric acid and ammonium nitrate, respectively.

4. *Hydroponic Systems for Treating Water Streams (Rhizofiltration).* Rhizofiltration is similar to phyto-accumulation, but the plants used for cleanup are raised in greenhouses with their roots in water. This system can be used for *ex-situ* groundwater treatment. That is, groundwater is pumped to the surface to irrigate these plants. Typically hydroponic systems utilize an artificial soil medium, such as sand mixed with perlite or vermiculite. As the roots become saturated with contaminants, they are harvested and disposed of.

5. *Phyto-volatilization.* In this process, plants take up water containing organic contaminants and release the contaminants into the air through their leaves.

6. *Phyto-degradation.* In this process, plants actually metabolize and destroy contaminants within plant tissues.

7. *Hydraulic Control.* In this process, trees indirectly remediate by controlling groundwater movement. Trees act as natural pumps when their roots reach down towards the water table and establish a dense root mass that takes up large quantities of water. A poplar tree, for example, pulls out of the ground 30 gallons of water per day, and a cottonwood can absorb up to 350 gallons per day.

The plants most used and studied are poplar trees. The U.S. Air Force has used poplar trees to contain trichloroethylene (TCE) in groundwater. In Iowa, EPA demonstrated that poplar trees acted as natural pumps to keep toxic herbicides, pesticides, and fertilizers out of the streams and groundwater. The US Army Corps of Engineers has experimented with wetland plants to destroy explosive compounds in the soil and groundwater. Submersed and floating-leafed species (coontail and pondweed, and arrowhead, respectively) decreased trinitrotoluene (TNT) to 5 per cent of original concentration. Submersed plants were able to decrease Royal Demolition Explosive (RDX) levels by 40 per cent, and when microbial degradation was added, RDX decreased by 80 per cent. Sunflowers, using rhizofiltration, were used successfully to remove radioactive contaminants from pond water in a test at Chernobyl, Ukraine.

Limitations and Concerns

The toxicity and bioavailability of biodegradation products is not always known.

Degradation by-products may be mobilized in groundwater or bio-accumulated in animals. Additional research is needed to determine the fate of various compounds in the plant metabolic cycle to ensure that plant droppings and products do not contribute toxic or harmful chemicals into the food chain.

Scientists need to establish whether contaminants that collect in the leaves and wood of trees are released when the leaves fall in the autumn or when firewood or mulch from the trees is used.

Disposal of harvested plants can be a problem if they contain high levels of heavy metals.

The depth of the contaminants limits treatment. The treatment zone is determined by plant root depth. In most cases, it is limited to shallow soils, streams, and groundwater. Pumping the water out of the ground and using it to irrigate plantations of trees may treat contaminated groundwater that is too deep to be reached by plant roots. Where practical, deep tilling, to bring heavy metals that may have moved downward in the soil closer to the roots, may be necessary.

Generally, the use of phytoremediation is limited to sites with lower contaminant concentrations and contamination in shallow soils, streams, and groundwater. However, researchers are finding that the use of trees (rather than smaller plants) allows them to treat deeper contamination because tree roots penetrate more deeply into the ground.

The success of phytoremediation may be seasonal, depending on location. Other climatic factors will also influence its effectiveness.

The success of remediation depends in establishing a selected plant community. Introducing new plant species can have widespread ecological ramifications. It should be studied beforehand and monitored. Additionally, the establishment of the plants may require several seasons of irrigation. It is important to consider extra mobilization of contaminants in the soil and groundwater during this start-up period.

If contaminant concentrations are too high, plants may die.

Some phytoremediation transfers contamination across media, (e.g., from soil to air).

Phytoremediation is not effective for strongly sorbed contaminants such as polychlorinated biphenyls (PCBs).

Phytoremediation requires a large surface area of land for remediation.

Applicability

Phytoremediation is used for the remediation of metals, radionuclides, pesticides, explosives, fuels, volatile organic compounds (VOCs) and semi-volatile organic compounds (SVOCs). Research is underway to understand the role of phytoremediation to remediate perchlorate, a contaminant that has been shown to be persistent in surface and groundwater systems. It may be used to cleanup contaminants found in soil and groundwater. For radioactive substances, chelating agents are sometimes used to make the contaminants amenable to plant uptake.

Technology Development Status

Phytoremediation is a broad technology type that has been successfully demonstrated for some contaminants and is experimental for others.

The Air Force has developed the Environmental Restoration Programme to identify, investigate, and cleanup environmental contamination associated with past activities. There are various technologies available to address remediation of contaminated sites, one such method is phytoremediation.

Phytoremediation is a biological process in which living plants are used to remove, accumulate, degrade, or contain environmental contaminants. This passive remediation technique is based on the natural ability of vegetation to utilize nutrients, which are transported by capillary action from the soil and groundwater through a plant's root system. The use of a plant's own biological mechanisms to contain and reduce concentrations of inorganic and organic contaminants in soils, sediments, and groundwater is a slow process relying on a plants growth rate. However, with advances in biological, chemical, and engineering technologies, phytoremediation has the potential to serve as a sustained, ecologically sound method to remediate contaminated soil and groundwater.

Biological Mechanisms

In addition to transporting nutrients, certain plants are capable of transporting environmental pollutants such as metals, radionuclides, chlorinated solvents, petroleum hydrocarbons, and ammunition wastes through biological means.

❖ *Hydraulic Control or phytohydraulics*: the use of a plant's root system to contain contaminant migration and infiltration by creating a hydraulic barrier, which also provides a vegetative cover that reduces contaminant exposure.

❖ *Phytoextraction or phytoaccumulation*: the use of a plant's root system to extract and transport inorganic contaminants such as metals or radionuclides from soils, which accumulate in aboveground plant tissue.

❖ *Rhizofiltration*: the use of a plant's root system to absorb inorganic contaminants dissolved in groundwater, which accumulate in a plant's roots.

❖ *Phytostabilization*: the use of a plant's root system to precipitate and absorb inorganic contaminants from soils and groundwater, which accumulate within the plant's roots.

❖ *Phytodegradation or phytotransformation*: the use of a plant's root system to extract organic contaminants such as solvents, pesticide and herbicide residues, and hydrocarbons from soils and groundwater, which degrade within the plant.

❖ *Rhizodegradation, also known as rhizosphere biodegradation, phytostimulation, or plant-assisted bioremediation*: is a result of a plant's root system releasing chemicals that enhance organic contaminant biodegradation by soil microorganisms.

❖ *Phytovolatilization*: the use of a plant's root system to extract and transport organic contaminants from groundwater to the leaves that transpire, evaporate, or volatilize the contaminants into the atmosphere.

Site & Plant Selection

Phytoremediation is not applicable at all contaminated sites. This remediation technique requires a large land area that has low to moderate contaminant concentrations within shallow soils and water table. Generally, phytoremediation is limited to within 3 feet of the surface for contaminated soils and within 10 feet of the surface for contaminated groundwater.

Aside from site conditions, phytoremediation is dependent on selecting the right plant for contaminants. Plants are complex systems whose biological mechanisms are dependent on various environmental factors; however, a fundamental trait in selecting a plant for phytoremediation is an expansive root system. Other plant selection parameters include the rate of plant and root growth, whether a plant freely transpires, and if a plant is suitable for the local environment and the contaminant of concern.

Fate of Contaminants

Once plants extract environmental contaminants, these contaminants are either broken down by the plant into non-hazardous substances or concentrated within plant tissues. Concentration of contaminants in plant material gives rise to the possibility of animals and insects consuming the contaminated material thereby introducing contaminants into the food chain.

To date, there is limited information regarding this issue due to the complex biological mechanisms involved. However, remedial sites implementing phytoremediation are taking measures to restrict grazing animals and birds by erecting fences and overhead nets. In addition, biodegradable pesticides are used to eliminate rodents and insects, and plants are harvested prior to seeding or flowering, which limits the availability of food.

Harvesting/Disposal

To sustain viable plant populations and operation of the system, plant material must occasionally be harvested. Harvested plant tissue, which has metabolized and accumulated environmental contaminants requires testing to determine if the tissue is hazardous. If testing confirms harvested material has metabolized the contaminant into non-hazardous substances, the plant material can be mulched or composted and reused on site. However, if the plant has accumulated contaminants and testing indicates plant tissues are hazardous, then those tissues which are hazardous (roots, stems, leaves, etc.)

must be disposed of as such. Although harvested plant material may be considered hazardous, phytoremediation is a cost effective remediation method, which causes little environmental disturbance while successfully reducing soil and groundwater contaminants.

Green Technology: Xenobiotics

21.1 Xenobiotic

A xenobiotic is a chemical which is found in an organism but which is not normally produced or expected to be present in it. It can also cover substances which are present in much higher concentrations than are usual. Specifically, drugs such as antibiotics are xenobiotics in humans because the human body does not produce them itself nor would they be expected to be present as part of a normal diet. However, the term is also used in the context of pollutants such as dioxins and polychlorinated biphenyls and their effect on the biota. Natural compounds can also become xenobiotics if they are taken up by another organism, such as the uptake of natural human hormones by fish found downstream of sewage treatment plant outfalls, or the chemical defenses produced by some organisms as protection against predators. Some xenobiotics can induce apoptosis at lower doses and necrosis at higher doses. Xenobiotics at low concentration can induce precommited reversible apoptotic phase which is followed by irreversible apoptotic phase.

Xenobiotic Metabolism

The body removes xenobiotics by xenobiotic metabolism. This consists of the deactivation and the secretion of xenobiotics, and happens mostly in the liver. Secretion routes are urine, feces, breath, and sweat. Hepatic enzymes are responsible for the metabolism of xenobiotics by first activating them (oxidation, reduction, hydrolysis and/or hydration of the xenobiotic), and then conjugating the active secondary metabolite with glucuronic or sulphuric acid, or glutathione, followed by excretion in bile or urine. An example of a group of enzymes involved in xenobiotic metabolism is hepatic microsomal cytochrome P450. These enzymes that metabolize xenobiotics are very important for the pharmaceutical industry, because they are responsible for the breakdown of medications.

Organisms can also evolve to tolerate xenobiotics. An example is the co-evolution of the production of tetrodotoxin in the rough-skinned newt and the evolution of tetrodotoxin resistance in its predator, the common garter snake. In this predator-prey pair, an evolutionary arms race has produced high levels of toxin in the newt and correspondingly high levels of resistance in the snake. This evolutionary response is based on the snake evolving modified forms of the ion channels that the toxin acts upon, so becoming resistant to its effects.

Xenobiotics in the Environment

Xenobiotic substances are becoming an increasingly large problem in Sewage Treatment systems, since they are relatively new substances and are very difficult to categorize. Antibiotics, for example, were derived from plants originally, and so mimic naturally occurring substances. This, along with the natural monopoly nature of municipal Waste Water Treatment Plants makes it nearly impossible to remove this new pollutant load.

Some xenobiotics are resistant to degradation. For example, they may be synthetic organochlorides such as plastics and pesticides, or naturally occurring organic chemicals such as polyaromatic hydrocarbons (PAHs) and some fractions of crude oil and coal. However, it is believed that microorganisms are capable of degrading all the different complex and resistant xenobiotics found on the earth.

Inter-Species Organ Transplantation

The term xenobiotic is also used to refer to organs transplanted from one species to another. For example, some researchers hope that hearts and other organs could be transplanted from pigs to humans. Many people die every year whose lives could have been saved if a critical organ had been available for transplant. Kidneys are currently the most commonly transplanted organ. Xenobiotic organs would need to be developed in such a way that they would not be rejected by the immune system. With the development of vitrification transplantable organs could be stored in organ banks for long periods.

21.2 Xenobiotics, Disease and Detoxification

Because of the significant overload of xenobiotics in our modern environment, you should undertake a light detoxification programme. You don't want to stir up too many of these toxins at one time.

Recently, scientists have coined a new term for foreign chemicals found in the body, xenobiotics. The term comes from the words xenos, meaning foreign, and bios, meaning life. Xenobiotics are found when the body absorbs chemicals that are not nutrients, or when normally occurring substances (nutrients or endogenously produced biochemicals) become denatured (modified from original molecular structure) or are in excess. Some authors suggest that we are burdened by as many as 100,000 man-made xenobiotics—and that doesn't take into account the natural ones.

A considerable number of clinical studies are demonstrating that xenobiotic chemicals, which were once thought to have a very precise and direct toxic effect, may actually affect different parts of the body in a variety of ways. One study has suggested a deficiency in detoxification pathways is involved in symptoms of Parkinson's disease. Other studies have associated the development of respiratory tract and food allergies with insufficient detoxification. Given the increasing number of individuals suffering from various recognized allergies, as well as the direct or indirect implication of allergies in a variety of disorders (they may cause the initial symptoms, or as is generally understood, exacerbate them), this has extraordinary ramifications. According to Dr. William Holub, "Treatment of allergy, then, requires far more than the identification and isolation from the allergen, but requires, in addition, an applicable programme of dietary regulation and food supplementation to insure adequate detoxification." Xenobiotics and food allergies are associated with many physical and psychological disorders.

Illnesses Directly Associated with Xenobiotics

- *Physical*: cardiac arrhythmia, eczema, edema, epilepsy, fatigue, headache, hypertension, multiple sclerosis, tinnitus, rheumatism, pain, psoriasis, vasculitis.
- *Psychological*: autism, aggressive behaviour, anxiety, fatigue, insomnia, organic mental disorders.

Illnesses That May be Triggered or Worsened by Food Allergies

- *Physical*: AIDS, alcoholism, bronchial asthma, constipation, Crohn's disease, dermatitis, diarrhoea, gallbladder disease, glaucoma, hypoglycaemia, irritable bowel syndrome, lupus, obesity, osteoarthritis, Raynaud's syndrome, rheumatoid arthritis, ulcerative colitis, ulcers (duodenal and gastric), urticaria.
- *Psychological*: attention deficit disorder, bipolar disorder (manic depression), learning disabilities, schizophrenia.

Xenobiotics, Allergies and Food Sensitivities

In order for foreign protein molecules to become antigens that cause food sensitivities, they must pass the digestive or respiratory tract mucosa relatively intact. Apart from acting as barriers to abnormally large molecules, Payer's patches (lymphocytes) in the gut wall promote secretion of antibodies called IgA into the gut. These antibodies neutralize, or "mop up," the antigens that are in the gut. Antigen molecules must be sufficiently complex for the body to recognize them as invader proteins, or the body to misinterprets them, identifying them either as endorphins or as a gamut of other amino-acid based complex molecules. Improperly digested, foreign molecules must also bypass the body's selective anti-toxin mechanisms, including detoxification and inactivation in the liver. Finally, they must either bypass or survive aggression by the immune system on a one-to-one basis. However, they might not be attacked if the body misinterprets them as "native" proteins. In some cases, the immune system may be unable to properly deal with these either because of a weakened condition or because of an excess of foreign substances.

21.3 Case Study: International Society for the Study of Xenobiotics

As we review the history of xenobiotic metabolism the rapid acceleration of the pace of discovery becomes quite clear. From the beginning of the 19th century reactions were enumerated at a slow deliberate pace. As the 20th century began the basics of transformation-oxidation, reduction, hydrolysis, and synthesis were understood but the coherence of applications was unclear.

It took another 50 years for the field to coalesce in the writings of R.T.Williams. From the early 1950s through the 60s the actual enzymes involved in metabolism as well as their intracellular locations came under study.

In the 80s the x-ray structure of key enzymes was unveiled and molecular biology began to explore the expression and synthesis of the individual proteins. The role of genetic polymorphisms in pharmacotherapy became apparent.

As the new millennium dawns we find xenobiotic disposition playing a major role in understanding individual response to drugs and to the chemical environment. It takes little imagination to see the day when a patient's knowledge of their genetic make-up and degree of expression is as important as the knowledge of the disease in shaping appropriate therapy. An understanding of how minor genetic changes can alter an individual response to the environment will build our database for controlling our response to new agents.

The combination of genomics and proteomics and powerful analytical techniques will allow us to probe human responses at the molecular level. The use of high powered computer programmes should allow us to assemble the molecular data into predictive paradigms relating to the individual organism's response to its surroundings. The dawn of the new millennium holds great promise and challenge for the field of xenobiotic metabolism.

21.4 Xenobiotic Metabolism

Xenobiotic metabolism is the set of metabolic pathways that modify the chemical structure of xenobiotics, which are compounds foreign to an organism's normal biochemistry, such as drugs and poisons. These pathways are a form of biotransformation present in all major groups of organisms, and are considered to be of ancient origin. These reactions often act to detoxify poisonous compounds; however, in some cases, the intermediates in xenobiotic metabolism can themselves be the cause of toxic effects.

Xenobiotic metabolism is divided into three phases. In phase I, enzymes such as cytochrome P450 oxidases introduce reactive or polar groups into xenobiotics. These modified compounds are then conjugated to polar compounds in phase II reactions. These reactions are catalysed by transferase enzymes such as glutathione S-transferases. Finally, in phase III, the conjugated xenobiotics may be further processed, before being recognised by efflux transporters and pumped out of cells.

The reactions in these pathways are of particular interest in medicine as part of drug metabolism and as a factor contributing to multidrug resistance in infectious diseases and cancer chemotherapy. The actions of some drugs as substrates or inhibitors of enzymes involved in xenobiotic metabolism are a common reason for hazardous drug interactions. These pathways are also important in environmental science, with the xenobiotic metabolism of microorganisms determining whether a pollutant will be broken down during bioremediation, or persist in the environment.

Permeability Barriers and Detoxification

That the exact compounds an organism is exposed to will be largely unpredictable, and may differ widely over time, is a major characteristic of xenobiotic toxic stress. The major challenge faced by xenobiotic detoxification systems is that they must be able to remove the almost-limitless number of xenobiotic compounds from the complex mixture of chemicals involved in normal metabolism. The solution that has evolved to address this problem is an elegant combination of physical barriers and low-specificity enzymatic systems.

All organisms use cell membranes as hydrophobic permeability barriers to control access to their internal environment. Polar compounds cannot diffuse across these cell membranes, and the uptake of useful molecules is mediated through transport proteins that specifically select substrates from the extracellular mixture. This selective uptake means that most hydrophilic molecules cannot enter cells, since they are not recognised by any specific transporters. In contrast, the diffusion of hydrophobic compounds across these barriers cannot be controlled, and organisms, therefore, cannot exclude lipid-soluble xenobiotics using membrane barriers.

However, the existence of a permeability barrier means that organisms were able to evolve detoxification systems that exploit the hydrophobicity common to membrane-permeable xenobiotics. These systems therefore solve the specificity problem by possessing such broad substrate specificities that they metabolise almost any non-polar compound. Useful metabolites are excluded since they are polar, and in general contain one or more charged groups.

The detoxification of the reactive by-products of normal metabolism cannot be achieved by the systems outlined above, because these species are derived from normal cellular constituents and usually share their polar characteristics. However, since these compounds are few in number, specific enzymes can recognize and remove them. Examples of these specific detoxification systems are the glyoxalase system, which removes the reactive aldehyde methylglyoxal, and the various antioxidant systems that eliminate reactive oxygen species.

Phases of Detoxification

The metabolism of xenobiotics is often divided into three phases: modification, conjugation, and excretion. These reactions act in concert to detoxify xenobiotics and remove them from cells.

Phase I—Modification

In phase I, a variety of enzymes acts to introduce reactive and polar groups into their substrates. One of the most common modifications is hydroxylation catalysed by the cytochrome P-450-dependent mixed-function oxidase system. These enzyme complexes act to incorporate an atom of oxygen into nonactivated hydrocarbons, which can result in either the introduction of hydroxyl groups or N-, O- and S-dealkylation of substrates. The reaction mechanism of the P-450 oxidases proceeds through the reduction of cytochrome-bound oxygen and the generation of a highly-reactive oxyferryl species, according to the following scheme:

Phase II—Conjugation

In subsequent phase II reactions, these activated xenobiotic metabolites are conjugated with charged species such as glutathione (GSH), sulfate, glycine, or glucuronic acid. These reactions are catalysed by a large group of broad-specificity transferases, which in combination can metabolise almost any hydrophobic compound that contains nucleophilic or electrophilic groups. One of the most important of these groups are the glutathione S-transferases (GSTs). The addition of large anionic groups (such as GSH) detoxifies reactive electrophiles and produces more polar metabolites that cannot diffuse across membranes, and may, therefore, be actively transported.

Phase III—Further Modification and Excretion

After phase II reactions, the xenobiotic conjugates may be further metabolised. A common example is the processing of glutathione conjugates to acetylcysteine (mercapturic acid) conjugates. Here, the ã-glutamate and glycine residues in the glutathione molecule are removed by Gamma-glutamyl transpeptidase and dipeptidases. In the final step, the cystine residue in the conjugate is acetylated.

Conjugates and their metabolites can be excreted from cells in phase III of their metabolism, with the anionic groups acting as affinity tags for a variety of membrane transporters of the multidrug resistance protein (MRP) family. These proteins are members of the family of ATP-binding cassette transporters and can catalyse the ATP-dependent transport of a huge variety of hydrophobic anions, and thus act to remove phase II products to the extracellular medium, where they may be further metabolised or excreted.

Endogenous Toxins

The detoxification of endogenous reactive metabolites such as peroxides and reactive aldehydes often cannot be achieved by the system described above. This is the result of these species' being derived from normal cellular constituents and usually sharing their polar characteristics. However, since these compounds are few in number, it is possible for enzymatic systems to utilize specific molecular recognition to recognize and remove them. The similarity of these molecules to useful metabolites therefore means that different detoxification enzymes are usually required for the metabolism of each group of endogenous toxins. Examples of these specific detoxification systems are the glyoxalase system, which acts to dispose of the reactive aldehyde methylglyoxal, and the various antioxidant systems that remove reactive oxygen species.

History

Studies on how people transform the substances that they ingest began in the mid-nineteenth century, with chemists discovering that organic chemicals such as benzaldehyde could be oxidized and conjugated to amino acids in the human body. During the remainder of the nineteenth century, several other basic detoxification reactions were discovered, such as methylation, acetylation, and sulfonation.

In the early twentieth century, work moved on to the investigation of the enzymes and pathways that were responsible for the production of these metabolites. This field became defined as a separate area of study with the publication by Richard Williams of the book *Detoxication mechanisms* in 1947. This modern biochemical research resulted in the identification of glutathione *S*-transferases in 1961, followed by the discovery of cytochrome P450s in 1962, and the realization of their central role in xenobiotic metabolism in 1963.

Green Technology:
Biosensors, Biochips and Biosurfactents

22.1 Biosensors and Other Medical and Environmental Probes

Bacteria. Our invisible friends, our invisible enemies. Some aid our digestion, others destroy our poisons. Still other "bugs" make us sick. Living inside and outside our bodies, natural bacteria are a fact of life. We have learned to live with them and they with us. Soldiers and sailors in action worry that even more dangerous bacteria are lurking in the environment. These are "killer" bacteria—the kind used as weapons. Such biological warfare agents are viewed by many to be as threatening to human life as nuclear weapons. In 1993, R.J. Wooley, director of the U.S. Central Intelligence Agency, proclaimed, "Proliferation (of nuclear, biological, and chemical weapons) poses one of the most complex challenges the intelligence community will face for the remainder of the century." General Colin Powell has stated that biological weapons worry him more than anything else. He has good reason to worry. A recent U.S. Arms Control and Disarmament Agency report on worldwide arms control compliance states that both Syria and Egypt have offensive biological warfare programmes. The key to protecting a military unit or community from dangerous bacteria is to detect them before they reach their intended victims. People can then be warned to leave an area or at least wear protective gear. Bacteria can be detected using "biosensors." A biosensor is a device that detects, records, and transmits information regarding a physiological change or the presence of various chemical or biological materials in the environment. More technically, a biosensor is a probe that integrates a biological component, such as a whole bacterium or a biological product (e.g., an enzyme or antibody) with an electronic component to yield a measurable signal. Biosensors, which come in a large variety of sizes and shapes, are used to monitor changes in environmental conditions. They can detect and measure concentrations of specific bacteria or hazardous chemicals; they can measure acidity levels (pH). In short, biosensors can use bacteria and detect them, too.

Genetically engineered bacteria can also be useful because of their ability to "tattle" on the environment. Such commonly used bacteria have been designed in Oak Ridge and Knoxville to give off a detectable signal, such as light, in the presence of a specific pollutant they like to eat. They may glow in the presence of toluene, a hazardous compound found in gasoline and other petroleum products. They can indicate whether an underground fuel tank is leaking or whether the site of an oil spill has b een cleaned up effectively. These informer bacteria are called bioreporters. In 1990, when a bioreporter of naphthalene was developed and tested at the University of Tennessee at Knoxville (UTK), the ability of these b acteria to glow was demonstrated for President George Bush during his visit to UTK.

Oak Ridge National Laboratory has been developing biosensors and bioreporters for almost a decade. Carl Gehrs, director of ORNL's Center for Biotechnology, says that ORNL's programme in biosensors is "a leader among DOE national labs, which is a best-kep t secret." He added that ORNL is proposing to develop biosensors that can detect the presence of biological and chemical warfare agents for military use and for determining the effectiveness of cleaning up waste sites.

It is clear that Oak Ridge can make some impressive, even revolutionary, contributions to military applications of biosensors. One of the rewards of developing biosensors for military use is that the resulting devices may eventually have applications in everyday life. As President Clinton recently told a group of military officials, "What you have done here is what I wish to do nationally: take some of the most talented people in the world who produce some of the most sophisticated military technol ogy and put that to work in the civilian economy."

ORNL's First Biosensor

The first biosensor developed at ORNL was intended for environmental monitoring. It used an antibody—a protein substance produced in the blood or tissues in response to a specific antigen, such as a bacterium or toxin normally foreign to the body. Antibodies destroy or weaken invading bacteria and neutralize organic poisons, forming the basis of immunity. Thus, ORNL's first biosensor was called an immunosensor.

In the mid-1980s, Tuan Vo-Dinh, Guy Griffin, and others in ORNL's Life Sciences Division were looking for a way to use light to detect cancer-causing agents in groundwater. So they attached to the end of an optical fi ber an antibody that reacts specifically with the carcinogen benzo(*a*)pyrene (B*a*P). The anti-B*a*P antibody on the end of the fiber was immersed in a sample of groundwater. The antibody was allowed to bind the B*a*P in the groun dwater sample. The antibody-B*a*P reaction product gives off light if illuminated by light of the right wavelength. So the right light was aimed through the fiber into the groundwater sample. After 5 to 10 minutes, the reaction product fluoresced, and the fluorescence was transmitted back up the fiber and measured. These successful results, reported in 1987 by Vo-Dinh and colleagues, brought the group a 1987 R&D 100 Award from *R&D* magazine and initiated the group's development of a seri es of fiber-optic-based biosensors.

Antibodies can be produced against bacteria, against complex carbohydrates, and even against smaller organic molecules that may cause cancer. To Vo-Dinh's group, the possibilities for applications of immunosensors seemed almost limitless. Indeed the po ssible uses for biosensors are limited only by our imagination. After all, there are many different ways to combine chemistry, physics, and biology with an electronic detector.

One type of biosensor has only five components: a biological sensing element, a transducer, a signal conditioner, a data processor, and a signal generator. The essential component must produce a signal that is related to the concentration of a specific chemical or biological substance in complex systems. This component takes advantage of the ability of a biomolecule, such as an antibody or enzyme, to specifically recognize the target substance.

Light emissions from microspheres and bacteria are seen through a fluorescence microscope. Shown are red-fluorescing *S. aureus* bacteria bound to 6.5-μm spheres and one yellow orange-fluorescing *E. coli* bound to the larger 10-μm sphere.

In another approach to the use of immunosensors, microspheres of different sizes are labeled with antibodies that bind to different bacteria; thus, microspheres of one size have one particular antibody and microspheres of another s ize have a different antibody. The sizes of the microspheres are identified by their "morphological resonances" (shape-based light emissions when excited by a laser), and the bacteria that become bound are detected by the colour of fluorescent dye with which they are stained. In a 1995 paper, Bill Whitten of ORNL's Chemical and Analytical Sciences Division suggests that up to 100 different types of bacteria can be identified simultaneous ly because the stained bacteria all would fluoresce at one wavelength of light and the diameters of the spheres could be illuminated at another wavelength.

Although all the spheres fluoresce when excited by one wavelength, the morphological resonances, which look like saw teeth superimposed on a fluorescence emission spectrum, can distinguish among diameters of many different-sized spheres. This approach satisfies one of today's challenges in biotechnology: multiplex biosensors to obtain more infor mation from one sample analysis.

Medical Telesensors

The goal is to develop an array of chips to collectively monitor bodily functions.

This "medical telesensor" chip on a fingertip can measure and transmit body temperature.

A chip on your fingertip may someday measure and transmit data on your body temperature. An array of chips attached to your body may provide additional information on blood pressure, oxygen level, and pulse rate. This type of medical telesensor, which is being developed at ORNL for military troops in combat zones, will report measurements of vital functions to remote recorders. The goal is to develop an array of chips to collectively monitor bodily functions. These chips may be attached at various points on a soldier using a nonirritating adhesive like that used in waterproof band-aids. These medical telesensors would send physiological data by wireless transmission to an intelligent monitor on another soldier's helmet. The monitor could alert medics if the data showed that the soldier's condition fit one of five levels of trauma. The monitor also would receive and transmit global satellite positioning data to help medics locate the wounded soldier.

Development of medical telesensors at ORNL is supported by the Defense Sciences Office of the Advanced Research Projects Agency, but the development is expected to have civilian applications. Wireless monitors a ttached to the skin could provide valuable information on the physiological condition of intensive-care patients in hospitals, high-risk outpatients, babies at risk of suffering sudden infant death syndrome, and police and firefighting personnel in ha zardous situations.

In ORNL's Life Sciences Division, Tom Ferrell has shown that a 2×2-millimeter (mm) siliconz chip attached to the skin can measure body temperature. The chip contains a temperature sensor in an integrated circuit, a lithium thin-film battery that supplies the very low level of power required by the circuit and signal processing and transmission electronics, and an antenna that sends the data by radio signals (radio-frequency transmission) to a monitor when the chip is queried. Ferrell calls thi s biosensor a "medical telesensor ASIC" because it uses an application-specific integrated circuit for telemetry—automatic measurement and transmission of data from remote sources to receivers for recording and analysis.

Ferrell also expects that a chip can be developed to measure blood oxygen level. As the blood oxygen level changes, the colour of hemoglobin in blood is altered. Such a chip would have a light source and light detector that could measure changes in the colour of hemoglobin transmitted when it is illuminated by light. The results are reported by wireless telemetry.

Blood pressure and pulse rate may be measured by chips designed to detect pressure changes. Indeed, Jeff Muhs and Steve Allison, both of ORNL's Engineering Technology Division, are working with optical fibers m ade of silicone to take advantage of the unique properties of this substance. Unlike a glass fiber, a silicone fiber is flexible—it can be squeezed or stretched, and the amount of compression or expansion can be measured by changes in light transm ission through the fiber. Thus, silicone fibers embedded in roads can be used to weigh trucks. If a silicone fiber on a chip can sense pressure at various positions in the body, it may be used for monitoring blood pressure, pulse rate, breathing (chest expansion), knee bending during physical rehabilitation, and foot pressure distribution.

Aging, diseases such as diabetes and Alzheimer's, and chemical warfare agents cause changes in metal ion concentrations in the body. If these changes could be detected and measured, the information could provide clues about changes in disease states a nd exposure to toxins. Tuan Vo-Dinh and his coworkers have developed a biosensor using a glass optical fiber and a hybrid molecule he synthesized. One half of the hybrid molecule binds calcium ions and the other half fluoresces when calcium ions are bo und to the molecule. By attaching this molecule to the end of a very small diameter optical fiber, Vo-Dinh measured the concentration of calcium ions in a solution. He plans to make a similar measurement within a single living cell!

Microcantilevers

Schematic of a microcantilever sensor, which can be adapted to detect physical, chemical, or biological activity.

An interesting alternative to the optical fiber is the microcantilever, which measures the presence of substances by nonoptical methods. It can act as a physical, chemical, or biological sensor by detecting changes in cantilever bending or vibrational frequency. Think of a diving board that wiggles up and down at a regular interval. This wiggling changes when someone steps on the board. Microcantilevers are a million times smaller but molecules adsorbed on a microcantilever cause vibrational freque ncy changes. Viscosity, density, and flow rate can also be measured by detecting the changes in vibrational frequency. Another way of detecting molecular adsorption is by measuring curling of the cantilever due to adsorption stress on just one side of the cantilever. Depending on the nature of chemical bonding of the molecule, the curling can be up or down. For example, if the microcantilever is bimetallic, just like the thermostat at home but a million times smaller, a temperature change as small as a millionth of a degree can be measured. There is much to learn about the basic mechanisms involved.

The microcantilever is ordinarily constructed of a silicon plank 100 micrometers (mm) long, 30 mm wide, and 3 to 4 mm thick (these dimensions are only approximate, and other geometries are sometimes used). When molecules are added to its surface, the extent to which the plank bends can be measured accurately by bouncing a light beam off the surface and measuring the extent to which the light beam is deflected. The vibrational frequency can be induced by piezoelectric transducers and measured with t he same laser beam that measured the deflection because it generates an alternating current in the detector.

Temperature is measured by coating the silicon surface with gold or aluminum, which expands at a different rate than silicon. Because the difference in heat expansion between silicon and gold affects bending of the microcantilever, temperature changes of a millionth of a degree can be measured. Chemical reactions generate heat, so this device can be used as a microcalorimeter to measure the heat of an enzyme-catalyzed reaction or a chemical reaction in a reaction volume of one microliter (1 µl).

Yet another mechanism of response was employed to measure proteins in solution. Antibodies were covalently attached to the silicon surface of a cantilever in such a way that the stresses induced in the antibody when it reacted with its antigen were det ected. Detection of biological warfare agents or bacteria and viruses in the hospital laboratory should be expedited with this stressed antibody technique. Additional experiments are under way to demonstrate the usefulness of the microcantilever as a b iosensor. Because of the small size and versatility of the microcantilever, arrays of sensors can be fabricated on a single chip to conceptually mimic the five sensory facilities: sight, hearing, smell, taste, and touch. ORNL researchers Thomas Thun dat, Bruce Warmack, Eric Wachter, Patrick Oden, and Panos Datskos received a 1996 R&D 100 Award for development of the microcantilever.

Detecting Cancer and Health Abnormalities

Another type of biosensor uses sophisticated technology to detect a specific trait or abnormality in a living organism. ORNL researchers have invented several biosensors of this type.

Tuan Vo-Dinh of ORNL (left) and Bergein Overholt and Masoud Panjehpour, both of Thompson Cancer Survival Center of Knoxville, have developed a new laser technique for nonsurgically determining whether tumors in the esophagus are cancerous or benign.

Of these biosensors, the most publicized is the optical biopsy sensor developed by Tuan Vo-Dinh in collaboration with medical researchers at Thompson Cancer Survival Center in Knoxville. This sensor can tell whether a tumor in the esophagus is cancerou s or benign. In the past, determining accurately whether a patient has cancer of the esophagus has required surgical biopsy. However, our laser-based fluorescence method has eliminated the need for biopsy, reducing pain and recovery time for patients.

Here's how it works. Laser light of the appropriate wavelength is directed to the inner surface of the esophagus by means of a fiber-optic device that is swallowed by the patient. The epithelial cells and tissue inside the esophagus fluoresce when exc ited by the laser light. When the esophagus interior is illuminated with blue light [410 nanometers (nm)], the normal tissue emits light at wavelengths different from those emitted by the cancer cells. Thanks to software developed by Vo-Dinh and colle agues, the spectral properties of the light at wavelengths ranging from 400 to 700 nm can be analyzed at various positions in the esophagus. Emissions from normal cells and cancer cells can be distinguished quite accurately; the difference is expressed as the differential normalized fluorescence index. Tests on more than 200 patients show that, compared with the results of surgical biopsies, laser fluorescence diagnosis is accurate in over 98 per cent of the cases.

Another biosensor from Vo-Dinh's laboratory provides a way to monitor the status of diabetes without using blood samples. In this case, light is used to illuminate the eyeball and stimulate certain substances, including proteins, to emit fluorescent li ght. This fluorescence changes in intensity and wavelength when the distribution and status of proteins in the eye change. This truly noninvasive method depends on a relatively new development for selecting the wavelength of light for illumination. Ins tead of prisms or gratings to refract the light into different wavelengths, a device called the acousto-optic tunable filter (AOTF) is used. One AOTF selects the wavelength of light to shine on the eyeball and another selects the wavelength of fluores cent light emitted from the eyeball. Both AOTF wavelengths are scanned simultaneously using the synchronous luminescence technique developed previously for environmental screening. The AOTFs, which are manipulated with a radio-frequency signal, can scan the entire visible spectrum and portions of the ultraviolet and infrared spectra in milliseconds to select the appropriate wavelengths to use to illuminate the eyeball. They can also select the correct wavelengt h to use to read the fluorescence signal from the patient's eye instantaneously. In this way many spectral scans can be taken and averaged in a computer to obtain the accuracy required to measure the status and changes in the eye proteins of diabetics. Vo-Dinh's group is also studying ways to apply optical techniques to detect skin, cervical, and colon cancers.

Bioreporters

Yet another example of a biosensor is based on detection of light emitted by a specially engineered microorganism that is involved in bioremediation. However, in this case the light originates from a particular protein that has been installed in certai n bacteria by modern molecular genetic methods. In one case, the gene for luciferase is placed in the operon (a sequence of genes that specify enzymes that carry out a related series of metabolic steps) that is responsible for degrading unwanted chemicals such as toluene, an organic solvent. When the bacteria are metabolizing the toluene, the genetic control mechanism also turns on the synthesis of the enzyme luciferase, which produces light in the presence of oxygen. A variation on this capability was invented at ORNL to deal with situations in which bacteria degrade an organic solvent under conditions o f very limited oxygen. In this case, a "green fluorescent protein," which emits green light (with a wavelength of 509 nm) when excited by blue light (395 nm), is installed in the operon. No oxygen is required for the light emission. Again, as the tolue ne is metabolized by the enzymes synthesized for that activity, the green fluorescent protein is also produced. Because it is active, it can be monitored by remote light activation and spectral emission analysis.

Bob Burlage of the Environmental Science Division has been installing these proteins and working with Vo-Dinh to create an optical biosensor to analyze the emitted light. Also, Michael Simpson of the Instrumentation and Controls Division Division is collaborating with Gary Sayler of UTK to produce a "critters on a chip" technology in which light sensor s pick up and transmit information from chip bacteria that glow in the presence of trace levels of poisons, explosives, or pollutants. These types of biosensors are useful for monitoring efforts to clean up industrial spills because these light-emitt ing bacteria can "report" continuously the progress of biodegradation. Such bioreporters have proven successful using trichloroethylene, toluene, and various petroleum products in laboratory tests. They are now being tested on a much larger scale usi ng a lysimeter.

In the late 1980s, a group at the Oak Ridge Y-12 Plant designed and built a set of 18 lysimeters. Each lysimeter is a cylinder 2.4 meters (8 feet) in diameter and 3 to 3.6 meters (10 to 12 feet) deep that is filled with soil and monitored for the prese nce of a variety of substances. The current experiment in which Sayler is collaborating with Burlage employs several of these cylinders so that they can observe the effectiveness of genetically engineered bacteria in degrading naphthalene.

John Hiller of the Oak Ridge Centers for Manufacturing Technology proposes to adapt a luminescence spectrometer, which was devised to measure uranium concentrations in groundwater, to detect bacteria and other organisms that may contaminate drinking water. Such organisms release small amounts of ATP (adensosine triphosphate, the universal bio-energy chemical), which, in the presence of oxygen, is converted into li ght energy by luciferase, which will be added to the assay system to detect the ATP. The luminescence spectrometer would detect and measure the intensity of light emitted by luciferase to determine the concentration of contaminants in water.

Miniaturized Devices

Another class of biosensors uses various techniques to turn a biological system into a tiny electronic device, to analyze biological or physiological processes, or to detect and identify bacteria. Some of these techniques produce or are carried out in miniaturized devices.

The site for photosynthesis in a green leaf contains a complex set of enzymes and proteins that capture light energy and convert carbon dioxide into compounds that help the plant grow. If a platinum salt in a certain oxidation state is supplied to one of two photosynthetic systems in plant chloroplasts, one photosynthetic reaction system will use light energy to provide electrons that will reduce platinum to the metal form. The metal is deposited on the photosystem complex to form a tiny platinum c enter that can be employed in sophisticated diode-based microelectronics for measurements at extremely high sensitivity, resolution, and speed. Such a biomolecular optoelectronic sensor has been demonstrated by Eli Greenbaum, James Lee, Ida Lee, and St eve Blankinship, all of ORNL's Chemical Technology Division.

The infrared microspectrometer developed at ORNL can be used for blood chemistry analysis, gasoline octane analysis, environmental monitoring, industrial process control, aircraft corrosion monitoring, and detection of chemical warfare agents.

ORNL researchers have made dramatic progress in miniaturizating clinical and chemical laboratories—another aspect of our biosensor work. One useful miniature device is an infrared microspectrometer the size of a sugar cube, constructed by Slo Raj ic and Chuck Egert, both of ORNL's Engineering Technology Division. Carved out of a solid block of plastic, the device measures 1.5 centimeters on a side and has no moving parts. It can be used for blood chemistry analysis, gasoline octane analysis, environmental monitoring, industrial process control, aircraft corrosion monitoring, and detection of chemical warfare agents. The plastic device uses a light source to excite certain types of compounds in gases, liquids, and solids. These excited com pounds give off infrared light of various wavelengths. Because the microspectrometer is precisely manufactured using single-point diamond turning, it can gather up the emitted light and channel it into an optical fiber for analysis. The measured emiss ions wavelengths are fed into a microchip, which identifies and determines the concentrations of chemicals in a sample. The device is inexpensive and performs well in different customized uses.

ORNL's best known miniaturization feat is the "lab on a chip." In 1996 this miniaturized chemical laboratory received a *Discover* magazine innovation award and an R&D 100 Award from *R&D* magazine. The chip consists of 50 to 100-mm channel s that are etched into the surface of a microscope slide. When reagents are mixed together and forced to migrate down these channels by differences in electrical potential, reaction rates can be measured and chemicals can be separated. Laser-induced f luorescence monitors the reaction's progress along the channel and measures concentrations of the separated products.

Mike Ramsey, Steve Jacobson, and colleagues in ORNL's Chemical and Analytical Sciences Division have demonstrated that these miniaturized devices are often more accurate than standard laboratory procedures. They have separated mixtures by performing ch romatography in these channels, demonstrated the separation of products of digestion of DNA with restriction enzymes, and demonstrated enzyme-catalyzed reactions. They are adapting such miniaturized devices for many clinical and military applications.

Biosensors and DNA Analysis

DNA can be used to identify organisms ranging from humans to bacteria and viruses. The identification consists of reading the sequence of the DNA letters (A, G, C, and T) that compose the alphabet used to describe the bases attached to the deoxyribose phosphate polymer that forms the backbone of the DNA helix. The bases join two strands of this polymer to form a spiral staircase, where the bases and the hydrogen bonds that join them are the steps on the stairs. Just as in the English alphabet, the letters are combined in various groupings and sequences to form words, sentences and paragraphs that organize information in a manner that can be read and utilized. Therefore, much effort has gone into ways to read the sequence of the four letters of t he DNA alphabet, and rather rudimentary methods have been developed that are reliable and useful. Using methods to form short sequences from the DNA of interest, DNA fragments are produced that have one of the letters at their end and that differ acco rding to their sizes. As a result, DNAs containing 400 to 600 letters can be sequenced accurately. However, many hours are required to prepare the fragments and separate them by size (using gel electrophoresis). Using this method, the sequences of mill ions of DNA letters have been determined, enabling the identification of the site of a genetic mutation that causes such diseases as sickle cell anemia, Huntington's disease, fragile X syndrome (a serious type of mental retardation), and several hundr ed other inherited diseases or traits. Furthermore, by combining sequences obtained ~400 at a time, the entire genome of a yeast was determined in 1996, a major accomplishment that led to the identification of an entire new group of genes that had not been recognized before. Many labs that were involved in the sequencing effort have now switched their attention to studying these new genes to learn of their function and significance. The Human Genome Project, which is sponsored by DOE and the Nation al Institutes of Health, plans to go well beyond yeast and determine the sequence of all DNA letters in the human genome. Undoubtedly, this information will lead to the discovery of many new human genes and a more sophisticated understanding of our he alth and disease states. More details may be found in the Primer for Molecular Genetics on the DOE World Wide Web site http://www.ornl.gov/TechResources/Human_Genome/pub licat/primer/ intro.html.

Because the methods used to sequence the yeast DNA required dozens of labs in Europe and the United States to work together for several years to determine the total sequence, new and faster methods to read the DNA sequence would clearly be advantage ous. If the methods were faster and less expensive, DNA sequence information would be as common as your blood type and would be much more informative. You could determine whether you are a carrier of a disease gene known to exist in your family, wheth er blood at a crime scene was that of the arrested suspect, or whether a biowarfare bacterium was descending on a battlefield. In some of these uses, the need to know the DNA sequence may be urgent. Because the current methods are not fast enough to sa tisfy such needs, many laboratories, including ORNL, are developing faster methods. In addition to sequence information, methods are also being developed to identify the site where a DNA sequence has been modified.

When a chemical reacts with DNA in the cell nucleus, possibly causing damage, a chemical compound called an adduct forms from the addition of the two species. ORNL researchers led by Bob Hettich and Michelle Buchanan in ORNL's Chemical and Analytical S ciences Division have used mass spectrometry to identify chemicals that form adducts with DNA. They have determined both the chemical identity of the adduct and the specific DNA site at which it occurs.

Laser desorption mass spectrometry has been used at ORNL to detect double-stranded and single-stranded DNA and to determine the sequence of bases in single-stranded DNA.

Simply speaking, DNA is like two strings of beads. One string contains beads of four different colours—red, blue, green and yellow—arranged in a particular order. The order of beads in the other string is governed by this rule: red must be pai red with blue, and yellow must be paired with green. Thus, if the order of the first four beads in the first string is red, green, blue and yellow, the order of the matching four in the second string is blue, yellow, red, and green. The DNA sequence consists of a string of chemical bases, known as A, T, C, and G, and the DNA molecule is folded, looped and coiled in various ways. For double-stranded DNA, the rule is that A on one strand must pair with T on the other, and C must pair with G.

Over the past four years ORNL's mass spectrometry groups have shown that DNA fragments of about 50 to 75 nucleotides (bases) can be sequenced to obtain the order of the A, G, C, and T letters. Because mass spectrometry accomplishes the sequence determ ination in less than a second, this speed supersedes that of gel electrophoresis by orders of magnitude. However, the gel method can do 400 to 600 letters on 50 samples at a time and is not in danger of being replaced yet. Other ways of characterizing DNA consist of determining the sizes of fragments produced by restriction endonucleases, enzymes that cut DNA when they find a specific sequence of letters. A physicist friend of mine calls these enzymes magic scissors because they not only cut the DNA but do so reliably at a sequence site that is specific to each of the restriction enzymes, of which there are over 400. The pattern of sizes of the DNA fragments produced by these magic scissors can be unique to the DNA from a biological species o r even an individual.

Gel electrophoresis is the usual method of classifying the size pattern but mass spectrometry can do that very well, and again in much shorter time. This application of mass spectrometry to DNA characterization will come much sooner than sequencing; i ndeed, one of the ORNL labs held the size record in 1996 by characterizing a 500-base fragment of DNA using matrix-assisted laser desorption ionization mass spectrometry. An ORNL group is also pursuing use of an electrospray method of mass spectrometry that has advantages such as simplicity of sample preparation and analysis. Hettich, Buchanan, Winston Chen, Scott McLuckey, Greg Hurst, and Mitch Doktycz, along with Rick Woychik, have made ma jor contributions to the use of mass spectrometry for DNA analysis.

Besides mass spectrometry, other methods exploit the characteristic reaction between two strands of DNA. Again, recalling that A pairs with T and G with C in opposite strands, a short strand of DNA, called an oligonucleotide, with a sequence of 5'-AGCTTTAACC will bind to 5'-GGTTAAAGCT (read it backwards and match it to the previous letter series) but not to 5'-GGTTAGGACT. This simple direct method of selecting DNA of complementary sequences has been used by Bob Foote and Mitch Doktycz, both of the Life Sciences Division.

Immobilized oligonucleotide arrays have been developed at ORNL to analyze DNA sequences. Above, the DNA sequence is amplified, labeled, and allowed to hybridize to the array of immobilized oligonucleotides. At right, the DNA sequence is deciphered by overlapping the hybridizing probes of similar sequence. The technique has applications in medicine, forensics, agriculture, and environmental bioremediation.

They synthesized a number of different DNA sequences directly on a glass surface and showed that the complementary sequence of the free DNA could be obtained by determining to which of the sequences it would bind. The free DNA was labeled with phospho rus-32 so that the pattern could be observed by autoradiography; in other words the glass was placed against a photographic film for a time, after which it was developed to determine the locations of the radioactive spots. In this case the immobile DNAs contained only eight bases and thus could determine the sequence of only eight bases in the free DNA. Longer immobile DNAs would give longer reads in the free DNA. This technique is often called sequencing by hybridization (SBH).

Interestingly, Doktycz and his colleagues have shown that the rules "A binds to T" and "G binds to C" are not completely reliable; mismatches can and do occur in hybridization experiments. If the mismatch is at or near the end of the DNA strand, the r ules are more likely to be broken than if the mismatch is at an interior position. The ORNL researchers developed a number of rules that predict when a mismatch is likely to occur. If these rules are ignored in an interpretation of binding patterns, t he reader could be misled as to the true sequence of the DNAs of interest. ORNL is collaborating with three other organizations that are developing this method of sequencing by hybridization. Because hybridization methods are so much faster than gel el ectrophoresis for sequence determination, it is likely that hybridization may become the method of choice for many applications for DNA fingerprinting. Ken Beattie, who recently joined the Life Sciences Division, brings his international experience wi th SBH to ORNL. He has proposed use of a flow-through SBH chip for multiple applications—genome sequencing, assessing the effectiveness of bioremediation, and evaluating the quality of agricultural microorganisms, plants, and animals, for examples.

The work of Foote and Doktycz used phosphorus-32 to label the DNA fragments. Other ways of labeling DNA are also employed. Currently, the most used are fluorescent labels, but Bruce Jacobson and Heinrich Arlinghaus of Atom Sciences, Inc. have been expl oring the use of enriched stable isotopes of tin and rare earth elements to label DNA fragments for sequencing by hybridization.

When a tin-labeled DNA segment hybridizes to its complementary DNA sequence in an array of short DNAs on a glass or nylon surface, the segment's location is determined by a laser-induced resonance ionization microprobe, which uses a laser beam to dissociate all atoms within several hundred monolayers of the sample surface. Those at oms that appear as ions are discarded; however, the neutral atoms of a selected element (in this case, tin) are converted to ions by illuminating them with a resonance laser that is tuned to the quantum energy of that element's electrons, causing only that element's atoms to be ionized. The ions produced in this way are introduced into a mass spectrometer, which separates them into isotopes according to differences in mass and measures the concentration of each tin isotope.

By working in a clean room, researchers using this highly selective and extremely sensitive technique have located DNAs that were hybridized to the complementary sequence with little background from either environmental tin or tin from noncomplementary tin-labeled DNAs, detectable on noncomplementary sequences. Because 10 stable isotopes of tin exist, it is possible to use all 10 labels at one time to "multiplex" the process. Ways of labeling specific DNA strands or fragments 8 to 20 bases long wit h tin enriched in a specific isotope have been devised by Rick Sachleben, Gil Brown, and Fred Sloop, all of ORNL's Chemical and Analytical Sciences Division. They have also developed a way to label specific DNAs with individual enriched stable isotope s of several rare earths, enabling the use of 20 or more labels at one time. Use of mass spectrometric detection of the stable isotopes increases the multiplexing possibilities greatly, allowing much more information to be obtained than is possible us ing only four fluorescent labels at a time. It is also an improvement because it overcomes the problem of the general fluorescent background when using nylon. Because phosphorus-32 decays with a half-life of about 14 days, the tin labeling approach al so looks attractive. Stable isotope labels may someday play a significant role in future DNA analyses. Imagine the SBH chip being interrogated by 10 or 20 stable isotope-labeled DNAs at one time. The laser atomization and mass spectrometry would identi fy each label in one pass, making the process much more efficient in time and materials. In addition, the laser atomization method would be accomplished in a few minutes, again much faster than present gel electrophoresis.

ORNL's surface-enhanced Raman gene (SERG) probes can locate free DNA molecules that have hybridized to other DNAs fixed on a surface. The technique has use in medi cine, forensics, agriculture, and environmental bioremediation.

Another hybridization method being developed at Oak Ridge uses a process called surface-enhanced Raman spectroscopy (SERS). Hybridized DNA is transferred from the nylon membrane to a glass strip coated with tiny silver spheres. The dye labels attached to the DNA bases have unique Raman infrared spectra, but the normally weak Raman lines are greatly enhanced by the presence of the silver spheres. This enhancement allows DNA bases to be detected at sufficient sensitivity to be useful for DNA sequenci ng studies. Earlier, Vo-Dinh, Mayo Uziel, and Alan Morrison, using SERS, detected a fluorescent carcinogen on DNA. Then, Dan Jacobson and Tom Ferrell demonstrated the power of SERS for DNA sequence analysis. Recently, Vo-Dinh, Kelly Houck, and David S tokes have developed it into a highly sensitive method that received an R&D 100 Award in 1996.

Yet another approach to DNA analysis taken by Vo-Dinh, in collaboration with members of the I&C Division, is to design an array of charge-coupled device (CCD) detectors and attach to their surface specific sequences of DNA. Free DNA strands are labeled with fluorescent molecules and hybridized to the bound DNA. When illuminated with laser light of the correct wavelength, the fluorescent tags of the hybridized DNA emit light signals that are detected by the individual CCD detectors behind each DNA site.

Because each of the CCD pixels has a different but known short DNA sequence bound to it, the sequence of the piece of the longer DNA strand that hybridizes with the short strand can be identified. By assembling all the information from the pixels, a l arger portion of the DNA sequence is obtained. Because bacteria have unique DNA sequences, these hybridization methods hold great promise as the basis for new techniques to rapidly analyze DNA, characterize its source, and identify bacteria.

DNA analysis of a different type depends on magic scissor enzymes, as described previously. For example, the enzyme EcoRI cuts DNA when 5'-GAATTC is on one strand of the DNA and its complement 5'-CTTAAG is on the other; the cut separates the G from th e A in both strands. This sequence occurs rather frequently in DNA, so the enzyme produces a large number of fragments. The enzyme Not I requires eight nucleotides in the sequence 5'-GCGGCCGC and its complement 5'GCGGCCGC; such a sequence occurs infreq uently, so fewer DNA fragments are produced, and they are generally much longer than those fragments cut by EcoRI. Because of the availability of more than 400 restriction enzymes that have unique and well-known sequence requirements, the DNA fragment patterns produced by the various enzymes can be used to characterize an individual's DNA. This is the basis for DNA fingerprints obtained through separation by size of DNA fragments by gel electrophoresis. Such fingerprints are used widely for eviden ce in research and in judicial courts.

To determine which DNA fragments make up a fingerprint, they must be separated and identified. The "lab on a chip" developed by Mike Ramsey and colleagues has been adapted to perform such separations within a few minutes—much faster than standard gel procedures. Like microcircuits and computers operating in parallel, these chemical separations chips can be used in parallel for DNA analysis. In one application, liquids containing DNA and a restriction enzyme are injected into different chambers etched into the chip. Electric fields pump the liquids through a microscopic channel into a reaction chamber, where the enzyme cuts the DNA into pieces of different lengths. The DNA snippets are then electrically pumped to the separation channel, wher e they are tagged with fluorescent dyes for detection.

The DNA fragments of various sizes are sorted in a liquid containing fibrous strands of a polymer. The DNA fragments get tangled with the polymer strands, which slow them down as they pass through. Small chunks of DNA find their way through the tangled web faster than the larger ones, so separation results. As the fragments ar e separated, they are illuminated with a laser light, causing them to fluoresce. The detected light intensities are fed to a computer, which sorts through signals from separated fragments to provide a sample analysis.

Automation and miniaturization are the two by-words in Ramsey's lab. To eliminate manual manipulation of DNA, Ramsey and his colleagues propose to build a more comprehensive chip that can prepare the DNA from a biological sample by lysing the cells, ex tracting the DNA, digesting it with restriction enzymes, amplifying it with the polymerase chain reaction, attaching fluorescent labels, and then obtaining the electrophoretic fingerprint. All steps would occur in a series of tiny chambers on the chip, and the results would be obtained in minutes rather than hours or days, as is the case now.

ORNL has developed a DNA mapping and sequencing technique using atomic force microscopy (AFM) to measure the length and width of double-stranded DNA molecules. The AFM is a sensitive device that traces the surface of a material, producing a three-dimen sional map. Certain viruses have circular DNA of known sizes, as characterized by various methods. Dave Allison, Doktycz, and Warmack, all in the Life Sciences Division, showed that the AFM can measure sizes of viral DNA circles.

The magic scissors enzyme, or restriction endonuclease, cuts a DNA sequence each time it occurs in the strand. But the mutant form of this enzyme simply binds to, rather than cleaves, the DNA strand. The AFM can image the resulting knot, or protein bum p, formed on the DNA thread. ORNL researchers can identify where the enzyme binds to the DNA within 100 base pairs, which is a very high resolution. They've shown that the distance from one bump to the next, as imaged by the AFM, was exactly as predicted from enzyme cutting experiments analyzed by gel electropho resis.

Through genetic engineering, other enzymes and proteins could be developed that bind to, rather than cut, particular DNA sequences. By using the AFM to view a DNA strand treated with different engineered enzymes, it may be possible to determine a certa in sequence of its chemical bases, a certain map position, or a spacing of genes and other biochemical landmarks on chromosomes. Using this biosensor, ORNL researchers have shown that the AFM can characterize a DNA strand according to distances betwee n particular restriction binding sites, eliminating the need to enzymatically cut the DNA into fragments for separation by gel electrophoresis. Because the AFM method is faster than gel electrophoresis, the genome sequencing community is already start ing to employ this technique for DNA characterization and mapping

Lipids in Bacteria and Human Fingerprints

Some bacteria can be identified by analyzing lipids and fatty acids with mass spectrometry, according to ORNL research. A lipid is a hydrocarbon compound in which the hydrogen and carbon atoms are linked in a long chain; a fatty acid is derived from a lipid. Bacteria are made of either polar or nonpolar lipids—that is, the lipids are soluble or not soluble in water depending on the ionic groups present. Using different solvents, ORNL researchers led by David White of the Environmental Sciences Division have extracted nonpolar and then polar lipids from a few dozen different types of bacteria. To identify each component of the mixture, they used gas chromatography and mass spectrometry. White has shown that each type of bacteria has an ident ifiable signature based on the unique chemical pattern in specific lipids. This technique could be used to identify rapidly many bacteria in the environment or those used during biological warfare.

Another bacteria detection method being developed by Bill Whitten does not require prior extraction. He has shown that, by illuminating airborne bacteria with laser light to obtain a mass spectrum, different bacteria can be distinguished by their indiv idual spectra. U.S. military organizations are seeking methods for identifying bacteria within five minutes to provide soldiers with enough warning about the hazard to employ effective countermeasures.

Our experience in analyzing lipids with mass spectrometry may have forensic applications. Lipids are the main components of human fingerprints. Recently, the Knoxville Police Department sought ORNL's assistance in determining why, under the same condit ions, fingerprints of young children disappear readily but adult fingerprints persist for weeks. Mass spectrometry of gas chromatographic analysis of lipids collected from fingers of children under the age of 10 shows that the spectra of children's li pids differ significantly from those of adult lipids. Lipids from children contain unesterified fatty acids, which disappear into the air from fingerprints exposed to the hot Tennessee sun. However, fatty acids in adult lipids are esterified with long-chain alcohols, making them much less volatile. Michelle Buchanan worked with Art Bohanan, a crime specialist in the Knoxville Police Department, to bring this finding at ORNL to the attention of the forensic community and to help stimulate support fo r forensics research at ORNL (see Pete Xiques' "ORNL's War on Crime, Technically Speaking"). Tuan Vo-Dinh and colleagues have developed computer software that can analyze the spectra of standard fingerprints very rapidly and enha nce spectra of fingerprint images too weak to be read visually, making possible identification of children's fingerprints

Anthropometry

Perhaps the most unusual of ORNL's biosensors is a new technique to measure human body surfaces. Such measurements, called anthropometry, are used by tailors, artists, and scientists. One of the finest minds in science to take a strong interest in anth ropometry was Leonardo da Vinci, who drew the famous *Proportions of the Human Figure* some 500 years ago. An ORNL scientist who has moved the field forward is Judson Jones of the Computer Science and Mathematic s Division. He has developed a technique using laser beams and mirrors to determine the shape of human body parts. He measures the topology of a solid surface, using amplitude-modulated laser radar, which measures arc lengths along complex and oft en inaccessible body contours. Because laser radar measures the phase and amplitude of a reflected, modulated laser beam, only one optical path is required between the sensor and the subject. Multiple images are combined by integrating information from different virtual viewpoints, any number of which can be created with strategically positioned mirrors. Arc lengths along arbitrary contours, surface areas, and volume estimates all become possible. The accuracy of the measurements is within 1 mm. Be cause creating "clothes that fit" may be accomplished using a single camera with several mirrors, a blue jeans manufacturer has shown interest in this methodology.

Oak Ridge scientists have been developing biosensors and bioreporters steadily for almost ten years. Now that the medical, military, and industrial communities are expressing more interest and providing more support, we can expect an accelerated pace i n new developments to use or evaluate living organisms, including bacteria, to gain valuable information about our bodies and the environment.

22.2 Biochip

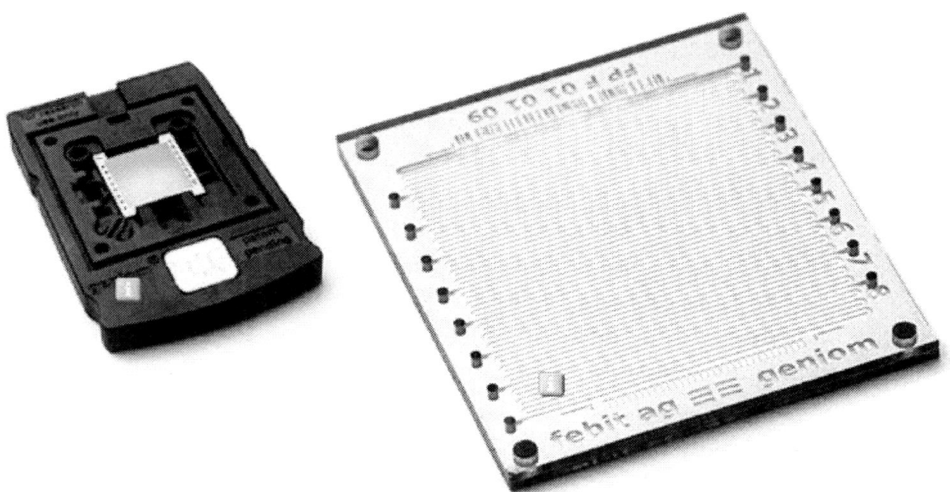

The development of *biochips* is a major thrust of the rapidly growing biotechnology industry, which encompasses a very diverse range of research efforts including genomics, proteomics, computational biology, and pharmaceuticals, among other activities. Advances in these areas are giving scientists new methods for unraveling the complex biochemical processes occurring inside cells, with the larger goal of understanding and treating human diseases. At the same time, the semiconductor industry has been steadily perfecting the science of microminiaturization. The merging of these two fields in recent years has enabled biotechnologists to begin packing their traditionally bulky sensing tools into smaller and smaller spaces, onto so-called biochips. These chips are essentially miniaturized laboratories that can perform hundreds or thousands of simultaneous biochemical reactions. Biochips enable researchers to quickly screen large numbers of biological analytes for a variety of purposes, from disease diagnosis to detection of bioterrorism agents.

Definition

A biochip is a collection of miniaturized test sites (microarrays) arranged on a solid substrate that permits many tests to be performed at the same time in order to achieve higher throughput and speed.

History

The development of biochips has a long history, starting with early work on the underlying sensor technology. One of the first portable, chemistry-based sensors was the glass pH electrode, invented in 1922 by Hughes (Hughes, 1922). Measurement of pH was accomplished by detecting the potential difference developed across a thin glass membrane selective to the permeation of hydrogen ions; this selectivity was achieved by exchanges between H^+ and SiO sites in the glass. The basic concept of using exchange sites to create permselective membranes was used to develop other ion sensors in subsequent years. For example, a K^+ sensor was produced by incorporating valinomycin into a thin membrane (Schultz, 1996). Over thirty years elapsed before the first true biosensor (*i.e.* a sensor utilizing biological molecules) emerged. In 1956, Leland Clark published a paper on an oxygen sensing electrode (Clark, 1956_41). This device became the basis for a glucose sensor developed in 1962 by Clark and colleague Lyons which utilized glucose oxidase molecules embedded in a dialysis membrane (Clark, 1962). The enzyme functioned in the presence of glucose to decrease the amount of oxygen available to the oxygen electrode, thereby relating oxygen levels to glucose concentration. This and similar biosensors became known as enzyme electrodes, and are still in use today.

In 1953, Watson and Crick announced their discovery of the now familiar double helix structure of DNA molecules and set the stage for genetics research that continues to the present day (Nelson, 2000). The development of sequencing techniques in 1977 by Gilbert (Maxam, 1977) and Sanger (Sanger, 1977) (working separately) enabled researchers to directly read the genetic codes that provide instructions for protein synthesis. This research showed how hybridization of complementary single oligonucleotide strands could be used as a basis for DNA sensing. Two additional developments enabled the technology used in modern DNA-based biosensors. First, in 1983 Kary Mullis invented the polymerase chain reaction (PCR) technique (Nelson, 2000), a method for amplifying DNA concentrations. This discovery made possible the detection of extremely small quantities of DNA in samples. Second, in 1986 Hood and coworkers devised a method to label DNA molecules with fluorescent tags instead of radiolabels (Smith, 1986), thus enabling hybridization experiments to be observed optically.

The rapid technological advances of the biochemistry and semiconductor fields in the 1980s led to the large scale development of biochips in the 1990s. At this time, it became clear that biochips were largely a "platform" technology which consisted of several separate, yet integrated components. The actual sensing component (or "chip") is just one piece of a complete analysis system. Transduction must be done to translate the actual sensing event (DNA binding, oxidation/reduction, *etc.*) into a format understandable by a computer (voltage, light intensity, mass, *etc.*), which then enables additional analysis and processing to produce a final, human-readable output. The multiple technologies needed to make a successful biochip—from sensing chemistry, to microarraying, to signal processing—require a true multidisciplinary approach, making the barrier to entry steep. One of the first commercial biochips was introduced by Affymetrix. Their "GeneChip" products contain thousands of individual DNA sensors for use in sensing defects, or single nucleotide polymorphisms (SNPs), in genes such as p53 (a tumor suppressor) and BRCA1 and BRCA2 (related to breast cancer) (Cheng, 2001). The chips are produced using microlithography techniques traditionally used to fabricate integrated circuits.

Today, a large variety of biochip technologies are either in development or being commercialized. Numerous advancements continue to be made in sensing research that enable new platforms to be developed for new applications. Cancer diagnosis through DNA typing is just one market opportunity. A variety of industries currently desire the ability to simultaneously screen for a wide range of chemical and biological agents, with purposes ranging from testing public water systems for disease agents to screening airline cargo for explosives. Pharmaceutical companies wish to combinatorially screen drug candidates against target enzymes. To achieve these ends, DNA, RNA, proteins, and even living cells are being employed as sensing mediators on biochips. Numerous transduction methods can be employed including surface plasmon resonance, fluorescence, and chemiluminescence. The particular sensing and transduction techniques chosen depend on factors such as price, sensitivity, and reusability.

Microarray Fabrication

The microarray—the dense, two-dimensional grid of biosensors—is the critical component of a biochip platform. Typically, the sensors are deposited on a flat substrate, which may either be passive (*e.g.* silicon or glass) or active, the latter consisting of integrated electronics or micromechanical devices that perform or assist signal transduction. Surface chemistry is used to the sensor molecules to the substrate medium. The fabrication of microarrays is non-trivial and is a major economic and technological hurdle that may ultimately decide the success of future biochip platforms. The primary manufacturing challenge is the process of placing each sensor at a specific position (typically on a Cartesian grid) on the substrate. Various means exist to achieve the placement, but typically robotic micro-pipetting (Schena, 1995) or micro-printing (MacBeath, 1999) systems are used to place tiny spots of sensor material on the chip surface. Because each sensor is unique, only a few spots can be placed at a time. The low-throughput nature of this process results in high manufacturing costs.

Fodor and colleagues developed a unique fabrication process (later used by Affymetrix) in which a series of microlithography steps is used to combinatorially synthesize hundreds of thousands of unique, single-stranded DNA sensors on a substrate one nucleotide at a time (Fodor, 1991; Pease, 1994). One lithography step is needed per base type; thus, a total of four steps is required per nucleotide level. Although this technique is very powerful in that many sensors can be created simultaneously, it is currently only feasible for creating short DNA strands (15-25 nucleotides). Reliability and cost factors limit the number of photolithography steps that can be done. Furthermore, light-directed combinatorial synthesis techniques are not currently possible for proteins or other sensing molecules.

As noted above, most microarrays consist of a Cartesian grid of sensors. This approach is used chiefly to map or "encode" the coordinate of each sensor to its function. Sensors in these arrays typically use a universal signaling technique (*e.g.* fluorescence), thus making coordinates their only identifying feature. These arrays must be made using a serial process (*i.e.* requiring multiple, sequential steps) to ensure that each sensor is placed at the correct position.

"Random" fabrication, in which the sensors are placed at arbitrary positions on the chip, is an alternative to the serial method. The tedious and expensive positioning process is not required, enabling the use of parallelized self-assembly techniques. In this approach, large batches of identical sensors can be produced; sensors from each batch are then combined and assembled into an array. A non-coordinate based encoding scheme must be used to identify each sensor. As the figure shows, such a design was first demonstrated (and later commercialized by Illumina) using functionalized beads placed randomly in the wells of an etched fiber optic cable (Steemers, 2000; Michael, 1998) Each bead was uniquely encoded with a fluorescent signature. However, this encoding scheme is limited in the number of unique dye combinations that be can be used and successfully differentiated.

Protein Biochip Array and Other Microarray Technologies

Microarrays are not limited to DNA analysis; protein microarrays, antibody microarray, Chemical Compound Microarray can also be produced using biochips. Randox Laboratories Ltd. launched Evidence, the first protein Biochip Array Technology analyzer in 2003. In protein Biochip Array Technology, the biochip replaces the ELISA plate or cuvette as the reaction platform. The biochip is used to simultaneously analyze a panel of related tests in a single sample, producing a patient profile. The patient profile can be used in disease screening, diagnosis, monitoring disease progression or monitoring treatment. Performing multiple analyses simultaneously, described as multiplexing, allows a significant reduction in processing time and the amount of patient sample required. Biochip Array Technology is a novel application of a familiar methodology, using sandwich, competitive and antibody-capture immunoassays. The difference from conventional immunoassays is that the capture ligands are covalently attached to the surface of the biochip in an ordered array rather than in solution.

In sandwich assays an enzyme-labelled antibody is used; in competitive assays an enzyme-labelled antigen is used. On antibody-antigen binding a chemiluminescence reaction produces light. Detection is by a charge-coupled device (CCD) camera. The CCD camera is a sensitive and high-resolution sensor able to accurately detect and quantify very low levels of light. The test regions are located using a grid pattern then the chemiluminescence signals are analysed by imaging software to rapidly and simultaneously quantify the individual analytes.

22.3 Biosurfactants: Properties, Commercial Production and Application

Biosurfactants or microbial surfactants are surface-active biomolecules that are produced by a variety of micro-organisms. Biosurfactants have gained importance in the fields of enhanced oil recovery, environmental bio-remediation, food processing and pharmaceuticals owing to their unique properties such as higher bio-degradability and lower toxicity. Interest in the production of biosurfactants has steadily increased during the past decade. However, large-scale production of these molecules has not been realized because of low yields in production processes and high recovery and purification costs. This article describes some practical approaches that have been adopted to make the bio-surfactant production process economically attractive. These include the use of cheaper raw materials, optimized and efficient bioprocesses and overproducing mutant and recombinant strains for obtaining maximum productivity. Here, we discuss the role and applications of biosurfactants focusing mainly on medicinal and therapeutic perspectives. With these specialized and cost-effective applications in biomedicine, we can look forward to biosurfactants as the molecules of the future.

Keywords: Biodegradability, biosurfactants, critical micelle concentration, cytotoxic, production process.

Biosurfactants are amphiphilic compounds produced on living surfaces, mostly on microbial cell surfaces, or excreted extracellularly and contain hydrophobic and hydrophilic moieties that confer the ability to accumulate between fluid phases, thus reducing surface and interfacial tension at the surface and interface respectively. They are a structurally diverse group of surface-active molecules synthesized by microorganisms. Rhamnolipids from *Pseudomonas aeruginosa*, surfactin from *Bacillus sub-tilis*, emulsan from *Acinetobacter calcoaceticus* and sophorolipids from *Candida bombicola* are some examples of microbial-derived surfactants. Originally, biosurfactants attracted attention as hydrocarbon dissolution agents in the late 1960s, and their applications have been greatly extended in the past five decades as an improved alternative to chemical surfactants (carboxylates, sulphonates and sulphate acid esters), especially in food, pharmaceutical and oil industry. The reason for their popularity as high value microbial products is primarily because of their specific action, low toxicity, higher biodegradability, effectiveness at extremes of temperature, pH, salinity and widespread applicability, and their unique structures which provide new properties that classical surfactants may lack. Biosurfactants possess the characteristic property of reducing the surface and interfacial tension using the same mechanisms as chemical surfactants. Unlike chemical surfactants, which are mostly derived from petroleum feedstock, these molecules can be produced by microbial fermentation processes using cheaper agro-based substrates and waste materials. During the past few years, bio-surfactant production by various microorganisms has been studied extensively. Also various aspects of biosurfactants, such as their biomedical and therapeutic properties, natural roles, production on cheap alternative substrates and commercial potential, have been recently reviewed. No attempt has been made, to the best of our knowledge, to describe the research and development strategies of making the biosurfactant production process cheaper and commercially attractive. The principle aim of the present article is to focus on such studies, with special emphasis on the development and use of mutant and recombinant hyperproducers of biosurfactants, and indication of direction towards their commercial production. Most of the work on biosurfactant applications has been focusing on bioremediation of pollutants and microbial enhanced oil recovery. However, these microbial compounds exhibit a variety of useful properties and applications in various fields. In this review, we discuss the potential roles and applications of biosurfactants mainly focusing on areas such as food and food-related industries (as emulsifiers, foaming, wetting, solubilizers, antiadhesive agents), biomedicine and therapeutics (as antimicrobial agents, immunoregulators and immunomodulators, their possible role in signalling and cytotoxic activity). With these specialized and cost-effective applications in biomedicine, we can look forward to biosurfactants as the molecules of the future.

Classification of Biosurfactants

Unlike chemically synthesized surfactants, which are usually classified according to the nature of their polar grouping, biosurfactants are generally categorized mainly by their chemical composition and microbial origin. Rosenberg and Ron suggested that biosurfactants can be divided into low-molecular-mass molecules, which efficiently lower surface and interfacial tension, and high molecular-mass polymers, which are more effective as emulsion-stabilizing agents. The major classes of low-mass surfactants include glycolipids, lipopeptides and phospholipids, whereas high-mass surfactants include polymeric and particulate surfactants. Most biosurfactants are either anionic or neutral and the hydrophobic moiety is based on long-chain fatty acids or fatty acid derivatives, whereas the hydrophilic portion can be a carbohydrate, amino acid, phosphate or cyclic peptide (Table 22.1). A brief discussion about each class of biosurfactant is given below.

Glycolipids

Most known biosurfactants are glycolipds. They are carbohydrates in combination with long-chain aliphatic acids or hydroxyaliphatic acids. The linkage is by means of ei ther ether or an ester group. Among the glycolipids, the best known are rhamnolipids, trehalolipids and sophorolipids.

Table 21.1: Major biosurfactant classes and microorganisms involved

Surfactant class	Microorganism
Glycolipids	
Rhamnolipids	*Pseudomonas aeruginosa*
Trehalose lipids	*Rhodococcus erithropolis*
	Arthobacter sp.
Sophorolipids	*Candida bombicola, C. apicola*
Mannosylerythritol lipids	*C. antartica*
Lipopeptides	
Surfactin/iturin/fengycin	*Bacillus subtilis*
Viscosin	*P. fluorescens*
Lichenysin	*B. licheniformis*
Serrawettin	*Serratia marcescens*
Phospholipids	*Acinetobacter* sp.
	Corynebacterium lepus
Surface-active antibiotics	
Gramicidin	*Brevibacterium brevis*
Polymixin	*B. polymyxa*
Antibiotic TA	*Myxococcus xanthus*
Fatty acids/neutral lipids	
Corynomicolic acids	*Corynebacterium insidibasseosum*
Polymeric surfactants	
Emulsan	*Acinetobacter calcoaceticus*
Alasan	*A. radioresistens*
Liposan	*C. lipolytica*
Lipomanan	*C. tropicalis*
Particulate biosurfactants	
Vesicles	*A. calcoaceticus*
Whole microbial cells	*Cyanobacteria*

❖ *Rhamnolipids:* These glycolipids, in which one or two molecules of rhamnose are linked to one or two molecules of ß-hydroxydecanoic acid, are the best studied.

   While the OH group of one of the acids is involved in glycosidic linkage with the reducing end of the rhamnose disaccharide, the OH group of the second acid is involved in ester formation. Production of rhamnose containing glycolipids was first described in *Pseudomonas aeruginosa* by Jarvis and Johnson. L-Rhamnosyl-L-rhamnosyl-ß-hydroxydecanoyl-ß-hydroxydecanoate and L-rhamnosyl- ß-hydroxydecanoyl-ß-hydrocydecanoate, referred to as rhamnolipids 1 and 2 respectively, are the principal glycolipids produced by *P. aeruginosa.*

❖ *Trehalolipids:* Several structural types of microbial tre-halolipid biosurfactants have been reported. Disaccharide trehalose linked at C-6 and C-6 'to mycolic acid is associated with most species of *Mycobacterium, Nocardia* and *Corynebacterium.* Mycolic acids are long-chain, a-branched-ß-hydroxy fatty acids. Trehalolipids from different organisms differ in the size and structure of mycolic acid, the number of carbon atoms and the degree of unsaturation. Trehalose lipids from *Rhodococcus erythropolis* and *Arthrobacter* sp. lowered the surface and interfacial tension in culture broth from 25 to 40 and 1 to 5 mN/m respectively.

❖ *Sophorolipids:* These glycolipids, which are produced mainly by yeast such as *Torulopsis bombicola T. petrophilum* and *T. apicola* consist of a dimeric carbohydrate sophorose linked to a long-chain hydroxyl fatty acid by glycosidic linkage. Generally, sophorolipids occur as a mixture of macrolactones and free acid form. It has been shown that the lactone form of the sophorolipid is necessary, or at least preferable, for many applications. These biosurfactants are a mixture of at least six to nine different hydrophobic sophorolipids.

Lipopeptides and Lipoproteins

A large number of cyclic lipopetides, including decapeptide antibiotics (gramicidins) and lipopeptide antibiotics (polymyxins) are produced. These consist of a lipid attached to a polypeptide chain.

❖ *Surfactin*: The cyclic lipopeptide surfactin, produced by *Bacillus subtilis* ATCC 21332, is one of the most powerful biosurfactants. It is composed of a seven amino-acid ring structure coupled to a fatty-acid chain via lactone linkage. It lowers the surface tension from 72 to 27.9 mN/m at concentrations as low as 0.005 per cent.

❖ *Lichenysin: Bacillus licheniformis* produces several biosurfacants which act synergistically and exhibit excellent temperature, pH and salt stability. These are also similar in structural and physio-chemical properties to the surfactin. The surfactants produced by *B. licheniformis* are capable of lowering the surface tension of water to 27 mN/m and the interfacial tension between water and *n*-hexadecane to 0.36 mN/m.

Fatty Acids, Phospholipids, and Neutral Lipids

Several bacteria and yeast produce large quantities of fatty acids and phospholipid surfactants during growth on n</i>-alkanes 27. The hydrophilic and lipophilic balance (HLB) is directly related to the length of the hydrocarbon chain in their structures. In Acinetobacter sp. strain HO1-N, phosphatidylethanolamine-rich vesicles are produced 28, which form optically clear microemulsions of alkanes in water. Phosphatidylethanolamine produced by R. erythropolis grown on n</i>-alkane causes a lowering of interfacial tension between water and hexadecane to less than 1 mN/m and a critical micelle concentration (CMC) of 30 mg/l (ref. 21).

Polymeric Biosurfactants

The best-studied polymeric biosurfactants are emulsan, liposan, alasan, lipomanan and other polysaccharide-protein complexes. Acinetobacter calcoaceticus RAG-1 produces an extracellular potent polyanionic amphipathics heteropolysaccharide bioemulsifier. Emulsan is an effective emisifying agent for hydrocarbons in water, even at a concentration as low as 0.001 to 0.01 per cent. Liposan is an extracellular water-souble emulsifier synthesized by Candida lipolytica and is composed of 83 per cent carbohydrate and 17 per cent protein.

Particulate Biosurfactants

Extracellular membrane vesicles partition hydrocarbons to from a microemulsion, which plays an important role in alkane uptake by microbial cells. Vesicles of Acinetobac ter sp. strain HO1-N with a diameter of 20-50 nm and a buoyant density of 1.158 cubic g/cm are composed of protein, phospholipids and lipopolysaccharide.

Properties of Biosurfactants

Biosurfactants are of increasing interest for commercial use because of the continually growing spectrum of available substances. There are many advantages of biosurfactants compared to their chemically synthesized counterpart. The main distinctive features of biosurfactants and a brief description of each property are given below.

Surface and Interface Activity

A good surfactant can lower surface tension of water from 72 to 35 mN/m and the interfacial tension of water/ hexadecane from 40 to 1 mN/m (ref. 14). Surfactin from B. subtilis can reduce the surface tension of water to 25 mN/m and interfacial tension of water/hexadecane to <1 mN/m (ref. 32). Rhamnolipids from P. aeruginosa decrease the surface tension of water to 26 mN/m and the interfacial tension of water/hexadecane to <1 mN/m (ref. 33). The sophorolipids from T. bombicola have been reported to reduce the surface tension to 33 m N/m and the interfacial tension to 5 mN/m (ref. 34). In general, biosurfactants are more effective and efficient and their CMC is about 10-40 times lower than that of chemical surfactants, i.e. less surfactant is necessary to get a maximum decrease in surface tension.

Temperature, pH and Ionic Strength Tolerance

Many biosurfactants and their surface activities are not affected by environmental conditions such as temperature and pH. McInerney *et al.* 26 reported that lichenysin from B. licheniformis JF-2 was not affected by temperature (up to 50 °C), pH (4.5-9.0) and by NaCl and Ca concentrations up to 50 and 25 g/l respectively. A lipopeptide from B. subtilis LB5a was stable after autoclaving (121 °C/20 min) and after 6 months at −18 °C; the surface activity did not change from pH 5 to 11 and NaCl concentrations up to 20 per cent (ref. 35).

Biodegradability

Unlike synthetic surfactants, microbial-produced compounds are easily degraded 36 and particularly suited for environmental applications such as bioremediation 14 and dispersion of oil spills.

Low Toxicity

Very little data are available in the literature regarding the toxicity of microbial surfactants. They are generally considered as low or non-toxic products and therefore, appropriate for pharmaceutical, cosmetic and food uses. A report suggested that a synthetic anionic surfactant (Corexit) displayed an LC50 (concentration lethal to 50% of test species) against Photobacterium phosphoreum ten times lower than rhamnolipids, demonstrating the higher toxicity of the chemical-derived surfactant. When comparing the toxicity of six biosurfactants, four synthetic surfactants and two commercial dispersants, it was found that most biosurfactants degraded faster, except for a synthetic sucrose-stearate that showed structure homology to glycolipids and was degraded more rapidly than the biogenic glycolipids. It was also reported that biosurfactants showed higher EC50 (effective concentration to decrease 50 per cent of test population) values than synthetic dispersants 37.

A biosurfactant from P. aeruginosa was compared with a synthetic surfactant (Marlon A-350) widely used in the industry, in terms of toxicity and mutagenic properties. Both assays indicated higher toxicity and mutagenic effect of the chemical-derived surfactant, whereas the biosurfactant was considered slightly non-toxic and non-mutagenic.

Emulsion Forming and Eemulsion Breaking

Stable emulsions can be produced with a lifespan of months and years 39. Biosurfactants may stabilize (emulsifiers) or destabilize (de-emulsifiers) the emulsion. High-molecular-mass biosurfactants are in general better emulsifiers than low-molecular-mass biosurfactants. Sophoro-lipids from T. bombicola have been shown to reduce surface and interfacial tension, but are not good emulsifiers. By contrast, liposan does not reduce surface tension, but has been used successfully to emulsify edible oils. Polymeric surfactants offer additional advantages because they coat droplets of oil, thereby forming stable emulsions. This property is especially useful for making oil/water emulsions for cosmetics and food.

Chemical Diversity

The chemical diversity of naturally produced biosurfactants offers a wide selection of surface-active agents with properties closely related to specific applications.

Medium Cheap Substrates: Economical and Promising Alternatives

Production economy is the major bottleneck in biosurfactant production, as in the case with most biotechnological processes. Often the amount and type of a raw material can contribute considerably to the production cost; it is estimated that raw materials account for 10-30 per cent of the total production cost in most biotechnological processes. Thus to reduce this cost it is desirable to use low-cost raw materials (Table 2) for the production of biosurfactants 10,40. One possibility explored extensively is the use of cheap and agro-based raw materials as substrates for biosurfactant production. A variety of cheap raw materials, including plant-derived oils, oil wastes, starchy substances, lactic whey and distillery wastes have been reported to support biosurfactant production.

Table 21.2: Use of inexpensive raw materials for the production of biosurfactants by various microbial strains.

Low Cost or Waste Raw Material	Biosurfactant Type	Maximum		
		Producer Microbial Strain	Yield (g/l)	Reference
Rapeseed oil	Rhamnolipids	Pseudomonas sp. DSM 2874	45	41
Babassu oil	Sophorolipids	Candida lipolytica IA 1055	11.72	42
Turkish corn oil	Sophorolipids	Candida bombicola ATCC 22214	400	43
Sunflower and soybean oil	Rhamnolipid	Pseudomonas aeruginosa DS10-129 4.31		44
Sunflower oil	Lipopeptide	Serratia marcescens	2.98 44	
Soybean oil	Mannosylerythritol lipid	Candida sp. SY16	95	45
Oil refinery waste	Glycolipids	Candida antarctica, Candida apicola	10.5 53	
Curd whey and distillery waste	Rhamnolipid	Pseudomonas aeruginosa strain BS2	0.92 48	
Potato process effluents	Lipopeptide	Bacillus subtilis 2.7		51
Cassava flour wastewater	Lipopeptide	B. subtilis ATCC 21332, B. subtilis LB5a	2.2 35	

Vegetable Oils and Oil Wastes

Several studies with plant-derived oils have shown that they can act as effective and cheap raw materials for biosurfactant production; for example, rapeseed oil, Babassu oil and corn oil 42,43. Similarly, vegetable oils such as sunflower and soybean oil were used for the production of rhamnolipid, sophorolipid and mannosylerythritol lipid biosurfactants by various microorganisms. Apart from various vegetable oils, oil wastes from vegetable-oil refineries and the food industry were also reported as good substrates for biosurfactant production. Furthermore, various waste oils with their origins at the domestic level, in vegetable-oil refineries or soap industries were found to be suitable for microbial growth and biosurfactant production.

Lactic Whey and Distillery Wastes

The effluent from the dairy industry, known as dairy waste-water, supports good microbial growth and is used as a cheap raw material for biosurfactant production. Dubey and Juwarkarcultivated P. aeruginosa BS2 on whey waste; within 48 h of incubation the yield of biosurfactant obtained was 0.92 g/l. Strain BS2 produced a crystalline biosurfactant as the secondary metabolites and its maximal production occurred after the onset of nitrogen-limiting conditions. The isolated biosurfactant possessed the potent surface-active properties, as it effectively reduced the surface tension of water from 72 to 27 mN/m and formed 100 per cent stable emulsion of a variety of water-insoluble compounds.

Starchy Substrates

Potato process effluents (waste from potato-processing industries) were used to produce biosurfactant by B. sub-tilis 50,51. Cassava wastewater, another carbohydrate-rich residue, which is generated in large amounts during the preparation of cassava flour, is also an attractive substrate and has been used for surfactin production by B. sub-tilis 35. Several other starchy waste substrates, such as rice water (effluent from rice processing industry and domestic cooking), cornsteep liquor and wastewater from the processing of cereals, pulses and molasses, have tremendous potential to support microbial growth and biosurfactant production.

Olive Oil Mill Effluent

Olive oil extraction involves an intensive consumption of water and produces large amounts of olive oil mill wastewater, thus causing deleterious environmental effects. Mercade *et al.* found that Pseudomonas sp. could reduce the surface tension in culture medium comprising olive oil mill effluent (OOME; 100 g/l) and NaNO 3 (2.5 g/l). Surface-active compounds produced from Pseudomonas sp. cultured in OOME medium included rhamnolipids biosurfactant, a total conversion yield was estimated to be 14 g of rhamnolipids per kg of OOME after 150 h of cultivation time.

Animal Fat

Animal fat and tallow can be obtained in large quantities from meat-processing industries and have been used as a cooking medium for food. Deshpande and Daniels used animal fat for the production of sophorolipid biosurfactant by yeast, C. bombicola. When only fat was provided as a sole carbon source, the growth was poor. A mixture of 10 per cent glucose and 10 per cent fat gave the highest level of growth. Sophorolipid was produced at levels of 97 and 12 g/l without and with pH control respectively.

Soapstock

Soapstock is a gummy, amber-coloured by-product of oil-seed processing. It is produced when hexane and other chemicals are used to extract and refine edible oil from the seeds. Shabtai reported the production of two extra-cellular capsular heteropolysaccharides, emulsan and biodispersan by A. calcoaceticus RAG-1 and A. calcoace-ticus A2 respectively, using soapstock as a carbon source. Emulsan forms and stabilizes the oil-water emulsion, whereas biodispersan disperses the large solid limestone granules, forming micrometre-sized water suspension.

Molasses

This is a co-product of sugar production, obtained from sugar cane as well as from sugar beet. Patel and Desai used molasses and cornsteep liquor as the primary carbon and nitrogen source to produce rhamnolipid biosurfactant from P. aeruginosa GS3. The biosurfactant production reached a maximum when 7 per cent (v/v) of molasses and 0.5 per cent (v/v) of cornsteep liquor were used. Maximal surfactant production occurred after 96 h of incubation, when cells reached the stationary phase of growth. A rhamnose concentration of 0.25 g/l and a reduction of interfacial tension between surfactant and crude oil of up to 0.47 mN/m were obtained.

Bioprocess Development: Optimum Production and Recovery

An efficient and economical bioprocess is the foundation for every profit-making biotechnology industry. Hence bioprocess development is the primary step towards commercialization of all biotechnological products, including biosurfactants. Any attempt to increase the yield of a biosurfactant demands optimal addition of media components and selection of the optimal culture conditions that will induce the maximum or optimum productivity. Similarly, efficient downstream processing techniques and methods are needed for maximum product recovery.

Process Optimization: The Best Combination of Essential Factors

Several elements, media components and precursors are reported to affect the process of biosurfactant production and the final quantity and quality. Different elements, such as nitrogen, iron and manganese are reported to affect the yield of biosurfactants; for example, the limitation of nitrogen is reported to enhance biosurfactant production in P. aeruginosa BS-2 (ref. 49) and Ustilago maydis 58. Similarly, addition of iron and manganese to the culture medium was reported to increase the production of biosurfactant by B. subtilis 59. The ratios of different elements such as C: N, C: P, C: Fe or C: Mg affected biosurfactant production and their optimization enhanced it.

Downstream Processing: Fast, Efficient and Cheap Product Recovery

Even if optimum production is obtained using optimal media and culture conditions, the production process is still incomplete without an efficient and economical means for recovery of the products. For many biotechnological products, the downstream processing costs account for ~60 per cent of the total production costs. Several conventional methods for the recovery of biosurfactants, such as acid precipitation, solvent extraction, crystallization, ammonium sulphate precipitation and centrifugation, have been widely reported in the literature 4. A few unconventional and interesting recovery methods have also been reported in recent years. Few examples of such biosurfactant recovery strategies (Table 22.3) include foam fractionation 60,61, ultrafiltration 62, adsorption-desorption on polystyrene resins and ion exchange chromatography 63, and adsorption-desorption on wood-based activated carbon 64. One of the main advantages of these methods is their ability to operate in a continuous mode for recovering biosurfactants with high level of purity. However, the solvents that are generally used for biosurfactant recovery, for example, acetone, methanol and chloroform, are toxic in nature and harmful to the environment. Cheap and less toxic solvents such as methyl tertiary-butyl ether have been successfully used in recent years to recover biosurfactants produced by Rhodococcus 65. These types of low cost, less toxic and readily available solvents can be used to cut the recovery expenses substantially and minimize environmental hazards. Often a single downstream processing technique is not enough for product recovery and purification. In such cases, a multi-step recovery strategy, using a sequence of concentration and purification steps, is more effective 63. In such a multi-step recovery for biosurfactants, it will be possible to obtain the product at any required degree of purity.

Mutant and Recombinant Strains: The Hyperproducers

The genetics of the producer organism is an important factor affecting the yield of all biotechnological products, because the capacity to produce a metabolite is bestowed by the genes of the organism. The bioindustrial production process is often dependent on the use of hyperproducing microbial strains, even with cheap raw materials, optimized medium and culture conditions, and efficient recovery processes.

Table 22.3: Physico-chemical property-based biosurfactant recovery methods and their relative advantages

Downstream recovery procedure	Biosurfactant property responsible for separation	Instrument/apparatus/ set-up required	Advantages
Acid precipitation	Biosurfactants become insoluble at low pH values	No set-up required	Low cost, efficient in crude biosurfactant recovery
Organic solvent extraction	Biosurfactants are soluble in organic solvents due to the presence of hydrophobic end	No set-up required	Efficient in crude biosurfactant recovery and partial purification, reusable nature
Ammonium sulphate precipitation	Salting-out of the polymeric or protein-rich biosurfactants	No set-up required	Effective in isolation of certain type of polymeric biosurfactants
Centrifugation	Insoluble biosurfactants get precipitated because of centrifugal force	Centrifuge required	Reusable, effective in crude biosurfactant recovery
Foam fractionation	Biosurfactants, due to surface activity, form and partition into foam	Specially designed bioreactors that facilitate foam recovery during fermentation	Useful in continuous recovery procedures, high purity of product
Membrane ultrafiltration	Biosurfactants form micelles above their critical micelle concentration, which are trapped by polymeric membranes	Ultrafiltration units with porous polymer membrane	Fast, one-step recovery, high level of purity
Adsorption on polystyrene resins	Biosurfactants are adsorbed on polymer resins and subsequently desorbed with organic solvents	Polystyrene resin packed in glass columns	Fast, one-step recovery, high level of purity, reusability
Adsorption on wood-activated carbon	Biosurfactants are adsorbed on activated carbon and can be desorbed using organic solvent	No set-up required. Can be added to culture broth. Can also be packed in glass columns	Highly pure biosurfactants, cheaper, reusability, recovery from continuous culture
Ion-exchange chromatography	Charged biosurfactants are attached to ion-exchange resins and can be eluted with proper buffer	Ion-exchange resins packed in columns	High purity, reusability, fast recovery
Solvent extraction (using methyl tertiary-butyl ether	Biosurfactants dissolve in organic solvents owing to the hydrophobic ends in the molecule	No set-up required	Less toxic than conventional solvents, reusability, cheap

A production process cannot be made commercially viable and profitable until the yield of the final product by the producer organisms is naturally high. Moreover, the industrial production process is dependent on the availability of recombinant and mutant hyperproducers if good yields are lacking from the natural producer strains. Even if high-yielding natural strains are available, the recombinant hyperproducers are always required to economize further the production process and to obtain products with better commercially important properties. Besides the natural biosurfactant producer strains, a few mutant and recombinant varieties with enhanced biosurfactant production characteristics are reported in the literature (Table 22.4). These mutant varieties were produced using various agents, for example, transposons, chemical mutagens such as N</i>-methyl-N '-nitro-N</i>-nitrosoguanidine, radiation 70 or by selection on the basis of resistance to ionic detergents such as CTAB 71. In addition to these mutant-hyperproducing varieties, several recombinant strains producing biosurfactants in better yields and showing improved production properties have been developed in recent years.

Table 22.4: Mutant and recombinant strains of microorganisms with enhanced biosurfactant yields and with improved product characteristics

Mutant and/or recombinant strain	Characteristic feature	Increased yield and/or improved production properties	Reference
P. aeruginosa 59C7	Transposon Tn5-GM-induced mutant of P. aeruginosa PG201	Two times more production	66
P. aeruginosa PTCC 1637	Random mutagenesis with N</i>-methyl-N'-nitro-N '-nitrosoguanidine	Ten times more production	67
B. licheniformis KGL11	Random mutagenesis with N</i>-methyl-'-nitro-N</i>-nitrosoguanidine	Twelve times more production	68
B. subtilis ATCC 55033	Random mutagenesis with N</i>-methyl-N'-nitro-N</i>-nitrosoguanidine	Approximately 4-6 times more production	69
P. aeruginosa EBN-8	Gamma ray-induced mutant of P. aeruginosa S8	2-3 times more production	70
A. calcoaceticus RAG-1	Mutant selection on basis of resistance to cationic detergent CTAB	2-3 times more production	71

Applications of Biosurfactants

All surfactants are chemically synthesized. Nevertheless, in recent years, much attention has been directed towards biosurfactants due to their broad range of functional properties and diverse synthetic capabilities of microbes. Most important is their environmental acceptability, because they are readily biodegradable and have low toxicity than synthetic surfactants. These unique properties of biosurfactants allow their use and possible replacement of chemically synthesized surfactants in a great number of industrial operations. Moreover, they are ecologically safe and can be applied in bioremediation and wastewater treatment. Some of the potential applications of biosurfactants in pollution and environmental control are microbial enhanced oil recovery, hydrocarbon degradation in soil environment and hexa-chloro cyclohexane degradation, heavy-metal removal from contaminated soil and hydrocarbon in aquatic environment (Table 22.5). In this review we discuss the potential roles and applications of biosurfactants, mainly focusing on areas such as food and food-related industries, biomedicine and therapeutics.

Table 22.5: Industrial applications of biosurfactants

Industry	Application	Role of biosurfactants
Petroleum	Enhanced oil recovery	Improving oil drainage into well bore, stimulating release of oil entrapped by capillaries, wetting of solid surfaces, reduction of oil viscosity and oil pour point, lowering of interfacial tension, dissolving of oil
	De-emulsification	De-emulsification of oil emulsions, oil solubilization, viscosity reduction, wetting agent
Environmental	Bioremediation	Emulsification of hydrocarbons, lowering of interfacial tension, metal sequestration
	Soil remediation and flushing	Emulsification through adherence to hydrocarbons, dispersion, foaming agent, detergent, soil flushing
Food	Emulsification and de-emulsification	Emulsifier, solubilizer, demulsifier, suspension, wetting, foaming, defoaming, thickener, lubricating agent
	Functional ingredient	Interaction with lipids, proteins and carbohydrates, protecting agent
Biological	Microbiological	Physiological behaviour such as cell mobility, cell communication, nutrient accession, cell-cell competition, plant and animal pathogenesis
	Pharmaceuticals and therapeutics	Antibacterial, antifungal, antiviral agents, adhesive agents, immunomodulatory molecules, vaccines, gene therapy
Agricultural	Biocontrol	Facilitation of biocontrol mechanisms of microbes such as parasitism, antibiosis, competition, induced systemic resistance and hypovirulence
Bioprocessing	Downstream processing	Biocatalysis in aqueous two-phase systems and microemulsions, biotransformations, recovery of intracellular products, enhanced production of extracellular enzymes and fermentation products
Cosmetic	Health and beauty products	Emulsifiers, foaming agents, solubilizers, wetting agents, cleansers, antimicrobial agents, mediators of enzyme action

Potential Food Applications

Biosurfactants can be explored for several food-processing applications. In this section we emphasize their potential as food-formulation ingredients and antiadhesive agents. Food-formulation ingredients: Apart from their obvious role as agents that decrease surface and interfacial tension, thus promoting the formation and stabilization of emulsions, surfactants can have several other functions in food. For example, to control the agglomeration of fat globules, stabilize aerated systems, improve texture and shelf-life of starch-containing products, modify rheological properties of wheat dough and improve consistency and texture of fat-based products. In bakery and icecream formulations biosurfactants act by controlling consistency, retarding staling and solubilizing flavour oils; they are also utilized as fat stabilizers and antispattering agents during cooking of oil and fats. Improvement in dough stability, texture, volume and conservation of bakery products is obtained by the addition of rhamnolipid surfactants. The study also suggested the use of rhamnolipids to improve the properties of butter cream, croissants and frozen confectionery products. L-Rhamnose has considerable potential as a precursor for flavouring. It is already used industrially as a precursor of high-quality flavour components like furaneol. Antiadhesive agents: A biofilm is described as a group of bacteria that have colonized a surface. The biofilm not only includes bacteria, but it also describes all the extracellular material produced at the surface and any material trapped within the resulting matrix. Bacterial biofilms present in the food industry surfaces are potential sources of contamination, which may lead to food spoilage and disease transmission. Thus controlling the adherence of microorganisms to food-contact surfaces is an essential step in providing safe and quality products to consumers. The involvement of biosurfactants in microbial adhesion and detachment from surfaces has been investigated. A surfactant released by Streptococcus thermophilus has been used for fouling control of heat-exchanger plates in pasteurizers, as it retards the colonization of other thermophilic strains of Streptococcus responsible for fouling. The preconditioning of stainless steel surfaces with a biosurfactant obtained from Pseudomonas fluorescens inhibits the adhesion of L. monocytogenes L028 strain. The bioconditioning of surfaces through the use of microbial surfactants has been suggested as a new strategy to reduce adhesion.

Therapeutic and Biomedical Applications

Antimicrobial activity: Several biosurfactants have shown antimicrobial action against bacteria, fungi, algae and viruses. The lipopeptide iturin from B. subtilis showed potent antifungal activity. Inactivation of enveloped virus such as herpes and retrovirus was observed with 80 mM of surfactin. Rhamnolipids inhibited the growth of harmful bloom algae species,

Heterosigma akashivo and Protocentrum dentatum at concentrations ranging from 0.4 to 10.0 mg/l. A rhamnolipid mixture obtained from P. aeruginosa AT10 showed inhibitory activity against the bacteria Escherichia coli, Micrococcus luteus, Alcaligenes faecalis (32 mg/ml), Serratia arcescens, Mycobacterium phlei (16 mg/ml) and Staphylococcus epidermidis (8 mg/ml) and excellent antifungal properties against Aspergillus niger (16 mg/ml), Chaetonium globosum, Enicillium crysogenum, Aureobasidium pullulans (32 mg/ml) and the phytopathogenic Botrytis cinerea and Rhizoctonia solani (18 mg/ml). Sophorolipids and rhamnolipids were found to be effective antifungal agents against plant and seed pathogenic fungi. The mannosylerythritol lipid (MEL), a glycolipid surfactant from Candida antartica, has demonstrated antimicrobial activity particularly against Gram-positive bacteria. Anticancer activity: The biological activities of seven microbial extracellular glycolipids, including manno-sylerythritol lipids-A, mannosylerythritol lipids-B, polyol lipid, rhamnolipid, sophorose lipid, succinoyl trehalose lipid (STL)-1 and succinoyl trehalose lipid-3 have been investigated. All these glycolipids, except rhamnolipid, were found to induce cell differentiation instead of cell proliferation in the human promyelocytic leukaemia cell line HL60. STL and MEL markedly increased common differentiation characteristics in monocytes and granulo-cytes respectively. Exposure of B16 cells to MEL resulted in the condensation of chromatin, DNA fragmentation and sub-G1 arrest (the sequence of events of apoptosis). This is the first evidence that growth arrest, apoptosis and differentiation of mouse malignant melanoma cells can be induced by glycolipids. In addition, exposure of PC12 cells to MEL enhanced the activity of acetylcholine esterase and interrupted the cell cycle at the G1 phase, with resulting outgrowth of neurites and partial cellular differentiation 82. This suggests that MEL induces neuronal differentiation in PC12 cells and provides the groundwork for the use of microbial extracellular glycolipids as novel reagents for the treatment of cancer cells. Another report suggested that the cytotoxic effects of sophorolipid on cancer cells of H7402, A549, HL60 and K562 were investigated by MTT assay. The results showed a dose-dependent inhibition ratio on cell viability according to the drug concentration <62.5 g/ml. These findings suggested that the sophorolipid produced by W. domercqiae have anticancer activity.

Immuno modulatory action: Sophorolipids are promising modulators of the immune response. It has been previously demonstrated that sophorolipids, (1) decreased sepsis related mortality at 36 h in vivo in a rat model of septic peritonitis by modulation of nitric oxide, adhesion molecules and cytokine production and (2) decreased IgE production in vitro in U266 cells possibly by affecting plasma cell activity. The results show that sophorolipids decrease

IgE production in U266 cells by downregulating important genes involved in IgE pathobiology in a synergistic manner. These data continue to support the utility of sophorolipids as an anti-inflammatory agent and a novel potential therapy in diseases of altered IgE regulation.

Anti-human immunodeficiency virus and sperm-immobilizing activity: The increased incidence of human immunodeficiency virus (HIV)/AIDS in women aged 15-49 years has identified the urgent need for a female-controlled, efficacious and safe vaginal topical microbicide. To meet this challenge, sophorolipid produced by C. bombicola Agents for stimulating skin fibroblast metabolism: The use of sophorolipids in lactone form comprises a major part of diacetyl lactones as agents for stimulating skin dermal fibroblast cell metabolism and more particularly, as agents for stimulating collagen neosynthesis, at a concentration of 0.01 ppm at 5 per cent (p/p) of dry matter in formulation. This is applicable in cosmetology and dermatology. The purified lactone sophorolipid product is of importance in the formulation of dermis anti-ageing, repair and restructuring products because of its effect on the stimulation of dermis cells. By encouraging the synthesis of new collagen fibres, purified lactone sophorolipids can be used both as a preventive measure against ageing of the skin and used in creams for the body, and in body milks, lotions and gels for the skin.

Antiadhesive agents in surgicals: Pre-treatment of silicone rubber with S. thermophilus surfactant inhibited by 85 per cent adhesion of C. albicans 87, whereas surfactants from L. fermentum and L. acidophilus adsorbed on glass, reduced by 77 per cent the number of adhering uropathogenic cells of Enterococcus faecalis. The biosurfactant from L. fermentum was reported to inhibit S. aureus infection and adherence to surgical implants 88. Surfactin decreased the amount of biofilm formation by Salmonella typhimurium, S. enterica, E. coli and Proteus mirabilis in PVC plates and vinyl urethral catheters.

Future Trends

Successful commercialization of every biotechnological product depends largely on its bioprocess economics. At present, the prices of microbial surfactants are not competitive with those of the chemical surfactants due to their high production costs and low yields. Hence, they have not been commercialized extensively. For the production of commercially viable biosurfactants, process optimization at the biological and engineering level needs to be improved. Improvement in the production technology of biosurfactants has already enabled a 10-20-fold increase in productivity, although further significant improvements are required. However, the use of cheaper substrates and optimal growth and production conditions coupled with novel and efficient multi-step downstream processing methods and the use of recombinant and mutant hyperproducing microbial strains can make bio-surfactant production economically feasible. Novel recombinant varieties of these microorganisms, which can grow on a wide range of cheap substrates and produce biosurfactants at high yields, can potentially bring the required breakthrough in the biosurfactant production process. Although a large number of biosurfactant producers have been reported in the literature, biosurfactant research, particularly related to production enhancement and economics, has been confined mostly to a few genera of microorganisms such as Bacillus, Pseudomonas and Candida. As documented in this review, biosurfactants are not only useful as antibacterial, antifungal and antiviral agents, they also have the potential for use as major immunomodulatory molecules, adhesive agents and even in vaccines and gene therapy. A judicial and effective combination of these strategies might, in the future, lead the way towards large-scale profitable production of biosurfactants. This will make biosurfactants highly sought after biomolecules for present and future applications as fine specialty chemicals, biological control agents, and new generation molecules for pharmaceutical, cosmetic and health care industries.

22.4 Microbial Production of Biosurfactants and their Importance

A large variety of microorganisms produce potent surface-active agents, biosurfactants, which vary in their chemical properties and molecular size. While the low molecular weight surfactants are often glycolipids, the high molecular weight surfactants are generally either polyanionic heteropolysaccharides containing covalently-linked hydrophobic side chains or complexes containing both polysaccharides and proteins. The yield of the biosurfactant greatly depends on the nutritional environment of the growing organism. The enormous diversity of biosurfactants makes them an interesting group of materials for application in many areas such as agriculture, public health, food, health care, waste utilization, and environmental pollution control such as in degradation of hydrocarbons present in soil.

Biosurfactants (BS) are amphiphilic compounds produced on living surfaces, mostly microbial cell surfaces, or excreted extracellularly and contain hydrophobic and hydrophilic moieties that reduce surface tension (ST) and interfacial tensions between individual molecules at the surface and interface, respectively. Since BS and bioemulsifiers both exhibit emulsification properties, bioemulsifiers are often categorized with BS, although emulsifiers may not lower surface tension. A biosurfactant may have one of the following structures: mycolic acid, glycolipids, polysaccharide—lipid complex, lipoprotein or lipopeptide, phospholipid, or the microbial cell surface itself.

Considerable attention has been given in the past to the production of surface-active molecules of biological origin because of their potential utilization in food-processing, pharmacology, and oil industry. Although the type and amount of the microbial surfactants produced depend primarily on the producer organism, factors like carbon and nitrogen, trace elements, temperature, and aeration also affect their production by the organism.

Hydrophobic pollutants present in petroleum hydrocarbons, and soil and water environment require solubilization before being degraded by microbial cells. Mineralization is governed by desorption of hydrocarbons from soil. Surfactants can increase the surface area of hydrophobic materials, such as pesticides in soil and water environment, thereby increasing their water solubility. Hence, the presence of surfactants may increase microbial degradation of pollutants. Use of biosurfactants for degradation of pesticides in soil and water environment has gained importance only recently. The identification and characterization of biosurfactant produced by various microorganisms have been extensively reviewed. Therefore, rather than describing the numerous types of biosurfactants and their properties, this article emphasizes the production of biosurfactants and their role in biodegradation of pesticides.

Microbiology

Microorganisms utilize a variety of organic compounds as the source of carbon and energy for their growth. When the carbon source is an insoluble substrate like a hydrocarbon (C_xH_y), microorganisms facilitate their diffusion into the cell by producing a variety of substances, the biosurfactants. Some bacteria and yeasts excrete ionic surfactants which emulsify the C_xH_y substrate in the growth medium. Some examples of this group of BS are rhamnolipids which are produced by different *Pseudomonas* sp. or the sophorolipids which are produced by several *Torulopsis* sp. Some other microorganisms are capable of changing the structure of their cell wall, which they achieve by synthesizing lipopolysaccharides or nonionic surfactants in their cell wall. Examples of this group are: *Candida lipolytica* and *C. tropicalis* which produce cell wall-bound lipopolysaccharides when growing on *n*-alkanes; and *Rhodococcus erythropolis*, and many *Mycobacterium* sp. and *Arthrobacter* sp. which synthesize nonionic trehalose corynomycolates. There are lipopolysaccharides, such as Emulsan, synthesized by *Acinetobacter* sp. and lipoproteins or lipopeptides, such as Surfactin and Subtilisin, produced by *Bacillus subtilis.* Other effective BS are: (i) Mycolates and Corynomycolates which are produced by *Rhodococcus* sp., *Corynebacteria* sp., *Mycobacteria* sp., and *Nocardia* sp.; and (ii) ornithinlipids, which are produced by *Pseudomonas rubescens, Gluconobacter cerinus,* and *Thiobacillus ferroxidans.* BS produced by various microorganisms together with their properties.

Classification and Chemical nature of Biosurfactants

The microbial surfactants (MS) are complex molecules covering a wide range of chemical types including peptides, fatty acids, phospholipids, glycolipids, antibiotics, lipopeptides, etc. Microorganisms also produce surfactants that are in some cases combination of many chemical types: referred to as the polymeric microbial surfactants (PMS). Many MS have been purified and their structures elucidated. While the high molecular weight MS are generally polyanionic heteropolysaccharides containing both polysaccharides and proteins, the low molecular weight MS are often glycolipids. The yield of MS varies with the nutritional environment of the growing microorganism. Intact microbial cells that have high cell surface hydrophobicity are themselves surfactants. In some cases, surfactants themselves play a natural role in growth of microbial cells on water-insoluble sub-strates like C_xH_y, sulphur, etc. Exocellular surfactants are involved in cell adhesion, emulsification, dispersion, flocculation, cell aggregation, and desorption phenomena. A very brief description of each group is given below.

Glycolipids

Glycolipids are the most common types of BS (ref. 32). The constituent mono-, di-, tri-and tetrasaccharides include glucose, mannose, galactose, glucuronic acid, rhamnose, and galactose sulphate. The fatty acid component usually has a composition similar to that of the phospholipids of the same microorganism. The glycolipids can be categorized as:

* ❖ *Trehalose lipids*: The serpentine growth seen in many members of the genus *Mycobacterium* is due to the presence of trehalose esters on the cell surface. Cord factors from different species of *Mycobacteria· Corynebacteria· Nocardia,* and *Brevibacteria* differ in size and structure of the mycolic acid esters.
* ❖ *Sophorolipids*: These are produced by different strains of the yeast, *Torulopsis.* The sugar unit is the disaccharide sophorose which consists of two b-1,2-linked glucose units. The 6 and 6¢ hydroxy groups are

generally acetylated. The sophorolipids reduce surface tensions between individual molecules at the surface, although they are effective emulsifying agents. The sophorolipids of *Torulopsis* have been reported to stimulate, inhibit, and have no effect on growth of yeast on water-insoluble substrates.

❖ *Rhamnolipids*: Some *Pseudomonas* sp. produce large quantities of a glycolipid consisting of two molecules of rhamnose and two molecules of b-hydroxydecanoic acid. While the OH group of one of the acids is involved in glycosidic linkage with the reducing end of the rhamnose disaccharide, the OH group of the second acids is involved in ester formation. Since one of the carboxylic acid is free, the rhamnolipids are anions above pH 4.0. Rhamnolipids are reported to lower surface tension, emulsify C_xH_y, and stimulate growth of *Pseudomonas* on *n*-hexadecane. Formation of rhamnolipids by *Pseudomonas* sp. MVB was greatly increased by nitrogen limitations. The pure rhamnolipid lowered the interfacial tension against *n*-hexadecane in water to about 1 mN/m and had a critical micellar concentration (cmc) of 10 to 30 mg/l depending on the pH and salt conditions.

Fatty Acids

The fatty acids produced from alkanes by microbial oxidations have received maximum attention as surfactants. Besides the straight-chain acids, microorganisms produce complex fatty acids containing OH groups and alkyl branches. Some of these complex acids, for example corynomucolic acids, are surfactants.

Phospholipids

These are major components of microbial membranes. When certain C_xH_y-degrading bacteria or yeast are grown on alkane substrates, the level of phospholipids increases greatly. Phospholipids from hexadecane-grown *Acinetobacter* sp. have potent surfactant properties. Phospholipids produced by *Thiobacillus thiooxidans* have been reported to be responsible for wetting elemental sulphur, which is necessary for growth.

Surface Active Antibiotics

❖ *Gramicidin S*: Many bacteria produce a cyclosymmetric decapeptide antibiotic, gramicidin S. Spore preparations of *Brevibacterium brevis* contain large amounts of gramicidin S bound strongly to the outer surface of the spores. Mutants lacking gramicidin S germinate rapidly and do not have a lipophilic surface. The antibacterial activity of gramicidin S is due to its high surface activity.

❖ *Polymixins*: These are a group of antibiotics produced by *Brevibacterium polymyxa* and related bacilli. Polymixin B is a decapeptide in which amino acids 3 through 10 form a cyclic octapeptide. A branched chain fatty acid is connected to the terminal 2,4-diaminobutyric acid (DAB). Polymixins are able to solubilize certain membrane enzymes.

❖ *Surfactin (subtilysin)*: One of the most active biosurfactants produced by *B. subtilis* is a cyclic lipopeptide surfactin. The yield of surfactin produced by *B. subtilis* can be improved to around 0.8 g/l by continuously removing the surfactant by foam fractionation and addition of either iron or manganese salts to the growth medium.

❖ *Antibiotic TA*: *Myxococcus xanthus* produces antibiotic TA which inhibits peptidoglycan synthesis by interfering with polymerization of the lipid disaccharide pentapeptide. Antibiotic TA has interesting chemotherapeutic applications.

Polymeric Microbial Surfactants

Most of these are polymeric heterosaccharide containing proteins.

❖ *Acinetobacter calcoaceticus RAG-1 (ATCC 31012) emulsan*: A bacterium, RAG-1, was isolated during an investigation of a factor that limited the degradation of crude oil in sea water. This bacterium efficiently emulsified C_xH_y in water. This bacterium, *Acinetobacter calcoaceticus*, was later successfully used to clear a cargo compartment of an oil tanker during its ballast voyage. The cleaning phenomenon was due to the production of an extracellular, high molecular weight emulsifying factor, emulsan.

❖ *The polysaccharide protein complex of Acinetobacter calcoaceticus BD413*: A mutant of *A. calcoaceticus* BD4, excreted large amounts of polysaccharide together with proteins. The emulsifying activity required the presence of both polysaccharide and proteins.

❖ *Other Acinetobacter emulsifiers*: Extracellular emulsifier production is widespread in the genus Acinetobacter. In one survey 8 to 16 strains of *A. calcaoceticus* produced high amounts of emulsifier following growth on ethanol medium. This extracellular fraction was extremely active in breaking (de-emulsifying) kerosene/water emulsion stabilized by a mixture of Tween 60 and Span 60.

❖ *Polysaccharide-lipid complexes from yeast*: The partially purified emulsifier, liposan, was reported to contain about 95 per cent carbohydrate and 5 per cent protein. A C_xH_y-degrading yeast, *Endomycopsis lipolytica* YM, produced an unstable alkane-solubilizing factor. *Torulopsis petrophilum* produced different types of surfactants depending on the growth medium. On water-insoluble substrates, the yeast produced glycolipids which were incapable of stabilizing emulsions. When glucose was the substrate, the yeast produced a potent emulsifier.

❖ *Emulsifying protein (PA) from Pseudomonas aeruginosa*: The bacterium *P. aeruginosa* has been observed to excrete a protein emulsifier. This protein PA is produced from long-chain *n*-alkanes, 1-hexadecane, and acetyl alcohol substrates; but not from glucose, glycerol or palmitic acid. The protein has a MW of 14,000 Da and is rich in serine and threonine.

❖ *Surfactants from Pseudomonas PG-1*: *Pseudomonas* PG-1 is an extremely efficient hydrocarbon-solubilizing bacterium. It utilizes a wide range of C_xH_y including gaseous volatile and liquid alkanes, alkenes, pristane, and alkyl benzenes.

❖ *Bioflocculant and emulcyan from the filamentous Cyanobacterium phormidium J-1*: The change in cell surface hydrophobicity of *Cyanobacterium phormidium* was correlated with the production of an emulsifying agent, emulcyan. The partially purified emulcyan has a MW greater than 10,000 Da and contains carbohydrate, protein and fatty acid esters. Addition of emulcyan to adherent hydrophobic cells resulted in their becomeing hydrophilic and detach from hexadecane droplets or phenyl sepharose beads.

Particulate Surfactants

❖ *Extracellular vesicles from Acinetobacter sp. H01-N*: *Acinetobacter* sp. when grown on hexadecane, accumulated extracellular vesicles of 20 to 50 mm diameter with a buoyant density of 1.158 g/cm. These vesicles appear to play a role in the uptake of alkanes by *Acinetobacter* sp. HO1-N. (refs 57, 84).

❖ *Microbial cells with high cell surface hydrophobicities*: Most hydrocarbon-degrading microorganisms, many nonhydrocarbon degraders, some species of *Cyanobacteria* and some pathogens have a strong affinity for hydrocarbon-water and air-water interfaces. In such cases, the microbial cell itself is a surfactant.

Factors Affecting Biosurfactant Production

Biosurfactants (BS) are amphiphilic compounds. They contain a hydrophobic and hydrophilic moiety. The polar moiety can be a carbohydrate, an amino acid, a phosphate group, or some other compound. The nonpolar moiety is mostly a long-carbon-chain fatty acid. Although the various BS possess different structures, there are some general phenomena concerning their biosynthesis. For example, BS production can be induced by hydrocarbons or other water-insoluble substrates. This effect, described by different authors, refers to many of the interfacially active compounds. Another striking phenomena is the catabolic repression of BS synthesis by glucose and other primary metabolites. For example, in the case of *Arthrobacter paraffineus*, no surface-active agent could be isolated from the medium when glucose was used as the carbon source instead of hexadecane. Similarly, a protein-like activator for *n*-alkane oxidation was formed by *P. aeruginosa* S7B1 from hydrocarbon, but not from glucose, glycerol, or palmitic acid. *Torulopsis petrophilum* did not produce any glycolipids when grown on a single-phase medium that contained water-soluble carbon source. When glycerol was used as substrate, rhamnolipid production by *P. aeruginosa* was sharply reduced by adding glucose, acetate, succinate or citrate to the medium.

Olive oil mill effluent, a major pollutant of the agricultural industry in mediterranian countries, has been used as raw material for rhamnolipid biosurfactant production by *Pseudomonas* sp. JAMM. Many microorganisms are known to synthesize different types of biosurfactants when grown on several carbon sources. However, there have been examples of the use of a water-soluble substrate for biosurfactant production by microorganisms. The type, quality and quantity of biosurfactant produced are influenced by the nature of the carbon substrate, the concentration of N, P, Mg, Fe, and Mn ions in the medium, and the culture conditions, such as pH, temperature, agitation and dilution rate in continous culture.

Biosurfactant production from *Pseudomonas* strains MEOR 171 and MEOR 172 are not affected by temperature, pH, and Ca, Mg, concentration in the ranges found in many oil reservior. Their production, on the other hand, in many cases improves with increased salinity. Thus, they are the biosurfactants of choice for the Venezuelan oil industry and in the cosmetics, food, and pharmaceutical markets.

The nitrogen source can be an important key to the regulation of BS synthesis. *Arthrobacter paraffineus* ATCC 19558 preferred ammonium to nitrate as inorganic nitrogen source for BS production. Urea also result in increased BS production. A change in growth rate of the concerned microorganisms is often sufficient to result in over production of BS (ref. 27). In some cases, addition of multivalent cations to the culture medium can have a positive effect on BS production. Besides the regulation of BS by chemicals indicated above, some compounds like ethambutol, penicillin, chloramphenicol, and EDTA influenced the formation of interfacially active compounds. The regulation of BS production by these compounds is either through their effect on solubilization of nonpolar hydrocarbon substrates or by increased production of water-soluble (polar) substrates. In some cases, BS synthesis is regulated by pH and temperature. For example in rhamnolipid production by *Pseudomonas* sp., in cellobioselipid formation by *Ustilago maydis,* and in sophorolipid formation by *Torulopsis bombicola,* pH played an important role, and in the case of *Arthrobacter paraffineus* ATCC 19558 (ref. 104), *Rhodococcus crythropolis,* and *Pseudomonas* sp. DSM 2874 (refs 47, 102) temperature was important. In all these cases however the yield of BS production was temperature dependent.

Applications of Biosurfactants in Pollution Control

The identification and characterization of microbial surfactants produced by various microorganisms have been extensively reviewed. Therefore rather than describing numeric types of MS, it is proposed to examine potential applications of MS.

Microbial Enhanced Oil Recovery

An area of considerable potential for BS application is microbial enhanced oil recovery (MEOR). In MEOR, microorganisms in reservoir are stimulated to produce polymers and surfactants which aid MEOR by lowering interfacial tension at the oil-rock interface. To produce MS *in situ,* microorganisms in the reservoir are usually provided with low-cost substrates, such as molasses and inorganic nutrients, to promote growth and surfactant production. To be useful for MEOR *in situ,* bacteria must be able to grow under extreme conditions encountered in oil reservoirs such as high temperature, pressure, salinity, and low oxygen level. Several aerobic and anaerobic thermophiles tolerant of pressure and moderate salinity have been isolated which are able to mobilize crude oil in the laboratory. Clark *et al.,* based on a computer search estimated that about 27 per cent of oil reservoirs in USA are amenable to microbial growth and MEOR. The effectiveness of MEOR has been reported in field studies carried out in US, Czechoslovakia, Romania, USSR, Hungary, Poland, and The Netherlands. Significant increase in oil recovery was noted in some cases.

Hydrocarbon Degradation

Hydrocarbon-utilizing microorganisms excrete a variety of biosurfactants. BS being natural products, are biodegradable and consequently environmentally safe. An important group of BS is mycolic acids which are the α-alkyl, β-hydroxy very long-chain fatty acids contributing to some characteristic properties of a cell such as acid fastness, hydrophobicity, adherability, and pathogenicity. Enriching waters and soils with long-and short-chain mycolic acids may be potentially hazardous. Daffe *et al.* reported trehalose polyphthienoylates as a specific glycolipid in virulent strains of *Mycobacterium tuberculosis*. Kaneda *et al.* reported that granuloma formation and hemopoiesis could be induced by C36-C48 mycolic acid-containing glycolipids from *Nocardia rubra*. Biolid extract (BE), obtained as a byproduct during the production of fodder yeast, is a dark brown heavy fluid with a characteristic odour and high interfacial activity. This product has many applications in agrochemistry, mineral flotation, and bitumen production and processing. Potentially, the product may be used as an emulsifying and dispersing agent while formulating herbicides, pesticides, and growth regulator preparations. Including phospholipids in formulations, facilitate penetration of active substances into the plant tissues, making it possible to apply only very low concentrations of the substances. The constituent fatty acids of biolipid extract have antiphytoviral and antifungal activities and therefore, can be applied in controlling plant diseases. These fatty acids also increase stress tolerance of plants, leading thereby to higher yields despite physiological drought.

Hydrocarbon Degradation in the Soil Environment

C_xH_y degradation in soil has been extensively studied. Degradation is dependent on presence in soil of hydrocarbon-degrading species of microorganisms, hydrocarbon composition, oxygen availability, water, temperature, pH, and inorganic nutrients. The physical state of C_xH_y can also affect biodegradation. Addition of synthetic surfactants or MS resulted in increased mobility and solubility of C_xH_y, which is essential for effective microbial degradation.

Use of MS in C_xH_y degradation has produced variable results. In the work of Lindley and Heydeman, the fungus *Cladosporium resiuae*, grown on alkane mixtures, produced extracellular fatty acids and phospholipids, mainly dodecanoic acid and phosphatidylcholine. Supplement of the growth medium with phosphatidylcholine enhanced the alkane degradation rate by 30 per cent. Foght *et al.* reported that the emulsifier, Emulsan, stimulated aromatic mineralization by pure bacterial cultures, but inhibited the degradation process when mixed cultures were used. Oberbremer and Muller-Harting used mixed soil population to assess C_xH_y degradation in model oil. Naphthalene was utilized in the first phase of C_xH_y degradation; other oil components were degraded during the second phase after the surfactants produced by concerned microorganisms lowered the interfacial tension. Addition of biosurfactants, such as some sophorolipids, increased both the extent of degradation and final biomass yield.

Biodetox (Germany) described a process to decontaminate soils, industrial sludges, and waste waters. They also described *in situ* bioreclamation of contaminated surface, deep ground and ground water. Microorganisms were added by means of a biodetox foam that contained bacteria, nutrients and surfactants; and was biodegradable. Another method to remove oil contaminants is to add BS into contaminated soil to increase C_xH_y mobility. The emulsified C_xH_y could then be recovered by using a production well, and subsequently degrading above ground in a bioreactor. *In situ* washing of soil was studied using two synthetic surfactants, Adsee 799 and Hyonic NP-90 (ref. 128). Removal of PCBs and petroleum C_xH_y from soil by adding surfactants to the wash water, has met with some success.

Several strains of anaerobic bacteria produce biosurfactants. However, the observed reduction in surface tension (45 to 50 mN/m) was not as large as the observed reduction in surface tension by anaerobic organisms (27 to 50 mN/m) (ref. 106). MS can also be used to enhance solubilization of toxic organic chemicals including xenobiotics. Berg *et al.* using the surfactant from *Pseudomonas aeruginosa* UG2, reported an increase in the solubility of hexachlorobiphenyl added to soil slurries, which resulted in a 31 per cent recovery of the compound in the aqueous phase. This was about 3-times higher than that solubilized by the chemical surfactant sodium ligninsulfonate (9.3%). When the *P. aeruginosa* bioemulsifier and sodium ligninsulphonate were used together, additive effect on solubilization (41.5%) was observed. *Pseudomonas ceparia* AC 1100 produced an emulsifier that formed a stable suspension with 2,4,5-T, and also exhibited some emulsifying activity against chlorophenols. Thus, this emulsifier can be used to enhance bacterial degradation of organochlorine compounds.

Hydrocarbon Degradation in Aquatic Environment

When oil is spilled in aquatic environment, the lighter hydrocarbon components volatilize while the polar hydrocarbon components dissolve in water. However, because of low solubility (< 1 ppm) of oil, most of the oil components will remain on the water surface. The primary means of hydrocarbon removal are photooxidation, evaporation, and microbial degradation. Since C_xH_y-degrading organisms are present in seawater, biodegradation may be one of the most efficient methods of removing pollutants. Surfactants enhance degradation by dispersing and emulsifying hydrocarbons. Microorganisms that are able to degrade C_xH_y have been isolated from aquatic environment. These microorganisms which exhibit emulsifying activity as well as the soil microorganisms which produced surfactants may be useful in aquatic environment. Chakrabarty reported that an emulsifier produced by *P. aeruginosa* SB30 was able to quickly disperse oil into fine droplets; therefore it may be useful in removing oil from contaminated beaches. BS produced by oil-degrading bacteria may be useful in cleaning oil tanks. When an oil tanker compartment containing oily ballast water was supplemented with urea and K_2HPO_4 and aerated for 4 days, the tanker was completely free of the thick layer of sludge that remained in the control tanker. Presumably this was owing to the surfactant produced, when growth of the natural bacterial population was enhanced.

Surfactants have been studied for their use in reducing viscosity of heavy oils, thereby facilitating recovery, transportation, and pipelining. Emulsan, a high MW lipopolysacharide produced by *A. calcaoceticus* RAG-1, has been proposed for a number of applications in the petroleum industry such as to clean oil and sludge from barges and tanks, reduce viscosity of heavy oils, enhance oil recovery, and stabilize water-in-oil emulsions in fuels. Specific solubilization of various C_xH_y types during growth of prokaryotic organism was demonstrated by Reddy *et al.* The specific solubilization of C_xH_y was strongly inhibited by EDTA which was overcome by excess Ca^{++}. It was concluded that specific solubilization of C_xH_y is an important mechanism in the microbial uptake of C_xH_y.

Pesticide-Specific Biosurfactants

Due to biodegradative property of biosurfactants, they are ideally suited for environmental applications, specially for removal of the pesticides—an important step in bioremediation. Survey of the literature reveals that application of biosurfactants in the field of pesticides is still in its infancy compared to the field of hydrocarbons. In India, a number of laboratories have initiated studies on BS. Some of the earlier works are by: (i) Banarjee *et al.* on 2,4,5-tricholoacetic acid, (ii) Patel and Gopinath on Fenthion, and (iii) Anu Appaiah and Karanth on alpha HCH. Very recently reports on production of microbial BS, based on preliminary studies by several groups, have appeared in posters/proceedings of symposia. The noteworthy feature being the increasing interest shown by the various researchers on: (i) degradation of pesticides, (ii) production and exploitation of BS for the removal of pesticides from the environment, and (iii) postulates on the possible replacement of synthetic surfactants with the biosurfactants in the pesticide formulation and clean-up.

Biosurfactant and HCH Degradation

Hexa-chlorocyclohexane (HCH) is still the highest ranking pesticide used in India and many other countries. Of the eight known isomers of HCH, the alpha-form constitutes more than 70 per cent of the technical product, which is not only noninsecticidal but also a suspected carcinogen. The use of technical HCH, which is a mixture of isomers, will continue in the Indian market because of their all-time availability with good insecticidal efficiency and at a price which is 10-12 times less than that of the pure gamma HCH (Lindane). It is pertinent to note that the environment burden of already-dumped HCH continues to pose threat to all forms of 'life'. The poor solubility is one of the limiting factors in the microbial degradation of alpha-HCH. Presence of six chlorines in the molecule is another factor that renders HCH lipophilic and persistent in the biosphere.

Even though several reports are available on biodegradation of specific isomers of HCH in animals, plants, soil and microbial systems, literature on metabolism of alpha-HCH by microorganisms is limited. Furthermore, the exact mechnism of translocation of HCH to the site of destruction and degradation of alpha-HCH in bacteria is not well understood.

During the course of our work at CFTRI on the bacterial degradation of alpha-HCH, we isolated several bacterial strains capable of degrading HCH. One of the strains efficient in HCH degradation was characterized as *Pseudomonas Ptm⁺* strain. The CFTRI isolate produced extracellular biosurfactant in a mineral medium containing HCH. While this BS emulsified the solid organochlorine-HCH to a higher extent, it emulsified other organochlorines such as DDT and cyclodienes to a lesser extent, implying thereby the specificity of the BS in dispersing HCH. It was also demonstrated that the peak in production of the emulsifier appeared before the onset of

HCH degradation by the *Pseudomonas* growing in liquid culture. The role of biosurfactant in the HCH degradation was ascertained using partially purified BS. The extracellular BS was a macro-molecule containing lipid, carbohydrate, and protein moieties. The carbohydrate part was identified as rhamnose by different analytical methods. The rhamnose part of the BS was stable and was necessary for the BS activity. Careful investigations revealed that the protein fraction represented the proximal enzymes of HCH metabolism. In the presence of BS, HCH was converted through the involvement of isomerase and dechlorinase to tertachlorohexenes and then to chlorophenols.

The BS acted by increasing the surface area of HCH, which accelerated this transformation. Hence, it is evident that extracellular BS has a definite role in HCH degradation by CFTRI strain of *Pseudomonas Ptm⁺*. Production of BS for Fenthion, a liqiud OP insecticide, has also received attention. *Bacillus subtilis* excreted the BS both in liqiud as well as in solid state fermentation system. The microbial surfactant produced by these two organisms also shows properties of a good cleansing agent for dislodging the pesticides from used containers, mixing tanks, cargo docks, etc. Attempts have also been made to standardize parameters for BS production both in liquid and solid state fermentations. A limited number of scale-up studies indicate good scope for expolitation of BS in industries.

In a separate study, it has been shown that addition of BS from *Pseudomonas Ptm⁺* strain facilitied 250-fold increase in dispersion of HCH in water. Addition of either this organism or BS dislodged surface-borne HCH residues from many types of fruits, seeds and vegetables as well. Laboratory-scale studies have revealed that BS is very efficient in cleaning the containers where HCH residues were sticking to the wall. Studies using fermentor for large-scale production of this BS from *Pseudomonas Ptm⁺* have been carried out. A bioformulation is planned from this BS for effective removal of HCH from contaminated soils.

Other Applications

By virtue of properties of biodegradability, substrate specificity, chemical and functional diversity, and rapid/ controlled inactivation, biosurfactants are gaining importance in various industires like agriculture, food, textiles, petrochemicals, etc. The potential applications of biosurfactants having desired functions and properties . The current consumption rate and estimated demand pattern for synthetic surfactants.

BS from some other bacterial taxa may be of public health concern.

Methylrhamnolipids from *Pseudomonas aeroginosa* have cytotoxic effects. Lipopolyglycans from mycoplasmas show endotoxic properties, potentially inducing procoagulant activity in human leukocytes. The toxicity and antigenic properties of mycobacterial glycolipids, produced by pathogenic mycobacteria such as *M. avium-intracellure, M. scrofulaceum,* and *M. fortulitum,* which are habitats of water polluted with industrial and domestic residues, are well known. The varied uses of BS also imply scope for MS, and the need to strengthen the research in this emerging area.

22.5 Biosurfactants and their Roles in Bioremediation

Introduction

The practical applications of biotechnology are vast and diverse; ranging from the cultivation of plant hybrids to the breeding of mice specifically designed as research instruments. It is a new world of possibilities, in which a spark of imagination one day can become the life-saving reality of the next. Many industries whose activities were traditionally remote from any living or biological systems have now recognised the tremendous potential of living cells in many areas, e.g. pre-treatment of raw material, product modifications, waste management, energy recycling and conservation and etc. Microbes and their products are nothing new to us as we have been using them since the ancient Egyptian era. Today, some of these products are used as specific food-stabilising agents, emulsifiers, vitamins, amino acids, proteins, food enzymes, specific fatty acids, water-binding agents, flavours, foaming agents and etc. Their applications also extend to the petroleum industries where various processes, like oil recovery, petroleum de-asphalting, viscosity control, desulphurization, oil-spill control, waste management and detoxification, are enhanced. Our reliance on microbial systems to produce alternative liquid and gaseous fuels from renewable world biomass will become greater, in view of the rapid deletion of the petroleum reserves at current stage.

Recent advances in industrial biotechnology have generated a lot of interest, from the perspectives of both research scientists and investors. One of the many reasons for the interest in biotechnology is the prediction of sales of by-products. The world market was about U.S. $25 million worth some twenty years back, but it was U.S. $1.7 billion in 1992 and reached U.S. $500 million just last year. That was an increase at an amazing rate of 3 to 5 per cent annually (Lin, 1996; Desai and Banat, 1997).

Surfactants constitute an important class of industrial chemicals that are widely used in almost every sector of modern industry. During the last decade, the growing demand of surfactants within the U.S. chemical industry was noted. The global production in 1999 At present, majority of the surfactant that one can find in the market is chemically synthesised. The manufacture of surfactants is on such a large scale that their production has always been considered to lie within the realms of organic chemistry and chemical engineering. Nevertheless, much interest and attention have been directed toward biosurfactants in recent years, due to their broad-range functional properties and the diverse synthetic capabilities of microbes. The structural analysis of biosurfactants has also opened possibilities for their chemical synthesis. They are likely to gain wide acceptance since they are readily biodegradable and have lower toxicity as compared to their chemically synthesised counterparts.

Some Facts about Surfactants and Biosurfactants

Surfactants are amphiphilic molecules consisting of a hydrophilic and a hydrophobic domain. The former can be non-ionic, positively or negatively charged or amphoteric, but the latter is usually a hydrocarbon (Georgiou *et al.*, 1992; Desai *et al.*, 1994). Common non-ionic surfactants include ethoxylates, ethylene and propylene oxide copolymers and sorbitan esters, while fatty acids, ester sulphonates or sulphates (anionic) and quaternary ammonium salts (cationic) are some examples of commercially available ionic surfactants. Since both the hydrophilic and hydrophobic groups reside within the same molecule, surfactants tend to partition preferentially at the interfaces between fluid phases with a different degree of polarity and hydrogen bonding (like that of oil/water or air/water interfaces). Unique properties of the surfactant molecules were driven by the reduction of interfacial energy (interfacial tension) and surface tension through the formation of an ordered molecular film at the interface. These properties allow them to be used extensively in industrial applications involving emulsification, foaming, detergence, wetting and phase dispersion or solubilization.

Many biological molecules are amphiphilic and partition preferentially at interphases. Microbial compounds that exhibit particularly high surface activity and emulsifying activity are classified as biosurfactants. At present, many types of biosurfactant are being utilised but they have been unable to compete economically with their chemically synthesised counterparts in the market, due to high production costs involved. However, this problem can be overcome by improving the efficiency of our current bioprocessing methodology and strain productivity, and the use of cost-effective substrates.

A variety of biosurfactants, of a broad spectrum of industrial applications, is synthesised by a wide range of microbes, particularly during their growth on water-immiscible substrates. They have definite structure: the lipophilic portion is usually the hydrocarbon (alkyl) tail of one or more fatty acids, which may be saturated, unsaturated, hydroxylated or branched; the fatty acid is linked to the hydrophilic group by a glycosidic ester or amide bond. Most of them are either neutral or negatively charged, the anionic character being due to carboxylate groups. According to Desai and Desai (1993), biosurfactants can be grouped into five major classes, namely glycolipids, phospholipids and fatty acids, lipopeptides or lipoproteins, polymeric surfactants and particulate surfactants. The types of biosurfactants, which are produced by microbes.

Physiological Roles of Biosurfactants

The main physiological role of biosurfactants is to permit microorganisms to grow on water-immiscible substrates by reducing the surface tension at the phase boundary, therefore making the substrate more readily available for uptake and metabolism. However, the molecular mechanisms of the uptake process of these substrates (e.g. alkanes) are still unclear. In addition to emulsification of the carbon source, they are also involved in the adhesion of microbial cells to the hydrocarbon. The cellular adsorption of the hydrocarbon-degrading microbes to water-immiscible substrates, and the excretion of surface-active compounds together allow growth on such carbon sources. In general, negatively charged biosurfactants inhibit, whereas positively charged biosurfactants promote microbial adhesion to hydrophobic phases. Furthermore, biosurfactants have been shown to be involved in antagonistic effect towards other microbes in the environment.

Biosynthesis of Biosurfactant and its Regulation

The hydrophilic and hydrophobic moieties of biosurfactants are synthesised by two metabolic pathways: the hydrocarbon and carbohydrate pathways (Desai and Banat, 1997). These pathways are usually diverse in nature and utilise specific sets of enzymes. According to Syldatk and Wagner (1987), who documented various possibilities for the synthesis of different moieties of biosurfactants and their association, any of the following possibilities could occur:

1. both hydrophilic and hydrophobic moieties are synthesised *de novo* by two independent pathways,

2. the hydrophilic moiety is synthesised de novo while synthesis of the hydrophobic moieties induced by the substrate,
3. the hydrophobic moiety is synthesised de novo while synthesis of the hydrophilic moieties induced by the substrate, and
4. both hydrophilic and hydrophobic moieties have substrate-dependent synthesis.

Generally, the regulation of biosurfactant production is governed by three main mechanisms, namely induction, repression and nitrogen and multivalent ions. Induction of biosurfactant synthesis can be achieved by addition of long chain fatty acids, hydrocarbons or glycerides to the growth medium; while addition of D-glucose, acetate or tricarboxylic acids represses their production. Nitrogen or metal ion-dependent regulation also played a prominent role in biosurfactant synthesis. It was observed that rhamnolipid synthesis in *Pseudomonas aeruginosa* occurs upon exhaustion of nitrogen and commencement of the stationary phase of growth (Guerra-Santos *et al.*, 1984; Ramana and Karanth, 1989). Lastly, the limitation of multivalent cations causes overproduction of biosurfactants.

Large Scale Production of Biosurfactants

Most of the biosurfactants are released into the culture medium either at the stationary phase or throughout the exponential phase. The production of biosurfactant has been carried out in batch fermentation or continuous fermentation at low dilution rates. Immobilised catalysis has been used for the biosurfactants that could be produced by resting cells. Alternatively, enhanced production of biosurfactants can be achieved by the application of an airlift fermentor and aqueous two-phase fermentation.

Application of Biosurfactants and Future Potential

A number of industrial of industrial applications of biosurfactants has been envisaged. One of the potential uses is in the oil industry with minimum purity specification so that whole cell-broth could be used. They are very selective, required in small amounts, effective under broad ranges of oil and reservoir conditions and environmentally friendly; especially in the safeguarding the coastal areas from additional damage caused by synthetic chemicals. An estimated 30 per cent increase in total oil recovery from underground sandstone by using trehalolipids from *Norcardia rhodochrous* has been documented.

Recently, Multi-biotech (a subsidiary of Geodyne Technology) has commercialised biosurfactants for enhanced oil recovery applications. *Bacillus licheniformis* JF-2, an isolate from oilfield injection water which, in addition to producing the most effective biosurfactants, has other characteristics like being anaerobic, halotolerant, and thermotolerant, produces biosurfactants that are potentially useful for in situ microbially enhanced oil recovery. As such, the viscosity of heavy crude oil can be greatly reduced when it is treated with this biosurfactant. In addition, the use of a biosurfactant for desludging of a crude oil storage tank for Kuwait Oil Co. has been reported (Banat *et al.*, 1991)

Apart from these, biosurfactants are also used in emerging technologies like microbial remediation of hydrocarbon and crude oil-contaminated soils. Hydrocarbon contaminants are removed from the environment, primarily, as a result of their biodegradation, which is performed by native microbial populations. Such biodegradation is known to be time-consuming and new technologies have been developed; for example bioaugmentation, biosurfactants and etc. While, the effectiveness of enhancing hydrocarbon degradation through addition of microbial inocula of nonindigenous populations (bioaugmentation in essence) has been ambiguous, the use of a combination of new approaches has shown much success. For instance, the addition of biosurfactant help to stimulate the indigenous microbial population to degrade hydrocarbons at rates higher than those which could be achieved through addition of nutrients alone. A good example of this is the use of rhamnolipid from *Pseudomonas aeruginosa* to remove large quantities of oil from contaminated Alaskan gravel a decade back, in the *Exxon Valdez* oil spill (Harvey *et al.*, 1990). Since then, Van Dyke *et al.* (1993) and Bragg *et al.* (1994) have demonstrated increased hydrocarbon recovery from the contaminated sites by rhamnolipid and the effectiveness of *in situ* bioremediation on the *Exxon Valdez* oil spill respectively.

The application of biosurfactant on the remediation of metal, phenanthrene and polychlorinated biphenyl (PCB) contaminated soil has recently been showed. For instance, Golyshin *et al.* (1999) conducted a study to determine the potential positive effect of novel biosurfactants on the enhancement of Aroclor 1248 metabolization in both *in vitro* and *in situ* experiments. Among two lipopeptides tested the highest activity was found in experiments with a hydrolytically opened form of lichenysin A. Lichenysin A itself did not enhance the degradation activity of chosen microorganism-degraders and in most cases inhibited their PCB mineralization rates. Glucolipid surfactant from marine bacterium *Alcanivorax borkumensis* (*A. borkumiensis*) showed in several tests a strong enhancing effect on microbial metabolization of Aroclor 1248 congeners. Biosurfactants appeared to act very specifically, i.e. depending on strain and concentration used. According to Zhang and Miller (1995), the three-way interaction between biosurfactant, substrate and cells is very critical to the enhanced rate of biodegradation. Experiments set up with soil samples did not give a clear answer whether bioemulsifiers applied at low concentration could sufficiently increase the rates of biodegradation *in situ*. Only *A. borkumiensis* glucose lipid caused the most marked enhancement of Aroclor 1248 metabolization in soil microcosm. Taking into account the specificity of surface-and biological activities of various biosurfactants, it was suggested that they might promote the mineralization of sorbed PCBs in polluted soils, when the optimised biosurfactant-degrader combination is used.

The ability to emulsify hydrocarbon-water mixtures is another established property of microbial biosurfactant. It has been shown to increase hydrocarbon degradation significantly and is therefore potentially applicable to oil spill management. Schulz *et al.* (1991) isolated biosurfactant-producing *Alcaligenes sp.* strain MM-1, *Arthrobacter sp.* strains EK1 and S-II from the North Sea, while screening for oil-degrading marine microorganisms. Dave *et al.* (1994) demonstrated that the use of a mixture of hydrocarbon-degrading microbes for bioaugmentation of soil contaminated with slop oil from a petrochemical industry resulted in the bioreclamation of soil, while Ghosh *et al.* (1995) showed an enhancement in bioremediation of polyaromatic hydrocarbons and soil contaminated with PCBs. In addition to these, Deziel *et al.* (1996) has shown biosurfactant production by polycyclic aromatic hydrocarbon-metabolising *Pseudomonas aeruginosa* 19SJ. Biosurfactants are also of great value to bioremediation of sites heavily contaminated with toxic heavy metals such as uranium, cadmium, lead and etc. (Miller, 1995). These observations suggest the application of natural microbial consortia and their products is a feasible approach to resolve certain environmental-related problems caused by mankind.

Moving away from the petrochemical industry, biosurfactants also found several applications in the food, healthcare and cosmetic industries. In the food industry, they are potential candidates in the search for functionally different products because they meet the requirements of functional food additives. Furthermore, lecithin and its derivatives, fatty acid esters containing glycerol, sorbitan, or ethylene glycol and ethoxylated derivatives of monoglycerides are currently in use as emulsifiers in the food industries globally. In the healthcare industry, biosurfactants are used commercially as a skin moisturiser due to their excellent skin compatibility (Yamane, 1987). Antibiotic effects and inhibition of growth of human immunodeficiency virus

(HIV), herpes simplex virus and influenza virus by biosurfactants have been documented (Uchida *et al.*, 1989; Kitamoto *et al.*, 1993). In the cosmetic industry, biosurfactants play a pivotal role in the production of many compounds for cosmetic applications. Kao Co. Ltd., for instance, uses sophorolipid as a humectant for cosmetic makeup brands such as Sofina. This organisation has developed a fermentation process for sophorolipid production, and after a two-step esterification process, the product can be used in lipsticks and as moisturiser for skin and hair products.

Other potential areas of application include that of the pulp and paper, coal, textile, ceramic processing and uranium ore-processing industries. For example, a heteropolysaccharide containing D-mannuronic acid and D-guluronate from *Macrocystis pyrifera* (kelp) and *Azobacter vinelandii* and polyglutamate from *Bacillus licheniformis* have been used as effective dispersants in the ceramic processing industry.

Conclusion

Recent interest has been generated in the production of biosurfactants due to their environment-friendly composition and wide range of applications. Their market is extremely competitive and manufacturers will have to focus on product development to capture the market for the year 2000 and beyond. Ultimately, it is the net economic balance between their production cost and functional benefits that controls the future of biosurfactant. To strike a fine balance between the two, one has to develop a cost-effective process that utilises low-cost materials and, on the other hand, achieving products of high yield and activity. Since biosurfactants and the microbes that produce them have numerous industrial, medical and environmental applications that often involve exposure to extremes of temperatures, pressure, pH and etc., there will be an urgent need to isolate extremophiles or to generate super-active genetically engineered bacteria that are capable of functioning under extreme conditions.

23

Bio-Polymers and Bio-Plastics

23.1 Biopolymer

Biopolymers are a class of polymers produced by living organisms. Starch, proteins and peptides, DNA, and RNA are all examples of biopolymers, in which the monomer units, respectively, are sugars, amino acids, and nucleotides.

Biopolymers versus Polymers

A major but defining difference between polymers and biopolymers can be found in their structures. Polymers, including biopolymers, are made of repetitive units called monomers. Biopolymers inherently have a well defined structure: The exact chemical composition and the sequence in which these units are arranged is called the primary structure. Many biopolymers spontaneously fold into characteristic compact shapes (see also "protein folding" as well as secondary structure and tertiary structure), which determine their biological functions and depend in a complicated way on their primary structures. Structural biology is the study of the structural properties of the biopolymers. In contrast most synthetic polymers have much simpler and more random (or stochastic) structures. This fact leads to a molecular mass distribution that is missing in biopolymers. In fact, as their synthesis is controlled by a template directed process in most in vivo systems all biopolymers of a type (say one specific protein) are all alike: they all contain the same sequence and number of monomers and thus all have the same mass. This phenomenon is called monodispersity in contrast to the polydispersity encountered in synthetic polymers. As a result biopolymers have a polydispersity index of 1.

Conventions and Nomenclature

Polypeptides

The convention for a polypeptide is to list its constituent amino acid residues as they occur from the amino terminus to the carboxylic acid terminus. The amino acid residues are always joined by peptide bonds. Protein, though used colloquially to refer to any polypeptide, refers to larger or fully functional forms and can consist of several polypeptide chains as well as single chains. Proteins can also be modified to include non-peptide components, such as saccharide chains and lipids.

Nucleic Acids

The convention for a nucleic acid sequence is to list the nucleotides as they occur from the 5' end to the 3' end of the polymer chain, where 5' and 3' refer to the numbering of carbons around the ribose ring which participate in forming the phosphate diester linkages of the chain. Such a sequence is called the primary structure of the biopolymer.

Sugars

Sugar-based biopolymers are often difficult with regards to convention. Sugar polymers can be linear or branched are typically joined with glycosidic bonds. However, the exact placement of the linkage can vary and the orientation of the linking functional groups is also important, resulting in á-and â-glycosidic bonds with numbering definitive of the linking carbons' location in the ring. In addition, many saccharide units can undergo various chemical modification, such as amination, and can even form parts of other molecules, such as glycoproteins.

Structural Characterization

There are a number of biophysical techniques for determining sequence information. Protein sequence can be determined by Edman degradation, in which the N-terminal residues are hydrolyzed from the chain one at a time, derivatized, and then identified. Mass spectrometer techniques can also be used. Nucleic acid sequence can be determined using gel electrophoresis and capillary electrophoresis. Lastly, mechanical properties of these biopolymers can often be measured using optical tweezers or atomic force microscopy.

Biopolymers as Materials

Some biopolymers—such as polylactic acid, naturally occurring zein, and poly-3-hydroxybutyrate can be used as plastics, replacing the need for polystyrene or polyethylene based plastics.

Some plastics are now referred to as being 'degradable', 'oxy-degradable' or 'UV-degradable'. This means that they break down when exposed to light or air, but these plastics are still primarily (as much as 98 per cent) oil-based and are not currently certified as 'biodegradable' under the European Union directive on Packaging and Packaging Waste (94/62/EC). Biopolymers, however, will break down and some are suitable for domestic composting.

Biopolymers as Packaging

Biopolymers (also called renewable polymers) are produced from biomass for use in the packaging industry. Biomass comes from crops such as sugar beet, potatoes or wheat: when used to produce biopolymers, these are classified as non food crops. These can be converted in the following pathways:

Sugar beet > Glyconic acid > Polyglonic acid
Starch > (fermentation) > Lactic acid > Polylactic acid (PLA)
Biomass > (fermentation) > Bioethanol > Ethene > Polyethylene

Many types of packaging can be made from biopolymers: food trays, blown starch pellets for shipping fragile goods, thin films for wrapping.

Biopolymers are Renewable, Sustainable, and can be Carbon Neutral

Biopolymers are renewable, because they are made from plant materials which can be grown year on year indefinitely. These plant materials come from agricultural non food crops. Therefore, the use of biopolymers would create a sustainable industry. In contrast, the feedstocks for polymers derived from petrochemicals will eventually run out. In addition, biopolymers have the potential to cut carbon emissions and reduce CO_2 quantities in the atmosphere: this is because the CO_2 released when they degrade can be reabsorbed by crops grown to replace them: this makes them close to carbon neutral.

Biopolymers are Biodegradable, and Some are Also Compostable

Some biopolymers are biodegradable: they are broken down into CO_2 and water by microorganisms. In addition, some of these biodegradable biopolymers are compostable: they can be put into an industrial composting process and will break down by 90 per cent within 6 months. Biopolymers that do this can be marked with a 'compostable' symbol, under European Standard EN 13432 (2000). Packaging marked with this symbol can be put into industrial composting processes and will break down within 6 months (or less). An example of a compostable polymer is PLA film under 20ìm thick: films which are thicker than that do not qualify as compostable, even though they are biodegradable. A home composting logo may soon be established: this will enable consumers to dispose of packaging directly onto their own compost heap. The standards for such a home composting logo have not yet been developed.

23.2 Tunable Biopolymers for Heavy Metal Removal

Introduction

Because of their intrinsically persistent nature, heavy metal ions are major contributors to pollution of the biosphere. These metals when discharged or transported into the environment may undergo transformations and can have a large environmental, public health, and economic impact. 3 Heavy metals such as Pb^{2+}, Hg^{2+}, and Cd^{2+} are currently ranked second, third, and seventh, respectively, on the EPA's priority list. Conventional physiochemical technologies are often inadequate to reduce metal concentration to acceptable regulatory standards. 4 The increasingly restrictive federal regulation of allowable levels of heavy metal discharge and accelerated requirements for the remediation of contaminated sites necessitate the development of highly efficient technologies to selectively remove dilute metal wastes to sub-part-per-billion levels.

The metal complexation behaviour of polymeric ligands has attracted much attention in recent years. "Polymer filtration" is an emerging cleanup technology, which employs water-soluble polymers for the chelation of heavy metals and ultrafiltration membranes to concentrate the polymer-metal complex and produce an effluent free of target metals. 5 Although this technology is highly effective, the cost of ultrafiltration and the toxic nature of the synthetic polymer make it undesirable for large-scale operation. Preferably, recovery of polymer-metal complexes can be achieved by simple changes in process condition, and environmentally benign materi-als can be employed. One alternative to synthetic polymers is the use of protein-based biopolymers. Advances in genetic engineering now enable us to specifically design metal-binding biopolymers with tunable properties that can be used to selectively remove heavy metals from dilute solutions.

Elastin (Ela)-based biopolymers consisting of the repeating pentapeptide (VPGVG) are structurally similar to the mammalian protein, elastin, that undergo a reversible phase transition from water-soluble forms into aggregates as the temperature increases. The transition from polymer solution into aggregates occurs over a wide range of transition temperatures (T t) and can be tuned by controlling the chain length and peptide sequence. The value of T t has also been shown to be sensitive to pH, ionic strength, pressure, and covalent modifications of the residues. Elastin oligomers with up to 180 repeats have been exploited for the thermally stimulated phase separation of enzymes. Elastin fusions exhibit a similar reversible, soluble-insoluble phase transition, and enzyme activity was unaffected even after repeated resolubilization. Moreover, Wang *et al.* have recently demonstrated that a genetically engineered His-tag coiled-coil protein domain can undergo repeated phase transitions without any effect on metal-binding via the histidine tag. These results suggested that thermally stimulated aggregation has no significant impact on the fusion domain. Because phase transition properties can be maintained as protein fusions and the fusion functionality can be maintained even after repeated phase transition, it is easy to envision that tunable biopolymers with metal-binding function could be similarly designed. This is a highly desirable property because the sequestered metals can be easily recovered as precipitates by inducing aggregation of the biopolymer-metal complex. In this paper, our goal was to develop tunable, metal-binding biopolymers and to investigate their metal-binding and reversible phase transition properties. Our results demonstrate the flexibility in engineering the phase transition and metal-binding functions independently for a wide range of applications.

Experimental Methods

Construction of Elastin-Based Biopolymers. DNA manipulations were performed according to standard methods. The high fidelity Pfu polymerase (Promega, Madison, WI) was used in PCR reactions. All cloning steps were carried out in E. coli JM109. DNA sequencing was carried out after each cloning step for verification.

The artificial genes were constructed as follows. Two oligonucletides (Loma Linda University, CA), CCCgaattcGTTC-CGGGTGTTGGT GTACCG-GGTGTTGGTGTGCCGGGTGT- TGGTGTTCCGGGT GTAGGCGTACCGGGGCGTAGGCG TGCCGGGGCG and CCCggatccTTA G T G G T G G T G G T G G T G G T G A C C T - A C A C - C C G G G ACGCCAACACCCGGCACGCCCACGCCCGGT-ACGCCCA CGCCCGG

AACG CCAACGCCCGGCACGCCTA-CGCCC, were annealed using a 20-base complementary region at the 3 'ends and extended with the Pfu polymerase to create the synthetic gene coding for (VPGVG) 10-H 6. The 189 bp fragment was digested with EcoRI and BamHI and inserted into the corresponding sites of plasmid pMAL-c2x (New England Biolabs, Beverly, MA). The resulting plasmid pEla10h6 allows expression of (VPGVG) 10-H 6 as a fusion to the maltose-binding protein (MalE). Proteins with higher number of VPGVG repeats were obtained by successive insertions of the (VPGVG) 10 repeating units. The VG(VPGVG) 10-H 6 sequence was PCR amplified using primers AGGCGTAGGCGTTC-CGGGTGTTGGTGTA and CCCGGATCCTTAGTG-GTGGTG-GTG and plasmid pEla10h6 as the template. After digestion with BamHI, the PCR product was inserted between the SmaI and BamHI sites located at the 3 ' end of the MalE-(VPGVG) n-H 6 fusion. This insert introduced 10 additional repeats of VPGVG and recreated the 3 ' terminal sequences including the H 6 tag. This extension step was repeated two times. DNA sequencing revealed that the resulting fusion gene was actually coding for MalE-(VPGVG) 38-H 6, instead of 40 repeats, which probably happened due to an error during PCR reactions.

The malE sequence was subsequently replaced with a short linker LIN1, resulting in a gene fusion coding for the MEF(VPGVG) 38 H 6 biopolymer. This fragment was subsequently excised with NdeI and BamHI and cloned into plasmid pET3a (Novagen, Madison, WI) under control of the T7 promoter. Further extension to 48, 58, 68, and 78 VPGVG repeats was accomplished using the same strategy described above. Modifications to the longest MEF(VPGVG) 78 H 6 fusion were made to construct biopolymers with either no histidine or 12 histidines. For the Ela78 biopolymers without histidine, plasmid pET-Ela78h6 was digested with SmaI-BamHI and ligated with the linker LIN2. For the Ela78H12 biopolymers containing two hexahistidines, plasmid pET-Ela78h6 was digested with BstEII and ligated with the linker LIN3, resulting in the production of MEF-(VPGVG) 78 H 6 GLQGH 6 fusions.

Production and Purification of Biopolymers. All cultivations were carried out in Luria-Bertani (LB) media supplemented with 100 µg/mL ampicillin. For protein expression, expression plasmids were introduced into E. coli BLR(DE3) (Novagen, Madison, WI). Cells were grown in 3 L of medium in a BIOFLO 3000 fermentor (New Brunswick Scientific, Edison, NJ) until late exponential phase when 1 mM IPTG was added. After 3 h, cells were harvested, washed in 0.9 per cent NaCl, and resuspended in 10 mM Tris pH 8.0 (Tb8) buffer. Cells were lysed with a French press, and cell debris was removed by centrifugation for 15 min at 30 000g. Purification of biopolymers by Ni-NTA agarose resin (QIAGEN, Valencia, CA) was done in small scale as described in the QUIAGEN manual. Purification of biopolymers by repeated temperature transition was achieved by modifying the procedures of McPherson *et al.* 13 To the cell free extracts, 5 M NaCl was added to bring a final concentration of 1 M. The sample was then heated to 30 °C and centrifuged at 30 000g at 30 °C. The pellet containing elastomeric protein was dissolved in cold Tb8 buffer. This temperature transition cycle (precipitation of the elastomeric protein with 1 M NaCl at 30 °C, centrifuging at 30 °C) was repeated two more times, and the pellet containng the biopolymers was finally dissolved in ice cold water. The protein concentration was determined as absorbance at 215 nm. The purity of the protein preparation was determined by SDS PAGE electrophoresis, 14 followed by silver staining (Bio-Rad, Hercules, CA).

Characterization of Metal-Binding Biopolymers. The transition temperature of the biopolymers was measured in a 96-well microplate. A 200 µL aliquot of protein solution was added in each well, and the optical density was followed at 655 nm in a microplate reader (BIO-RAD 3550-UV) with temperature control. The temperature was raised from 20 to 40 °C, in 2 °C increments each 5 min. The transition temperature was determined as the temperature at which the optical density reaches half of its maximum. 13,15 Metal-binding experiments were performed in 1 mL of Tb8 buffer at a final protein concentration of 44.5 µM and a final $CdCl_2$ concentration of 100 µM. After incubating for 30 min at room temperature, the biopolymers were precipitated by raising the temperature to 30 °C and by NaCl addition to 1 M. The solution was then centrifuged at 14 000g for 1 min in a benchtop microcentrifuge to separate the pellet. The pellet was washed with Tb8 buffer with the same NaCl concentration, redissolved in 1 mL of ice cold Tb8 buffer, and diluted in 2 per cent HNO_3. The amount of Cd^{2+} bound to the biopolymers was analyzed by flame atomic absorption spectrophotometry (Shi-madzu AA6701). In the isothermic or isotonic mode, metal binding was carried out in Tb8 buffer, and precipitation was achieved either by the addition of NaCl to a final concentration 1 M or by addition of 0.1 M NaCl and raising the temperatue *to 37°C.*

For the metal-binding experiments, 5 µg of cadmium was added in each cycle. After the precipitated biopolymers were recovered by centrifugation, they were incubated with a "stripping buffer", which was composed of either 50 mM sodium acetate pH 4 with 0.5 M NaCl (acid stripping) or Tb8 buffer containing 0.1 M NaCl and 100 mM EDTA (EDTA stripping). The regenerated biopolymers were recovered by centrifugation and redissolved in ice cold water. Subsequent cycles were repeated using the same procedure.

Design and Production of Tunable Biopolymers. Since the chain length of the elastin repeats has been shown to affect the transition properties, 6 initial efforts were focused on designing tunable biopolymers with varying VPGVG repeats. Only shorter biopolymers ranging from 30 to 80 repeats with predicted transition phase temperatures 8 around 30 °C in water were investigated because aggregation and recovery of biopoly-mer-metal complexes can be achieved with only a small increase in process temperature.

The strategy used for constructing the tunable biopolymers was similar to that reported by McPherson *et al.* The basic biopolymer was comprised of repeating VPGVG and a hexahistidine tail as the metal-binding moiety. The synthetic gene coding for (VPGVG) 10-(H) 6 was generated by annealing two synthetic oligos. Repeating VPGVG units were added until a chain length of 78 repeats was achieved. A SmaI site was specifically designed at the end of the elastin repeats to enable the easy replacement of different metal-binding moieties. Six different biopolymers consisting of a N-terminal 38-, 48-, 58-, 68-, or 78-repeating VPGVG domain and either one hexahistidine or two hexahistidines at the C-terminal were constructed. A similar Ela78 biopolymer without the hexahistidine tail was also constructed. Using this iterative approach of cloning repetitive elastin sequences, a series of specially designed biopolymers with precisely defined chain length and metal-binding properties were obtained.

All biopolymers were easily produced in E. coli BLR(DE3) and purified taking advantage of their inherent temperature responsive properties. The apparent molecular mass of the biopolymers was slightly higher than the calculated size, a phenomenon observed for other elastin-based proteins. 15 In addition, except for Ela78, which has no histidines, all other biopolymers were also purified by Ni-nitrilotriacetic acid (NTA) affinity chromatography, demonstrating the presence and functionality of the polyhistidine tags. These results clearly demonstrate that the metal-binding domain and the elastin domain are fully functional without interfering with each other. Selected biopolymers were produced in larger scale and purified by repeated temperature transition, with yields ranging from 148 to 236 mg in a 3 L culture. Shorter biopolymers were obtained in larger quantities as was reported for the production of other elastin-based proteins in E. coli.

Phase Transition Properties. The reversible phase transition behaviours of the biopolymers were investigated. The temperature profile for turbidity formation was used to measure the onset of folding and aggregation of the biopolymers. 17 The value of T t, which is defined as the temperature at which 50 per cent turbidity occurred, was used to indicate the phase transition properties of the different biopolymers. In agreement with other reports, the transition temperature decreased with increasing protein concentration (data not shown), protein size, and NaCl concentration. The Ela38H6 biopolymers exhibited a T t value >40 °C in Tris (pH 8) buffer with 0.25 M NaCl, while the value of T t decreased to 30 °C for biopolymers with 78 elastin repeats. The transition profile was also dependent on the polyhistidine tails; small differences in T t values were detected between the Ela78, Ela78H6, and Ela78H12 biopolymers, indicating that the presence of a small metal-binding moiety has some effect on the phase transition behaviour of the elastin domain. Aggregation of the biopolymers was reversible, and the precipitates were resolubilized completely when the temperature was decreased below T t. These data were used to select appropriate experimental conditions for the metal-binding experiments.

The phase transition behaviour could be influenced by other process parameters such as ionic strength. 18 As shown in Table 1, the value of T t can be tuned from >40 to 26 °C for the Ela38H6 biopolymers simply by adjusting the salt concentration. Similar behaviours were observed with the other biopolymers. In addition to temperature, this result suggests that biopolymermetal complexes could be recovered by salt addition under isothermal conditions. This operation may be preferred for large-scale processes.

Heavy Metal Removal by Tunable Biopolymers. The metal-binding capability of the biopolymers was demonstrated by incubating the biopolymers with excess cadmium. After 30 min incubation, the elastin biopolymers were recovered by precipitation, and the amount of Cd 2+ bound to the aggregates was measured. Independent of the number of elastin repeats, all biopolymers containing a hexahistidine domain bound Cd 2+ at a 1:1 ratio, an observation in line with the binding affinity of the hexahistidine tag. 19 Increasing the concentration of biopolymers resulted in a corresponding increase in Cd 2+ removal (data not shown). In contrast, a similar elastin biopolymer (Ela78) without the hexahistidine tag was unable to remove Cd 2+ from the solution even after induced precipitation. This result demonstrates that the elastin moiety itself does not bind or entrap Cd 2+. To demonstrate the possibility of fine-tuning the metal-binding capability of the biopolymers, metal-binding experiments were performed with the Ela78H12 biopolymer. The addition of six extra histidines increased the Cd 2+ binding capacity of the Ela78H12 biopolymers to a ratio of 1.5:1. Clearly, the metal-binding capability of the biopolymers can be tailored specifically for any target metal of interest simply by employing the appropriate metal-binding domain.

Beside temperature, another easy way to achieve precipitation of the biopolymers is by increasing the salt concentration at constant temperature (isothermal precipitation). On the basis of the temperature transition profiles reported in Table 23.1, the possibility of using either isotonic or isothermic conditions in the metalbinding removal process was investigated. Table 23.2 shows that both conditions were almost equally effective for Cd 2+ binding and removal by the Ela78H12 biopolymers.

Table 23.1: Transition Temperatures (C) of Biopolymers at Different NaCl Concentrations[a]

	NaCl (M)				
biopolymer	0.0	0.25	0.5	0.75	1.0
Ela38H6	>40	>40	>40	34	26
Ela58H6	>40	37	35	27	<20
Ela78H6	>40	33	29	22	<20
Ela78	>40	35	29	23	<20
Ela78H12	35	30	22	<20	<20

[a] Biopolymers were dissolved in 10 mM Tris (pH 8.0) buffer.

Table 23.2: Binding of Cadmium in Tb8 Buffer[a]

sbiopolymer	Cd^{2+} binding (mol protein: mol Cd^{2+} (SD^b))	biopolymer	Cd^{2+} binding (mol protein: mol Cd^{2+} (SD))
Ela38H6	1:1.04 (0.04)	Ela78H12	1:1.50 (0.02
Ela58H6	1:0.97 (0.01)	Ela78H12c	1:1.39 (0.01
Ela78H6	1:0.95 (0.01)	Ela78H12d	1:1.50 (0.03
Ela78	1:0.01 (0.00)		

[a] A 44.5 nmol sample of biopolymers was incubated with excess cadmium (100 nmol of Cd^{2+}). Aggregation of the biopolymers was induced by raising the temperature to 30 °C and by NaCl addition to 1 M. The amount of Cd^{2+} bound to the precipitates was measured. [b] SD) standard deviation. [c] Precipitation by increasing temperature only. [d] Precipitation by salt addition only.

Regeneration of the metal-binding sites for repeated usage is an important criterion for any successful application. Previous reports have demonstrated that sequestered metals can be removed from polyhistidine either by lowing the pH 20 or by treating with chelators such as EDTA. To demonstrate this possibility, precipitated Ela78H12-Cd 2+ complexes were removed from solution and treated with a "stripping buffer", which was composed of either a pH 4.0 acetate buffer or a solution containing 100 mM EDTA. Although both stripping buffers were effective in removing virtually all the bound Cd 2+, only the acidic stripping conditions could be used for repeated cycles.

Traces of EDTA prevented further binding of Cd 2+ to the biopolymers (data not shown). The regenerated biopolymer aggregates were resolubilized below 25 °C and remained fully functional even after four repeating cycles.

Conclusions

The tunable biopolymers presented here extended on ideas from nature toward entirely new objectives. Protein-protein interaction is tailored specifically into tunable metal-binding biopolymers. The biosynthetic approach is environmentally friendly and allows precise and independent control of the length, composition, and charge density of the interacting end blocks and metalbinding domains. The net result is the flexibility in designing tunable biopolymers that can undergo a transition from water-soluble forms into aggregates under a wild range of conditions, and such precise control is valuable to satisfy the needs of different process conditions. In addition, these tunable biopolymers offer the advantages of being easily regenerated and reused for many repeating cycles. Production and purification of biopolymers are based on the same phase transition principle and could be easily scaled up to kilogram quantity, therefore providing a low-cost and environmentally benign technology for heavy metal removal.

Although the results reported here are for biopolymers with polyhistidine as the metal-binding moiety, other metal-binding domains with high affinity toward heavy metals may be similarly applied. The ability to incorporate multiple binding domains with different selectivity for different metals within a single polypeptide may prove to be a versatile strategy for the removal of mixed metal wastes.

Acknowledgment. This work was supported in part by the UC Toxic Substances Teaching and Research Programme and a grant (BES0086698) from the National Science Foundation (New Technologies for the Environment). We thank Tammy Chen for her help with DNA sequencing and oligo synthesis.

23.3 Biopolymers and Bioplastics

Many everyday products used by Canadians are made of plastic. Pens, computers, cars, food packaging, and clothing are all examples of products which contain plastic. Plastic is a material made up of one or more polymers. The main source of the chemicals needed to manufacture plastics are fossil fuels. These petrochemical plastics are very durable, but take a long time to biodegrade when disposed. Rising concern about the cost of fossil fuels, and their impact on the environment has resulted in a search for alternatives to petrochemical plastics, namely biopolymers and bioplastics.

What are Biopolymers and Bioplastics?

Biopolymers and bioplastics go by many different names. They are often referred to as bio-based plastics and polymers, or as biodegradable plastics or polymers. They are defined below:

- ❖ *Biopolymers* are polymers which are present in, or created by, living organisms. These include polymers from renewable resources that can be polymerized to create bioplastics.
- ❖ *Bioplastics* are plastics manufactured using biopolymers, and are biodegradable.
- ❖ Biopolymers and bioplastics are not new products. Henry Ford developed a method of manufacturing plastic car parts from soybeans in the mid-1900s. However, World War II side-tracked the production of bioplastic cars. Today, bioplastics are gaining popularity once again as new manufacturing techniques developed through biotechnology are being applied to their production.

Types of Biopolymers

There are two main types of biopolymers: those that come from living organisms; and, those which need to be polymerized but come from renewable resources. Both types are used in the production of bioplastics.

Biopolymers from Living Organisms

These biopolymers are present in, or created by, living organisms. These include carbohydrates and proteins. These can be used in the production of plastic for commercial purposes. Examples are listed in the table 23.3.

Polymerizable Molecules

These molecules come from renewable natural resources, and can be polymerized to be used in the manufacture of biodegradable plastics.

Biopolymer	Natural Source	What is it?
Lactic Acid	Beets, corn, potatoes, and others	Produced through fermentation of sugar feedstocks, such as beets, and by converting starch in corn, potatoes, or other starch sources. It is polymerized to produce polylactic acid—a polymer that is used to produce plastic.
Triglycerides	Vegetable oils	These form a large part of the storage lipids found in plant and animal cells. Vegetable oils are one possible source of triglycerides that can be polymerized into plastics.

The Science—How are Biopolymers and Bioplastics Made?

There are two methods being researched and used to produce plastics from plants. The first uses fermentation, and the second relies on the plant to become the factory for plastic production. These two methods are outlined below.

Table 23.3: Biopolymers from living organisms

Biopolymer	Natural Source	What is it?
Cellulose	Wood, cotton, corn, wheat, and others	This polymer is made up of glucose. It is the main component of plant cell walls.
Soy protein	Soybeans	Protein which naturally occurs in the soy plant.
Starch	Corn, potatoes, wheat, tapioca, and others	This polymer is one way carbohydrates are stored in plant tissue. It is a polymer made up of glucose. It is not found in animal tissues.
Polyesters	Bacteria	These polyesters are created through naturally occurring chemical reactions that are carried out by certain types of bacteria.

Using Fermentation to Produce Plastics

Fermentation, used for hundreds of years by humans, is even more powerful when coupled with new biotechnology techniques. Fermentation is the use of microorganisms to break down organic substances in the absence of oxygen. Today, fermentation can be carried out with genetically engineered microorganisms, specially designed for the conditions under which fermentation takes place, and for the specific substance that is being broken down by the microorganism. There are two ways fermentation can be used to create biopolymers and bioplastics:

❖ *Bacterial Polyester Fermentation*: Bacteria are one group of microorganisms that can be used in the fermentation process. Fermentation, in fact, is the process by which bacteria can be used to create polyesters. Bacteria called Ralstonia eutropha are used to do this. The bacteria use the sugar of harvested plants, such as corn, to fuel their cellular processes. The by-product of these cellular processes is the polymer. The polymers are then separated from the bacterial cells.

❖ *Lactic Acid Fermentation:* Lactic acid is fermented from sugar, much like the process used to directly manufacture polymers by bacteria. However, in this fermentation process, the final product of fermentation is lactic acid, rather than a polymer. After the lactic acid is produced, it is converted to polylactic acid using traditional polymerization processes.

Growing Plastics in Plants

Plants are becoming factories for the production of plastics. Researchers created a Arabidopis thaliana plant through genetic engineering. The plant contains the enzymes used by bacteria to create plastics. Bacteria create the plastic through the conversion of sunlight into energy. The researchers have transferred the gene that codes for this enzyme into the plant, as a result the plant produces plastic through its cellular processes. The plant is harvested and the plastic is extracted from it using a solvent. The liquid resulting from this process is distilled to separate the solvent from the plastic.

Biotechnology and Biopolymers and Bioplastics

Biotechnology is driving the production of new bioplastics. Biotechnology techniques used to produce bioplastics include fermentation, and genetic engineering. For example, fermentation is used to release the cellulose from plants, so the cellulose can be used to create plastics. Also, genetic engineering can be used to create plants, such as soybean, specifically designed to be used as a raw material for the production of bioplastics.

Current Research Areas in Biopolymers and Bioplastics

Improving efficiency is a major concern for the production of plastics and bioplastics. Currently, fossil fuel is still used as an energy source during the production process. This has raised questions by some regarding how much fossil fuel is actually saved by manufacturing bioplastics. Only a few processes have emerged that actually use less energy in the production process. Therefore, researchers are still working on refining the processes used in order to make bioplastics viable alternatives to petrochemical plastics.

Energy use is not the only concern when it comes to biopolymers and bioplastics. There are also concerns about how to balance the need to grow plants for food, and the need to grow plants for use as raw materials.

Agricultural space needs to be shared. Researchers are looking into creating a plant that can be used for food, but also as feedstock for plastic production. One group is attempting to genetically engineer corn to contain the bacterial enzyme responsible for plastic production. Eventually, they are hoping to create the plant in a way which would restrict the plastic production to the stem, and leaves of the plant. This would leave the edible part of the corn plastic free. The edible part of the corn would be used as food, or as livestock feed. The plastic would be removed from the remaining part of the corn plant.

Sustainable Development and Biopolymers and Bioplastics

Biopolymers and bioplastics are the main components in creating a sustainable plastics industry. These products reduce the dependence on non-renewable fossil fuels, and are easily biodegradable. Together, this greatly limits the environmental impacts of plastic use and manufacture. Also, characteristics such as being biodegradable make plastics more acceptable for long term use by society. It is likely that in the long term, these products will mean plastics will remain affordable, even as fossil fuel reserves diminish.

23.4 Bioplastic

Bioplastics are a form of plastics derived from renewable biomass sources, such as vegetable oil, corn starch, pea starch or microbiota, rather than traditional plastics which are derived from petroleum. They are used either as a direct replacement for traditional plastics or as blends with traditional plastics. There is no international agreement on how much bio-derived content is required to use the term bioplastic.

Bioplastics and Biodegradation

The terminology used in the bioplastics sector is sometimes misleading. Most in the industry use the term bioplastic to mean a plastic produced from a biological source. One of the oldest plastics, cellulose film, is made from wood cellulose. All bio-and petroleum-based plastics are technically biodegradable, meaning they can be degraded by microbes under suitable conditions. However many degrade at such slow rates as to be considered non-biodegradable. Some petrochemical-based plastics are considered biodegradable, and may be used as an additive to improve the performance of many commercial bioplastics. Non-biodegradable bioplastics are referred to as durable. The degree of biodegradation varies with temperature, polymer stability, and available oxygen content. Consequently, most bioplastics will only degrade in the tightly controlled conditions of commercial composting units. An internationally agreed standard, EN13432, defines how quickly and to what extent a plastic must be degraded under commercial composting conditions for it to be called biodegradable. This is published by the International Organisation for Standardization ISO and is recognised in many countries, including all of Europe, Japan and the US. However, it is designed only for the aggressive conditions of commercial composting units. There is no standard applicable to home composting conditions.

The term "biodegradable plastic" is often also used by producers of specially modified petrochemical-based plastics which appear to biodegrade. Traditional plastics such as polyethylene are degraded by ultra-violet (UV) light and oxygen. To prevent this process manufacturers add stabilising chemicals. However with the addition of a degradation initiator to the plastic, it is possible to achieve a controlled UV/oxidation disintegration process. This type of plastic may be referred to as *degradable plastic* or *oxy-degradable plastic* or *photodegradable plastic* because the process is not initiated by microbial action. While some degradable plastics manufacturers argue that degraded plastic residue will be attacked by microbes, these degradable materials do not meet the requirements of the EN13432 commercial composting standard.

Environmental Impacts

The production and use of bioplastics is generally regarded as a more sustainable activity when compared with plastic production from petroleum, because it relies less on fossil fuel as a carbon source and also introduces less, net-new greenhouse emissions if it biodegrades. However, manufacturing of bioplastic materials is often still reliant upon petroleum as an energy and materials source. This comes in the form of energy required to power farm machinery and irrigate growing crops, to produce fertilisers and pesticides, to transport crops and crop products to processing plants, to process raw materials, and ultimately to produce the bioplastic.

Italian bioplastic manufacturer Novamont states in its own environmental audit that producing one kilogram of its starch-based product uses 500g of petroleum and consumes almost 80 per cent of the energy required to produce a traditional polyethylene polymer. Environmental data from NatureWorks, the only commercial manufacturer of PLA (polylactic acid) bioplastic, says that making its plastic material delivers a fossil fuel saving of between 25 and 68 per cent compared with polyethylene, in part due to its purchasing of renewable energy certificates for its manufacturing plant.

A detailed study examining the process of manufacturing a number of common packaging items in several traditional plastics and polylactic acid carried out by US-group and published by the Athena Institute shows the bioplastic to be less environmentally damaging for some products, but more environmentally damaging for others.

While production of most bioplastics results in reduced carbon dioxide emissions compared to traditional alternatives, there are some real concerns that the creation of a global bioeconomy could contribute to an accelerated rate of deforestation if not managed effectively. There are associated concerns over the impact on water supply and soil erosion.

Cost

With the exception of cellulose, most bioplastic technology is relatively new and is currently not as cost competitive with petroleum-based plastics. Many bioplastics are reliant on fossil fuel-derived energy for their manufacturing, reducing the cost advantage over petroleum-based plastic.

Performance and Usage

Many bioplastics also lack the performance and ease of processing of traditional materials albeit materials such as Bioplast from Stanelco have closed this performance gap. Polylactic acid plastic is being used by a handful of small companies for water bottles. But shelf life is limited because the plastic is permeable to water—the bottles lose their contents and slowly deform. However, bioplastics are seeing some use in Europe, where they account for 60 per cent of the biodegradable materials market. The most common end use market is for packaging materials. Japan has also been a pioneer in bioplastics, incorporating them into electronics and automobiles.

Recycling

There are also fears that bioplastics will damage existing recycling projects. Packaging such as HDPE milk bottles and PET water and soft drinks bottles is easily identified and hence setting up a recycling infrastructure has been quite successful in many parts of the world. Polylactic acid and PET do not mix—as bottles made from polylactic acid cannot be distinguished from PET bottles by the consumer there is a risk that recycled PET could be rendered unusable. This could be overcome by ensuring distinctive bottle types or by investing in suitable sorting technology. However, the first route is unreliable and the second costly.

Genetically Modified Bioplastics

Genetic modification (GM) is also a challenge for the bioplastics industry. None of the currently available bioplastics—which can be considered first generation products—require the use of GM crops. However, it is not possible to ensure corn used to make bioplastic in North America is GM-free.

European consumers are hostile to any products that are linked to the GM industry. As a result, some UK retailers such as Sainsbury's will not use bioplastic manufactured in the US, such as Natureworks polylactic acid. There is currently no commercial European source of polylactic acid bioplastic.

There is also concern that the route from corn to bioplastics is not the most efficient. Looking further ahead, some of the second generation bioplastics manufacturing technologies under development employ the "plant factory" model, using genetically modified crops or genetically modified bacteria to optimise efficiency. However, a change in consumer perception of GM technology in Europe will be required for these to be widely accepted.

Market Size

Because of the fragmentation in the market it is difficult to estimate the total market size for bioplastics, but estimates by SRI Consulting put global consumption in 2006 at around 85,000 tonnes. In contrast, global consumption of all flexible packaging is estimated at around 12.3 million tonnes.

COPA (Committee of Agricultural Organisation in the European Union) and COGEGA (General Committee for the Agricultural Cooperation in the European Union) have made an assessment of the potential of bioplastics in different sectors of the European economy:

- ❖ Catering products: 450,000 tonnes per year
- ❖ Organic waste bags: 100,000 tonnes per year
- ❖ Biodegradable mulch foils: 130,000 tonnes per year
- ❖ Biodegradable foils for diapers 80,000 tonnes per year
- ❖ Diapers, 100 per cent biodegradable: 240,000 tonnes per year
- ❖ Foil packaging: 400,000 tonnes per year
- ❖ Vegetable packaging: 400,000 tonnes per year
- ❖ Tyre components: 200,000 tonnes per year
- ❖ Total 2,000,000 tonnes per year

Certification

Biodegradability—EN 13432, ASTM D6400

The EN 13432 industrial standard is arguably the most international in scope and compliance with this standard is required to claim that a product is compostable in the European marketplace. In summary, it requires biodegradation of 90 per cent of the materials in a commercial composting unit within 90 days. The ASTM 6400 standard is the regulatory framework for the United States and sets a less stringent threshold of 60 per cent biodegradation within 180 days, again within commercial composting conditions.

The "compostable" marking found on many items of packaging indicates that the package complies with either of the two standards mentined above. However, the marking is not owned by either regulatory body but by third party trade associations representing companies making or selling biodegradable plastics. In Europe, this is European Bioplastics, in the U.S. it is the Biodegradable Products Institute.

Many starch based plastics, PLA based plastics and certain aliphatic-aromatic co- compounds such as succinates and adipates, have obtained these certificates. Additivated plastics sold as fotodegradable or oxobiodegradable do not comply with these standards in their current form.

Biobased—ASTM D6866

The ASTM D6866 method has been developed to certify the biologically derived content of bioplastics. There is an important difference between biodegradability and biobased content. A bioplastic such as high density polyethylene (HDPE) can be 100 per cent biobased (i.e. contain 100 per cent renewable carbon), yet be non-biodegradable. These bioplastics such HDPE play nonetheless an important role in greenhouse gas abatement, particularly when they are combusted for energy production. The biobased component of these bioplastics is considered carbon-neutral since their origin is from biomass.

Applications

Because of their biological biodegradability, the use of bioplastics is especially popular for disposable items, such as packaging and catering items (crockery, cutlery, pots, bowls, straws). The use of bioplastics for shopping bags is already very common. After their initial use they can be reused as bags for organic waste and then be composted. Trays and containers for fruit, vegetables, eggs and meat, bottles for soft drinks and dairy products and blister foils for fruit and vegetables are also already widely manufactured from bioplastics.

Non-disposable applications include mobile phone casings (NEC), carpet fibres (Dupont Sorona), and car interiors (Mazda). The French company, Arkema, produces a grade of bioplastic called Rilsan, which is being used in fuel line and plastic pipe applications. In these areas, the goal is obviously not biodegradability, but to create items from sustainable resources.

Plastic Types

Starch Based Plastics

Constituting about 50 per cent of the bioplastics market, thermoplastic starch, such as Plastarch Material, currently represents the most important and widely used bioplastic. Pure starch possesses the characteristic of being able to absorb humidity and is thus being used for the production of drug capsules in the pharmaceutical sector. Flexibiliser and plasticiser such as sorbitol and glycerine are added so that starch can also be processed thermo-plastically. By varying the amounts of these additives, the characteristic of the material can be tailored to specific needs (also called "thermo-plastical starch").

Polylactide Acid (PLA) Plastics

Polylactide acid (PLA) is a transparent plastic produced from cane sugar or corn starch. It not only resembles conventional petrochemical mass plastics (like PE or PP) in its characteristics, but it can also be processed easily on standard equipment that already exists for the production of conventional plastics. PLA and PLA-Blends (such as the CompostablesTM by Cereplastnc.) generally come in the form of granulates with various properties and are used in the plastic processing industry for the production of foil, moulds, tins, cups, bottles and other packaging.

Poly-3-Hydroxybutyrate (PHB)

The biopolymer poly-3-hydroxybutyrate (PHB) is a polyester produced produced by certain bacteria processing glucose or starch. Its characteristics are similar to those of the petrochemical-produced plastic polypropylene. The South American sugar industry, for example, has decided to expand PHB production to an industrial scale. PHB is distinguished primarily by its physical characteristics. It produces transparent film at a melting point higher than 130 degrees Celsius, and is biodegradable without residue.

Polyamide 11 (PA 11)

PA 11 is a biopolymer derived from natural oil. It is also known under the tradename Rilsan B commercialized by Arkema. PA 11 belongs to the technical polymers family and is not biodegradable. Its properties are similar than PA 12 although emissions of greenhouse gases and consumption of non-renewable resources are reduced during its production. Its thermal resistance is also superior than PA 12. It is used in high performance applications as automotive fuel lines, pneumatic airbrake tubing, electrical anti-termite cable sheathing, oil & gas flexible pipes & control fluid umbilicals, sports shoes, electronic device components, catheters, etc.

Bio-Derived Polyethylene

The basic building block (monomer) of polyethylene is ethylene. This is just one small chemical step from ethanol, which can be produced by fermentation of agricultural feedstocks such as sugar cane or corn. Bio-derived polyethylene is chemically and physically identical to traditional polyethylene—it does not biodegrade but can be recycled. It can also considerably reduce greenhouse gas emissions. Brazilian chemicals group Braskem claims that using its route from sugar cane ethanol to produce one tonne of polyethylene captures (removes from the environment) 2.5 tonnes of carbon dioxide while the traditional petrochemical route results in emissions of close to 3.5 tonnes.

Braskem plans to introduce commercial quantities of its first bio-derived high density polyethylene, used in a packaging such as bottles and tubs, in 2010 and has developed a technology to produce bio-derived butene, required to make the linear low density polethylene types used in film production.

Developments

- ❖ In the early 1950s, Amylomaize (>50% starch content corn) was successfully bred and commercial bioplastics applications started to be explored.
- ❖ In 2004, NEC developed a flame retardant plastic, polylactic acid, without using toxic chemicals such as halogens and phosphorus compounds.
- ❖ In 2005, Fujitsu became one of the first technology company to make

personal computer cases from bioplastics, which are featured in their FMV-BIBLO NB80K line.

❖ In 2007 Braskem of Brazil announced it had developed a route to manufacture high density polyethylene (HDPE) using ethylene derived from sugar cane.

Bioleaching and Biomining

24.1 Bioleaching

Bioleaching is the extraction of specific metals from their ores through the use of bacteria. Bioleaching is one of several applications within biohydrometallurgy and several methods are used to recover copper, zinc, lead, arsenic, antimony, nickel, , gold, cobalt.

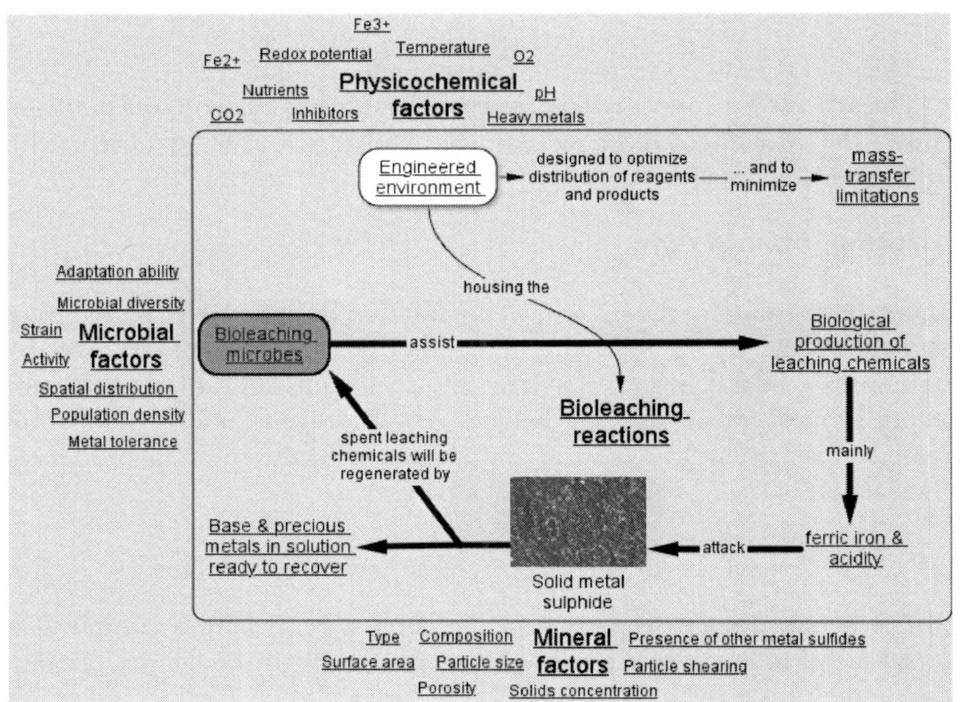

The extraction of gold from its ore can involve numerous ferrous and sulfur oxidizingbacteria, including *Acidithiobacillus ferrooxidans* and

Acidithiobacillus thiooxidans (formerly known as *Thiobacillus*). For example, bacteria catalyse the breakdown of the mineral arsenopyrite (FeAsS) by oxidising the sulfur and metal (in this case arsenic ions) to higher oxidation states whilst reducing dioxygen by H_2 and Fe^{3+}. This allows the soluble products to dissolve.

$$FeAsS_{(s)} \rightarrow Fe^{2+}_{(aq)} + As^{3+}_{(aq)} + S^{6+}_{(aq)}$$

This process actually occurs at the cell membrane of the bacteria. The electrons pass into the cells and are used in biochemical processes to produce energy for the bacteria to reduce oxygen molecules to water.

In stage 2, bacteria oxidise Fe^{2+} to Fe^{3+} (whilst reducing O_2).

$$Fe^{2+} \rightarrow Fe^{3+}$$

They then oxidise the metal to a higher positive oxidation state. With the electrons gained, they reduce Fe^{3+} to Fe^{2+} to continue the cycle.

$$M^{3+} \rightarrow M^{5+}$$

The gold is now separated from the ore and in solution.

The process for copper is very similar. The mineral chalcopyrite ($CuFeS_2$) follows the two stages of being dissolved and then further oxidised, with Cu^{2+} ions being left.

Extraction from Mixture

Copper (Cu^{2+}) ions are removed from the solution by ligand exchange solvent extraction which leaves other ions in the solution. The copper is removed by bonding to a ligand, which is a large molecule consisting of a number of smaller groups each possessing a lone pair. The ligand is dissolved in an organic solvent such as kerosene and shaken with the solution producing this reaction:

$$Cu^{2+}_{(aq)} + 2LH(organic) \rightarrow CuL_2(organic) + 2H^+_{(aq)}$$

The ligand donates electrons to the copper, producing a complex—a central metal atom (copper) bonded to 2 molecules of the ligand. Because this complex has no charge, it is no longer attracted to polar water molecules and dissolves in the kerosene, which is then easily separated from the solution. Because the initial reaction is reversible, it is determined by pH. Adding concentrated acid reverses the equation, and the copper ions go back into an aqueous solution.

Then the copper is passed through an electro-winning process to increase its purity: an electric current is passed through the resulting solution of copper ions. Because copper ions have a 2+ charge, they are attracted to the negative cathodes and collect there.

The copper can also be concentrated and separated by displacing the copper with Fe from scrap iron:

$$Cu^{2+}_{(aq)} + Fe_{(s)} \rightarrow Cu_{(s)} + Fe^{2+}_{(aq)}$$

The electrons lost by the iron are taken up by the copper. Copper is the oxidising agent (it accepts electrons), and iron is the reducing agent (it loses electrons).

Traces of precious metals such as gold may be left in the original solution. Treating the mixture with sodium cyanide in the presence of free oxygen dissolves the gold. The gold is removed from the solution by adsorbing (taking it up on the surface) to charcoal.

Bioleaching with Fungi

Several species of fungi can be used for bioleaching. Fungi can be grown on many different strata, as with electronic scrap, catalytic converters, and fly ash from municipal waste incineration. Experiments have shown that two fungal strains (*Aspergillus Niger, Penicillium simplicissimum*) were able to mobilize Cu and Sn by 65 per cent, and Al, Ni, Pb, and Zn by more than 95 per cent. *As pergillus Niger* can produce some organic acids such as citric acid. So it can be used for bioleaching sulfides.

Bioleaching Compared with other Extraction Techniques

Traditional extractions involve many expensive steps such as roasting and smelting, which require sufficient concentrations of elements in ores and are environmentally unfriendly. Low concentrations are not a problem for bacteria because they simply ignore the waste which surrounds the metals, attaining extraction yields of over 90 per cent in some cases. These microorganisms actually gain energy by breaking down minerals into their constituent elements. The company simply collects the ions out of the solution after the bacteria have finished.

Some advantages associated with bioleaching are:

❖ *Economical*: bioleaching is generally simpler and therefore cheaper to operate and maintain than traditional processes, since fewer specialists are needed to operate complex chemical plants.

❖ *Environmental*: The process is more environmentally friendly than traditional extraction methods. For the company this can translate into profit, since the necessary limiting of sulfur dioxide emissions during smelting is expensive. Less landscape damage occurs, since the bacteria involved grow naturally, and the mine and surrounding area can be left relatively untouched. As the bacteria breed in the conditions of the mine, they are easily cultivated and recycled.

Some disadvantages associated with bioleaching are:

❖ *Economical*: the bacterial leaching process is very slow compared to smelting. This brings in less profit as well as introducing a significant delay in cash flow for new plants.

❖ *Environmental*: Toxic chemicals are sometimes produced in the process. Sulfuric acid and H^+ ions which have been formed can leak into the ground and surface water turning it acidic, causing environmental damage. Heavy ions such as iron, zinc, and arsenic leak during acid mine drainage. When the pH of this solution rises, as a result of dilution by fresh water, these ions precipitate, forming "Yellow Boy" pollution. For these reasons, a setup of bioleaching must be carefully planned, since the process can lead to a biosafety failure.

Currently it is more economical to smelt copper ore rather than to use bioleaching, since the concentration of copper in its ore is generally quite high. The profit obtained from the speed and yield of smelting justifies its cost. However, the concentration of gold in its ore is generally very low. The lower cost of bacterial leaching in this case outweighs the time it takes to extract the metal.

24.2 Bioleaching Microbes

The microbes that are found in bioleaching environments are of two types: those that produce leaching chemicals ("true" bioleaching microbes) and those that do not produce leaching chemicals but support "true" bioleaching microbes—bioleaching supporters. In a consortium, bioleaching supporters assist those microbes producing leaching reagents by for example feeding on, and thereby removing, organic waste products that otherwise would build up and be detrimental to said "true" bioleaching microbes.

Prokaryotes are Common while Eukaryotes are Rare

Even though all (?) microbes used in commercial plants for bioleaching are prokaryotes (bacteria and archaea), bioleaching eukaryotes exist too. These organisms often excrete organic acids which may act as . Research on metal leaching of electronic scrap by bioleaching fungi has been performed with a certain degree of success. {Brandl, 2001 #9}

Reasons why "true" Bioleachers may not Thrive

Sub-optimal growth conditions are attributable to at least the following:

1. incorrect pH conditions;
2. a lack of critical macro-and micro-nutrients;
3. a high ionic strength or total salt content
4. the presence of dissolved or entrained organic compounds with inhibitory effects on microbial growth; and
5. carbon-or oxygen-limiting conditions.

Bioleaching uses bacterial microorganisms to extract precious metals, such as gold, from ore in which it is embedded. As an alternative to smelting or roasting, miners use bioleaching when there are lower concentrations of metal in ore and they need an efficient, environmentally responsible method. The bacteria feeds on nutrients in minerals, thereby separating the metal that leaves the organism's system; then the metal can be collected in a solution.

BioleachingBioleachingBioleaching works because of how special microorganisms act on mineral deposits. They are a catalyst to speed up natural processes inside ore. The bacteria uses a chemical reaction called oxidation to turn metal sulphide crystals into sulfates and pure metals. These constituent parts of ore are separated into valuable metal and leftover sulphur and other acidic chemicals. Eventually, enough material builds up in the waste solution to filter and concentrate it into metal.

For some types of metal, such as copper, bioleaching is not always economically feasible or fast enough, even with its low cost. However, in certain areas of the world or with other metals, this simple, effective, and low cost method offers a smart choice. For example, developing countries often do not have the infrastructure or capital investment to begin smelting, yet their land contains enough ore that its extraction can significantly improve their national economy. One day we may use bioleaching to mine other metals, such as zinc and nickel, on the Moon.

After gaining popularity, about 20 per cent of the extracted copper in the world currently comes from bioleachingbioleachingbioleaching. Mining companies must be careful of pollution that might result from solutions reaching a groundwater source. Yet overall, bioleaching produces less air pollution and little damage to geological formations, since the bacteria occurs there naturally. An ideal metal deposit must allow a certain amount of water into the rock to carry the bacteria. However, it should be surrounded by rock that is impermeable to water to make sure no ground water gets polluted with sulphur.

24.3 BioHeap—A Bioleaching Process for Nickel Extraction

Background

Titan Resources, the Australian-based nickel producer, has recently confirmed the economics of its BioHeap process, following a pre-feasibihty cost analysis of the exploration of two small low-grade nickel sulphide deposits in northern Western Australia.

Who Developed the Process?

The process has been developed over the past four years by a 75 per cent-owned subsidiary of Titan, Pacific Ore Technology Ltd, with capital to help the project from the Australian Government's R & D Start Grant programme.

Bioleaching

The BioHeap process is a form of bioleaching. Traditionally, this is carried out by naturally occurring micro organisms, usually thermophiles, which are micro organisms usually found in acidic environments produced by the oxidation of sulphur—for example in and around hot springs, volcanic regions and sulphide-rich areas. While most living organisms derive energy for growth and reproduction from organic carbon, thermophiles grow on inorganic matter and are harmless to living creatures. Their diet consists of pyrite, arsenopyrite and other metal suphides, such as chalcocite and chalcopyrite.

How Does the Bioleaching Work?

The exact mechanisms by which these micro-organisms oxidise sulphidised materials is not precisely known, although both chemical and biological forces work together to oxidise the metal sulphide to form acid-soluble sulphates. Precious metals, which are not soluble, remain with the residue. Iron, arsenic and base metals, such as copper, cobalt and zinc pass into solution, which can then be separated from the residue and be treated by conventional processing methods. The residue created by the process may contain precious metals, which can be recovered by cyanidation.

Types of Bioleaching

There are currently two methods of bioleaching—tank bioleaching and heap bioleaching. The latter involves crushing the ore, stacking it on plastic mats and spraying it with a dilute sulphuric acid solution containing bacteria and nutrients. The solution drains through the heap and is recovered and resprayed over the heap. When the solution is considered rich enough, it is drained off and the metals extracted by conventional processes.

Advantages of the BioHeap Process

Although both bacterial oxidation and non-bacterial heap leaching are established in the gold and copper industries, their combined use in nickel extraction is unique. The BioHeap process treats crushed ore, so avoiding the need for fine grinding and concentration steps usually needed for most bacterial oxidation techniques. This in turn leads to cost savings which allows the processing of lower grade nickel sulphide resources, which may currently be uneconomic to exploit.

How does the BioHeap Process Perform?

Small samples of the disseminated nickel sulphide ore from the company's Radio Hill mine were subjected to bacterial leach amenability treating during 1998. The results demonstrated that greater than 90 per cent of the contained nickel, copper and cobalt could be extracted. The BioHeap process was then tested using disseminated ore—ore that had a content of 0.73 per cent Ni and 0.87 per cent Cu.

How does the Bacteria Work?

The bacterial culture works in the temperature range 45-60°C, at pH levels of less than 2 and requires air, from which it fixes carbon from carbon dioxide and oxygen for oxidation reactions. This occurs both directly and indirectly— in the direct method the bacteria attaches to the sulphide mineral and oxidises Fe and S moieties in the mineral to release metal ions into solution. The indirect bacterial attack on the mineral involves the bacteria in solution oxidising ferrous ion to ferric ion. Ferric ion is a strong oxidising agent and this in turn attacks the sulphide mineral.

Evaluation of the BioHeap Process

The actual BioHeap test was constructed on the apron of the Radio Hill tailings dam using ore from the Mount Sholl deposit, and consists of three pads, sand filtration bunds, drainage gutters from each pad and a holding pond. The entire area is lined with HPDE.

Two heaps have initially been erected, one a 5,000 tonne ore heap and the second a 5,000 tonne waste heap. The former was inoculated with the bacterial cell culture, while the later was used in conditioning the pregnant liquor. Aeration pipes are located one metre from the base of the heaps and run the length of the heap (figure 24.1). These are connected to a manifold and low-pressure blowers used to aerate the heaps. Covering the heaps is a network of irrigation pipes and drippers used for acid irrigation, bacterial inoculation and nutrient irrigation of both heaps.

Figure 24.1: The BioHeap trial, showing aeration pipes.

Following bacterial inoculation, the heaps are continually irrigated with

the pregnant liquor stream and with a nutrient solution. As the oxidation of the ore progresses, so the base metal tenure in the pregnant liquor stream increases. Once a desired level of metal in solution has been achieved, a bleed stream is removed from the system and the dissolved metals removed using conventional precipitation, electrowinning and solvent extraction methods. By 22 weeks of operation, nickel recovery was 74 per cent, well above the company's estimate of 70 per cent after nine months.

Establishment Costs

Total capital costs for the venture were estimated at Aus$15 million, including preparation of the leach pads, bacterial breeding facilities and ion exchange and precipitation equipment. This capital cost equates to a unit comparison figure of less than US$2 per pound of nickel produced per annum over the two year life of the project.

Summary

While the actual production costs of nickel using BioHeap will vary for different ore deposits and for different scales of operation, the results of the study are consistent with Titan's projections for BioHeap as a process with low capital and operating expenditures and inherent technical and environmental advantages over conventional flotation and smelting.

Biomethanation

25.1 Methanogenesis

Methanogenesis or biomethanation is the formation of methane by microbes known as methanogens. Organisms capable of producing methane have been identified only from the kingdom Archaea, a group phylogenetically distinct from both eukaryotes and bacteria, although many live in close association with anaerobic bacteria. The production of methane is an important and widespread form of microbial metabolism. In most environments, it is the final step in the decomposition of biomass.

Recently, it has been demonstrated that leaf tissues of living plants emit methane. Although the mechanism by which such methane production occurs is, as yet, unknown, the implications are far-reaching; this is an example of methanogenesis occurring in non-microbes, presumably under aerobic conditions. Most of what is known about methanogenesis comes from microbial studies.

Biochemistry of Methanogenesis

Methanogenesis in microbes is a form of anaerobic respiration. Methanogens do not use oxygen to breathe; in fact, oxygen inhibits the growth of methanogens. The terminal electron acceptor in methanogenesis is not oxygen, but carbon. The carbon can occur in a small number of organic compounds, all with low molecular weights. The two best described pathways involve the use of carbon dioxide and acetic acid as terminal electron acceptors:

$$CO_2 + 4\,H_2 \rightarrow CH_4 + 2H_2O$$
$$CH_3COOH \rightarrow CH_4 + CO_2$$

However, methanogenesis has been shown to use carbon from other small organic compounds, such as formic acid (formate), methanol, methylamines, dimethyl sulfide, and methanethiol.

The biochemistry of methanogenesis is relatively complex, involving the following coenzymes and cofactors: F430, coenzyme B, coenzyme M, methanofuran, and methanopterin.

Importance in Carbon Cycle

Methanogenesis is the final step in the decay of organic matter. During the decay process, electron acceptors (such as oxygen, ferric iron, sulfate, nitrate, and manganese) become depleted, while hydrogen (H_2) and carbon dioxide accumulate. Light organics produced by fermentation also accumulate. During advanced stages of organic decay, all electron acceptors become depleted except carbon dioxide. Carbon dioxide is a product of most catabolic processes, so it is not depleted like other potential electron acceptors.

Only methanogenesis and fermentation can occur in the absence of electron acceptors other than carbon. Fermentation only allows the breakdown of larger organic compounds, and produces small organic compounds. Methanogenesis effectively removes the semi-final products of decay: hydrogen, small organics, and carbon dioxide. Without methanogenesis, a great deal of carbon (in the form of fermentation products) would accumulate in anaerobic environments.

In Ruminants

Methanogenesis occurs in the guts of humans and other animals, especially ruminants. In the rumen, anaerobic organisms including methanogens digest cellulose into forms usable by the animal, without them, livestock such as cattle would not be able to graze grass. The useful products of methanogenesis are absorbed by the gut, but the methane is released from the animal mainly by belching (eructation). The average cow emits around 600 litres of methane per day.

Role in Global Warming

Methane in the Earth's atmosphere is an important greenhouse gas with a global warming potential 25 times greater than carbon dioxide (averaged over 100 years), and methanogenesis in livestock and the decay of organic material is thus a considerable contributor to global warming. It may not be a net contributor in the sense that it works on organic material which used up atmospheric carbon dioxide when it was created, but its overall effect is to convert the carbon dioxide into methane which is a much more potent greenhouse gas.

Methanogenesis can also be beneficially exploited, to treat organic waste, to produce useful compounds, and the methane can be collected and used as biogas, a fuel.

Methanogenesis and Extra-Terrestrial Life

The presence of atmospheric methane has a role in the scientific search for extra-terrestrial life . The argument being that methane in the atmosphere will eventually dissipate, unless something is replenishing it. This something could then be the decomposition of organic matter (Methanogenesis). So, if it can be detected (by using a spectrometer for example) then that means there is, or relatively recently was, life present. This was debated when methane was discovered in the Martian atmosphere by (among others) the Mars Express Orbiter (2004) and in Titan's atmosphere by the Huygens probe (2005) . It is also argued that atmospheric methane can come from volcanoes or other fissures in the planet's crust and that without an Isotopic signature it is difficult to say what exactly was the origin.

Biofuel and Biodiesel

26.1 Biofuel

Biofuel (if cultivated, then also called agrofuel or agrifuel) can be broadly defined as solid, liquid, or gas fuel consisting of, or derived from recently dead biological material, most commonly plants. This distinguishes it from fossil fuel, which is derived from long dead biological material.

Biofuel can be theoretically produced from any (biological) carbon source. The most common by far is photosynthetic plants that capture solar energy. Many different plants and plant-derived materials are used for biofuel manufacture.

Biofuels are used globally and biofuel industries are expanding in Europe, Asia and the Americas. The most common use for biofuels is as liquid fuels for automotive transport. The use of renewable biofuels provides increased independence from petroleum and enhances energy security.

There are various current issues with biofuel production and use, which are presently being discussed in the popular media and scientific journals. These include: the effect of moderating oil prices, the "food vs fuel" debate, carbon emissions levels, sustainable biofuel production, deforestation and soil erosion, impact on water resources, human rights issues, poverty reduction potential, biofuel prices, energy balance and efficiency, and centralised versus decentralised production models.

One of the greatest technical challenges is to develop ways to convert biomass energy specifically to liquid fuels for transportation. To achieve this, the two most common strategies are:

1. To grow sugar crops (sugar cane, sugar beet, and sweet sorghum), or starch (corn/maize), and then use yeast fermentation to produce ethanol (ethyl alcohol).
2. To grow plants that (naturally) produce oils, such as oil palm, soybean,

algae, or jatropha. When these oils are heated, their viscosity is reduced, and they can be burned directly in a diesel engine, or the oils can be chemically processed to produce fuels such as biodiesel.

Wood and its byproducts can be converted into biofuels such as woodgas, methanol or ethanol fuel. Some researchers are working to improve these processes.

History and Policy

Humans have used biomass fuels in the form of solid biofuels for heating and cooking since the discovery of fire. Following the discovery of electricity, it became possible to use biofuels to generate electrical power as well. However, the discovery and use of fossil fuels: coal, gas and oil, have dramatically reduced the amount of biomass fuel used in the developed world for transport, heat and power. /biofuels-p2.html National Geographic, Green Dreams, Oct 2007]</ref> However, when large supplies of crude oil were discovered in Pennsylvania and Texas, petroleum based fuels became inexpensive, and soon were widely used. Cars and trucks began using fuels derived from mineral oil/petroleum: gasoline/petrol or diesel.

Nevertheless, before World War II, and during the high demand wartime period, biofuels were valued as a strategic alternative to imported oil. Wartime Germany experienced extreme oil shortages, and many energy innovations resulted. This include the powering of some of its vehicles using a blend of gasoline with alcohol fermented from potatoes, called Reichskraftsprit. In Britain, grain alcohol was blended with petrol by the Distillers Company Limited under the name Discol, and marketed through Esso's affiliate Cleveland.

During the peacetime post-war period, inexpensive oil from the Middle East contributed in part to the lessened economic and geopolitical interest in biofuels. Then in 1973 and 1979, geopolitical conflict in the Middle East caused OPEC to cut exports, and non-OPEC nations experienced a very large decrease in their oil supply. This "energy crisis" resulted in severe shortages, and a sharp increase in high demand oil-based products, notably petrol/gasoline. There was also increased interest from governments and academics in energy issues and biofuels. Throughout history, the fluctuations of supply and demand, energy policy, military conflict, and the environmental impacts, have all contributed to a highly complex and volatile market for energy and fuel.

In the year 2000 and beyond, renewed interest in biofuels has been seen.

The drivers for biofuel research and development include rising oil prices, concerns over the potential oil peak, greenhouse gas emissions (causing global warming and climate change), rural development interests, and instability in the Middle East.

Biomass

Biomass is material derived from recently living organisms. This includes plants, animals and their by-products. For example, manure, garden waste and crop residues are all sources of biomass. It is a renewable energy source based on the carbon cycle, unlike other natural resources such as petroleum, coal, and nuclear fuels.

Animal waste is a persistent and unavoidable pollutant produced primarily by the animals housed in industrial sized farms. Researchers from Washington University have figured out a way to turn manure into biomass. In April 2008 with the help of imaging technology they noticed that vigorous mixing helps microorganisms turn farm waste into alternative energy, providing farmers with a simple way to treat their waste and convert it into energy.

There are also agricultural products specifically grown for biofuel production include corn, switchgrass, and soybeans, primarily in the United States; rapeseed, wheat and sugar beet primarily in Europe; sugar cane in Brazil; palm oil and miscanthus in South-East Asia; sorghum and cassava in China; and jatropha in India. Hemp has also been proven to work as a biofuel. Biodegradable outputs from industry, agriculture, forestry and households can be used for biofuel production, either using anaerobic digestion to produce biogas, or using second generation biofuels; examples include straw, timber, manure, rice husks, sewage, and food waste. The use of biomass fuels can therefore contribute to waste management as well as fuel security and help to prevent climate change, though alone they are not a comprehensive solution to these problems.

Bioenergy from Waste

Using waste biomass to produce energy can reduce the use of fossil fuels, reduce greenhouse gas emissions and reduce pollution and waste management problems. A recent publication by the European Union highlighted the potential for waste-derived bioenergy to contribute to the reduction of global warming. The report concluded that 19 million tons of oil equivalent is available from biomass by 2020, 46 per cent from bio-wastes: municipal solid waste (MSW), agricultural residues, farm waste and other biodegradable waste streams.

Landfill sites generate gases as the waste buried in them undergoes anaerobic digestion. These gases are known collectively as landfill gas (LFG). This can be burned and is considered a source of renewable energy, even though landfill disposal are often non-sustainable. Landfill gas can be burned either directly for heat or to generate electricity for public consumption. Landfill gas contains approximately 50 per cent methane, the same gas that is found in natural gas.

Biomass can come form waste plant material.If landfill gas is not harvested, it escapes into the atmosphere: this is not desirable because methane is a greenhouse gas, with more global warming potential than carbon dioxide. Over a time span of 100 years, methane has a global warming potential of 23 relative to CO_2. Therefore, during this time, one ton of methane produces the same greenhouse gas (GHG) effect as 23 tons of CO_2. When methane burns the formula is $CH_4 + 2O_2 = CO_2 + 2H_2O$ So by harvesting and burning landfill gas, its global warming potential is reduced a factor of 23, in addition to providing energy for heat and power.

Frank Keppler and Thomas Rockmann discovered that living plants also produce methane CH_4. The amount of methane produced by living plants is 10 to 100 times greater than that produced by dead plants (in an aerobic environment) but does not increase global warming because of the carbon cycle.

Anaerobic digestion can be used as a distinct waste management strategy to reduce the amount of waste sent to landfill and generate methane, or biogas. Any form of biomass can be used in anaerobic digestion and will break down to produce methane, which can be harvested and burned to generate heat, power or to power certain automotive vehicles.

A 3 MW landfill power plant would power 1,900 homes. It would eliminate 6,000 tons per year of methane from getting into the environment. It would eliminate 18,000 tons per year of CO_2 from fossil fuel replacement.[citation needed] This is the same as removing 25,000 cars from the road, or planting 36,000 acres (146 km²) of forest, or not using 305,000 barrels (48,500 m³) of oil per year.

Liquid Fuels for Transportation

Most transportation fuels are liquids, because vehicles usually require high energy density, as occurs in liquids and solids. Vehicles usually need high power density as can be provided most inexpensively by an internal combustion engine. These engines require clean burning fuels, in order to keep the engine clean and minimize air pollution. The fuels that are easier to burn cleanly are typically liquids and gases. Thus only liquids meet the requirements of being both portable and clean burning. Also, liquids can be pumped, which means handling is easily mechanized, and thus less laborious.

First Generation Biofuels

'First-generation biofuels' refer to biofuels made from sugar, starch, vegetable oil, or animal fats using conventional technology. The basic feedstocks for the production of first generation biofuels are often seeds or grains such as wheat, which yields starch that is fermented into bioethanol, or sunflower seeds, which are pressed to yield vegetable oil that can be used in biodiesel. These feedstocks could also enter the animal or human food chain, and as the global population has risen their use in producing biofuels has been criticised for diverting food away from the human food chain, leading to food shortages and price rises.

The most common first generation biofuels are listed below:

Vegetable Oil

Vegetable oil can be used for either food or fuel; the quality of the oil may be lower for fuel use. Vegetable oil can be used in many older diesel engines (equipped with indirect injection systems), but only in warm climates. In most cases, vegetable oil is used to manufacture biodiesel, which is compatible with most diesel engines when blended with conventional diesel fuel. MAN B&W Diesel, Wartsila and Deutz AG offer engines that are compatible with straight vegetable oil. Used vegetable oil is increasingly being processed into biodiesel, and at a smaller scale, cleaned of water and particulates and used as a fuel.

Biodiesel

Biodiesel is the most common biofuel in Europe. It is produced from oils or fats using transesterification and is a liquid similar in composition to mineral diesel. Its chemical name is fatty acid methyl (or ethyl) ester (FAME). Oils are mixed with sodium hydroxide and methanol (or ethanol) and the chemical reaction produces biodiesel (FAME) and glycerol. 1 part glycerol is produced for every 10 parts biodiesel.

Biodiesel can be used in any diesel engine when mixed with mineral diesel. In some countries manufacturers cover their diesel engines under warranty for 100 per cent biodiesel use, although Volkswagen of Germany, for example, asks drivers to make a telephone check with the VW environmental services department before switching to 100 per cent biodiesel (see biodiesel use). Many people have run their vehicles on biodiesel without problems, although it can become thick/viscous at lower temperatures, depending on the feedstock used, and vehicles may require fuel line heaters. However, the majority of vehicle manufacturers limit their recommendations to 15 per cent biodiesel blended with mineral diesel. Many newer diesel engines are made so that they can run with 100 per cent biodiesel fuel without altering the engine itself, although this can be dependent on the fuel rail design. Since biodiesels burn cleaner than regular mineral diesel, filters may need to be replaced more often, especially as the biofuel dissolves old deposits in the fuel tank and pipes. In many European countries, a 5 per cent biodiesel blend is widely used and is available at thousands of gas stations.

In the USA, more than 80 per cent of commercial trucks and city buses run on diesel. Therefore "the nascent U.S. market for biodiesel is growing at a staggering rate—from 25 million gallons per year in 2004 to 78 million gallons by the beginning of 2005. By the end of 2006 biodiesel production was estimated to increase fourfold to more than 1 billion gallons," energy expert Will Thurmond writes in an article for the July-August 2007 issue of THE FUTURIST magazine.

Bioalcohols

Biologically produced alcohols, most commonly ethanol, and less commonly propanol and butanol, are produced by the action of microorganisms and enzymes through the fermentation of sugars or starches (easiest), or cellulose (which is more difficult). Biobutanol (also called biogasoline) is often claimed to provide a direct replacement for gasoline, because it can be used directly in a gasoline engine (in a similar way to biodiesel in diesel engines).

Butanol is formed by ABE fermentation (acetone, butanol, ethanol) and experimental modifications of the process show potentially high net energy gains with butanol as the only liquid product. Butanol will produce more energy and allegedly can be burned "straight" in existing gasoline engines (without modification to the engine or car), and is less corrosive and less water soluble than ethanol, and could be distributed via existing infrastructures. DuPont and BP are working together to help develop Butanol.

Ethanol fuel is the most common biofuel worldwide, particularly in Brazil. Alcohol fuels are produced by fermentation of sugars derived from wheat, corn, sugar beets, sugar cane, molasses and any sugar or starch that alcoholic beverages can be made from (like potato and fruit waste, etc.). The ethanol production methods used are enzyme digestion (to release sugars from stored starches, fermentation of the sugars, distillation and drying. The distillation process requires significant energy input for heat (often unsustainable natural gas fossil fuel, but cellulosic biomass such as bagasse, the waste left after sugar cane is pressed to extract its juice, can also be used more sustainably).

Ethanol can be used in petrol engines as a replacement for gasoline; it can be mixed with gasoline to any percentage. Most existing automobile petrol engines can run on blends of up to 15 per cent bioethanol with petroleum/gasoline. Gasoline with ethanol added has higher octane, which means that your engine can typically burn hotter and more efficiently. In high altitude (thin air) locations, some states mandate a mix of gasoline and ethanol as a winter oxidizer to reduce atmospheric pollution emissions.

Ethanol fuel has less BTU energy content, which means it takes more fuel (volume and mass) to go the same distance. More-expensive premium fuels contain less, or no, ethanol. In high-compression engines, less ethanol, slower-burning premium fuel is required to avoid harmful pre-ignition (knocking). Very-expensive aviation gasoline (Avgas) is 100 octane made from 100 per cent petroleum. The high price of zero-ethanol Avgas does not include federal-and-state road-use taxes.

Ethanol is very corrosive to fuel systems, rubber hoses-and-gaskets, aluminum, and combustion chambers. It is therefore illegal to use fuels containing alcohol in aircraft (although at least one model of ethanol-powered aircraft has been developed, the Embraer EMB 202 Ipanema). Ethanol is incompatible with marine fiberglass fuel tanks (it makes them leak). For higher ethanol percentage blends, and 100 per cent ethanol vehicles, engine modifications are required.

Corrosive ethanol cannot be transported in petroleum pipelines, so more-expensive over-the-road stainless-steel tank trucks increase the cost and energy consumption required to deliver ethanol to the customer at the pump.

In the current alcohol-from-corn production model in the United States, considering the total energy consumed by farm equipment, cultivation, planting, fertilizers, pesticides, herbicides, and fungicides made from petroleum, irrigation systems, harvesting, transport of feedstock to processing plants, fermentation, distillation, drying, transport to fuel terminals and retail pumps, and lower ethanol fuel energy content, the net energy content value added and delivered to consumers is very small. And, the net benefit (all things considered) does little to reduce un-sustainable imported oil and fossil fuels required to produce the ethanol.

Many car manufacturers are now producing flexible-fuel vehicles (FFV's), which can safely run on any combination of bioethanol and petrol, up to 100 per cent bioethanol. They dynamically sense exhaust oxygen content, and adjust the engine's computer systems, spark, and fuel injection accordingly. This adds initial cost and ongoing increased vehicle maintenance[citation needed]. Efficiency falls and pollution emissions increase when FFV system maintenance is needed (regardless of the 0 per cent-to-100 per cent ethanol mix being used), but not performed (as with all vehicles). FFV internal combustion engines are becoming increasingly complex, as are multiple-propulsion-system FFV hybrid vehicles, which impacts cost, maintenance, reliability, and useful lifetime longevity.

Alcohol mixes with both petroleum and with water, so ethanol fuels are often diluted after the drying process by absorbing environmental moisture from the atmosphere. Water in alcohol-mix fuels reduces efficiency, makes engines harder to start, causes intermittent operation (sputtering), and oxidizes aluminum (carburetors) and steel components (rust).

Even dry ethanol has roughly one-third lower energy content per unit of volume compared to gasoline, so larger/heavier fuel tanks are required to travel the same distance, or more fuel stops are required. With large current un-sustainable, non-scalable subsidies, ethanol fuel still costs much more per unit of distance traveled than current high gasoline prices in the United States.

Methanol is currently produced from natural gas, a non-renewable fossil fuel. It can also be produced from biomass as biomethanol. The methanol economy is an interesting alternative to the hydrogen economy, compared to today's hydrogen produced from natural gas, but not hydrogen production directly from water and state-of-the-art clean solar thermal energy processes.

BioGas

Biogas is produced by the process of anaerobic digestion of organic material by anaerobes. It can be produced either from biodegradable waste materials or by the use of energy crops fed into anaerobic digesters to supplement gas yields. The solid byproduct, digestate, can be used as a biofuel or a fertilizer. In the UK, the National Coal Board experimented with microorganisms that digested coal in situ converting it directly to gases such as methane.

Biogas contains methane and can be recovered from industrial anaerobic digesters and mechanical biological treatment systems. Landfill gas is a less clean form of biogas which is produced in landfills through naturally occurring anaerobic digestion. If it escapes into the atmosphere it is a potent greenhouse gas.

Oils and gases can be produced from various biological wastes:
   Thermal depolymerization of waste can extract methane and other oils similar to petroleum.

GreenFuel Technologies Corporation developed a patented bioreactor system that uses nontoxic photosynthetic algae to take in smokestacks flue gases and produce biofuels such as biodiesel, biogas and a dry fuel comparable to coal.

Solid Biofuels

Examples include wood, grass cuttings, domestic refuse, charcoal, and dried manure.

Syngas

Syngas is produced by the combined processes of pyrolysis, combustion, and gasification. Biofuel is converted into carbon monoxide and energy by pyrolysis. A limited supply of oxygen is introduced to support combustion. Gasification converts further organic material to hydrogen and additional carbon monoxide.

The resulting gas mixture, syngas, is itself a fuel. Using the syngas is more efficient than direct combustion of the original biofuel; more of the energy contained in the fuel is extracted.

Syngas may be burned directly in internal combustion engines. The wood gas generator is a wood-fueled gasification reactor mounted on an internal combustion engine. Syngas can be used to produce methanol and hydrogen, or converted via the Fischer-Tropsch process to produce a synthetic petroleum substitute. Gasification normally relies on temperatures >700°C. Lower temperature gasification is desirable when co-producing biochar.

Second Generation Biofuels

Supporters of biofuels claim that a more viable solution is to increase political and industrial support for, and rapidity of, second-generation biofuel implementation from non food crops, including cellulosic biofuels. Second-generation biofuel production processes can use a variety of non food crops. These include waste biomass, the stalks of wheat, corn, wood, and special-energy-or-biomass crops (e.g. Miscanthus). Second generation (2G) biofuels use biomass to liquid technology, including cellulosic biofuels from non food crops. Many second generation biofuels are under development such as biohydrogen, biomethanol, DMF, Bio-DME, Fischer-Tropsch diesel, biohydrogen diesel, mixed alcohols and wood diesel.

Cellulosic ethanol production uses non food crops or inedible waste products and does not divert food away from the animal or human food chain. Lignocellulose is the "woody" structural material of plants. This feedstock is abundant and diverse, and in some cases (like citrus peels or sawdust) it is a significant disposal problem.

Producing ethanol from cellulose is a difficult technical problem to solve. In nature, Ruminant livestock (like cattle) eat grass and then use slow enzymatic digestive processes to break it into glucose (sugar). In cellulosic ethanol laboratories, various experimental processes are being developed to do the same thing, and then the sugars released can be fermented to make ethanol fuel.

Scientists also work on experimental recombinant DNA genetic engineering organisms that could increase biofuel potential.

Third Generation Biofuels

Algae fuel, also called oilgae or third generation biofuel, is a biofuel from algae. Algae are low-input/high-yield (30 times more energy per acre than land) feedstocks to produce biofuels and algae fuel are biodegradable:

- ❖ With the higher prices of fossil fuels (petroleum), there is much interest in algaculture (farming algae).
- ❖ One advantage of many biofuels over most other fuel types is that they are biodegradable, and so relatively harmless to the environment if spilled.
- ❖ The United States Department of Energy estimates that if algae fuel replaced all the petroleum fuel in the United States, it would require 15,000 square miles (38,849 square kilometers), which is a few thousand miles larger than Maryland.

Second and third generation biofuels are also called advanced biofuels.

On the other hand, an appearing fourth generation is based in the conversion of vegoil and biodiesel into gasoline.

Fourth Generation Biofuels

Craig Venter's company Synthetic Genomics is genetically engineering microorganisms to produce fuel directly from carbon dioxide on an industrial scale.

Biofuels by Country

Recognizing the importance of implementing bioenergy, there are international organizations such as IEA Bioenergy, established in 1978 by the OECD International Energy Agency (IEA), with the aim of improving cooperation and information exchange between countries that have national programmes in bioenergy research, development and deployment. The U.N. International Biofuels Forum is formed by Brazil, China, India, South Africa, the United States and the European Commission. The world leaders in biofuel development and use are Brazil, United States, France, Sweden and Germany.

China

In China, the government is making E10 blends mandatory in five provinces that account for 16 per cent of the nation's passenger cars. In Southeast Asia, Thailand has mandated an ambitious 10 per cent ethanol mix in gasoline starting in 2007. For similar reasons, the palm oil industry plans to supply an increasing portion of national diesel fuel requirements in Malaysia and Indonesia.[citation needed] In Canada, the government aims for 45 per cent of the country's gasoline consumption to contain 10 per cent ethanol by 2010.

India

In India, a bioethanol programme calls for E5 blends throughout most of the country targeting to raise this requirement to E10 and then E20.

Europe

The European Union in its biofuels directive (updated 2006) has set the goal that for 2010 that each member state should achieve at least 5.75 per cent biofuel usage of all used traffic fuel. By 2020 the figure should be 10 per cent. As of January 2008 these aims are being reconsidered in light of certain environmental and social concerns associated with biofuels such as rising food prices and deforestation.

France

France is the second largest biofuel consumer among the EU States in 2006. According to the Ministry of Industry, France's consumption increased by 62.7 per cent to reach 682,000 toe (i.e. 1.6% of French fuel consumption). Biodiesel represents the largest share of this (78%, far ahead of bioethanol with 22%). The unquestionable biodiesel leader in Europe is the French company Diester Industrie. In bioethanol, the French agro-industrial group

Téréos is increasing its production capacities. Germany itself remained the largest European biofuel consumer, with a consumption estimate of 2.8 million tons of biodiesel (equivalent to 2,408,000 toe), 0.71 million ton of vegetable oil (628.492 toe) and 0.48 million ton of bioethanol (307,200 toe).

Germany

The biggest biodiesel German company is ADM Ölmühle Hamburg AG, which is a subsidiary of the American group Archer Daniels Midland Company. Among the other large German producers, MUW (Mitteldeutsche Umesterungswerke GmbH & Co KG) and EOP Biodiesel AG. A major contender in terms of bioethanol production is the German sugar corporation, Südzucker.

Spain

The Spanish group Abengoa, via its American subsidiary Abengoa Bioenergy, is the European leader in production of bioethanol.

Sweden

The government in Sweden has together with BIL Sweden, the national association for the automobile industry, that are the automakers in Sweden started the work to end oil dependency. One-fifth of cars in Stockholm can run on alternative fuels, mostly ethanol fuel. Also Stockholm will introduce a fleet of Swedish-made hybrid ethanol-electric buses. In 2005, oil phase-out in Sweden by 2020 was announced.

United Kingdom

In the United Kingdom the Renewable Transport Fuel Obligation (RTFO) (announced 2005) is the requirement that by 2010 5 per cent of all road vehicle fuel is renewable. In 2008 a critical report by the Royal Society stated that biofuels risk failing to deliver significant reductions in greenhouse gas emissions from transport and could even be environmentally damaging unless the Government puts the right policies in place.

Brazil

In Brazil, the government hopes to build on the success of the Proálcool ethanol programme by expanding the production of biodiesel which must contain 2 per cent biodiesel by 2008, increasing to 5 per cent by 2013.

Colombia

Colombia mandates the use of 10 per cent ethanol in all gasoline sold in cities with populations exceeding 500,000. In Venezuela, the state oil company is supporting the construction of 15 sugar cane distilleries over the next five years, as the government introduces a E10 (10% ethanol) blending mandate.

USA

In 2006, the United States president George W. Bush said in a State of the Union speech that the US is "addicted to oil" and should replace 75 per cent of imported oil by 2025 by alternative sources of energy including biofuels.

Essentially all of the ethanol fuel in the US is produced from corn. Corn is a very energy intensive crop, which requires one unit of fossil-fuel energy to create just 0.9 to 1.3 energy units of ethanol. A senior member of the House Energy and Commerce Committee Congressman Fred Upton has introduced legislation to use at least E10 fuel by 2012 in all cars in the USA.

The 2007-12-19 U.S. Energy Independence and Security Act of 2007 requires American "fuel producers to use at least 36 billion gallons of biofuel in 2022. This is nearly a fivefold increase over current levels." This is causing a significant agricultural resource shift away from food production to biofuels. American food exports have decreased (increasing grain prices worldwide), and US food imports have increased significantly.

Most biofuels are not currently cost-effective without significant subsidies. "America's ethanol programme is a product of government subsidies. There are more than 200 different kinds, as well as a 54 cents-a-gallon tariff on imported ethanol. This prices Brazilian ethanol out of an otherwise competitive market. Brazil makes ethanol from sugarcane rather than corn (maize), which has a better EROEI. Federal subsidies alone cost $7 billion a year (equal to around $1.90 a gallon)."

General Motors is starting a project to produce E85 fuel from cellulose ethanol for a projected cost of $1 a gallon. This is optimistic however, because $1/gal equates to $10/MBTU which is comparable to woodchips at $7/MBTU or cord wood at $6-$12/MBTU, and this does not account for conversion losses and plant operating and capital costs which are significant. The raw materials can be as simple as corn stalks and scrap petroleum-based vehicle tires, but used tires are an expensive feedstock with other more-valuable uses. GM has over 4 million E85 cars on the road now, and by 2012 half of the production cars for the U.S. will be capable of running on E85 fuel, however by 2012 the supply of ethanol will not even be close to supplying this much E85. Coskata Inc. is building two new plants for the ethanol fuel. Theoretically, the process is claimed to be five times more energy efficient than corn based ethanol, however it is still in development and has not been proven to be cost effective in a free market.

The greenhouse gas emissions are reduced by 86 per cent for cellulose compared to corn's 29 per cent reduction.

Biofuels in Developing Countries

Biofuel industries are becoming established in many developing countries. Many developing countries have extensive biomass resources that are becoming more valuable as demand for biomass and biofuels increases. The approaches to biofuel development in different parts of the world varies. Countries such as India and China are developing both bioethanol and biodiesel programmes. India is extending plantations of jatropha, an oil-producing tree that is used in biodiesel production. The Indian sugar ethanol programme sets a target of 5 per cent bioethanol incorporation into transport fuel. China is a major bioethanol producer and aims to incorporate 15 per cent bioethanol into transport fuels by 2010. Costs of biofuel promotion programmes can be very high, though.

Amongst rural populations in developing countries, biomass provides the majority of fuel for heat and cooking. Wood, animal dung and crop residues are commonly burned. Figures from the International Energy Agency show that biomass energy provides around 30 per cent of the total primary energy supply in developing countries; over 2 billion people depend on biomass fuels as their primary energy source.

The use of biomass fuels for cooking indoors is a source of health problems and pollution. 1.3 million deaths were attributed to the use of biomass fuels with inadequate ventilation by the International Energy Agency in its World Energy Outlook 2006. Proposed solutions include improved stoves and alternative fuels. However, fuels are easily damaged, and alternative fuels tend to be expensive. Very low cost, fuel efficient, low pollution biomass stove designs have existed since 1980 or earlier. Issues are a lack of education, distribution, excess corruption, and very low levels of foreign aid. People in developing countries are often unable to afford these solutions without assistance or financing such as microloans. Organizations such as Intermediate Technology Development Group work to make improved facilities for biofuel use and better alternatives accessible to those who cannot get them.

Current Issues in Biofuel Production and Use

Biofuels are proposed as having such benefits as: reduction of greenhouse gas emissions, reduction of fossil fuel use, increased national energy security, increased rural development and a sustainable fuel supply for the future.

However, biofuel production is questioned from a number of angles. The chairman of the International Panel on Climate Change, Rajendra Pachauri, notably observed in March 2008 that questions arise on the emissions implications of that route, and that biofuel production has clearly raised prices of corn, with an overall implication for food security.

Biofuels are also seen as having limitations. The feedstocks for biofuel production must be replaced rapidly and biofuel production processes must be designed and implemented so as to supply the maximum amount of fuel at the cheapest cost, while providing maximum environmental benefits. Broadly speaking, first generation biofuel production processes cannot supply us with more than a few percent of our energy requirements sustainably. The reasons for this are described below. Second generation processes can supply us with more biofuel, with better environmental gains. The major barrier to the development of second generation biofuel processes is their capital cost: establishing second generation biodiesel plants has been estimated at 500 million.

Recently, an inflexion point about advantages/disadvantages of biofuels seems to be gaining momentum. The March 27, 2008 TIME magazine cover features the subject under the title "The Clean Energy Myth":

> Politicians and Big Business are pushing biofuels like corn-based ethanol as alternatives to oil. All they're really doing is driving up world food prices, helping to destroy the Amazon jungle, and making global warming worse.

In the June, 2008 issue of the journal Conservation Biology, scientists argue that because such large amounts of energy are required to grow corn and convert it to ethanol, the net energy gain of the resulting fuel is modest. Using a crop such as switchgrass, common forage for cattle, would require much less energy to produce the fuel, and using algae would require even less. Changing direction to biofuels based on switchgrass or algae would require significant policy changes, since the technologies to produce such fuels are not fully developed.

Oil Price Moderation

The International Energy Agency's World Energy Outlook 2006 concludes that rising oil demand, if left unchecked, would accentuate the consuming countries' vulnerability to a severe supply disruption and resulting price shock. The report suggested that biofuels may one day offer a viable alternative, but also that "the implications of the use of biofuels for global security as well as for economic, environmental, and public health need to be further evaluated".

Economists disagree on the extent that biofuel production affects crude oil prices. According to the Francisco Blanch, a commodity strategist for Merrill Lynch, crude oil would be trading 15 per cent higher and gasoline would be as much as 25 per cent more expensive, if it were not for biofuels. Gordon Quaiattini, president of the Canadian Renewable Fuels Association, argued that a healthy supply of alternative energy sources will help to combat gasoline price spikes. However, the Federal Reserve Bank of Dallas concluded that "Biofuels are too limited in scale and currently too costly to make much difference to crude oil pricing."

Rising Food Prices—the "Food vs. Fuel" Debate

This topic is internationally controversial. There are those, such as the National Corn Growers Association, who say biofuel is not the main cause. Some say the problem is a result of government actions to support biofuels. Others say it is just due to oil price increases. The impact of food price increases is greatest on poorer countries. Some have called for a freeze on biofuels. Some have called for more funding of second generation biofuels which should not compete with food production so much. In May 2008 Olivier de Schutter, the United Nations food adviser, called for a halt on biofuel investment. In an interview in Le Monde he stated: "The ambitious goals for biofuel production set by the United States and the European Union are irresponsible. I am calling for a freeze on all investment in this sector." 100 million people are currently at risk due to the food price increases.

Carbon Emissions

Biofuels and other forms of renewable energy aim to be carbon neutral or even carbon negative. Carbon neutral means that the carbon released during the use of the fuel, e.g. through burning to power transport or generate electricity, is reabsorbed and balanced by the carbon absorbed by new plant growth. These plants are then harvested to make the next batch of fuel. Carbon neutral fuels lead to no net increases in human contributions to atmospheric carbon dioxide levels, reducing the human contributions to global warming. A carbon negative aim is achieved when a portion of the biomass is used for carbon sequestration. Calculating exactly how much greenhouse gas (GNG) is produced in burning biofuels is a complex and inexact process, which depends very much on the method by which the fuel is produced and the assumptions made in the calculation.

The carbon emissions (Carbon footprint) produced by biofuels are calculated using a technique called Life Cycle Analysis (LCA). This uses a "cradle to grave" or "well to wheels" approach to calculate the total amount of carbon dioxide and other greenhouse gases emitted during biofuel production, from putting seed in the ground to using the fuel in cars and trucks. Many different LCAs have been done for different biofuels, with widely differing results. The majority of LCA studies show that biofuels provide significant greenhouse gas emissions savings when compared to fossil fuels such as petroleum and diesel.[citation needed] Therefore, using biofuels to replace a proportion of the fossil fuels that are burned for transportation can reduce overall greenhouse gas emissions. The well-to-wheel analysis for biofuels has shown that first generation biofuels can save up to 60 per cent carbon emission and second generation biofuels can save up to 80 per cent as opposed to using fossil fuels. However these studies do not take into account emissions from nitrogen fixation, deforestation, land use, or any indirect emissions.

In October 2007, a study was published by scientists from Britain, U.S., Germany and Austria, including Professor Paul Crutzen, who won a Nobel Prize for his work on ozone. They reported that the burning of biofuels derived from rapeseed and corn (maize) can contribute as much or more to global warming by nitrous oxide emissions than cooling by fossil fuel savings. Nitrous oxide is both a potent greenhouse gas and a destroyer of atmospheric ozone. But they also reported that crops with lower requirements for nitrogen fertilizers, such as grasses and woody coppicing will result in a net absorption of greenhouse gases.

In February 2008, two articles were published in Science which investigated the GHG emissions effects of the large amount of natural land that is being converted to cropland globally to support biofuels development. The first of these studies, conducted at the University of Minnesota, found that:

> ...converting rainforests, peatlands, savannas, or grasslands to produce food-based biofuels in Brazil, Southeast Asia, and the United States creates a 'biofuel carbon debt' by releasing 17 to 420 times more CO_2 than the annual greenhouse gas (GHG) reductions these biofuels provide by displacing fossil fuels.

This study not only takes into account removal of the original vegetation (as timber or by burning) but also the biomass present in the soil, for example roots, which is released on continued plowing. It also pointed out that:

> ...biofuels made from waste biomass or from biomass grown on degraded and abandoned agricultural lands planted with perennials incur little or no carbon debt and can offer immediate and sustained GHG advantages.

The second study, conducted at Princeton University, used a worldwide agricultural model to show that:

> ...corn-based ethanol, instead of producing a 20 per cent savings, nearly doubles greenhouse emissions over 30 years and increases greenhouse gases for 167 years.

Both of the Science studies highlight the need for sustainable biofuels, using feedstocks that minimize competition for prime croplands. These include farm, forest and municipal waste streams; energy crops grown on marginal lands, and algaes. These second generation biofuels feedstocks "are expected to dramatically reduce GHGs compared to first generation biofuels such as corn ethanol". In short, biofuels done unsustainably could make the climate problem worse, while biofuels done sustainably could play a leading role in solving the carbon challenge.

Sustainable Biofuel Production

Responsible policies and economic instruments would help to ensure that biofuel commercialization, including the development of new cellulosic technologies, is sustainable. Sustainable biofuel production practices would not hamper food and fibre production, nor cause water or environmental problems, and would actually enhance soil fertitlity. Responsible commercialization of biofuels represents an opportunity to enhance sustainable economic prospects in Africa, Latin America and impoverished Asia.

Soil Erosion, Deforestation, and Biodiversity

It is important to note that carbon compounds in waste biomass that is left on the ground are consumed by other microorganisms. They break down biomass in the soil to produce valuable nutrients that are necessary for future crops. On a larger scale, plant biomass waste provides small wildlife habitat, which in turn ripples up through the food chain. The widespread human use of biomass (which would normally compost the field) would threaten these organisms and natural habitats. When cellulosic ethanol is produced from feedstock like switchgrass and saw grass, the nutrients that were required to grow the lignocellulose are removed and cannot be processed by microorganisms to replenish the soil nutrients. The soil is then of poorer quality. Loss of ground cover root structures accelerates unsustainable soil erosion.

Large-scale deforestation of mature trees (which help remove CO2 through photosynthesis—much better than does sugar cane or most other biofuel feedstock crops do) contributes to un-sustainable global warming atmospheric greenhouse gas levels, loss of habitat, and a reduction of valuable biodiversity. Demand for biofuel has led to clearing land for Palm Oil plantations.

A portion of the biomass should be retained onsite to support the soil resource. Normally this will be in the form of raw biomass, but processed biomass is also an option. If the exported biomass is used to produce syngas, the process can be used to co-produce biochar, a low-temperature charcoal used as a soil amendment to increase soil organic matter to a degree not practical with less recalcitrant forms of organic carbon. For co-production of biochar to be widely adopted, the soil amendment and carbon sequestration value of co-produced charcoal must exceed its net value as a source of energy.

Impact on Water Resources

Increased use of biofuels puts increasing pressure on water resources in at least two ways: water use for the irrigation of crops used as feedstocks for biodiesel production; and water use in the production of biofuels in refineries, mostly for boiling and cooling.

In many parts of the world supplemental or full irrigation is needed to grow feedstocks. For example, if in the production of corn (maize) half the water needs of crops are met through irrigation and the other half through rainfall, about 860 liters of water are needed to produce one liter of ethanol.

In the United States, the number of ethanol factories has almost tripled from 50 in 2000 to about 140 in 2008. A further 60 or so are under construction, and many more are planned. Projects are being challenged by residents at courts in Missouri (where water is drawn from the Ozark Aquifer), Iowa, Nebraska, Kansas (all of which draw water from the non-renewable

Ogallala Aquifer), central Illinois (where water is drawn from the Mahomet Aquifer) and Minnesota.

Burning Biofuels Produces Unhealthy Aldehydes

Deadly Formaldehyde, Acetaldehyde and other Aldehydes are produced when biofuels are oxidized (such as burning them in internal combustion engines). The European Union has banned products that contain Formaldehyde, due to its documented carcinogenic characteristics. The U.S. Environmental Protection Agency has labeled Formaldehyde as a probable cause of cancer in humans.

At higher concentrations, atmospheric aldehydes can be significant respiratory irritants causing nose bleeds, respiratory distress, lung disease, and persistent headaches.

When only a 10 per cent mixture of ethanol is added to gasoline (as is common in America and elsewhere), aldehyde emissions increase 40 per cent. The more ethanol, the more aldehydes are emitted (as in E85 and E100 biofuels). Biodiesel also produces aldehydes, to a degree that depends in part on the type of engine.

Many aldehydes are highly toxic to living cells. Formaldehyde irreversibly cross-links protein amino acids, which produces the hard flesh of embalmed bodies.

Highly-reactive organic aldehyde compounds like Acetaldehyde are carcinogenic and mutagenic. They bind to DNA and protein, destroy folate and result in secondary cellular hyperregeneration (i.e., they mutate cells, causing cancer). Acetaldehyde (from oxidized ethanol) is particularly harmful to liver and brain tissues, causing cancer in both. The effect is cumulative over time, particularly to susceptible children and the elderly. The development of cancer from atmospheric aldehydes can take 10 to 15 years.

Brazil burns significant amounts of ethanol biofuel. Gas chromatograph studies were performed of ambient air in Sao Paulo Brazil, and compared to Osaka Japan, which does not burn ethanol fuel. Atmospheric Formaldehyde was 260 per cent higher in Brazil, and Acetaldehyde was 360 per cent higher. For a variety of complex reasons, the average life expectancy in Japan is 12 years longer than in Brazil.

Social and Water impact in Indonesia

In some locations such as Indonesia deforestation for Palm Oil plantations is leading to displacement of Indigenous peoples. Also, extensive use of pesticide for biofuel crops is reducing clean water supplies.

Environmental Organizations Stance

Some mainstream environmental groups support biofuels as a significant step toward slowing or stopping global climate change.[citation needed] However, biofuel production can threaten the environment if it is not done sustainably. This finding has been backed by reports of the UN, the IPCC, and some other smaller environmental and social groups as the EEB and the Bank Sarasin, which generally remain negative about biofuels.

As a result, governmental and environmental organisations are turning against biofuels made at a non-sustainable way (hereby preferring certain oil sources as jatropha and lignocellulose over palm oil) and are asking for global support for this. Also, besides supporting these more sustainable biofuels, environmental organisations are redirecting to new technologies that do not use internal combustion engines such as hydrogen and compressed air.

The "Roundtable on Sustainable Biofuels" is an international initiative which brings together farmers, companies, governments, non-governmental organizations, and scientists who are interested in the sustainability of biofuels production and distribution. During 2008, the Roundtable is developing a series of principles and criteria for sustainable biofuels production through meetings, teleconferences, and online discussions.

The increased manufacture of biofuels will require increasing land areas to be used for agriculture. Second and third generation biofuel processes can ease the pressure on land, because they can use waste biomass, and existing (untapped) sources of biomass such as crop residues and potentially even marine algae.

In some regions of the world, a combination of increasing demand for food, and increasing demand for biofuel, is causing deforestation and threats to biodiversity. The best reported example of this is the expansion of oil palm plantations in Malaysia and Indonesia, where rainforest is being destroyed to establish new oil palm plantations. It is an important fact that 90 per cent of the palm oil produced in Malaysia is used by the food industry; therefore biofuels cannot be held solely responsible for this deforestation. There is a pressing need for sustainable palm oil production for the food and fuel industries; palm oil is used in a wide variety of food products. The Roundtable on Sustainable Biofuels is working to define criteria, standards and processes to promote sustainably produced biofuels. Palm oil is also used in the manufacture of detergents, and in electricity and heat generation both in Asia and around the world (the UK burns palm oil in coal-fired power stations to generate electricity).

Significant area is likely to be dedicated to sugar cane in future years as demand for ethanol increases worldwide. The expansion of sugar cane plantations will place pressure on environmentally-sensitive native ecosystems including rainforest in South America. In forest ecosystems, these effects themselves will undermine the climate benefits of alternative fuels, in addition to representing a major threat to global biodiversity.

Although biofuels are generally considered to improve net carbon output, biodiesel and other fuels do produce local air pollution, including nitrogen oxides, the principal cause of smog.[citation needed]

Potential for Poverty Reduction

Researchers at the Overseas Development Institute have argued that biofuels could help to reduce poverty in the developing world, through increased employment, wider economic growth multipliers and energy price effects. However, this potential is described as 'fragile', and is reduced where feedstock production tends to be large scale, or causes pressure on limited agricultural resources: capital investment, land, water, and the net cost of food for the poor.

With regards to the potential for poverty reduction or exacerbation, biofuels rely on many of the same policy, regulatory or investment shortcomings that impede agriculture as a route to poverty reduction. Since many of these shortcomings require policy improvements at a country level rather than a global one, they argue for a country-by-country analysis of the potential poverty impacts of biofuels. This would consider, among other things, land administration systems, market coordination and prioritising investment in biodiesel, as this 'generates more labour, has lower transportation costs and uses simpler technology'.

Biofuel Prices

Retail, at the pump prices, including U.S. subsidies, Federal and state motor taxes, B2/B5 prices for low-level Biodiesel (B2-B5) are lower than petroleum diesel by about 12 cents, and B20 blends are the same per unit of volume as petrodiesel.

Due to the 1/3 lower energy content of ethanol fuel, even the heavily-subsidized net cost to drive a specific distance in flexible-fuel vehicles is higher than current gasoline prices.

Energy Efficiency and Energy Balance of Biofuels

Production of biofuels from raw materials requires energy (for farming, transport and conversion to final product, and the production/application of fertilizers, pesticides, herbicides, and fungicides), and has environmental consequences.

The energy balance of a biofuel is determined by the amount of energy put into the manufacture of fuel compared to the amount of energy released when it is burned in a vehicle. This varies by feedstock and according to the assumptions used. Biodiesel made from sunflowers may produce only 0.46 times the input rate of fuel energy. Biodiesel made from soybeans may produce 3.2 times the input rate of fossil fuels. This compares to 0.805 for gasoline and 0.843 for diesel made from petroleum. Biofuels may require higher energy input per unit of BTU energy content produced than fossil fuels: petroleum can be pumped out of the ground and processed more efficiently than biofuels can be grown and processed. However, this is not necessarily a reason to use oil instead of biofuels, nor does it have an impact on the environmental benefits provided by a given biofuel.

Studies have been done that calculate energy balances for biofuel production. Some of these show large differences depending on the biomass feedstock used and location.

To explain one specific example, a June 17, 2006 editorial in the Wall. St. Journal stated, "The most widely cited research on this subject comes from Cornell's David Pimental and Berkeley's Ted Patzek. They've found that it takes more than a gallon of fossil fuel to make one gallon of ethanol— 29 per cent more. That's because it takes enormous amounts of fossil-fuel energy to grow corn (using fertilizer and irrigation), to transport the crops and then to turn that corn into ethanol."

Life cycle assessments of biofuel production show that under certain circumstances, biofuels produce only limited savings in energy and greenhouse gas emissions. Fertiliser inputs and transportation of biomass across large distances can reduce the GHG savings achieved. The location of biofuel processing plants can be planned to minimize the need for transport, and agricultural regimes can be developed to limit the amount of fertiliser used for biomass production. A European study on the greenhouse gas emissions found that well-to-wheel (WTW) CO_2 emissions of biodiesel from seed crops such as rapeseed could be almost as high as fossil diesel. It showed a similar result for bio-ethanol from starch crops, which could have almost as many WTW CO_2 emissions as fossil petrol. This study showed that second generation biofuels have far lower WTW CO_2 emissions.

Other independent LCA studies show that biofuels save around 50 per cent of the CO_2 emissions of the equivalent fossil fuels. This can be increased to 80-90 per cent GHG emissions savings if second generation processes or reduced fertiliser growing regimes are used. Further GHG savings can be achieved by using by-products to provide heat, such as using bagasse to power ethanol production from sugarcane.

Collocation of synergistic processing plants can enhance efficiency. One example is to use the exhaust heat from an industrial process for ethanol production, which can then recycle cooler processing water, instead of evaporating hot water that warms the atmosphere.

Biofuels and Solar Energy Efficiency

Biofuels from plant materials convert energy that was originally captured from solar energy via photosynthesis. A comparison of conversion efficiency from solar to usable energy (taking into account the whole energy budgets) shows that photovoltaics are 100 times more efficient than corn ethanol and 10 times more efficient than the best biofuel.

Centralised vs. Decentralised Production

There is debate around the best model for production.

One side sees centralised vegetable oil fuel production offering:

❖ efficiency,
❖ greater potential for fuel standardisation,
❖ ease of administrating taxes,
❖ possibility for rapid expansion.

The other side of the argument points to:

❖ increased fuel security,
❖ rural job creation,
❖ less of a 'monopolistic' or 'oligopolistic' market due to the increased number of producers,
❖ benefits to local economy as a greater part of any profits stay in the local economy,
❖ decreased transportation and greenhouse gases of feedstock and end product, and
❖ consumers close to and able to observe the effects of production.

The majority of established biofuel markets have followed the centralised model with a few small or micro producers holding a minor segment of the market. A noticeable exception to this has been the pure plant oil (PPO) market in Germany which grew exponentially until the beginning of 2008 when increasing feedstock prices and the introduction of fuel duty combined to stifle the market. Fuel was produced in hundreds of small oil mills distributed throughout Germany often run as part of farm businesses.

Initially fuel quality could be variable but as the market matured new technologies were developed that made significantly improvements. As the technologies surrounding this fuel improved usage and production rapidly increased with rapeseed oil PPO forming a significant segment of transportation biofuels consumed in 2007.

26.2 Biogas

Biogas typically refers to a gas produced by the biological breakdown of organic matter in the absence of oxygen. Biogas originates from biogenic material and is a type of. One type of biogas is produced by anaerobic digestion or fermentation of biodegradable materials such as biomass, manure or sewage, municipal waste, and energy crops. This type of biogas comprises primarily methane and carbon dioxide. The other principal type of biogas is wood gas which is created by gasification of wood or other biomass. This type of biogas is comprised primarily of nitrogen, hydrogen, and carbon monoxide, with trace amounts of methane.

The gases methane, hydrogen and carbon monoxide can be combusted or oxidized with oxygen. Air contains 21 per cent oxygen. This energy release allows biogas to be used as a fuel. Biogas can be used as a low-cost fuel in any country for any heating purpose, such as cooking. It can also be utilized in modern waste management facilities where it can be used to run any type of heat engine, to generate either mechanical or electrical power. Biogas is a renewable fuel and electricity produced from it can be used to attract renewable energy subsidies in some parts of the world.

Production

Depending on where it is produced, biogas can also be called swamp, marsh, landfill or digester gas. A biogas plant is the name often given to an anaerobic digester that treats farm wastes or energy crops.

Biogas can be produced utilizing anaerobic digesters. These plants can be fed with energy crops such as maize silage or biodegradable wastes including sewage sludge and food waste.

Landfill gas is produced by organic waste decomposing under anaerobic conditions in a landfill. The waste is covered and compressed mechanically and by the weight of the material that is deposited from above. This material prevents oxygen from accessing the waste and anaerobic microbes thrive. This gas builds up and is slowly released into the atmosphere if the landfill site has not been engineered to capture the gas. Landfill gas is hazardous for three key reasons. Landfill gas becomes explosive when it escapes from the landfill and mixes with oxygen within lower and higher explosive limits. The methane in biogas forms explosive mixtures in air. The lower explosive limit is 5 per cent methane and the upper explosive limit is 15 per cent methane. The methane contained within biogas is 20 times more potent as a greenhouse gas than carbon dioxide. Therefore uncontained landfill gas which escapes into the atmosphere may significantly contribute to the effects of global warming. In addition to this volatile organic compounds (VOCs) contained within landfill gas contribute to the formation of photochemical smog.

Sweden produces biogas from confiscated alcoholic beverages.

Composition

Typical composition of biogas:

Matter	%
Methane, CH_4	50-75
Carbon dioxide, CO_2	25-50
Nitrogen, N_2	0-10
Hydrogen, H_2	0-1
Hydrogen sulfide, H_2S	0-3
Oxygen, O_2	0-2

The composition of biogas varies depending upon the origin of the anaerobic digestion process. Landfill gas typically has methane concentrations around 50 per cent. Advanced waste treatment technologies can produce biogas with 55-75 per cent CH_4.

In some cases biogas contains siloxanes. These siloxanes are formed from the anaerobic decomposition of materials commonly found in soaps and detergents. During combustion of biogas containing siloxanes, silicon is released and can combine with free oxygen or various other elements in the combustion gas. Deposits are formed containing mostly silica (SiO_2) or silicates (Si_xO_y) and can also contain calcium, sulfur, zinc, phosphorus. These white mineral deposits build to a surface thickness of several millimetres and must be removed by chemical or mechanical means.

Applications

Biogas can be utilized for electricity production, cooking, space heating, water heating and process heating. If compressed, it can replace compressed natural gas for use in vehicles, where it can fuel an internal combustion engine or fuel cells.

Methane within biogas can be concentrated to the same standards as natural gas, when it is, it is called biomethane. If the local gas network permits it the producer of the biogas may be able to utilize the local gas distribution networks. Gas must be very clean to reach pipeline quality, and must be of the correct composition for the local distribution network to accept. Carbon dioxide, Water, hydrogen sulfide and particulates must be removed if present. If concentrated and compressed it can also be used in vehicle transportation. Compressed biogas is becoming widely used in Sweden, Switzerland and Germany. A biogas-powered train has been in service in Sweden since 2005.

Bates' and his biogas car were the subject of a short documentary film called 'Sweet as a Nut' in 1974, at which point he had run his car for 17 years on gas he had produced by processing pig manure. Bates, an inventor, lived in Devon, UK and in the film talks through the simple process and benefits of running a car on biogas. The conversion was simply made with an adapter attached to any combustion engine.

Biogas in Developing Nations

In India biogas produced from the anaerobic digestion of manure in small-scale digestion facilities is called Gober gas. In India biogas is generated at an estimated 2 million+ household facilities. The digester is an airtight circular pit made of concrete with a pipe connection. The manure is directed to the pit, usually directly from the cattle shed. The pit is then filled with a required quantity of wastewater. The gas pipe is connected to the kitchen fire place through control valves. The combustion of this biogas has very little odour or smoke. Owing to simplicity in implementation and use of cheap raw materials in villages, it is one of the most environmentally sound energy sources for rural needs.

Biogas is used extensively throughout rural China and where wastewater treatment and industry coincide.

The Biogas Support Programme in Nepal has installed over 100,000 biogas plants in rural areas.

Vietnam's Biogas Programme for Animal Husbandry Sector has led to the installation of over 20,000 plants throughout that country.

Biogas is also in use in rural Costa Rica.

In Colombia experiments with diesel engines-generator sets partially fuelled by biogas demonstrated that biogas could be used for power generation, reducing elecricity costs by 40 per cent compared with purchase from the regional utility.

Legislation

The European Union presently has some of the strictest legislation regarding waste management and landfill sites called the Landfill Directive. The United States legislates against landfill gas as it contains these VOCs. The United States Clean Air Act and Title 40 of the Code of Federal Regulations (CFR) requires landfill owners to estimate the quantity of non-methane organic compounds (NMOCs) emitted. If the estimated NMOC emissions exceeds 50 tonnes per year the landfill owner is required to collect the landfill gas and treat it to remove the entrained NMOCs. Treatment of the landfill gas is usually by combustion. Because of the remoteness of landfill sites it is sometimes not economically feasible to produce electricity from the gas.

Gober Gas

Gober gas is a *biogas* generated out of cow dung. In India, gober gas is generated at the countless number of micro plants (an estimated more than 2 million) attached to households. The gober gas plant is basically an airtight circular pit made of concrete with a pipe connection. The manure is directed to the pit (usually directed from the cattle shed). The pit is then filled with a required quantity of water (usually waste water). The gas pipe is connected to the kitchen fire place through control valves. The flammable methane gas generated out of this is practically odorless and smokeless. The residue left after the extraction of the gas is used as biofertiliser. Owing to its simplicity in implementation and use of cheap raw materials in the villages, it is often quoted as one of the most environmentally sound energy source for the rural needs.

26.3 Butanol Fuel

Butanol may be used as a fuel in an internal combustion engine. Because its longer hydrocarbon chain causes it to be fairly non-polar, it is more similar to gasoline than is ethanol. Butanol has been demonstrated to work in some vehicles designed for use with gasoline without any modification. It can be produced from biomass (as "biobutanol") as well as fossil fuels (as "petrobutanol"); both biobutanol and petrobutanol have the same chemical properties.

Production of Butanol from Biomass

Butanol from biomass is called biobutanol. It can be produced by fermentation of biomass by the A.B.E. process. The process uses the bacterium *Clostridium acetobutylicum*, also known as the *Weizmann organism*. It was Chaim Weizmann who first used this bacteria for the production of acetone from starch (with the main use of acetone being the making of Cordite) in 1916. The butanol was a by-product of this fermentation (twice as much butanol was produced). The process also creates a recoverable amount of H_2 and a number of other by-products: acetic, lactic and propionic acids, acetone, isopropanol and ethanol.

The difference from ethanol production is primarily in the fermentation of the feedstock and minor changes in distillation. The feedstocks are the same as for ethanol: energy crops such as sugar beets, sugar cane, corn grain, wheat and cassava as well as agricultural byproducts such as straw and corn stalks (reference needed). According to DuPont, existing bioethanol plants can cost-effectively be retrofitted to biobutanol production.

Algae Butanol

Biobutanol can be made entirely with solar energy, from algae (called Solalgal Fuel) or diatoms.

Centia Process

Centia is based on a three-step thermal, catalytic, and reforming process that has the potential to turn virtually any lipidic compound—e.g., vegetable oils, oils from animal fat and oils from algae—into 1-for-1 replacements for petroleum jet fuel, diesel, and gasoline. The three steps are:

- ❖ Hydrolytic conversion.
- ❖ Decarboxylation.
- ❖ Reforming long-chain alkanes.

Distribution

Butanol better tolerates water contamination and is less corrosive than ethanol and more suitable for distribution through existing pipelines for gasoline. In blends with diesel or gasoline, butanol is less likely to separate from this fuel than ethanol if the fuel is contaminated with water. There is also a vapor pressure co-blend synergy with butanol and gasoline containing ethanol, which facilitates ethanol blending. This facilitates storage and distribution of blended fuels.

Properties of Common Fuels

Fuel	Energy density	Air-fuel ratio	Specific energy	Heat of vaporization	RON	MON
Gasoline and biogasoline	32 MJ/L	14.6	2.9 MJ/kg air	0.36 MJ/kg	91-99	81-89
Butanol fuel	29.2 MJ/L	11.2	3.2 MJ/kg air	0.43 MJ/kg	96	78
Ethanol fuel	19.6 MJ/L	9.0	3.0 MJ/kg air	0.92 MJ/kg	129	102
Methanol	16 MJ/L	6.5	3.1 MJ/kg air	1.2 MJ/kg	136	104

Energy Content and Effects on Fuel Economy

Switching a gasoline engine over to butanol would in theory result in a fuel consumption penalty of about 10 per cent but butanol's effect on mileage is yet to be determined by a scientific study. While the energy density for any mixture of gasoline and butanol can be calculated, tests with other alcohol fuels have demonstrated that the effect on fuel economy is not proportional to the change in energy density.

Octane Rating

The octane rating of n-butanol is similar to that of gasoline but lower than that of ethanol and methanol. n-Butanol has a RON (Research Octane number) of 96 and a MON (Motor octane number) of 78 while t-butanol has octane ratings of 105 RON and 89 MON. t-Butanol is used as an additive in gasoline but cannot be used as a fuel in its pure form because its relatively high melting point of 25.5 °C causes it to gel and freeze near room temperature.

A fuel with a higher octane rating is less prone to knocking (extremely rapid and spontaneous combustion by compression) and the control system of any modern car engine can take advantage of this by adjusting the ignition timing. This will improve energy efficiency, leading to a better fuel economy than the comparisons of energy content different fuels indicate. By increasing the compression ratio, further gains in fuel economy, power and torque can be achieved. Conversely, a fuel with lower octane rating is more prone to knocking and will lower efficiency. Knocking can also cause engine damage.

Air-fuel Ratio

Alcohol fuels, including butanol and ethanol, are partially oxidized and therefore need to run at richer mixtures than gasoline. Standard gasoline engines in cars can adjust the air-fuel ratio to accommodate variations in the fuel, but only within certain limits depending on model. If the limit is exceeded by running the engine on pure butanol or a gasoline blend with a high percentage of butanol, the engine will run lean, something which can damage it. Compared to ethanol, butanol can be mixed in higher ratios with gasoline for use in existing cars without the need for retrofit as the air-fuel ratio and energy content are closer to that of gasoline.

Specific Energy

Alcohol fuels have less energy per unit weight and unit volume than gasoline. To make it possible to compare the net energy released per cycle a measure called the fuels specific energy is sometimes used. It is defined as the energy released per air fuel ratio. The net energy released per cycle is higher for butanol than ethanol or methanol and about 10 per cent higher than for gasoline.

Viscosity

Substance	*Kinematic viscosity at 20°C*
Butanol	3.64 cSt
Ethanol	1.52 cSt
Methanol	0.64 cSt
Gasoline	0.4-0.8 cSt
Diesel	>3 cSt
Water	1.0 cSt

The viscosity of alcohols increase with longer carbon chains. For this reason, butanol is used as an alternative to shorter alcohols when a more viscous solvent is desired. The kinematic viscosity of butanol is several times higher than that of gasoline and about as viscous as high quality diesel fuel.

Heat of Vaporization

The fuel in an engine has to be vaporized before it will burn. Insufficient vaporization is a known problem with alcohol fuels during cold starts in cold weather. As the latent heat of vaporization of butanol is less than half of that of ethanol, an engine running on butanol should be easier to start in cold weather than one running on ethanol or methanol.

Potential Problems with the use of Butanol Fuel

The potential problems with the use of butanol are similar to those of ethanol:

❖ To match the combustion characteristics of gasoline, the utilization of butanol fuel as a substitute for gasoline requires fuel-flow increases (though butanol has only slightly less energy than gasoline, so the fuel-flow increase required is only minimal, maybe 10 per cent, compared to 40 per cent for ethanol.).

❖ Alcohol-based fuels are not compatible with some fuel system components.

❖ Alcohol fuels may cause erroneous gas gauge readings in vehicles with capacitance fuel level gauging.

❖ While ethanol and methanol have lower energy densities than butanol, their higher octane number allows for greater compression ratio and efficiency. Higher combustion engine efficiency allows for lesser greenhouse gas emissions per unit motive energy extracted.

As an advantage, butanol production from biomass could be more efficient (i.e. unit engine motive power delivered per unit solar energy consumed) than ethanol or methanol routes. Also, some bacteria that produce butanol are able to digest cellulose, not just starch and sugars.

Possible Butanol Fuel Mixtures

Standards for the blending of ethanol and methanol in gasoline exist in many countries, including the EU, the US and Brazil. Approximate equivalent butanol blends can be calculated from the relations between the stochiometric fuel-air ratio of butanol, ethanol and gasoline. Common ethanol fuel mixtures for fuel sold as gasoline currently range from 5 per cent to 10 per cent. The share of butanol can be 60 per cent greater than the equivalent ethanol share, which gives a range from 8 to 32 per cent. "Equivalent" in this case refers only to the vehicle's ability to adjust to the fuel. Other properties such as energy density, viscosity and heat of vaporisation will vary and may further limit the percentage of butanol that can be blended with gasoline.

Current use of Butanol in Vehicles

Currently no production vehicle is known to be approved by the manufacturer for use with 100 per cent butanol, though any model that is able to run 10 per cent ethanol blends should be able to use butanol without any problems.

David Ramey drove from Blacklick, Ohio to San Diego, California using butanol in an unmodified 1992 Buick Park Avenue. Although further long term testing must be done, it is highly likely that most late model cars can run on 100 per cent butanol safely with no modifications. Justification for this conclusion is based on data for RON in comparison of n-Butanol with Gasoline. Also, modern ECU-injected motorcar piston engines are designed to be flexible enough to deliver good performance with 91-RON fuels, which n-Butanol exceeds in RON rating.

Research Challenges

The key research challenge that must be resolved is that butanol production inhibits microbial growth even at low concentrations. The result is that the product of the fermentation is less than 2 per cent butanol. The overwhelming majority of the fermentation broth is water, so an energy-intensive distillation step is required for purification. This may be acceptable if the goal is to produce butanol for use as a solvent, but if butanol is to gain traction as a motor fuel, energy inputs into the process need to be minimized.

The Swiss company Butalco GmbH uses a special technology to modify yeasts in order to produce butanol instead of ethanol. Yeasts as production organisms for butanol have decisive advantages compared to bacteria.

26.4 Biodiesel

Biodiesel refers to a non-petroleum-based diesel fuel consisting of short chain alkyl (methyl or ethyl) esters, made by transesterification of vegetable oil, which can be used (alone, or blended with conventional petrodiesel) in unmodified diesel-engine vehicles. Biodiesel is distinguished from the *straight vegetable oil* (SVO) (aka "waste vegetable oil", "WVO", "used vegetable oil", "UVO", "unwashed biodiesel", "pure plant oil", "PPO") used (alone, or blended) as fuels in some *converted* diesel vehicles. "Biodiesel" is standardized as mono-alkyl ester and other kinds of diesel-grade fuels of biological origin are not included.

Blends

Blends of biodiesel and conventional hydrocarbon-based diesel are products most commonly distributed for use in the retail diesel fuel marketplace. Much of the world uses a system known as the "B" factor to state the amount of biodiesel in any fuel mix: fuel containing 20 per cent biodiesel is labeled *B20*, while pure biodiesel is referred to as *B100*. It is common to see *B99*, since 1 per cent petrodiesel is sufficiently toxic to retard mold. Blends of 20 percent biodiesel with 80 percent petroleum diesel (B20) can generally be used in unmodified diesel engines. Biodiesel can also be used in its pure form (B100), but may require certain engine modifications to avoid maintenance and performance problems. Blending B100 with petro diesel may be accomplished by:

- ❖ Mixing in tanks at manufacturing point prior to delivery to tanker truck.
- ❖ Splash mixing in the tanker truck (adding specific percentages of Biodiesel and Petro Diesel).
- ❖ In-line mixing, two components arrive at tanker truck simultaneously.

Origin

On August 31, 1937, G. Chavanne of the University of Brussels (Belgium) was granted a patent for a 'Procedure for the transformation of vegetable oils for their uses as fuels' (fr. 'Procédé de Transformation d'Huiles Végétales en Vue de Leur Utilisation comme Carburants') Belgian Patent 422,877. This patent described the alcoholysis (often referred to as transesterification) of vegetable oils using ethanol (and mentions methanol) in order to separate the fatty acids from the glycerol by replacing the glycerol with short linear alcohols. This appears to be the first account of the production of what is known as 'biodiesel' today.

Applications

Biodiesel can be used in pure form (B100) or may be blended with petroleum diesel at any concentration in most modern diesel engines. Biodiesel has different solvent properties than petrodiesel, and will degrade natural rubber gaskets and hoses in vehicles (mostly found in vehicles manufactured before 1992), although these tend to wear out naturally and most likely will have already been replaced with FKM, which is nonreactive to biodiesel. Biodiesel has been known to break down deposits of residue in the fuel lines where petrodiesel has been used. As a result, fuel filters may become clogged with particulates if a quick transition to pure biodiesel is made. Therefore, it is recommended to change the fuel filters on engines and heaters shortly after first switching to a biodiesel blend.

Distribution

Biodiesel use and production are increasing rapidly. Fueling stations make biodiesel readily available to consumers across Europe, and increasingly in the USA and Canada. A growing number of transport fleets use it as an additive in their fuel. Biodiesel is often more expensive to purchase than petroleum diesel but this is expected to diminish due to economies of scale and agricultural subsidies versus the rising cost of petroleum as reserves are depleted.

Vehicular use and Manufacturer Acceptance

In 2005, DaimlerChrysler released Jeep Liberty CRD diesels from the factory into the American market with 5 per cent biodiesel blends, indicating at least partial acceptance of biodiesel as an acceptable diesel fuel additive. In 2007, DiamlerChrysler indicated intention to increase warranty coverage to 20 per cent biodiesel blends if biofuel quality in the United States can be standardized.

Railroad Use

The British businessman Richard Branson's Virgin Voyager train, number 220007 *Thames Voyager*, billed as the world's first "biodiesel train" was converted to run on 80 per cent petrodiesel and only 20 per cent biodiesel, and it is claimed it will save 14 per cent on direct emissions.

Aircraft Use

Aircraft manufacturers are understandably even more cautious, but a test flight has been performed by an ex Soviet Aircraft (completely powered on biofuel); testing has been announced by Rolls Royce plc, Air New Zealand and Boeing (one engine out of four on a Boeing 747); and commercial passenger jet testing has also been announced by Virgin Atlantic's Richard Branson.

The world's first biofuel-powered commercial aircraft took off from London's Heathrow Airport on February 24, 2008 and touched down in Amsterdam on a demonstration flight hailed as a first step towards "cleaner" flying. The "BioJet" fuel for this flight was produced by Seattle based Imperium Renewables, Inc.

As a Heating Oil

Biodiesel can also be used as a heating fuel in domestic and commercial boilers, sometimes known as bioheat. Older furnaces may contain rubber parts that would be affected by biodiesel's solvent properties, but can otherwise burn biodiesel without any conversion required. Care must be taken at first, however, given that varnishes left behind by petrodiesel will be released and can clog pipes-fuel filtering and prompt filter replacement is required. Another approach is to start using biodiesel as blend, and decreasing the petroleum proportion over time can allow the varnishes to come off more gradually and be less likely to clog. Thanks to its strong solvent properties, however, the furnace is cleaned out and generally becomes more efficient. A technical research paper   describes laboratory research and field trials project using pure biodiesel and biodiesel blends as a heating fuel in oil fired boilers. During the Biodiesel Expo 2006 in the UK, Andrew J. Robertson presented his biodiesel heating oil research from his technical paper and suggested that B20 biodiesel could reduce UK household CO_2 emissions by 1.5 million tons per year.

Historical Background

Transesterification of a vegetable oil was conducted as early as 1853 by scientists E. Duffy and J. Patrick, many years before the first diesel engine became functional. Rudolf Diesel's prime model, a single 10 ft (3 m) iron cylinder with a flywheel at its base, ran on its own power for the first time in Augsburg, Germany, on August 10, 1893. In remembrance of this event, August 10 has been declared "International Biodiesel Day".

Rudolf Diesel demonstrated a Diesel engine running on peanut oil (at the request of the French government) built by the French Otto Company at the World Fair in Paris, France in 1900, where it received the *Grand Prix* (highest prize).

This engine stood as an example of Diesel's vision because it was powered by peanut oil—a biofuel, though not *biodiesel*, since it was not transesterified. He believed that the utilization of biomass fuel was the real future of his engine. In a 1912 speech Diesel said, "the use of vegetable oils for engine fuels may seem insignificant today but such oils may become, in the course of time, as important as petroleum and the coal-tar products of the present time."

During the 1920s, diesel engine manufacturers altered their engines to utilize the lower viscosity of petrodiesel (a fossil fuel), rather than vegetable oil (a biomass fuel). The petroleum industries were able to make inroads in fuel markets because their fuel was much cheaper to produce than the biomass alternatives. The result, for many years, was a near elimination of the biomass fuel production infrastructure. Only recently, have environmental impact concerns and a decreasing price differential made biomass fuels such as biodiesel a growing alternative.

Despite the widespread use of fossil petroleum-derived diesel fuels, interest in vegetable oils as fuels in internal combustion engines is reported in several countries during the 1920s and 1930's and later during World War II. Belgium, France, Italy, the United Kingdom, Portugal, Germany, Brazil, Argentina, Japan and China have been reported to have tested and used vegetable oils as diesel fuels during this time. Some operational problems were reported due to the high viscosity of vegetable oils compared to petroleum diesel fuel, which result in poor atomization of the fuel in the fuel spray and often leads to deposits and coking of the injectors, combustion chamber and valves. Attempts to overcome these problems included heating of the vegetable oil, blending it with petroleum-derived diesel fuel or ethanol, pyrolysis and cracking of the oils.

On August 31, 1937, G. Chavanne of the University of Brussels (Belgium) was granted a patent for a "Procedure for the transformation of vegetable oils for their uses as fuels" (fr. 'Procédé de Transformation d'Huiles Végétales en Vue de Leur Utilisation comme Carburants') Belgian Patent 422,877. This patent described the alcoholysis (often referred to as transesterification) of vegetable oils using methanol and ethanol in order to separate the fatty acids from the glycerol by replacing the glycerol by short linear alcohols. This appears to be the first account of the production of what is known as "biodiesel" today.

More recently, in 1977, Brazilian scientist Expedito Parente produced biodiesel using transesterification with ethanol, and again filed a patent for the same process. This process is classified as biodiesel by international norms, conferring a "standardized identity and quality. No other proposed biofuel has been validated by the motor industry." Currently, Parente's company Tecbio is working with Boeing and NASA to certify bioquerosene (bio-kerosene), another product produced and patented by the Brazilian scientist.

Research into the use of transesterified sunflower oil, and refining it to diesel fuel standards, was initiated in South Africa in 1979. By 1983, the process for producing fuel-quality, engine-tested biodiesel was completed and published internationally. An Austrian company, Gaskoks, obtained the technology from the South African Agricultural Engineers; the company erected the first biodiesel pilot plant in November 1987, and the first industrial-scale plant in April 1989 (with a capacity of 30,000 tons of rapeseed per annum).

Throughout the 1990s, plants were opened in many European countries, including the Czech Republic, Germany and Sweden. France launched local production of biodiesel fuel (referred to as *diester*) from rapeseed oil, which is mixed into regular diesel fuel at a level of 5 per cent, and into the diesel fuel used by some captive fleets (e.g. public transportation) at a level of 30 per cent. Renault, Peugeot and other manufacturers have certified truck engines for use with up to that level of partial biodiesel; experiments with 50 per cent biodiesel are underway. During the same period, nations in other parts of the world also saw local production of biodiesel starting up: by 1998, the Austrian Biofuels Institute had identified 21 countries with commercial biodiesel projects. 100 per cent Biodiesel is now available at many normal service stations across Europe.

In September 2005 Minnesota became the first U.S. state to mandate that all diesel fuel sold in the state contain part biodiesel, requiring a content of at least 2 per cent biodiesel.

Properties

Biodiesel *has better lubricant than that of today's diesel fuels*. During the manufacture of these, to comply with low SO_2 engine emission limits set in modern standards, severe hydrotreatment is included. Biodiesel addition reduces wear increasing the life of the fuel injection equipment that relies on the fuel for its lubrication, such as high pressure injection pumps, pump injectors (also called *unit injectors*) and fuel injectors.

The calorific value of biodiesel is about 33 MJ/L. This is 9 per cent lower than regular Number 2 petrodiesel. Variations in biodiesel energy density is more dependent on the feedstock used than the production process. Still these variations are less than for petrodiesel. It has been claimed biodiesel gives better lubricity and more complete combustion thus increasing the engine energy output and partially compensating for the higher energy density of petrodiesel.

Biodiesel is a liquid which varies in colour—between golden and dark brown—depending on the production feedstock. It is immiscible with water, has a high boiling point and low vapor pressure. *The flash point of biodiesel (>130 °C, >266 °F) is significantly higher than that of petroleum diesel (64 °C, 147 °F) or gasoline (-45 °C, -52 °F). Biodiesel has a density of ~ 0.88 g/cm³, less than that of water.

Biodiesel has a viscosity similar to petrodiesel, the current industry term for diesel produced from petroleum. Biodiesel has high lubricity and virtually no sulfur content, and it is often used as an additive to Ultra-Low Sulfur Diesel (ULSD) fuel.

Biodiesel is 3½ times more energy efficient than bioethanol.

Technical Standards

The European standard for biodiesel is EN 14214, which is translated into the respective national standards for each country that forms the CEN (European Committee for Standardization) area e.g., for the United Kingdom, BS EN 14214 and for Germany DIN EN 14214. It may be used outside the CEN area as well.

There are other national specifications. ASTM D6751 is the most common standard referenced in the United States and Canada. There are also DIN standards for three different varieties of biodiesel, which are made of different oils:

- ❖ RME (rapeseed methyl ester, according to DIN E 51606).
- ❖ PME (vegetable methyl ester, purely vegetable products, according to DIN E 51606).
- ❖ FME (fat methyl ester, vegetable and animal products, according to DIN V 51606).

The standards ensure that the following important factors in the fuel production process are satisfied:

- ❖ Acid value
- ❖ Complete reaction
- ❖ Removal of glycerin
- ❖ Removal of catalyst
- ❖ Removal of alcohol
- ❖ Absence of free fatty acids
- ❖ Low sulfur content
- ❖ Cold Filter Plugging point
- ❖ Cloud Point

Basic industrial tests to determine whether the products conform to the standards typically include , a test that verifies only the more important of the variables above. Tests that are more complete are more expensive. Fuel meeting the quality standards is very non-toxic, with a toxicity rating (LD_{50}) of greater than 50 mL/kg.

Gelling

The cloud point, or temperature at which pure (B100) biodiesel starts to gel, varies significantly and depends upon the mix of esters and therefore the feedstock oil used to produce the biodiesel. For example, biodiesel produced from low erucic acid varieties of canola seed (RME) starts to gel at approximately -10 °C (14 °F). Biodiesel produced from tallow tends to gel at around +16 °C (61 °F). As of 2006, there are a very limited number of products that will significantly lower the gel point of straight biodiesel. A study carried out by Assiniboine Community College in Manitoba, Canada managed to produce B100 biodiesel that was a clear flowing liquid at -38° by using a commercially available additive, Wintron XC30, in addition to low temperature filtration. A number of studies have shown that winter operation is possible with biodiesel blended with other fuel oils including #2 low sulfur diesel fuel and #1 diesel/kerosene. The exact blend depends on the operating environment: successful operations have run using a 65 per cent LS #2, 30 per cent K #1, and 5 per cent bio blend. Other areas have run a 70 per cent Low Sulfur #2, 20 per cent Kerosene #1, and 10 per cent bio blend or an 80 per cent K#1, and 20 per cent biodiesel blend. According to the National Biodiesel Board (NBB), B20 (20 per cent biodiesel, 80 per cent petrodiesel) does not need any treatment in addition to what is already taken with petrodiesel.

To permit the use of biodiesel without mixing and without the possibility of gelling at low temperatures, some people modify their vehicles with a second fuel tank for biodiesel in addition to the standard fuel tank. Alternately, a vehicle with two tanks is chosen. The second fuel tank is insulated and a heating coil using engine coolant is run through the tank. When a temperature sensor indicates that the fuel is warm enough to burn, the driver switches from the petrodiesel tank to the biodiesel tank. This is similar to the method used for running straight vegetable oil.

Contamination by Water

Biodiesel may contain small but problematic quantities of water. Although it is hydrophobic (non-miscible with water molecules), it is said to be, at the same time, hygroscopic to the point of attracting water molecules from atmospheric moisture; one of the reasons biodiesel can absorb water is the persistence of mono and diglycerides left over from an incomplete reaction. These molecules can act as an emulsifier, allowing water to mix with the biodiesel. In addition, there may be water that is residual to processing or resulting from storage tank condensation. The presence of water is a problem because:

❖ Water reduces the heat of combustion of the bulk fuel. This means more smoke, harder starting, less power.

- ❖ Water causes corrosion of vital fuel system components: fuel pumps, injector pumps, fuel lines, etc.
- ❖ Water & microbes cause the paper element filters in the system to fail (rot), which in turn results in premature failure of the fuel pump due to ingestion of large particles.
- ❖ Water freezes to form ice crystals near 0 °C (32 °F). These crystals provide sites for nucleation and accelerate the gelling of the residual fuel.
- ❖ Water accelerates the growth of microbe colonies, which can plug up a fuel system. Biodiesel users who have heated fuel tanks therefore face a year-round microbe problem.
- ❖ Additionally, water can cause pitting in the pistons on a diesel engine.

Previously, the amount of water contaminating biodiesel has been difficult to measure by taking samples, since water and oil separate. However, it is now possible to measure the water content using water-in-oil sensors.

Water contamination is also a potential problem when using certain chemical catalysts involved in the production process, substantially reducing catalytic efficiency of base (high pH) catalysts such as KOH. However, the super-critical methanol production methodology, whereby the transesterification process of oil feedstock and methanol is effectuated under high temperature and pressure, has been shown to be largely unaffected by the presence of water contamination during the production phase.

Availability and Prices

Global biodiesel production reached 3.8 million tons in 2005. Approximately 85 per cent of biodiesel production came from the European Union.

In the United States, average retail (at the pump) prices, including Federal and state fuel taxes, of B2/B5 are lower than petroleum diesel by about 12 cents, and B20 blends are the same as petrodiesel. B99 and B100 generally cost more than petrodiesel except where local governments provide a subsidy.

Production

Biodiesel is commonly produced by the transesterification of the vegetable oil or animal fat feedstock. There are several methods for carrying out this transesterification reaction including the common batch process, supercritical processes, ultrasonic methods, and even microwave methods.

Chemically, transesterified biodiesel comprises a mix of mono-alkyl esters of long chain fatty acids. The most common form uses methanol (converted to sodium methoxide) to produce methyl esters as it is the cheapest alcohol available, though ethanol can be used to produce an ethyl ester biodiesel and higher alcohols such as isopropanol and butanol have also been used. Using alcohols of higher molecular weights improves the cold flow properties of the resulting ester, at the cost of a less efficient transesterification reaction. A lipid transesterification production process is used to convert the base oil to the desired esters. Any Free fatty acids (FFAs) in the base oil are either converted to soap and removed from the process, or they are esterified (yielding more biodiesel) using an acidic catalyst. After this processing, unlike straight vegetable oil, biodiesel has combustion properties very similar to those of petroleum diesel, and can replace it in most current uses.

A byproduct of the transesterification process is the production of glycerol. For every 1 tonne of biodiesel that is manufactured, 100 kg of glycerol are produced. Originally, there was a valuable market for the glycerol, which assisted the economics of the process as a whole. However, with the increase in global biodiesel production, the market price for this crude glycerol (containing 20% water and catalyst residues) has crashed. Research is being conducted globally to use this glycerol as a chemical building block. One initiative in the UK is The Glycerol Challenge.

Usually this crude glycerol has to be purified, typically by performing vacuum distillation. This is rather energy intensive. The refined glycerol (98%+ purity) can then be utilised directly, or converted into other products. The following announcements were made in 2007: A joint venture of Ashland Inc. and Cargill announced plans to make propylene glycol in Europe from glycerol and Dow Chemical announced similar plans for North America. Dow also plans to build a plant in China to make epichlorhydrin from glycerol. Epichlorhydrin is a raw material for epoxy resins.

Production Levels

Biodiesel production capacity is growing rapidly, with an average annual growth rate from 2002-2006 of over 40 per cent. For the year 2006, the latest for which actual production figures could be obtained, total world biodiesel production was about 5-6 million tonnes, with 4.9 million tonnes processed in Europe (of which 2.7 million tonnes was from Germany) and most of the rest from the USA. The capacity for 2007 in Europe totalled 10.3 million tonnes. This compares with a total demand for diesel in the US and Europe of approximately 490 million tonnes (147 billion gallons). Total world production of vegetable oil for all purposes in 2005/06 was about 110 million tonnes, with about 34 million tonnes each of palm oil and soybean oil.

Biodiesel Feedstocks

A variety of oils can be used to produce biodiesel. These include:

❖ Virgin oil feedstock; rapeseed and soybean oils are most commonly used, soybean oil alone accounting for about ninety percent of all fuel stocks in the US. It also can be obtained from field pennycress and Jatropha other crops such as mustard, flax, sunflower, palm oil, hemp;

❖ Waste vegetable oil (WVO);

❖ Animal fats including tallow, lard, yellow grease, chicken fat, and the by-products of the production of Omega-3 fatty acids from fish oil; and

❖ Algae, which can be grown using waste materials such as sewage and without displacing land currently used for food production.

Many advocates suggest that waste vegetable oil is the best source of oil to produce biodiesel, but since the available supply is drastically less than the amount of petroleum-based fuel that is burned for transportation and home heating in the world, this local solution does not scale well.

Animal fats are similarly limited in supply, and it would not be efficient to raise animals (or catch fish) simply for their fat. However, producing biodiesel with animal fat that would have otherwise been discarded could replace a small percentage of petroleum diesel usage. Currently, a 5-million dollar plant is being built in the USA, with the intent of producing 11.4 million litres (3 million gallons) biodiesel from some of the estimated 1 billion kg (2.3 billion pounds) of chicken fat produced annually the local Tyson poultry plant.. Similarly, some small-scale biodiesel factories use waste fish oil as feedstock.

Quantity of Feedstocks Required

Worldwide production of vegetable oil and animal fat is not yet sufficient to replace liquid fossil fuel use. Furthermore, some object to the vast amount of farming and the resulting fertilization, pesticide use, and land use conversion that would be needed to produce the additional vegetable oil. The estimated transportation diesel fuel and home heating oil used in the United States is about 160 million tonnes (350 billion pounds) according to the Energy Information Administration, US Department of Energy. In the United States, estimated production of vegetable oil for all uses is about 11 million tonnes

(24 billion pounds) and estimated production of animal fat is 5.3 million tonnes (12 billion pounds).

If the entire arable land area of the USA (470 million acres, or 1.9 million square kilometers) were devoted to biodiesel production from soy, this would just about provide the 160 million tonnes required (assuming an optimistic 98 gpa of biodiesel). This land area could in principle be reduced significantly using algae, if the obstacles can be overcome. The US DOE estimates that if algae fuel replaced all the petroleum fuel in the United States, it would require 15,000 square miles (38,849 square kilometers), which is a few thousand square miles larger than Maryland, or 1.3 Belgiums,   assuming a yield of 15000 gpa. The advantages of algae are that it can be grown on non-arable land such as deserts or in marine environments, and the potential oil yields are much higher than from plants.

Yield

Feedstock yield efficiency per acre affects the feasibility of ramping up production to the huge industrial levels required to power a significant percentage of national or world vehicles. Some typical yields in US gallons of biodiesel per acre are:

- ❖ Algae: 1800 gpa or more (est. see soy figures and DOE quote below)
- ❖ Palm oil: 508 gpa
- ❖ Coconut: 230 gpa
- ❖ Rapeseed: 102 gpa
- ❖ Soy: 59.2-98.6 gpa in Indiana (Soy is used in 80% of USA biodiesel)
- ❖ Peanut: 90 gpa
- ❖ Sunflower: 82 gpa

Algae fuel yields have not yet been accurately determined, but DOE is reported as saying that algae yield 30 times more energy per acre than land crops such as soybeans, and some estimate even higher yields up to 15000 gpa.

The Jatropha plant has been cited as a high-yield source of biodiesel but such claims have also been exaggerated. The more realistic estimates put the yield at about 200 gpa (1.5-2 tonnes per hectare). It is grown in the Philippines, Mali and India, is drought-resistant, and can share space with other cash crops such as coffee, sugar, fruits and vegetables. It is well-suited to semi-arid lands and can contribute to slow down desertification, according to its advocates.

Efficiency and Economic Arguments

According to a study written by Drs. Van Dyne and Raymer for the Tennessee Valley Authority, the average US farm consumes fuel at the rate of 82 litres per hectare (8.75 US gallons per acre) of land to produce one crop. However, average crops of rapeseed produce oil at an average rate of 1,029 L/ha (110 US gal/acre), and high-yield rapeseed fields produce about 1,356 L/ha (145 US gal/acre). The ratio of input to output in these cases is roughly 1:12.5 and 1:16.5. Photosynthesis is known to have an efficiency rate of about 3-6 per cent of total solar radiation and if the entire mass of a crop is utilized for energy production, the overall efficiency of this chain is currently about 1 per cent While this may compare unfavourably to solar cells combined with an electric drive train, biodiesel is less costly to deploy (solar cells cost approximately US$1,000 per square meter) and transport (electric vehicles require batteries which currently have a much lower energy density than liquid fuels).

However, these statistics by themselves are not enough to show whether such a change makes economic sense. Additional factors must be taken into account, such as: the fuel equivalent of the energy required for processing, the yield of fuel from raw oil, the return on cultivating food, the effect biodiesel will have of food prices and the relative cost of biodiesel versus petrodiesel.

The debate over the energy balance of biodiesel is ongoing. Transitioning fully to biofuels could require immense tracts of land if traditional food crops are used (although non food crops can be utilized). The problem would be especially severe for nations with large economies, since energy consumption scales with economic output.

If using only traditional food plants, most such nations do not have sufficient arable land to produce biofuel for the nation's vehicles. Nations with smaller economies (hence less energy consumption) and more arable land may be in better situations, although many regions cannot afford to divert land away from food production.

For third world countries, biodiesel sources that use marginal land could make more sense, e.g. honge oil nuts grown along roads or jatropha grown along rail lines.

In tropical regions, such as Malaysia and Indonesia, oil palm is being planted at a rapid pace to supply growing biodiesel demand in Europe and other markets. It has been estimated in Germany that palm oil biodiesel has less than 1/3 the production costs of rapeseed biodiesel. The direct source of the energy content of biodiesel is solar energy captured by plants during photosynthesis. Regarding the positive energy balance of biodiesel:

When straw was left in the field, biodiesel production was strongly energy positive, yielding 1 GJ biodiesel for every 0.561 GJ of energy input (a yield/cost ratio of 1.78).

When straw was burned as fuel and oilseed rapemeal was used as a fertilizer, the yield/cost ratio for biodiesel production was even better (3.71). In other words, for every unit of energy input to produce biodiesel, the output was 3.71 units (the difference of 2.71 units would be from solar energy).

Biodiesel is becoming of interest to companies interested in commercial scale production as well as the more usual home brew biodiesel user and the user of straight vegetable oil or waste vegetable oil in diesel engines. Homemade biodiesel processors are many and varied.

Energy Security

One of the main drivers for adoption of biodiesel is energy security. This means that a nations dependence on oil is reduced, and substituted with use of locally available sources, such as coal, gas, or renewable sources. Thus significant benefits can accrue to a country from adoption of biofuels, even without a reduction in greenhouse gas emissions. Whilst the total energy balance is debated, it is clear that the dependence on oil is reduced. One example is the energy used to manufacture fertilizers, which could come from a variety of sources other than petroleum. The the US NREL says that energy security is the number one driving force behind the US biofuels programme. and the White House "Energy Security for the 21st Century" makes clear that energy security is a major reason for promoting biodiesel. The EU commission president, Jose Manuel Barroso, speaking at a recent EU biofuels conference, stressed that properly managed biofuels have the potential to reinforce the EU's security of supply through diversification of energy sources.

Environmental Effects

Greenhouse Gas Emissions

An often mentioned incentive for using biofuel is its capacity to lower greenhouse gas emissions compared to those of fossil fuels. If this is true or not depends on many factors. Especially the effects from land use change have potential to cause even more emissions than what would be caused by using fossil fuels alone.

Carbon dioxide is one of the major greenhouse gases. Although the burning of biodiesel produces carbon dioxide emissions similar to those from ordinary fossil fuels, the plant feedstock used in the production absorbs carbon dioxide from the atmosphere when it grows. Plants absorb carbon dioxide through a process known as photosynthesis which allows it to store energy from sunlight in the form of oil. After the oil is converted into biodiesel and burnt as fuel the energy and carbon is released again. Some of that energy can be used to power an engine while the carbon dioxide is released back into the atmosphere.

When considering the total amount of greenhouse gas emissions it is therefore important to consider the whole production process and what indirect effects such production might cause. The effect on carbon dioxide emissions is highly dependent on production methods and the type of feedstock used. Calculating the carbon intensity of biofuels is a complex and inexact process, and is highly dependent on the assumptions made in the calculation. A calculation usually includes:

- ❖ Emissions from growing the feedstock (e.g. Petrochemicals used in fertilizers).
- ❖ Emissions from transporting the feedstock to the factory.
- ❖ Emissions from processing the feedstock into biodiesel.

Other factors can be very significant but are sometimes not considered. These include:

- ❖ Emissions from the change in land use of the area where the fuel feedstock is grown.
- ❖ Emissions from transportation of the biodiesel from the factory to its point of use.
- ❖ The efficiency of the biodiesel compared with standard diesel.
- ❖ The amount of Carbon Dioxide produced at the tail pipe. (Biodiesel can produce 4.7% more).
- ❖ The benefits due to the production of useful by-products, such as cattle feed or glycerine.

If land use change is not considered and assuming todays production methods, biodiesel from rapeseed and sunflower oil produce 45 to 65 per cent lower greenhouse gas emissions than petrodiesel. However, there is ongoing research to improve the efficiency of the production process. Biodiesel produced from used cooking oil or other waste fat could reduce CO_2 emissions by as much as 85 per cent. As long as the feedstock is grown on existing cropland, land use change has little or no effect on greenhouse gas emissions.

However, there is concern that increased feedstock production directly affect the rate of deforestation. Such clearcutting cause carbon stored in the forest, soil and peat layers to be released. The amount of greenhouse gas emissions from deforestation is so large that the benefits from lower emissions (caused by biodiesel use alone) would be negligible for hundreds of years. Biofuel produced from feedstocks such as palm oil could therefore cause much higher carbon dioxide emissions than ordinary fossil fuels.

Deforestation

If deforestation, and monoculture farming techniques were used to grow biofuel crops, biodiesel may become a serious threat to the environment:

- Increasing the emission of climate change gases rather than helping curb them,
- Damaging ecosystems and biodiversity, and
- Exacerbating social conflict.

The demand for cheap oil from the tropical regions is of rising concern. In order to increase production, the amount of arable land is being expanded at the cost of tropical rainforest. Feedstock oils produced in Asia, South America and Africa are currently less expensive than those produced in Europe and North America suggesting that imports to these wealthier nations are likely to increase in the future.

In the Philippines and Indonesia forest clearing is already underway for the production of palm oil. Indigenous people are forced to move and their livelihood is destroyed when forest is cleared to make room for oil palm plantations. In some areas the use of pesticides for biofuel crops are disrupting clean water supplies, and the loss of habitat caused by deforestation is threatening many species of unique plants and animals. One example is the already-shrinking populations of orangutans on the Indonesian islands of Borneo and Sumatra, which face extinction if deforestation continue at it's projected rate.

Pollution

In the United States, biodiesel is the only alternative fuel to have successfully completed the Health Effects Testing requirements (Tier I and Tier II) of the Clean Air Act (1990).

Biodiesel can reduce the direct tailpipe-emission of particulates, small particles of solid combustion products, on vehicles with particulate filters by as much as 20 percent compared with low-sulfur (< 50 ppm) diesel. Particulate emissions as the result of production are reduced by around 50 percent compared with fossil-sourced diesel. (Beer *et al*, 2004). Biodiesel has a higher cetane rating than petrodiesel, which can improve performance and clean up emissions compared to crude petro-diesel (with cetane lower than 40). Biodiesel contains fewer aromatic hydrocarbons: benzofluoranthene: 56 per cent reduction; Benzopyrenes: 71 per cent reduction.

If burned without additives, Biodiesel (B100) is estimated to produce about 10 per cent more nitrogen oxide NO_x tailpipe-emissions than petrodiesel. As biodiesel has a low sulfur content, NO_x emissions can be reduced through the use of catalytic converters to less than the NO_x emissions from conventional diesel engines. However, modern diesel engines already use exhaust aftertreatment and EGR to reduce NO_x emissions. These systems add complexity, increase costs, and reduce fuel economy (leading to higher CO_2 emissions). As a transportation fuel, biodiesel is in its infancy in terms of additives which are capable of improving energy density, resistance to gelling, and NO_x emissions. Debate continues over NO_x, particulates, smog, and greenhouse gas emissions from biodiesel and all other new transportation fuels, biofuels in particular. Ultimately, greater clarity on the fundamental distinctions between smog and other local pollution issues vs. greenhouse gas emissions will be essential for both well founded public policy as well as well informed consumer choices.

Biodegradable

A University of Idaho study compared biodegradation rates of biodiesel, neat vegetable oils, biodiesel and petroleum diesel blends, and neat 2-D diesel fuel. Using low concentrations of the product to be degraded (10 ppm) in nutrient and sewage sludge amended solutions, they demonstrated that biodiesel degraded at the same rate as a dextrose control and 5 times as quickly as petroleum diesel over a period of 28 days, and that biodiesel blends doubled the rate of petroleum diesel degradation through co-metabolism. The same study examined soil degradation using 10 000 ppm of biodiesel and petroleum diesel, and found biodiesel degraded at twice the rate of petroleum diesel in soil. In all cases, it was determined biodiesel also degraded more completely than petroleum diesel, which produced poorly degradable undetermined intermediates. Toxicity studies for the same project demonstrated no mortalities and few toxic effects on rats and rabbits with up to 5000 mg/kg of biodiesel. Petroleum diesel showed no mortalities at the same concentration either, however toxic effects such as hair loss and urinary discolouring were noted with concentrations of greater than 2000 mg/l in rabbits.

Food vs. Fuel

Food quality vegetable oil has become so expensive there is no longer a profit viability for its use. Food grade vegetable oil pricing is on a similar upward ramp as food in general. Accessing food stuffs in poor countries has always been problematic for the inhabitants. Non food grade vegetable oils are under use or consideration for use to make biodiesel and have been so during the entire history of biodiesel.

In some poor countries the rising price of vegetable oil is causing problems. There are those that say using a food crop for fuel sets up competition between food in poor countries and fuel in rich countries. Some propose that fuel only be made from non-edible vegetable oils like jatropha oil and algal oil.

Others argue that the problem is more fundamental. Farmers can switch from producing food crops to producing biofuel crops to make more money, even if the new crops are not edible. The law of supply and demand predicts that if fewer farmers are producing food the price of food will rise. It may take some time, as farmers can take some time to change which things they are growing, but increasing demand for first generation biofuels is likely to result in price increases for many kinds of food. Some have pointed out that there are poor farmers and poor countries making more money because of the higher price of vegetable oil. . In any case, algae biodiesel would not displace land currently used for food production and new algaculture jobs could be created.

Current Research

There is ongoing research into finding more suitable crops and improving oil yield. Using the current yields, vast amounts of land and fresh water would be needed to produce enough oil to completely replace fossil fuel usage. It would require twice the land area of the US to be devoted to soybean production, or two-thirds to be devoted to rapeseed production, to meet current US heating and transportation needs.

Specially bred mustard varieties can produce reasonably high oil yields and are very usefull in crop rotation with cereals, and have the added benefit that the meal leftover after the oil has been pressed out can act as an effective and biodegradable pesticide.

Algaculture

From 1978 to 1996, the U.S. National Renewable Energy Laboratory experimented with using algae as a biodiesel source in the "Aquatic Species Programme". A self-published article by Michael Briggs, at the UNH Biodiesel Group, offers estimates for the realistic replacement of all vehicular fuel with biodiesel by utilizing algae that have a natural oil content greater than 50 per cent, which Briggs suggests can be grown on algae ponds at plants. This oil-rich algae can then be extracted from the system and processed into biodiesel, with the dried remainder further reprocessed to create ethanol.

The production of algae to harvest oil for biodiesel has not yet been undertaken on a commercial scale, but feasibility studies have been conducted to arrive at the above yield estimate. In addition to its projected high yield, algaculture—unlike crop-based biofuels—does not entail a decrease in food production, since it requires neither farmland nor fresh water. Many companies are pursuing algae bio-reactors for various purposes, including scaling up biodiesel production to commercial levels.

26.5 Ethanol Fuel

Ethanol fuel is ethanol (ethyl alcohol), the same type of alcohol found in alcoholic beverages. It can be used as a fuel, mainly as a biofuel alternative to gasoline, and is widely used in cars in Brazil. Because it is easy to manufacture and process, and can be made from very common crops, such as sugar cane and maize (corn), it is an increasingly common alternative to gasoline in some parts of the world.

Anhydrous ethanol (ethanol with less than 1% water) can be blended with gasoline in varying quantities up to pure ethanol (E100), and most spark-ignited gasoline style engines will operate well with mixtures of 10 per cent ethanol (E10). Most cars on the road today in the U.S. can run on blends of up to 10 per cent ethanol, and the use of 10 per cent ethanol gasoline is mandated in some cities where harmful levels of auto emissions are possible.

Ethanol can be mass-produced by fermentation of sugar or by hydration of ethylene from petroleum and other sources. Current interest in ethanol mainly lies in bio-ethanol, produced from the starch or sugar in a wide variety of crops, but there has been considerable debate about how useful bio-ethanol will be in replacing fossil fuels in vehicles. Concerns relate to the large amount of arable land required for crops, as well as the energy and pollution balance of the whole cycle of ethanol production. Recent developments with cellulosic ethanol production and commercialization may allay some of these concerns.

According to the International Energy Agency, cellulosic ethanol could allow ethanol fuels to play a much bigger role in the future than previously thought. Cellulosic ethanol offers promise as resistant cellulose fibers, a major component in plant cells walls, can be used to generate ethanol. Dedicated energy crops, such as switchgrass, are also promising cellulose sources that can be produced in many regions of the United States.

Chemistry

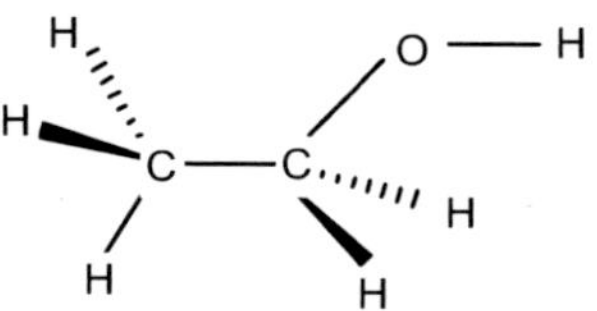

In this 3-d diagram of ethanol, the lines represent single bonds.

During ethanol fermentation, glucose is decomposed into ethanol and carbon dioxide.

$$C_6H_{12}O_6 \rightarrow 2C_2H_6O + 2CO_2$$

During combustion ethanol reacts with oxygen to produce carbon dioxide, water, and heat: (other air pollutants are also produced when ethanol is burned in the atmosphere rather than in pure oxygen)

$$C_2H_6O + 3O_2 \rightarrow 2CO_2 + 3H_2O$$

Together, they add up to:

$$C_6H_{12}O_6 + 6O_2 \rightarrow 6CO_2 + 6H_2O + heat$$

Also harmful nitrous oxide gases are produced. Nitrogen dioxide is one of the harmful gases as is a major contributor to the formation of "brown smog".

Sources

Ethanol is considered "renewable" because it is primarily the result of conversion of the sun's energy into usable energy. Creation of ethanol starts with photosynthesis causing the feedstocks such as switchgrass, sugar cane, or corn to grow. These feedstocks are processed into ethanol.

About 5 per cent of the ethanol produced in the world in 2003 was actually a petroleum product. It is made by the catalytic hydration of ethylene with sulfuric acid as the catalyst. It can also be obtained via ethylene or acetylene, from calcium carbide, coal, oil gas, and other sources. Two million tons of petroleum-derived ethanol are produced annually. The principal suppliers are plants in the United States, Europe, and South Africa. Petroleum derived ethanol (synthetic ethanol) is chemically identical to bio-ethanol and can be differentiated only by radiocarbon dating.

Bio-ethanol is obtained from the conversion of carbon based feedstock. Agricultural feedstocks are considered renewable because they get energy from the sun using photosynthesis, provided that all minerals required for growth (such as nitrogen and phosphorus) are returned to the land. Ethanol can be produced from a variety of feedstocks such as sugar cane, bagasse, miscanthus, sugar beet, sorghum, grain sorghum, switchgrass, barley, hemp, kenaf, potatoes, sweet potatoes, cassava, sunflower, fruit, molasses, corn, stover, grain, wheat, straw, cotton, other biomass, as well as many types of cellulose waste and harvestings, whichever has the best well-to-wheel assessment.

Current, first generation processes for the production of ethanol from corn use only a small part of the corn plant: the corn kernels are taken from the corn plant and only the starch, which represents about 50 per cent of the dry kernel mass, is transformed into ethanol. Two types of second generation processes are under development. The first type uses enzymes and yeast to convert the plant cellulose into ethanol while the second type uses pyrolysis to convert the whole plant to either a liquid bio-oil or a syngas. Second generation processes can also be used with plants such as grasses, wood or agricultural waste material such as straw.

Production Process

The basic steps for large scale production of ethanol are: microbial (yeast) fermentation of sugars, distillation, dehydration (requirements vary, see Ethanol fuel mixtures, below), and denaturing (optional). Prior to fermentation, some crops require saccharification or hydrolysis of carbohydrates such as cellulose and starch into sugars. Saccharification of cellulose is called cellulolysis. Enzymes are used to convert starch into sugar.

Fermentation

Ethanol is produced by microbial fermentation of the sugar. Microbial fermentation will currently only work directly with sugars. Two major components of plants, starch and cellulose, are both made up of sugars, and can in principle be converted to sugars for fermentation. Currently, only the sugar (e.g. sugar cane) and starch (e.g. corn) portions can be economically converted. However, there is much activity in the area of cellulosic ethanol, where the cellulose part of a plant is broken down to sugars and subsequently converted to ethanol.

Distillation

For the ethanol to be usable as a fuel, water must be removed. Most of the water is removed by distillation, but the purity is limited to 95-96 per cent due to the formation of a low-boiling water-ethanol azeotrope. The 95.6 per cent m/m (96.5% v/v) ethanol, 4.4 per cent m/m (3.5% v/v) water mixture may be used as a fuel alone, but unlike anhydrous ethanol, is immiscible in gasoline, so the water fraction is typically removed in further treatment in order to burn with in combination with gasoline in gasoline engines.

Dehydration

Currently, the most widely used purification method is a physical absorption process using a molecular sieve, for example, ZEOCHEM Z3-03 (a special 3A molecular sieve for EtOH dehydration). Another method, azeotropic distillation, is achieved by adding the hydrocarbon benzene which also denatures the ethanol (to render it undrinkable for duty purposes). A third method involves use of calcium oxide as a desiccant.

Technology

Ethanol-based Engines

Ethanol is most commonly used to power automobiles, though it may be used to power other vehicles, such as farm tractors and airplanes. Ethanol (E100) consumption in an engine is approximately 34 per cent higher than that of gasoline (the energy per volume unit is 34% lower). However, higher compression ratios in an ethanol-only engine allow for increased power output and better fuel economy than would be obtained with the lower compression ratio. In general, ethanol-only engines are tuned to give slightly better power and torque output to gasoline-powered engines. In flexible fuel vehicles, the lower compression ratio requires tunings that give the same output when using either gasoline or hydrated ethanol. For maximum use of ethanol's benefits, a much higher compression ratio should be used, which would render that engine unsuitable for gasoline use. When ethanol fuel availability allows high-compression ethanol-only vehicles to be practical, the fuel efficiency of such engines should be equal or greater than current gasoline engines. However, since the energy content (by volume) of ethanol fuel is less than gasoline, a larger volume of ethanol fuel (151%) would still be required to produce the same amount of energy.

A 2004 MIT study, and an earlier paper published by the Society of Automotive Engineers, describing tests, identify a method to exploit the characteristics of fuel ethanol that is substantially better than mixing it with gasoline. The method presents the possibility of leveraging the use of alcohol to even achieve definite improvement over the cost-effectiveness of hybrid electric. The improvement consists of using dual-fuel direct-injection of pure alcohol (or the azeotrope or E85) and gasoline, in any ratio up to 100 per cent of either, in a turbocharged, high compression-ratio, small-displacement engine having performance similar to an engine having twice the displacement. Each fuel is carried separately, with a much smaller tank for alcohol. The high-compression (which increases efficiency) engine will run on ordinary gasoline under low-power cruise conditions. Alcohol is directly injected into the cylinders (and the gasoline injection simultaneously reduced) only when necessary to suppress 'knock' such as when significantly accelerating. Direct cylinder injection raises the already high octane rating of ethanol up to an effective 130. The calculated over-all reduction of gasoline use and CO2 emission is 30 per cent. The consumer cost payback time shows a 4:1 improvement over turbo-diesel and a 5:1 improvement over hybrid. In addition, the problems of water absorption into pre-mixed gasoline (causing phase separation), supply issues of multiple mix ratios and cold-weather starting are avoided.

Ethanol's higher octane rating allows an increase of an engine's compression ratio for increased thermal efficiency. In one study, complex engine controls and increased exhaust gas recirculation allowed a compression ratio of 19.5 with fuels ranging from neat ethanol to E50. Thermal efficiency up to approximately that for a diesel was achieved. This would result in the MPG (miles per gallon) of a dedicated ethanol vehicle to be about the same as one burning gasoline.

Engines using fuel with 30 per cent to 100 per cent ethanol also need a cold-starting system. For E85 fuel at temperatures below 11 °C (52 °F) a cold-starting system is required for reliable starting and to meet EPA emissions standards. However, the EPA does not require cold start systems on E85 vehicles. No current production E85 vehicles in the USA are equipped with these cold start systems, and they meet EPA emission guidelines.

Ethanol Fuel Mixtures

To avoid engine stall due to "slugs" of water in the fuel lines interrupting fuel flow, the fuel must exist as a single phase. The fraction of water that an ethanol-gasoline fuel can contain without phase separation increases with the percentage of ethanol. . This shows, for example, that E30 can have up to about 2 per cent water. If there is more than about 71 per cent ethanol, the remainder can be any proportion of water or gasoline and phase separation will not occur. However, the fuel mileage declines with increased water content. The increased solubility of water with higher ethanol content permits E30 and hydrated ethanol to be put in the same tank since any combination of them always results in a single phase. Somewhat less water is tolerated at lower temperatures. For E10 it is about 0.5 per cent v/v at 70 F and decreases to about 0.23 per cent v/v at -30 F.

In many countries cars are mandated to run on mixtures of ethanol. Brazil requires cars be suitable for a 25 per cent ethanol blend, and has required various mixtures between 22 per cent and 25 per cent ethanol, since of July 2007 25 per cent is required. The United States allows up to 10 per cent blends, and some states require this (or a smaller amount) in all gasoline sold. Other countries have adopted their own requirements. Beginning with the model year 1999, an increasing number of vehicles in the world are manufactured with engines which can run on any fuel from 0 per cent ethanol up to 100 per cent ethanol without modification. Many cars and light trucks (a class containing minivans, SUVs and pickup trucks) are designed to be flexible-fuel vehicles (also called *dual-fuel* vehicles). In older model years, their engine systems contained alcohol sensors in the fuel and/or oxygen sensors in the exhaust that provide input to the engine control computer to adjust the fuel injection to achieve stochiometric (no residual fuel or free oxygen in the exhaust) air-to-fuel ratio for any fuel mix. In newer models, the alcohol sensors have been removed, with the computer using only oxygen and airflow sensor feedback to estimate alcohol content. The engine control computer can also adjust (advance) the ignition timing to achieve a higher output without pre-ignition when it predicts that higher alcohol percentages are present in the fuel being burned. This method is backed up by advanced knock sensors—used in most high performance gasoline engines regardless if they're designed to use ethanol or not—that detect pre-ignition and detonation.

Fuel Economy

In theory, all fuel-driven vehicles have a fuel economy (measured as miles per US gallon, or liters per 100 km) that is directly proportional to the fuel's energy content. In reality, there are many other variables that come in to play that affect the performance of a particular fuel in a particular engine. Ethanol contains approx. 34 per cent less energy per unit volume than gasoline, and therefore in theory, burning pure ethanol in a vehicle will result in a 34 per cent reduction in miles per US gallon, given the same fuel economy, compared to burning pure gasoline. This assumes that the octane ratings of the fuels, and the thus the engine's ability to extract energy from the fuels, are the same. For E10 (10% ethanol and 90 per cent gasoline), the effect is small (– 3%) when compared to conventional gasoline, and even smaller (1-2%) when compared to oxygenated and reformulated blends. However, for E85 (85% ethanol), the effect becomes significant. E85 will produce lower mileage than gasoline, and will require more frequent refueling. Actual performance may vary depending on the vehicle. The EPA-rated mileage of current USA flex-fuel vehicles should be considered when making price comparisons, but it must be noted that E85 is a high performance fuel, with an octane rating of about 104, and should be compared to premium. In one estimate the US retail price for E85 ethanol is 2.62 US dollar per gallon or 3.71 dollar corrected for energy equivalency compared to a gallon of gasoline priced at 3.03 dollar. Brazilian cane ethanol (100%) is priced at 3.88 dollar against 4.91 dollar for E25 (figures July 2007).

Experience by Country

The top five ethanol producers in 2006 were the United States with 4.855 billion U.S. liquid gallons (bg), Brazil (4.49 bg), China (1.02 bg), India (0.50 bg) and France (0.25 bg). Brazil and the United States accounted for 70 percent of all ethanol production, with total world production of 13.5 billion US gallons (40 million tonnes). When accounting just for fuel ethanol production in 2007, the U.S. and Brazil are responsible for 88 per cent of the 13.1 billion gallons total world production. Strong incentives, coupled with other industry development initiatives, are giving rise to fledgling ethanol industries in countries such as Thailand, Colombia, and some Central American countries. Nevertheless, ethanol has yet to make a dent in world oil consumption of approximately 4000 million tonnes/yr (84 million barrels/ day).

Total Annual Ethanol Production (All Grades) by Country (2004-2006) Top 15 countries (Millions of U.S. liquid gallons)					Annual Fuel Ethanol Production by Country (2004-2006) Top 15 countries/blocks (Millions of U.S. liquid gallons)		
World rank	Country	2006	2005	2004	World rank	Country/Region	2007
1	United States	4,855	4,264	3,535	1	United States	6,498.6
2	Brazil	4,491	4,227	3,989	2	Brazil	5,019.2
3	China	1,017	1,004	964	3	European Union	570.3
4	India	502	449	462	4	China	486.0
5	France	251	240	219	5	Canada	211.3
6	Germany	202	114	71	6	Thailand	79.2
7	Russia	171	198	198	7	Colombia	74.9
8	Canada	153	61	61	8	India	52.8
9	Spain	122	93	79	9	Central America	39.6
10	South Africa	102	103	110	10	Australia	26.4
11	Thailand	93	79	74	11	Turkey	15.8
12	United Kingdom	74	92	106	12	Pakistan	9.2
13	Ukraine	71	65	66	13	Peru	7.9
14	Poland	66	58	53	14	Argentina	5.2
15	Saudi Arabia	52	32	79	15	Paraguay	4.7
	World Total	13,489	12,150	10,770		World Total	13,101.7

Brazil

Brazil has ethanol fuel available throughout the country. A typical Petrobras filling station at São Paulo with dual fuel service, marked A for alcohol (ethanol) and G for gasoline.

Brazil has the largest and most successful bio-fuel programmes in the world, involving production of ethanol fuel from sugar cane, and it is considered to have the world's first sustainable biofuels economy. In 2006 Brazilian ethanol provided around 20 per cent of the country's road transport sector fuel consumption needs, and more than 40 per cent of fuel consumption for the light vehicle fleet. As a result of the increasing use of ethanol, together with the exploitation of domestic deep water oil sources, Brazil, which years ago had to import a large share of the petroleum needed for domestic consumption, in 2006 reached complete self-sufficiency in oil supply.

Together, Brazil and the United States lead the industrial world in global ethanol production, accounting together for 70 per cent of the world's production and nearly 90 per cent of ethanol used for fuel. In 2006 Brazil produced 16.3 billion liters (4.3 billion U.S. liquid gallons), which represents 33.3 per cent of the world's total ethanol production and 42 per cent of the world's ethanol used as fuel. Sugar cane plantations cover 3.6 million hectares of land for ethanol production, representing just 1 per cent of Brazil's arable land, with a productivity of 7,500 liters of ethanol per hectare, as compared with the U.S. maize ethanol productivity of 3,000 liters per hectare.

Production and use of ethanol has been stimulated through:

* ❖ Low-interest loans for the construction of ethanol distilleries
* ❖ Guaranteed purchase of ethanol by the state-owned oil company at a reasonable price
* ❖ Retail pricing of neat ethanol so it is competitive if not slightly favourable to the gasoline-ethanol blend
* ❖ Tax incentives provided during the 1980s to stimulate the purchase of neat ethanol vehicles.

Guaranteed purchase and price regulation were ended some years ago, with relatively positive results. In addition to these other policies, ethanol producers in the state of São Paulo established a research and technology transfer center that has been effective in improving sugar cane and ethanol yields.

There are no longer light vehicles in Brazil running on pure gasoline. Since 1977 the government made mandatory to blend 20 per cent of ethanol (E20) with gasoline (gasohol), requiring just a minor adjustment on regular gasoline motors. Today the mandatory blend is allowed to vary nationwide between 20 per cent to 25 per cent ethanol (E25) and it is used by all regular gasoline vehicles, plus three million cars running on 100 per cent anhydrous ethanol and five million of dual or flexible-fuel vehicles. The Brazilian car manufacturing industry developed full flexible-fuel vehicles that can run on any proportion of gasoline and ethanol. Introduced in the market in 2003, these vehicles became a commercial success. On March 2008, the fleet of "flex" cars and light commercial vehicles had reached 5 million new vehicles sold. which represents around 10 per cent of Brazil's motor vehicle fleet and 15.6 per cent of all light vehicles. The ethanol-powered and "flex" vehicles, as they are popularly known, are manufactured to tolerate hydrated ethanol, an azeotrope comprised of 95.6 per cent ethanol and 4.4 per cent water.

The first vehicle in the world, moved by pure ethanol, was the Fiat 147, built in July 1979. In the late 1988, ethanol vehicles held almost 90 per cent of the Brazilian's market, but an strong crisis in ethanol supplies in early 1990 left thousands of vehicles out of fuel in their garages, and makes the ethanol vehicles production fall to less than 20 per cent of the total fuel produced in one year. Recently, in early 2003, Volkswagen started the production of the Volkswagen Gol Total Flex, the first full flexible-fuel vehicle, that supports any percentage of ethanol and gasoline used as fuel. Because of the flexible-fuel vehicles, the ethanol production in 2006, reached more than 60 per cent of the total fuel production in 2006.

United States

United States fuel ethanol production and imports (2001-2007) (Millions of U.S. liquid gallons)			
Year	Production	Imports	Demand
2001	1,770	n/a	n/a
2002	2,130	46	2,085
2003	2,800	61	2,900
2004	3,400	161	3,530
2005	3,904	135	4,049
2006	4,855	653	5,377
2007	6,485	435	6,847

Note: Demand figures includes stocks change and small exports in 2005

The United States produces and consumes more ethanol fuel than any other country in the world. Most cars on the road today in the U.S. can run on blends of up to 10 per cent ethanol, and motor vehicle manufacturers already produce vehicles designed to run on much higher ethanol blends. In 2007 Portland, Oregon, became the first city in the United States to require all gasoline sold within city limits to contain at least 10 per cent ethanol. As of January 2008, three states—Missouri, Minnesota, and Hawaii—require ethanol to be blended with gasoline motor fuel. Many cities are also required to use an ethanol blend due to non-attainment of federal air quality goals.

Several motor vehicle manufacturers, including Ford, DaimlerChrysler, and GM, sell flexible-fuel vehicles that can use gasoline and ethanol blends ranging from pure gasoline all the way up to 85 per cent ethanol (E85). By mid-2006, there were approximately six million E85-compatible vehicles on U.S. roads.

In the USA there are currently 1,587 stations distributing ethanol, although most stations are in the corn belt area. One of the debated methods for distribution in the US is using existing oil pipelines, which raises concerns over corrosion. In any case, some companies proposed building a 1,700-mile pipeline to carry ethanol from the Midwest through Central Pennsylvania to New York.

The production of fuel ethanol from corn in the United States is controversial for a few reasons. Production of ethanol from corn is 5 to 6 times less efficient than producing it from sugarcane. Ethanol production from corn is highly dependent upon subsidies and it consumes a food crop to produce fuel. The subsidies paid to fuel blenders and ethanol refineries have often been cited as the reason for driving up the price of corn, and in farmers planting more corn and the conversion of considerable land to corn (maize) production which generally consumes more fertilizers and pesticides than many other land uses. This is at odds with the subsidies actually paid directly to farmers that are designed to take corn land out of production and pay farmers to plant grass and idle the land, often in conjunction with soil conservation programmes, in an attempt to boost corn prices. Recent developments with cellulosic ethanol production and commercialization may allay some of these concerns. A theoretically much more efficient way of ethanol production has been suggested to use sugar beets which make about the same amount of ethanol as corn without using the corn food crop especially since sugar beats can grow in less tropical conditions than sugar cane.

Europe

Production of Bioethanol in the European Union (GWh)				Consumption of Bioethanol in the European Union (GWh)			
No	Country	2006	2005	No	Country	2006	2005
1	Germany	2,554	978	1	Germany	3,573	1,682
2	Spain	2,382	1,796	2	Sweden	1,895	1,681
3	France	1,482	853	3	France	1,747	871
4	Sweden	830	907	4	Spain	1,332	1,314
5	Italy	759	47	5	Poland	611	329
6	Poland	711	379	6	United Kingdom	561	502
7	Hungary	201	207	7	Netherlands	238	0
8	Lithuania	107	47	8	Hungary	125	28
9	Netherlands	89	47	9	Lithuania	99	10
10	Czech Republic	89	0	10	Czech Republic	14	0
11	Latvia	71	71	11	Finland	9	0
12	Finland	0	77	12	Ireland	8	0
27	**Total**	**9,274**	**5,411**	13	Italy	0	59
				14	Latvia	0	5
				27	**EU**	**10,210**	**6,481**

100 l bioethanol = 79,62 kg,

1 tonne bioethanol = 0,64 toe

1 toe = 11,63 MWh

The consumption of bioethanol is largest in Europe in Germany, Sweden, France and Spain. Europe produces equivalent to 90 per cent of its consumption (2006). Germany produced ca 70 per cent of its consumption, Spain 60 per cent and Sweden 50 per cent (2006). In Sweden there are 792 E85 filling stations and in France 131 E85 service stations with 550 more under construction.

On Monday, September 17, 2007 the first ethanol fuel pump was opened in Reykjavik, Iceland. This pump is the only one of its kind in Iceland. The fuel is imported by Brimborg, a Volvo dealer, as a pilot to see how ethanol fueled cars work in Iceland. In a few weeks, the pump will be opened for public use.

In The Netherlands regular petrol with no bio-additives is slowly outphased, since EU-legislation has been passed that requires the fraction of nonmineral origin to become minimum 5,75 per cent of the total fuel consumption volume in 2010. This can be realised by substitutions in diesel or in petrol of any biological source; or fuel sold in the form of pure biofuel. (2007:) There are only a few gas stations where E85 is sold, which is an 85 per cent ethanol, 15 per cent petrol mix. Directly neighbouring country Germany is reported to have a much better biofuel infrastructure and offers both E85 and E50. Biofuel is taxed equally as regular fuel. However, fuel tanked abroad cannot be taxed and a recent payment receipt will in most cases suffice to prevent fines if customs check tank contents. (Authorities are aware of high taxation on fuels and cross-border fuel refilling is a well-known practice.)

All Swedish gas stations are required by an act of parliament to offer at least one alternative fuel, and every fifth car in Stockholm now drives at least partially on alternative fuels, mostly ethanol. The number of bioethanol stations in Europe is highest in Sweden, with 792 stations.

Stockholm will introduce a fleet of Swedish-made electric hybrid buses in its public transport system on a trial basis in 2008. These buses will use ethanol-powered internal-combustion engines and electric motors. The vehicles' diesel engines will use ethanol.

Bioethanol stations European Union		
Country	*Stations*	*No/10^6 persons*
Sweden	792	86.6
Germany	73	0.89
France	36	0.56
United Kingdom	14	0.24
Ireland	13	3.07
Switzerland	6	0.8

Asia

China

China is promoting ethanol-based fuel on a pilot basis in five cities in its central and northeastern region, a move designed to create a new market for its surplus grain and reduce consumption of petroleum. The cities include Zhengzhou, Luoyang and Nanyang in central China's Henan province, and

Harbin and Zhaodong in Heilongjiang province, northeast China. Under the programme, Henan will promote ethanol-based fuel across the province by the end of this year. Officials say the move is of great importance in helping to stabilize grain prices, raise farmers' income and reducing petrol—induced air pollution.

Thailand

Thailand already use 10 per cent ethanol (E10) widely on big scale on the local market. Beginning in 2008 Thailand started with the sale of E20 and the in the third quarter of 2008 E85 will come on the mark.

Australia

Legislation in Australia imposes a 10 per cent cap on the concentration of fuel ethanol blends. Blends of 90 per cent unleaded petrol and 10 per cent fuel ethanol are commonly referred to as E10. E10 is available through service stations operating under the BP, Caltex, Shell and United brands as well as those of a number of smaller independents. Not surprisingly, E10 is most widely available closer to the sources of production in Queensland and New South Wales. E10 is most commonly blended with 91 RON "regular unleaded" fuel. There is a requirement that retailers label blends containing fuel ethanol on the dispenser.

Caribbean Basin

United States fuel ethanol imports by country (2002-2007) (Millions of U.S. liquid gallons)						
Country	2007*	2006	2005	2004	2003	2002
Brazil	188.8	433.7	31.2	90.3	0	0
Jamaica	75.2	66.8	36.3	36.6	39.3	29.0
El Salvador	73.3	38.5	23.7	5.7	6.9	4.5
Trinidad and Tobago	42.7	24.8	10.0	0	0	0
Costa Rica	39.3	35.9	33.4	25.4	14.7	12.0
*Note: 2007 figures through November only.						

*Note: 2007 figures through November only.

All countries in Central America, northern South America and the Caribbean are located in a tropical zone with suitable climate for growing sugar cane. In fact, most of these countries have a long tradition of growing sugar cane mainly for producing sugar and alcoholic beverages.

As a result of the guerilla movements in Central America, in 1983 the United States unilateral and temporarily approved the Caribbean Basin

Initiative, allowing most countries in the region to benefit from several tariff and trade benefits. These benefits were made permanent in 1990 and more recently, these benefits were replaced by the Caribbean Basin Trade and Partnership Act, approved in 2000, and the Dominican Republic-Central America Free Trade Agreement that went to effect in 2008. All these agreements have allowed several countries in the region to export ethanol to the U.S free of tariffs. Until 2004, the countries that benefited the most were Jamaica and Costa Rica, but as the U.S. began demanding more fuel ethanol, the two countries increased their exports and two others began exporting. In 2007, Jamaica, El Salvador, Trinidad & Tobago and Costa Rica exported together to the U.S. a total of 230.5 million gallons of ethanol, representing 54.1 per cent of U.S. fuel ethanol imports. Brasil began exporting ethanol to the U.S. in 2004 and exported 188.8 million gallons representing 44.3 per cent of U.S. ethanol imports in 2007. The remaining imports that year came from Canada and China.

In March 2007, "ethanol diplomacy" was the focus of President George W. Bush's Latin American tour, in which he and Brazil's president, Luiz Inacio Lula da Silva, were seeking to promote the production and use of sugar cane based ethanol throughout Latin America and the Caribbean. The two countries also agreed to share technology and set international standards for biofuels. The Brazilian sugar cane technology transfer would allow several Central American, Caribbean and Andean countries to take advantage of their tariff-free trade agreements to increase or become exporters to the United States in the short-term. Also, in August 2007, Brazil's President toured Mexico and several countries in Central America and the Caribbean to promote Brazilian ethanol technology. The ethanol alliance between the U.S. and Brazil generated some negative reactions from Venezuela's President Hugo Chavez, and by then Cuba's President, Fidel Castro, who wrote that *"you will see how many people among the hungry masses of our planet will no longer consume corn." "Or even worse,"* he continued, *"by offering financing to poor countries to produce ethanol from corn or any other kind of food, no tree will be left to defend humanity from climate change."* Daniel Ortega, Nicaragua's President, and one of the preferencial recipients of Brazilian technical aid also voiced critics to the Bush plan, but he vowed support for sugar cane based ethanol during Lula's visit to Nicaragua.

Colombia

Colombia's ethanol programme began in 2002, based on a law approved in

2001 mandating a mix of 10 per cent ethanol with regular gasoline. Sugar cane-based ethanol production began in 2005, and as local production was not enough to supply enough ethanol to the entire country's fleet, the programme was implemented only on cities with more than 500,000 inhabitants, such as Cali, Pereira, and the capital city of Bogotá. All of the ethanol production comes from the Department of Valle del Cauca, Colombia's traditional sugar cane region.

Costa Rica

Starting in October 2008, all gasoline sold in Costa Rica will be blended with 7.5 per cent ethanol. This follows a two year trial that took place in the provinces of Guanacaste and Puntarenas. The government expects to increase the percent of ethanol mixed with gasoline to 12 per cent in the next 4 to 5 years. The Costa Rican government is pursuing this policy to lower the country's dependency of foreign oil and to reduce the amount of greenhouse gases produced. The plan also calls for an increase in ethanol producing crops and tax breaks for flex-fuel vehicles.

El Salvador

As a result of the cooperation agreement between the United States and Brazil, El Salvador was chosen in 2007 to lead a pilot experience to introduce state-of-the-art technology for growing sugar cane for production of ethanol fuel in Central America, as this technical bilateral cooperation is looking for helping Central American countries to reduce their dependence on foreign oil.

Comparison between Brazil and the U.S.

Brazil's sugar cane-based industry is far more efficient than the U.S. corn-based industry. Brazilian distillers are able to produce ethanol for 22 cents per liter, compared with the 30 cents per liter for corn-based ethanol. Sugarcane cultivation requires a tropical or subtropical climate, with a minimum of 600 mm (24 in) of annual rainfall. Sugarcane is one of the most efficient photosynthesizers in the plant kingdom, able to convert up to 2 per cent of incident solar energy into biomass. Ethanol is produced by yeast fermentation of the sugar extracted from sugar cane.

Sugarcane production in the United States occurs in Florida, Louisiana, Hawaii, and Texas. In prime growing regions, such as Hawaii, sugarcane can produce 20 kg for each square meter exposed to the sun. The first three plants to produce sugar cane-based ethanol are expected to go online in Louisiana by mid 2009. Sugar mill plants in Lacassine, St. James and Bunkie were converted to sugar cane-based ethanol production using Colombian technology in order to make possible a profitable ethanol production. These three plants will produce 100 million gallons of ethanol within five years.

Comparison of key characteristics between the ethanol industries in the United States and Brazil			
Characteristic	*Brazil*	*U.S.*	*Units/comments*
Feedstock	Sugar cane	Maize	Main cash crop for ethanol production, the US has less than 2% from other crops.
Total ethanol production (2007)	5,019.2	6,498.6	Million U.S. liquid gallons
Total arable land	355	270	Million hectares.
Total area used for ethanol crop	3.6 (1%)	10 (3.7%)	Million hectares (% total arable)
Productivity per hectare	7,500	4,000	Liters of ethanol per hectare. Brazil is 727 to 870 gal/acre (2006), US is 424 gal/acre (2006)
Energy balance (input energy productivity)	8.3 to 10.2 times	1.3 to 1.6 times	Ratio of the energy obtained from ethanol to the energy expended in its production
Estimated greenhouse gas emission reduction	86-90%	10-30%	% GHGs avoided by using ethanol instead of gasoline, using existing crop land.
Ethanol fueling stations in the counrty	33,000 (100%)	873 (0,5%)	As % of total fueling gas stations in the country. U.S. has 170,000 (see Inslee, *op cit.,* pp. 161)
Fuel ethanol used by the road transport sector	20%	3.6%	As % of the sector's total on a volumetric basis for 2006.
Cost of production (USD/gallon)	0.83	1.14	2006/2007 for Brazil (22¢/liter), 2004 for U.S. (35¢/liter)
Government subsidy (in USD)	0	0.51/gallon	U.S. as of 2008-04-30. Brazilian ethanol production is no longer subsidized.
Import tariffs (in USD)	0	0.54/gallon	As of April 2008, Brazil does not import ethanol, the U.S. does

Notes: (1) Only contiguous U.S., excludes Alaska. (2) Assuming no land use change. (3) Excluding diesel-powered vehicles, ethanol consumption in the road sector is more than 40%

U.S. corn-derived ethanol costs 30 per cent more because the corn starch must first be converted to sugar before being distilled into alcohol. Unfortunately, despite this cost differential in production, in contrast to Japan and Sweden, the U.S. does not import much of Brazilian ethanol because of U.S. trade barriers corresponding to a tariff of 54-cent per gallon—a levy designed to offset the 51-cent per gallon blender's federal tax credit that is applied to ethanol no matter its country of origin. One advantage U.S. corn-derived ethanol offers is the ability to return 1/3 of the feedstock back into the market as a replacement for the corn used in the form of Distillers Dried Grain.

Country	Type	Energy balance
United States	Corn ethanol	1.3
Brazil	Sugarcane ethanol	8
Germany	Biodiesel	2.5
United States	Cellulosic ethanol	†2-36

† depending on production method.

All biomass goes through at least some of these steps: it needs to be grown, collected, dried, fermented, and burned. All of these steps require resources and an infrastructure. The total amount of energy input into the process compared to the energy released by burning the resulting ethanol fuel is known as the *energy balance*. Figures compiled in a 2007 by *National Geographic Magazine* point to modest results for corn ethanol produced in the US: one unit of fossil-fuel energy is required to create 1.3 energy units from the resulting ethanol. The energy balance for sugarcane ethanol produced in Brazil is more favourable, 1:8. Energy balance estimates are not easily produced, thus numerous such reports have been generated that are contradictory. For instance, a separate survey reports that production of ethanol from sugarcane, which requires a tropical climate to grow productively, returns from 8 to 9 units of energy for each unit expended, as compared to corn which only returns about 1.34 units of fuel energy for each unit of energy expended.

Carbon dioxide, a greenhouse gas, is emitted during fermentation and combustion. However, this is canceled out by the greater uptake of carbon dioxide by the plants as they grow to produce the biomass. When compared to gasoline, depending on the production method, ethanol releases less greenhouse gases.

Air Pollution

Compared with conventional unleaded gasoline, ethanol is a particulate-free burning fuel source that combusts with oxygen to form carbon dioxide, water and aldehydes (a contraction of alcohol dehydrogenated). Gasoline produces 2.44 CO2 equivalent kg/l and ethanol 1.94 (this is -21% CO2). The Clean Air Act requires the addition of oxygenates to reduce carbon monoxide emissions in the United States. The additive MTBE is currently being phased out due to ground water contamination, hence ethanol becomes an attractive alternative additive. Current production methods include air pollution from the manufacturer of macronutrient fertilizers such as ammonia.

A study by atmospheric scientists at Stanford University found that E85 fuel would increase the risk of air pollution deaths relative to gasoline. levels are significantly increased, thereby increasing photochemical smog and aggravating medical problems such as asthma.

Burning Ethanol Produces Unhealthy Aldehydes

Deadly Formaldehyde, Acetaldehyde and other Aldehydes are produced when alcohol is oxidized (such as burning ethanol in internal combustion engines). The European Union has banned products that contain Formaldehyde, due to its documented carcinogenic characteristics. The U.S. Environmental Protection Agency has labeled Formaldehyde as a probable cause of cancer in humans.

At higher concentrations, atmospheric aldehydes can be significant respiratory irritants causing nose bleeds, respiratory distress, lung disease, and persistent headaches.

When only a 10 per cent mixture of ethanol is added to gasoline (as is common in America and elsewhere), aldehyde emissions increase 40 per cent. The more ethanol, the more aldehydes are emitted (as in E85 and E100 biofuels).

Many aldehydes are highly toxic to living cells. Formaldehyde irreversibly cross-links protein amino acids, which produces the hard flesh of embalmed bodies.

Highly-reactive organic aldehyde compounds like Acetaldehyde are carcinogenic and mutagenic. They bind to DNA and protein, destroy folate and result in secondary cellular hyperregeneration (i.e., they mutate cells, causing cancer). Acetaldehyde (from oxidized ethanol) is particularly harmful to liver and brain tissues, causing cancer in both. The effect is cumulative over time, particularly to susceptible children and the elderly. The development of cancer from atmospheric aldehydes can take 10 to 15 years.

Brazil burns significant amounts of ethanol biofuel. Gas chromatograph studies were performed on ambient air in Sao Paulo Brazil, and compared to Osaka Japan, which does not burn ethanol fuel. Atmospheric Formaldehyde was 260 per cent higher in Brazil, and Acetaldehyde was 360 per cent higher. For a variety of complex reasons, the average life expectancy in Japan is 12 years longer than in Brazil.

Manufacture

In 2002, monitoring of ethanol plants revealed that they released VOCs (volatile organic compounds) at a higher rate than had previously been disclosed. The Environmental Protection Agency (EPA) subsequently reached settlement with Archer Daniels Midland and Cargill, two of the largest producers of ethanol, to reduce emission of these VOCs. VOCs are produced when fermented corn mash is dried for sale as a supplement for livestock feed. Devices known as thermal oxidizers or catalytic oxidizers can be attached to the plants to burn off the hazardous gases.

Carbon Dioxide

The calculation of exactly how much carbon dioxide is produced in the manufacture of bioethanol is a complex and inexact process, and is highly dependent on the method by which the ethanol is produced and the assumptions made in the calculation. A calculation should include:

- The *cost* of growing the feedstock.
- The *cost* of transporting the feedstock to the factory.
- The *cost* of processing the feedstock into bioethanol.

Such a calculation may or may not consider the following effects:

- The *cost* of the change in land use of the area where the fuel feedstock is grown.
- The cost of transportation of the bioethanol from the factory to its point of use.
- The efficiency of the bioethanol compared with standard gasoline.
- The amount of Carbon Dioxide produced at the tail pipe.
- The benefits due to the production of useful bi-products, such as cattle feed or electricity.

The graph on the right shows figures calculated by the UK government for the purposes of the Renewable transport fuel obligation.

The January 2006 Science article from UC Berkeley's ERG, estimated reduction from corn ethanol in GHG to be 13 per cent after reviewing a large number of studies. However, in a correction to that article released shortly after publication, they reduce the estimated value to 7.4 per cent. A National Geographic Magazine overview article (2007) puts the figures at 22 per cent less CO_2 emissions in production and use for corn ethanol compared to gasoline and a 56 per cent reduction for cane ethanol. Carmaker Ford reports a 70 per cent reduction in CO_2 emissions with bioethanol compared to petrol for one of their flexible-fuel vehicles.

An additional complication is that production requires tilling new soil which produces a one-off release of GHG that it can take decades or centuries of production reductions in GHG emissions to equalize. As an example, converting grass lands to corn production for ethanol takes about a century of annual savings to make up for the GHG released from the initial tilling.

Change in Land Use

Large-scale farming is necessary to produce agricultural alcohol and this requires substantial amounts of cultivated land. University of Minnesota researchers report that if all corn grown in the U.S. were used to make ethanol it would displace 12 per cent of current U.S. gasoline consumption. There are claims that land for ethanol production is acquired through deforestation, while others have observed that areas currently supporting forests are usually not suitable for growing crops. In any case, farming may involve a decline in soil fertility due to reduction of organic matter, a decrease in water availability and quality, an increase in the use of pesticides and fertilizers, and potential dislocation of local communities. However, new technology enables farmers and processors to increasingly produce the same output using less inputs.

There is a concern that as demand for ethanol fuel increases, food crops are replaced by fuel crops, driving food supply down and food prices up. Growing demand for ethanol in the United States has been discussed as a factor in the increased corn prices in Mexico. Average barley prices in the United States rose 17 per cent from January to June 2007 to the highest in 11 years. However, some commentators suggest that recent food price increases mainly reflect high oil prices in recent years, not specific pressures associated with ethanol production.

Cellulosic ethanol production is a new approach which may alleviate land use and related concerns. Cellulosic ethanol can be produced from any plant material, potentially doubling yields, in an effort to minimize conflict between food needs versus fuel needs. Instead of utilizing only the starch by-products from grinding wheat and other crops, cellulosic ethanol production maximizes the use of all plant materials, including gluten. This approach would have a smaller carbon footprint because the amount of energy-intensive fertilisers and fungicides remain the same for higher output of usable material. The technology for producing cellulosic ethanol is currently in the commercialization stage.

Many analysts suggest that, whichever ethanol fuel production strategy is used, fuel conservation efforts are also needed to make a large impact on reducing petroleum fuel use.

Efficiency of Common Crops

As ethanol yields improve or different feedstocks are introduced, ethanol production may become more economically feasible in the US. Currently, research on improving ethanol yields from each unit of corn is underway using biotechnology. Also, as long as oil prices remain high, the economical use of other feedstocks, such as cellulose, become viable. By-products such as straw or wood chips can be converted to ethanol. Fast growing species like switchgrass can be grown on land not suitable for other cash crops and yield high levels of ethanol per unit area.

sCrop	Annual yield (Liters/hectare)	Annual yield (US gal/acre)	Greenhouse-gas savings (% vs. petrol)	Comments
Miscanthus	7300	780	37-73	Low-input perennial grass. Ethanol production depends on development of cellulosic technology.
Switchgrass	3100-7600	330-810	37-73	Low-input perennial grass. Ethanol production depends on development of cellulosic technology. Breeding efforts underway to increase yields. Higher biomass production possible with mixed species of perennial grasses.
Poplar	3700-6000	400-640	51-100	Fast-growing tree. Ethanol production depends on development of cellulosic technology. Completion of genomic sequencing project will aid breeding efforts to increase yields.
Sugar cane	5300-6500	570-700	87-96	Long-season annual grass. Used as feedstock for most bioethanol produced in Brazil. Newer processing plants burn residues not used for ethanol to generate electricity. Only grows in tropical and subtropical climates.
Sweet sorghum	2500-7000	270-750	No data	Low-input annual grass. Ethanol production possible using existing technology. Grows in tropical and temperate climates, but highest ethanol yield estimates assume multiple crops per year (only possible in tropical climates). Does not store well.
Corn	3100-3900	330-420	10-20	High-input annual grass. Used as feedstock for most bioethanol produced in USA. Only kernels can be processed using available technology; development of commercial cellulosic technology would allow stover to be used and increase ethanol yield by 1,100-2,000 litres/ha.

Source: (except sorghum): *Nature* 444 (December 7, 2006): 670-654.

Reduced Petroleum Imports and Costs

One rationale given for extensive ethanol production in the U.S. is its benefit to energy security, by shifting the need for some foreign-produced oil to domestically-produced energy sources. Production of ethanol requires significant energy, but current U.S. production derives most of that energy from coal, natural gas and other sources, rather than oil. Because 66 per cent of oil consumed in the U.S. is imported, compared to a net surplus of coal and just 16 per cent of natural gas (2006 figures), the displacement of oil-based fuels to ethanol produces a net shift from foreign to domestic U.S. energy sources.

According to a 2008 analysis by Iowa State University, the growth in US ethanol production has caused retail gasoline prices to be US $0.29 to US $0.40 per gallon lower than would otherwise have been the case.

Recent Patents

In 2006-2-23, Veridium Corporation announced the technology to convert exhaust carbon dioxide from the fermentation stage of ethanol production facilities back into new ethanol and biodiesel. The bioreactor process is based on a new strain of iron-loving blue-green algae discovered thriving in a hot stream at Yellowstone National Park.

In 2006-11-14, US Patent Office approved Patent 7135308, a process for the production of ethanol by harvesting starch-accumulating filament-forming or colony-forming algae to form a biomass, initiating cellular decay of the biomass in a dark and anaerobic environment, fermenting the biomass in the presence of a yeast, and the isolating the ethanol produced.

Criticism and Controversy

In 2007, biofuels consumed one third of America's corn (maize) harvest. Filling up one large vehicle fuel tank one time with 100 per cent ethanol uses enough corn to feed one person for a year. Thirty million tons of U.S. corn going to ethanol in 2007 greatly reduces the world's overall supply of grain. However, 31 per cent of the corn put into the process comes out as distiller's grain, or DDGS, which is very high in protein, and is used to feed livestock.

Jean Ziegler, the United Nations Special Rapporteur on the Right to Food, called for a five-year moratorium on biofuel production to halt the increasing catastrophe for the poor. He proclaimed that the rising practice of converting food crops into biofuel is "A Crime Against Humanity," saying it is creating food shortages and price jumps that cause millions of poor people to go hungry.

The European Organisation for Economic Co-operation and Development warns that "the current push to expand the use of biofuels is creating unsustainable tensions that will disrupt markets without generating significant environmental benefits."

When all 200 American ethanol subsidies are considered, they cost about $7 billion USD per year (equal to roughly $1.90 USD total for each a gallon of ethanol). When the price of one agricultural commodity increases, farmers are motivated to quickly shift finite land and water resources to it, away from traditional food crops.

The 2007-12-19 U.S. Energy Independence and Security Act of 2007 requires American "fuel producers to use at least 36 billion gallons of biofuel in 2022. This is nearly a fivefold increase over current levels."

When cellulosic ethanol is produced from feedstock like switchgrass and sawgrass, the nutrients required to grow the cellulose are removed and cannot decay and replenish the soil. The soil is of poorer quality, and unsustainable soil erosion occurs.

Ethanol production consumes large quantities of unsustainable petroleum and natural gas. Even with the most-optimistic energy return on investment claims, in order to use 100 per cent solar energy to grow corn and produce ethanol (fueling farm-and-transportation machinery with ethanol, distilling with heat from burning crop residues, using NO fossil fuels), the consumption of ethanol to replace current U.S. petroleum use alone would require about 75 per cent of all cultivated land on the face of the Earth, with no ethanol for other countries, or sufficient food for humans and animals.

Fuel System Problems

Several of the outstanding ethanol fuel issues are linked specifically to fuel systems. Fuels with more than 10 per cent ethanol are not compatible with non E85-ready fuel system components and may cause corrosion of ferrous components. Ethanol fuel can negatively affect electric fuel pumps by increasing internal wear, cause undesirable spark generation, and is not compatible with capacitance fuel level gauging indicators and may cause erroneous fuel quantity indications in vehicles that employ that system. It is also not always compatible with marine craft, especially those that use fiberglass fuel tanks.

Using 100 per cent ethanol fuel decreases fuel-economy by 15-30 per cent over using 100 per cent gasoline; this can be avoided using certain modifications that would, however, render the engine inoperable on regular petrol without the addition of an adjustable ECU. Tough materials are needed to accommodate a higher compression ratio to make an ethanol engine as efficient as it would be on petrol; these would be similar to those used in diesel engines which typically run at a CR of 20:1, versus about 8-12:1 for petrol engines.

In April 2008 the German environmental minister cancelled a proposed 10 per cent ethanol fuel scheme citing technical problems: too many older cars in Germany are unequipped to handle this fuel. Ethanol levels in fuel will remain at 5 per cent.

Sewage Treatment

27.1 Sewage Treatment

Sewage treatment, or domestic wastewater treatment, is the process of removing contaminants from wastewater, both runoff (effluents) and domestic. It includes physical, chemical and biological processes to remove physical, chemical and biological contaminants. Its objective is to produce a waste stream (or treated effluent) and a solid waste or sludge suitable for discharge or reuse back into the environment. This material is often inadvertently contaminated with many toxic organic and inorganic compounds.

Sewage is created by residences, institutions, hospitals and commercial and industrial establishments. It can be treated close to where it is created (in septic tanks, biofilters or aerobic treatment systems), or collected and transported via a network of pipes and pump stations to a municipal treatment plant (see sewerage and pipes and infrastructure). Sewage collection and treatment is typically subject to local, state and federal regulations and standards. Industrial sources of wastewater often require specialized treatment processes (see Industrial wastewater treatment).

The sewage treatment involves three stages, called *primary*, *secondary* and *tertiary treatment*. First, the solids are separated from the wastewater stream. Then dissolved biological matter is progressively converted into a solid mass by using indigenous, water-borne microorganisms. Finally, the biological solids are neutralized then disposed of or re-used, and the treated water may be disinfected chemically or physically (for example by lagoons and micro-filtration). The final effluent can be discharged into a stream, river, bay, lagoon or wetland, or it can be used for the irrigation of a golf course, green way or park. If it is sufficiently clean, it can also be used for groundwater recharge.

Description

Raw influent (sewage) includes household waste liquid from toilets, baths, showers, kitchens, sinks, and so forth that is disposed of via sewers. In many areas, sewage also includes liquid waste from industry and commerce. The draining of household waste into greywater and blackwater is becoming more common in the developed world, with greywater being permitted to be used for watering plants or recycled for flushing toilets. A lot of sewage also includes some surface water from roofs or hard-standing areas. Municipal wastewater therefore includes residential, commercial, and industrial liquid waste discharges, and may include stormwater runoff. Sewage systems capable of handling stormwater are known as combined systems or combined sewers. Such systems are usually avoided since they complicate and thereby reduce the efficiency of sewage treatment plants owing to their seasonality. The variability in flow also leads to often larger than necessary, and subsequently more expensive, treatment facilities. In addition, heavy storms that contribute more flows than the treatment plant can handle may overwhelm the sewage treatment system, causing a spill or overflow (called a combined sewer overflow, or CSO, in the United States). It is preferable to have a separate storm drain system for stormwater in areas that are developed with sewer systems.

As rainfall runs over the surface of roofs and the ground, it may pick up various contaminants including soil particles and other sediment, heavy metals, organic compounds, animal waste, and oil and grease. Some jurisdictions require stormwater to receive some level of treatment before being discharged directly into waterways. Examples of treatment processes used for stormwater include sedimentation basins, wetlands, buried concrete vaults with various kinds of filters, and vortex separators (to remove coarse solids).

The site where the raw wastewater is processed before it is discharged back to the environment is called a wastewater treatment plant (WWTP). The order and types of mechanical, chemical and biological systems that comprise the wastewater treatment plant are typically the same for most developed countries:

❖ *Mechanical treatment*

- Influx (Influent)
- Removal of large objects
- Removal of sand and grit

- ● Pre-precipitation

- ❖ *Biological treatment*

  - ● Oxidation bed (oxidizing bed) or aeration system
  - ● Post precipitation

- ❖ *Chemical treatment* (this step is usually combined with settling and other processes to remove solids, such as filtration. The combination is referred to in the U.S. as physical-chemical treatment.

Treatment Stages

Primary Treatment

Primary treatment removes the materials that can be easily collected from the raw wastewater and disposed of. The typical materials that are removed during primary treatment include fats, oils, and greases (also referred to as FOG), sand, gravels and rocks (also referred to as grit), larger settleable solids and floating materials (such as rags and flushed feminine hygiene products). This step is done entirely with machinery.

Removal of Large Objects from Influent Sewage

In primary treatment, the influent sewage water is strained to remove all large objects that are deposited in the sewer system, such as rags, sticks, tampons, cans, fruit, etc. This is most commonly done with a manual or automated mechanically raked screen. The raking action of a mechanical bar screen is typically paced according to the accumulation on the bar screens and/or flow rate. The bar screen is used because large solids can damage or clog the equipment used later in the sewage treatment plant. The solids are collected in a dumpster and later disposed in a landfill.

Sand and Grit Removal

Primary treatment also typically includes a sand or grit channel or chamber where the velocity of the incoming wastewater is carefully controlled to allow sand grit rand stones to settle, while keeping the majority of the suspended organic material in the water column. This equipment is called a detritor or sand catcher. Sand, grit, and stones need to be removed early in the process to avoid damage to pumps and other equipment in the remaining treatment stages. Sometimes there is a sand washer (grit classifier) followed by a conveyor that transports the sand to a container for disposal. The contents from the sand catcher may be fed into the incinerator in a sludge processing plant, but in many cases, the sand and grit is sent to a landfill.

Sedimentation

Many plants have a sedimentation stage where the sewage is allowed to pass slowly through large tanks, commonly called "primary clarifiers" or "primary sedimentation tanks". The tanks are large enough that sludge can settle and floating material such as grease and oils can rise to the surface and be skimmed off. The main purpose of the primary clarification stage is to produce both a generally homogeneous liquid capable of being treated biologically and a sludge that can be separately treated or processed. Primary settling tanks are usually equipped with mechanically driven scrapers that continually drive the collected sludge towards a hopper in the base of the tank from where it can be pumped to further sludge treatment stages.

Secondary Treatment

Secondary treatment is designed to substantially degrade the biological content of the sewage such as are derived from human waste, food waste, soaps and detergent. The majority of municipal and industrial plants treat the settled sewage liquor using aerobic biological processes. For this to be effective, the biota require both oxygen and a substrate on which to live. There are number of ways in which this is done. In all these methods, the bacteria and protozoa consume biodegradable soluble organic contaminants (e.g. sugars, fats, organic short-chain carbon molecules, etc.) and bind much of the less soluble fractions into floc. Secondary treatment systems are classified as *fixed film* or suspended growth. Fixed-film treatment process including trickling filter and rotating biological contactors where the biomass grows on media and the sewage passes over its surface. In *suspended growth systems*—such as activated sludge—the biomass is well mixed with the sewage and can be operated in a smaller space than fixed-film systems that treat the same amount of water. However, fixed-film systems are more able to cope with drastic changes in the amount of biological material and can provide higher removal rates for organic material and suspended solids than suspended growth systems.

Roughing filters are intended to treat particularly strong or variable organic loads, typically industrial, to allow them to then be treated by conventional secondary treatment processes. Characteristics include typically tall, circular filters filled with open synthetic filter media to which wastewater is applied at a relatively high rate. They are designed to allow high hydraulic loading and a high flow-through of air. On larger installations, air is forced through the media using blowers. The resultant wastewater is usually within the normal range for conventional treatment processes.

Activated Sludge

In general, activated sludge plants encompass a variety of mechanisms and processes that use dissolved oxygen to promote the growth of biological floc that substantially removes organic material.

The process traps particulate material and can, under ideal conditions, convert ammonia to nitrite and nitrate and ultimately to nitrogen gas, (see also denitrification).

Surface-Aerated Basins

Most biological oxidation processes for treating industrial wastewaters have in common the use of oxygen (or air) and microbial action. Surface-aerated basins achieve 80 to 90 per cent removal of BOD with retention times of 1 to 10 days. The basins may range in depth from 1.5 to 5.0 metres and use motor-driven aerators floating on the surface of the wastewater.

In an aerated basin system, the aerators provide two functions: they transfer air into the basins required by the biological oxidation reactions, and they provide the mixing required for dispersing the air and for contacting the reactants (that is, oxygen, wastewater and microbes). Typically, the floating surface aerators are rated to deliver the amount of air equivalent to 1.8 to 2.7 kg O_2/kWh. However, they do not provide as good mixing as is normally achieved in activated sludge systems and therefore aerated basins do not achieve the same performance level as activated sludge units.

Biological oxidation processes are sensitive to temperature and, between 0 °C and 40 °C, the rate of biological reactions increase with temperature. Most surface aerated vessels operate at between 4 °C and 32 °C.

Fluidized Bed Reactors

The carbon absorption following biological treatment is particularly effective in reducing both the BOD and COD to low levels. A fluidized bed reactor is a combination of the most common stirred tank packed bed, continuous flow reactors. It is very important to chemical engineering because of its excellent heat and mass transfer characteristics. In a fluidized bed reactor, the substrate is passed upward through the immobilized enzyme bed at a high velocity to lift the particles. However the velocity must not be so high that the enzymes are swept away from the reactor entirely. This causes low mixing; these type of reactors are highly suitable for the exothermic reactions. It is most often applied in immobilized enzyme catalysis.

Filter Beds (Oxidising Beds)

In older plants and plants receiving more variable loads, trickling filter beds are used where the settled sewage liquor is spread onto the surface of a deep bed made up of coke (carbonised coal), limestone chips or specially fabricated plastic media. Such media must have high surface areas to support the biofilms that form. The liquor is distributed through perforated rotating arms radiating from a central pivot. The distributed liquor trickles through this bed and is collected in drains at the base. These drains also provide a source of air which percolates up through the bed, keeping it aerobic. Biological films of bacteria, protozoa and fungi form on the media's surfaces and eat or otherwise reduce the organic content. This biofilm is grazed by insect larvae and worms which help maintain an optimal thickness. Overloading of beds increases the thickness of the film leading to clogging of the filter media and ponding on the surface.

Biological Aerated Filters

Biological Aerated (or Anoxic) Filter (BAF) or Biofilters combine filtration with biological carbon reduction, nitrification or denitrification. BAF usually includes a reactor filled with a filter media. The media is either in suspension or supported by a gravel layer at the foot of the filter. The dual purpose of this media is to support highly active biomass that is attached to it and to filter suspended solids. Carbon reduction and ammonia conversion occurs in aerobic mode and sometime achieved in a single reactor while nitrate conversion occurs in anoxic mode. BAF is operated either in upflow or downflow configuration depending on design specified by manufacturer.

Membrane Bioreactors

Membrane bioreactors (MBR) combines activated sludge treatment with a membrane liquid-solid separation process. The membrane component uses low pressure microfiltration or ultra filtration membranes and eliminates the need for clarification and tertiary filtration. The membranes are typically immersed in the aeration tank (however, some applications utilize a separate membrane tank). One of the key benefits of a membrane bioreactor system is that it effectively overcomes the limitations associated with poor settling of sludge in conventional activated sludge (CAS) processes. The technology permits bioreactor operation with considerably higher mixed liquor suspended solids (MLSS) concentration than CAS systems, which are limited by sludge settling. The process is typically operated at MLSS in the range of 8,000-12,000 mg/L, while CAS are operated in the range of 2,000-3,000 mg/L. The elevated biomass concentration in the membrane bioreactor process allows for very effective removal of both soluble and particulate biodegradable materials at higher loading rates. Thus increased Sludge Retention Times (SRTs)—usually exceeding 15 days—ensure complete nitrification even in extremely cold weather.

The cost of building and operating a MBR is usually higher than conventional wastewater treatment, however, as the technology has become increasingly popular and has gained wider acceptance throughout the industry, the life-cycle costs have been steadily decreasing. As well, in developed urban areas where the footprint of the treatment plant is considered a limiting factor MBR facilities can be considered a desirable option.

Secondary Sedimentation

The final step in the secondary treatment stage is to settle out the biological floc or filter material and produce sewage water containing very low levels of organic material and suspended matter.

Rotating Biological Contactors

Rotating biological contactors (RBCs) are mechanical secondary treatment systems, which are robust and capable of withstanding surges in organic load. RBCs were first installed in Germany in 1960 and have since been developed and refined into a reliable operating unit. The rotating disks support the growth of bacteria and micro-organisms present in the sewage, which breakdown and stabilise organic pollutants. To be successful, micro-organisms need both oxygen to live and food to grow. Oxygen is obtained from the atmosphere as the disks rotate. As the micro-organisms grow, they build up on the media until they are sloughed off due to shear forces provided by the rotating discs in the sewage. Effluent from the RBC is then passed through final clarifiers where the micro-organisms in suspension settle as a sludge. The sludge is withdrawn from the clarifier for further treatment.

Tertiary Treatment

Tertiary treatment provides a final stage to raise the effluent quality before it is discharged to the receiving environment (sea, river, lake, ground, etc.). More than one tertiary treatment process may be used at any treatment plant. If disinfection is practiced, it is always the final process. It is also called "effluent polishing".

Filtration

Sand filtration removes much of the residual suspended matter. Filtration over activated carbon removes residual toxins.

Lagooning

Lagooning provides settlement and further biological improvement through storage in large man-made ponds or lagoons. These lagoons are highly aerobic and colonization by native macrophytes, especially reeds, is often encouraged. Small filter feeding invertebrates such as Daphnia and species of Rotifera greatly assist in treatment by removing fine particulates.

Constructed Wetlands

Constructed wetlands include engineered reedbeds and a range of similar methodologies, all of which provide a high degree of aerobic biological improvement and can often be used instead of secondary treatment for small communities, also see phytoremediation. One example is a small reedbed used to clean the drainage from the elephants' enclosure at Chester Zoo in England.

Nutrient Removal

Wastewater may contain high levels of the nutrients nitrogen and phosphorus. Excessive release to the environment can lead to a build up of nutrients, called eutrophication, which can in turn encourage the overgrowth of weeds, algae, and cyanobacteria (blue-green algae). This may cause an algal bloom, a rapid growth in the population of algae. The algae numbers are unsustainable and eventually most of them die. The decomposition of the algae by bacteria uses up so much of oxygen in the water that most or all of the animals die, which creates more organic matter for the bacteria to decompose. In addition to causing deoxygenation, some algal species produce toxins that contaminate drinking water supplies. Different treatment processes are required to remove nitrogen and phosphorus.

Nitrogen removal: The removal of nitrogen is effected through the biological oxidation of nitrogen from ammonia (nitrification) to nitrate, followed by denitrification, the reduction of nitrate to nitrogen gas. Nitrogen gas is released to the atmosphere and thus removed from the water.

Nitrification itself is a two-step aerobic process, each step facilitated by a different type of bacteria. The oxidation of ammonia (NH_3) to nitrite (NO_2) is most often facilitated by *Nitrosomonas* spp. (nitroso referring to the formation of a nitroso functional group). Nitrite oxidation to nitrate (NO_3), though traditionally believed to be facilitated by *Nitrobacter* spp. (nitro referring the formation of a nitro functional group), is now known to be facilitated in the environment almost exclusively by *Nitrospira* spp.

Denitrification requires anoxic conditions to encourage the appropriate biological communities to form. It is facilitated by a wide diversity of bacteria. Sand filters, lagooning and reed beds can all be used to reduce nitrogen, but the activated sludge process (if designed well) can do the job the most easily. Since denitrification is the reduction of nitrate to dinitrogen gas, an electron donor is needed. This can be, depending on the wastewater, organic matter (from faeces), sulfide, or an added donor like methanol.

Sometimes the conversion of toxic ammonia to nitrate alone is referred to as tertiary treatment.

Phosphorus removal: Phosphorus can be removed biologically in a process called enhanced biological phosphorus removal. In this process, specific bacteria, called polyphosphate accumulating organisms, are selectively enriched and accumulate large quantities of phosphorus within their cells (up to 20% of their mass). When the biomass enriched in these bacteria is separated from the treated water, these biosolids have a high fertilizer value.

Phosphorus removal can also be achieved by chemical precipitation, usually with salts of iron (e.g. ferric chloride) or aluminum (e.g. alum). The resulting chemical sludge is difficult to handle and the added chemicals can be expensive. Despite this, chemical phosphorus removal requires significantly smaller equipment footprint than biological removal, is easier to operate and can be more reliable in areas that have wastewater compositions that make biological phosphorus removal difficult.

Disinfection

The purpose of disinfection in the treatment of wastewater is to substantially reduce the number of microorganisms in the water to be discharged back into the environment. The effectiveness of disinfection depends on the quality of the water being treated (e.g., cloudiness, pH, etc.), the type of disinfection being used, the disinfectant dosage (concentration and time), and other environmental variables. Cloudy water will be treated less successfully since solid matter can shield organisms, especially from ultraviolet light or if contact times are low. Generally, short contact times, low doses and high flows all militate against effective disinfection. Common methods of disinfection include ozone, chlorine, or ultraviolet light. Chloramine, which is used for drinking water, is not used in wastewater treatment because of its persistence.

Chlorination remains the most common form of wastewater disinfection in North America due to its low cost and long-term history of effectiveness. One disadvantage is that chlorination of residual organic material can generate chlorinated-organic compounds that may be carcinogenic or harmful to the environment. Residual chlorine or chloramines may also be capable of chlorinating organic material in the natural aquatic environment. Further, because residual chlorine is toxic to aquatic species, the treated effluent must also be chemically dechlorinated, adding to the complexity and cost of treatment.

Ultraviolet (UV) light can be used instead of chlorine, iodine, or other chemicals. Because no chemicals are used, the treated water has no adverse effect on organisms that later consume it, as may be the case with other methods. UV radiation causes damage to the genetic structure of bacteria, viruses, and other pathogens, making them incapable of reproduction. The key disadvantages of UV disinfection are the need for frequent lamp maintenance and replacement and the need for a highly treated effluent to ensure that the target microorganisms are not shielded from the UV radiation (i.e., any solids present in the treated effluent may protect microorganisms from the UV light). In the United Kingdom, light is becoming the most common means of disinfection because of the concerns about the impacts of chlorine in chlorinating residual organics in the wastewater and in chlorinating organics in the receiving water. Edmonton, Alberta, Canada also uses UV light for its water treatment.

Ozone O_3 is generated by passing oxygen O_2 through a high voltage potential resulting in a third oxygen atom becoming attached and forming O_3. Ozone is very unstable and reactive and oxidizes most organic material it comes in contact with, thereby destroying many pathogenic microorganisms. Ozone is considered to be safer than chlorine because, unlike chlorine which has to be stored on site (highly poisonous in the event of an accidental release), ozone is generated onsite as needed. Ozonation also produces fewer disinfection by-products than chlorination. A disadvantage of ozone disinfection is the high cost of the ozone generation equipment and the requirements for special operators.

Package Plants and Batch Reactors

In order to use less space, treat difficult waste, deal with intermittent flow or achieve higher environmental standards, a number of designs of hybrid treatment plants have been produced. Such plants often combine all or at least two stages of the three main treatment stages into one combined stage. In the UK, where a large number of sewage treatment plants serve small populations, package plants are a viable alternative to building discrete structures for each process stage.

One type of system that combines secondary treatment and settlement is the sequencing batch reactor (SBR). Typically, activated sludge is mixed with raw incoming sewage and mixed and aerated. The resultant mixture is then allowed to settle producing a high quality effluent. The settled sludge is run off and re-aerated before a proportion is returned to the head of the works. SBR plants are now being deployed in many parts of the world including North Liberty, Iowa, and Llanasa, North Wales.

The disadvantage of such processes is that precise control of timing, mixing and aeration is required. This precision is usually achieved by computer controls linked to many sensors in the plant. Such a complex, fragile system is unsuited to places where such controls may be unreliable, or poorly maintained, or where the power supply may be intermittent.

Package plants may be referred to as *high charged* or *low charged*. This refers to the way the biological load is processed. In high charged systems, the biological stage is presented with a high organic load and the combined floc and organic material is then oxygenated for a few hours before being charged again with a new load. In the low charged system the biological stage contains a low organic load and is combined with floculate for a relatively long time.

Sludge Treatment and Disposal

The sludges accumulated in a wastewater treatment process must be treated and disposed of in a safe and effective manner. The purpose of digestion is to reduce the amount of organic matter and the number of disease-causing microorganisms present in the solids. The most common treatment options include anaerobic digestion, aerobic digestion, and composting.

The choice of a wastewater solid treatment method depends on the amount of solids generated and other site-specific conditions. However, in general, composting is most often applied to smaller-scale applications followed by aerobic digestion and then lastly anaerobic digestion for the larger-scale municipal applications.

Anaerobic Digestion

Anaerobic digestion is a bacterial process that is carried out in the absence of oxygen. The process can either be *thermophilic* digestion, in which sludge is fermented in tanks at a temperature of 55°C, or *mesophilic*, at a temperature of around 36°C. Though allowing shorter retention time (and thus smaller tanks), thermophilic digestion is more expensive in terms of energy consumption for heating the sludge.

One major feature of anaerobic digestion is the production of biogas, which can be used in generators for electricity production and/or in boilers for heating purposes.

Aerobic Digestion

Aerobic digestion is a bacterial process occurring in the presence of oxygen. Under aerobic conditions, bacteria rapidly consume organic matter and convert it into carbon dioxide. The operating costs are characteristically much greater for aerobic digestion because of the energy costs needed to add oxygen to the process.

Composting

Composting is also an aerobic process that involves mixing the sludge with sources of carbon such as sawdust, straw or wood chips. In the presence of oxygen, bacteria digest both the wastewater solids and the added carbon source and, in doing so, produce a large amount of heat.

Thermal Depolymerization

Thermal depolymerization uses hydrous pyrolysis to convert reduced complex organics to oil.

Sludge Disposal

When a liquid sludge is produced, further treatment may be required to make it suitable for final disposal. Typically, sludges are thickened (dewatered) to reduce the volumes transported off-site for disposal. There is no process which completely eliminates the need to dispose of biosolids. There is, however, an additional step some cities are taking to superheat the wastewater sludge and convert it into small pelletized granules that are high in nitrogen and other organic materials. This product is then sold to local farmers and turf farms as a soil amendment or fertilizer, reducing the amount of space required to dispose of sludge in landfills.

Treatment in the Receiving Environment

Many processes in a wastewater treatment plant are designed to mimic the natural treatment processes that occur in the environment, whether that environment is a natural water body or the ground. If not overloaded, bacteria in the environment will consume organic contaminants, although this will reduce the levels of oxygen in the water and may significantly change the overall ecology of the receiving water. Native bacterial populations feed on the organic contaminants, and the numbers of disease-causing microorganisms are reduced by natural environmental conditions such as predation exposure to ultraviolet radiation, for example. Consequently, in cases where the receiving environment provides a high level of dilution, a high degree of wastewater treatment may not be required. However, recent evidence has demonstrated that very low levels of certain contaminants in wastewater, including hormones (from animal husbandry and residue from human hormonal contraception methods) and synthetic materials such as phthalates that mimic hormones in their action, can have an unpredictable adverse impact on the natural biota and potentially on humans if the water is re-used for drinking water. In the US and EU, uncontrolled discharges of wastewater to the environment are not permitted under law, and strict water quality requirements are to be met. A significant threat in the coming decades will be the increasing uncontrolled discharges of wastewater within rapidly developing countries.

Sewage Treatment in Developing Countries

There are few reliable figures on the share of the wastewater collected in sewers that is being treated in the world. In many developing countries the bulk of domestic and industrial wastewater is discharged without any treatment or after primary treatment only. In Latin America about 15 per cent of collected wastewater passes through treatment plants (with varying levels of actual treatment). In Venezuela, a below average country in South America with respect to wastewater treatment, 97 per cent of the country's sewage is discharged raw into the environment. In a relatively developed Middle Eastern country such as Iran, Tehran's majority of population has totally untreated sewage injected to the city's groundwater. Most of sub-Saharan Africa is without wastewater treatment.

Water utilities in developing countries are chronically underfunded because of low water tariffs, the inexistence of sanitation tariffs in many cases, low billing efficiency (i.e. many users that are billed do not pay) and poor operational efficiency (i.e. there are overly high levels of staff, there are high physical losses, and many users have illegal connections and are thus not being billed). In addition, wastewater treatment typically is the process within the utility that receives the least attention, partly because enforcement of environmental standards is poor. As a result of all these factors, operation and maintenance of many wastewater treatment plants is poor. This is evidenced by the frequent breakdown of equipment, shutdown of electrically operated equipment due to power outages or to reduce costs, and sedimentation due to lack of sludge removal. Developing countries as diverse as Egypt, Algeria, China or Colombia have invested substantial sums in wastewater treatment without achieving a significant impact in terms of environmental improvement. Even if wastewater treatment plants are properly operating, it can be argued that the environmental impact is limited in cases where the assimilative capacity of the receiving waters (ocean with strong currents or large rivers) is high, as it is often the case.

Benefits of Wastewater Treatment Compared to Benefits of Sewage Collection in Developing Countries

Waterborne diseases that are prevalent in developing countries, such as typhus and cholera, are caused primarily by poor hygiene practices and the absence of improved household sanitation facilities. The public health impact of the discharge of untreated wastewater is comparatively much lower. Hygiene promotion, on-site sanitation and low-cost sanitation thus are likely to have a much greater impact on public health than wastewater treatment.

27.2 Hyperion Sewage Treatment Plant

The Hyperion Wastewater Treatment plant is located in southwest Los Angeles, California next to Dockweiler State Beach on Santa Monica Bay. The largest such facility in the Los Angeles Metropolitan Area, Hyperion is operated by the City of Los Angeles, Department of Public Works, Bureau of Sanitation.

History

Until 1925, raw sewage from the city of Los Angeles was discharged untreated directly into Santa Monica Bay in the region of today's Hyperion Treatment Plant.

With the population increase, the amount of sewage became a major problem to the beaches, so in 1925 the city of Los Angeles built a simple screening plant in the 200 acres the city had acquired in 1892.

Even with the screening plant, the quality of the water in the Santa Monica Bay was unacceptable, and in 1950 the city of Los Angeles opened the Hyperion Treatment Plant with full secondary treatment processes. In addition, the new plant included capture of biogas from anaerobic digesters to produce heat dried fertilizer.

In order to keep up with the increase of influent wastewater produced by the ever growing city of Los Angeles, by 1957 the plant engineers had cut back treatment levels and increased the discharge of a blend of primary and secondary effluent through a five-mile pipe into the ocean. They also opted to halt the production of fertilizers and started discharging digested sludge into the Santa Monica Bay through a seven-mile pipe.

Environmental Forces

The discharge of sewage effluent into Santa Monica Bay changed the ecosystem so dramatically that by the 1970s, only worms and hardy clams existed in the ocean floor of the bay. The increasing volume of treated sewage being dumped into Santa Monica Bay captured the attention of a group of concerned users of the bay, who started to observe a decrease in the number and quality of fish in the bay, sick dolphins and people. From this movement, Heal the Bay was founded by Dorothy Green in 1985.

Heal the Bay joined the Environmental Protection Agency in a lawsuit against the city of Los Angeles to force the sewage treatment done by Hyperion to comply with the Clean Water Act. Heal the Bay and EPA succeeded and by November 1987, Hyperion stopped discharging sludge into the Santa Monica Bay and by December 1998, Hyperion had a full secondary treatment plant. For its significance and impact to quality of life, in 2001 the Hyperion Treatment Plan was named one of the Top Ten Public Works project of the 20th century by the American Public Works Association.

Treatment System

The plant treats approximately 350-450 million gallons per day of raw sewage. The sewage undergoes both primary and secondary treatment.

The wastewater produced by houses, businesses, and industries, flows in the sewer system, separately from the storm drain system to the Hyperion treatment plant.

Once the wastewater reached the plant, the first treatment stage is to remove large objects such as plastic, rags, metals, and wood. The wastewater then flows into sedimentation tanks were about 15 tons of sand and other materials settle to the bottom of the tank and are removed every day.

The chemical treatment starts with the addition of coagulants to the wastewater to improve the settling of small particles. At Hyperion, the wastewater flows during one hour in underground tanks after the addition of coagulants. The settled solids are pumped to digesters and oil and grease are skimmed off the top of the wastewater.

After the removal of solids, the wastewater is pumped to tanks where aerators deliver 96 per cent oxygen to the wastewater, providing microorganisms with the necessary conditions to decompose organic solids in the wastewater. The wastewater spends about one to two hours in the oxygen reactors.

Now that the microorganisms fed on most of the organics, they are able to settle to the bottom in quiescent conditions. The wastewater is pumped into settling, clarifying tanks where it stays for about four hours. After this time, about 90-95 per cent of the solids have been removed from the wastewater and the effluent is clean enough to be discharged or recycled. About 94 per cent of the water is discharged into the Santa Monica Bay through a pipe 5 miles out from the shore and 190 feet deep. Approximately 30 million gallons of secondary effluent per day is pumped to the West Basin Municipal Water District where the water is filtered further and reused for various purposes.

All the solids settled throughout the several stages in the treatment are treated in anaerobic digesters. The idea is to destroy pathogens before the sludge can be dumped in landfills. At Hyperion, the biosolids stay about 15 days in the digesters—enough time for the anaerobic bacteria to consume pathogens. One by-product of this process is the production of methane gas, which is captured and processed by a nearby power plant. After the "sterilization", the biosolids have their water removed to reduce volume and thus reduce transportation costs. The dewatering is achieved by using centrifuges with variable settings and addition of polymers (Horenstein *et al.*, 1990). Each 1 per cent decrease in volume results in an economy of $1 million in disposal costs (Horenstein *et al.*, 1990). The dewatered biosolids are then transported to landfills or to locations where it can be reused.

Future Challenges

Recently, the public has become aware of the presence of pharmaceutical products in the nation's waterways. Several studies, such as (Jones-Lepp and Stevens, 2007) have shown the presence of pharmaceuticals and personal care products in biosolids/sewage sludge from wastewater treatment plant. How to reduce the presence of pharmaceuticals and personal care products may become a major challenge in the future not only for Hyperion, but for all wastewater treatment plants. However, currently there are no requirements for Hyperion to remove such products.

27.3 Sewage Treatment

Introduction

There are few aspects that are more responsible for moving civilization out of the Middle Ages and into the Modern Age than is the treatment of our municipal and industrial sewage. This and the treatment of our drinking water (potable water) comprise THE two major advances in public health that have contributed to greatly extended life expectancies due to the control of infectious diseases. (What *non*-infectious disease kills the most people worldwide?) Anyway, if you do not recognize the significance of this technology to your life, go hit your head on a brick wall until you do appreciate its importance. IT IS VERY IMPORTANT, and, as a citizen, you should know about it, unpleasant as it may be to your nose.

In essence, sewage treatment has two functions:

❖ to kill pathogens (whatever they are), and
❖ to eliminate harmful chemicals from the water.

- chemicals that are toxic per se, and
- the bulk of organic chemicals that cause the depletion of life-supporting oxygen in the receiving waters of a stream, river or lake.

Let's take a look at the process and equipment your locality possesses. While you follow the steps, take note of what "primary treatment" is, and what "secondary treatment" is:

1. The first part of sewage treatment is concerned with the removal of large items that enter the treatment plant. These items can jam the system, can consist of twigs, Volkswagens, gravel, animals, tampons and condoms.

The raw influent first goes through a self-cleaning screen (B = Bar screen), and then into one end of a shallow and rather fast moving basin (S = Sand basin) so that sand and gravel can settle out. Often skimmers rotate around the surface of the basin to remove oils that may have been flushed into the system.

2. In order to provide for the further processes—such as the addition of flocculants, the volume of the incoming raw sewage water must be measured (M = Meter). One of the most common types of meters that has no moving parts to become clogged or jammed is the kind shown here, and is known as a Parshal Flume. As the volume of water increases, the difference in water level on each side of the constriction increases. This difference is then translated into flow-through volume. With today's electronics, signals are transferred downstream for the needed adjustments for chlorine, flocculants, etc.

3. Finally the waste water enters the first step of sanitization and is directed into the Primary Clarifier (PC). The theory behind this device is that it creates an ecosystem that is very alien to cellular pathogens (e.g.: bacteria) of humans. It does this by supporting anaerobic growth of microbes that also digest a small amount of the dissolved organic material (BOD, see later) while they produce lots of sulfides and other chemicals that are noxious to both noses and to the pathogens of humans. The construction is a huge circular vat with a metal curtain that does not quite reach the bottom. The water comes in between the curtain and the outer wall; gently and slowly flows under the curtain and rises to spill into a receiving trough at the top of the tank. As microbes digest the organic materials they grow and floc together. Large quantities settle to the bottom of this tank. If you are lucky to find the treatment plant, which you are visiting, has spare, empty tanks, you can see scrapers on the bottom that slowly sweep the sludge into a hole. The sludge is then augered into a sludge holding tank. You have now finished what is called "Primary Treatment." The effluent is more or less safe with regard to bacterial pathogens, but it is a disaster to dump this into a stream or river because it contains so much dissolved organic material. In the receiving body of water, which is filled with lots of its own microbes, all that organic material is digested at the expense of the oxygen that is dissolved in that river or lake. This is called biological oxygen demand (BOD) which is directly correlated with the amount of rapidly biodegradable organic material. It is disasterous to dump this "high BOD" water

into a lake because soon all the fish and other aerobic creatures will die, and you will get a fetid, stagnant cess pool.

4. Rather than dumping the "primary effluent" into a beautiful natural stream, let us make an artificial lake and vigorously aerate the water to promote the rapid growth of microbes that will "eat up" the BOD before the water gets to the river. Hence we have an "aeration chamber" (AC). In the days before electric power, the water was allowed to sprinkle out over the surface of a basin that was filled with rocks. The water would trickle down through the rocks before exiting at the bottom. The water was thus exposed to a great amount of surface area, and the rocks became coated with a slime of all sorts of aerobic microbes that rapidly dropped the BOD. On the sunlit top rocks, algae vigorously grew, but just below them the slime turned to a vast microbial dog-eat-dog ecosystem of bacteria, fungi, grazing amoebae, snails.
An alternative way of doing this is by making huge silos filled with plastic honeycombs. The water is sprayed on top to trickle down through the honeycomb while air is forced upwards through the honeycomb. Yet a third way of promoting aeration is shown in this diagram: giant "egg beaters" churn and splash the water. A fourth way is even more direct: bubble air up through the water from the bottom. All these alternative devices are with the purpose of converting the BOD to either carbon dioxide (which wafts away) or to cell matter (for the next step).

5. We take the outflow from the aeration chamber, and try to recover as much of the microbial growth as possible to feed it back into the beginning of the aeration process. This is done with another clarifier (SC = Secondary Clarifier). Many treatment plants have secondary clarifiers that are almost identical to the primary clarifier's design. But, for change of pace, here is shown one where the incoming water flows down a central shaft and slowly works its way back up to the top, where is spills over at the edges into a circular trough. Often alum or other flocculant is added to the incoming water to assist in the aggregation of the microbial cells into globs that a big enough to sink. The water that spills out of this tank has passed "secondary treatment." It is surprisingly clear water.

6. Before allowing the effluent to reach the river or other receiving natural body of water, a final safety step is taken—the disinfection of the water. Shown here is a maze into which chlorine is added at the beginning, and remains in "contact" until it reaches the end of the tank. (CC = Contact Chamber)

Because excess chlorine is harmful to the receiving body of water, the amount added must be carefully monitored at the outfall. Because chlorine is so extremely toxic, alternative methods are used elsewhere. Ozone is commonly used in Europe, which is generally more advanced in its treatment of sewage than is North America. (The reason is that Europe is much more crowded and the tolerance of receiving bodies of water cannot absorb overages there, while much less crowded North American watersheds can absorb and "mistakes." Other processes are being considered: ultraviolet radiation; high intensity sound (sonic disruption of cells), and very intense 1,000 cps electrical current. It is always better to add "physics" to a system because there is no residue to be monitored. It is easy to get rid of "physics" —just turn it off!

There are two other components of sewage that must be considered that often escape the above "secondary treatment" plants.

❖   Nitrogen compounds, and
❖   Viruses.

"Tertiary Treatment Plants" do one further step after the foregoing: they allow the secondary effluent to flow into large ponds where algae grows and uses up the ammonia and nitrates in the water. This is a slow process and requires very large holding ponds. The algal growth is harvested and used as a nitrogen-rich organic fertilizer.

The elimination of viruses from sewage is problematic. Addition of chlorine is known to have little effect on many types of viruses. The European usage of ozone is known to damage and thus inactivate most viruses, and so that process is better than the use of chlorine. Ultraviolet irradiation of the effluent should be highly effective, but there is a problem that even the slightest amount of any organic chemical containing double-bonds absorbs and renders UV rather ineffective.

A class project to test one's local sewage treatment plant's efficacy in eliminating viruses is rather simple. Just as public health use usually harmless E.coli as an indicator of fecal pollution, you can use those viruses that infect only E.coli as your indicator viruses. Of course, the treatment plant's raw influent waters contain huge amounts of E.coli. Therefore, these waters ought also contain even greater amounts of those viruses that infect E.coli. Those viruses have a nickname "coliphages" (after the bacteriophages of E.coli, where the term bacteriophage means a virus of bacteria).

1.   Obtain 5 ml samples of treatment plant waters taken at various stages along the process.

2. Add 1 ml of chloroform to the samples; shake vigorously for a few minutes. The chloroform emulsion kills all the cells in the samples, but does not harm the coliphages.
3. Allow the chloroform to settle to the bottom of the tubes.
4. Make some dilutions of the overlying aqueous phases of these tubes.
5. Mix 0.05 ml of those dilutions with 0.05 ml of an overnight culture of a lab strain of E.coli (such as K-12, C600, B, etc.).
6. Transfer to the middle of the agar in a nutrient agar filled petri plate, and spread the droplet evenly over the surface of the agar. (Use a presterilized bent glass or metal rod for this purpose.) The droplet will soon be absorbed into the agar leaving the bacteria and coliphages stranded high and dry on the surface.
7. Incubate at body temperature; results often appear within 6 to 9 hours.

As the thousands of stranded bacteria begin to grow, some cells come in contact with a coliphage, which infects the contacting cell. In about 40 minutes, that cell bursts releasing a hundred or so progeny coliphages, which infect surrounding cells. The infection and bursting continues upon the "lawn" of growing E.coli resulting in holes or "plaques" in the lawn. By holding the plate up to the light, the plaques become visible. You should see plaques in the samples derived from raw influent water. Do you see any decrease in numbers of plaques as the water moved through the treatment process? Probably. But is there complete removal of "plaque-forming units" (pfu's)? If not, what does that mean? While coliphages are not a health hazard to you, they do indicate that other viruses—flu, hepatitis, HIV—might also escape the treatment process. If you find that your treatment plant is not removing the coliphages, it is something to think about very seriously. But do not run off to the news reporters with your information. You will not accomplish anything constructive. Instead, try to work with the plant's engineers to discover a way to treat the viruses. You will make history in a constructive way. Society needs this discovery.

27.4 Sewage Treatment

The wastes generated by some 60 per cent of the U.S. population are collected in sewer systems and carried along by some 14 billion gallons of water a day. Of this enormous volume, some 10 per cent is allowed to pass untreated into rivers, streams, and the ocean. The rest receives some form of treatment to improve the quality of the water (which makes up 99.9% of sewage) before it is released for reuse.

Biochemical Oxygen Demand (BOD)

The BOD is an important measure of water quality. It is a measure of the amount of oxygen needed (in milligrams per liter or parts per million) by bacteria and other microorganisms to oxidize the organic matter present in a water sample over a period of 5 days. The BOD of drinking water should be less than 1. That of raw sewage may run to several hundred. It is also called the "biological" oxygen demand.

Primary Treatment

The simplest, and least effective, method of treatment is to allow the undissolved solids in raw sewage to settle out of suspension forming sludge. Such primary treatment removes only one-third of the BOD and virtually none of the dissolved minerals.

Attempts to use digested sludge as a fertilizer have been hampered by its frequent contamination by toxic chemicals derived from industrial wastes.

Secondary Treatment

However, many treatment plants in North America then pass the effluent from primary treatment to secondary treatment. Here the effluent is brought in contact with oxygen and aerobic microorganisms. They break down much of the organic matter to harmless substances such as carbon dioxide.

Primary and secondary treatment together can remove up to 90 per cent of the BOD. After chlorination to remove its content of bacteria, the effluent from secondary treatment is returned to the local surface water.

Advanced Waste Treatment

The combination of primary and secondary treatment removes most of the organic matter in sewage and thus lowers the BOD. However, most of the nitrogen and phosphorus in sewage remains in the effluent from secondary treatment. These inorganic nutrients can cause eutrophication of surface water receiving the effluent causing blooms of algae. To avoid this, a few communities add a third stage of treatment called tertiary or advanced waste treatment. Several techniques are available to remove dissolved salts from sewage effluent, but all are quite expensive.

28

Bioethics and Biotechnology in Australia

Bioethics is the study of the ethical, social, legal, philosophical and other related issues arising in healthcare, the biological sciences and from biotechnology. Biotechnology operates in an environment on which past experiences and current norms may be insufficient to guide our moral reasoning. The ever-expanding possible applications of modern biotechnology and the often uncertain consequences make it imperative to discuss the implications and issues arising from the science. Australian Government has been making all efforts to increase the awareness of bioethics and improve the Australian public's access to information about bioethics. The Bioethics Portal acts as an information portal to bioethics and biotechnology-related information in all State and Territory Governments and the Australian Government. The site is part of a National Approach to Ethics in Biotechnology established by the Biotechnology Liaison Committee (BLC). The BLC was established in late 2001 with agreement by all States and Territories and the Australian Government, to participate in the development of strategies for coordinating Australia's emerging biotechnology policies. The Committee, comprised of nominated senior government officials, does following works:

- ❖ Encourages greater harmonization and coordination between all key government biotechnology agencies;
- ❖ Exchanges information and raises awareness of biotechnology strategies and initiatives;
- ❖ Explores opportunities for collaborating, developing synergies and avoiding duplication in the implementation of the Australian Government's National Biotechnology Strategy and State and Territory biotechnology policies/strategies; and
- ❖ As necessary, develops coordinated recommendations to Australian

Government and State/Territory Ministers on policies for a harmonized Australian approach to biotechnology and related issues.

The other important portal called "BioRegs Online" offers a comprehensive information about the regulation and development of biotechnology and related products in Australia. One is now able to find out which Commonwealth Government agencies regulate one's product, what compliance steps are involved and where to go for more information. Also find out about technology licensing, business planning, good manufacturing practice, exporting, government assistance and many other useful business topics of relevance to biotechnology. Regarding the Biotechnology Regulation Webtool, one may follow a pathway through the BioRegs Online decision tree. The researchers, companies and other users can develop a customised package of the key regulatory and business development requirements to take a specific biotechnology product from concept to market.

Case Study: Environmental Biotech, Australia

Environmental Biotech, Australia, offers unique and effective environmental solutions systems. We manufacture various strains of bacteria that have been trained to "remediate" (eat) the waste other industries produce—waste that causes these industries monumental operational and financial concerns. Bioremediation provides a healthy and safe solution to the environmental waste problems of the new millenium. Years of research and development have led to the manufacture and application of live bacteria that actually "eat" (bioremediate) grease, oil, sugar, starch, and gelatin. Our research lab proudly leads the industry in environmental biotechnology services. Our global network of franchises provides superior systems and service to many municipalities and a variety of industries, including commercial food service, hospitals, manufacturing, x-ray processors, and food processing.

The Problem

The highly detrimental impact of these residual substances carry a heavy and expensive toll on both mankind and Mother Earth. The safe and proper disposal of grease, sugar, starch, and gelatin accumulations has become a major concern for councils and industries alike. These buildups are the direct cause of many daily problems, such as:

- ❖ Stinky kitchen odours,
- ❖ Nasty drain line backups,

❖ Filthy grease interceptor cleaning,
❖ Clogged sewers and pump stations,
❖ Environmental problems, and
❖ Expensive grease trap pumping.

The Solution

The right solution is Environmental Biotech's scientifically developed systems of waste removal. Through the use of our live, vegetative systems, bioremediation uses natural bacteria (our "Bug Team®") to turn these harmful wastes into harmless carbon dioxide and water. Our highly effective, scientifically proven systems can eliminate those waste problems, saving our clients money, time, and the headaches which come with ineffective waste treatment. Our solutions help both their "bottom line" and the environment.

Company Profile

Environmental Biotech, was founded with the purpose of providing environmentally safe, economically feasible, and truly beneficial solutions to problems associated with commercial food service, industrial processing, manufacturing, and wastewater treatment facilities. To accomplish this, we apply naturally occurring, non-pathogenic, vegetative bacteria that "eat" (bioremediate) environmental wastes, leaving behind harmless products such as carbon dioxide and water.

E.B. is the world leader in waste bioremediation technology. Our diverse systems, global network, and effective service provide satisfaction and savings for our customers and a healthier environment for us all. And our cutting-edge technology has been combined with a proven business format of franchising to meet the growing demand for our services, which include:

(a) GES Grease Eradication System™
(b) SES® Sugar/Starch Eradication System
(c) GEL-OUT® Gelatin Eradication System
(d) Odour Elimination Systems

(a) (GES) Grease Eradication System

The Problem: Grease and Oil

Grease and oil are natural by-products of the food processing and preparation industries. Unfortunately, these by-products often lead to clogged drains, backups, and rancid odours. Since grease floats on water, it accumulates inside pipes (1 and 2) from the top down, eventually clogging the drain (3) and creating a backup.

There are NO quick fixes! Caustic chemicals, enzymes, surfactants (soaps), solvents, or mechanical routing (snaking) cannot solve this problem. Instead they simply relocate or actually compound it. Chemicals and routing can cause significant damage to pipes, resulting in costly downtime and repairs. Enzymes, solvents, and surfactants simply move the grease problem further down the line, where it will repeat itself.

The Proven Solution

The *(GES) Grease Eradication System* is a cost-effective, environmentally conscious system that "remediates" grease problems for good. Our state-of-the-art technology harnesses the power of bioremediation, a natural process in which bacteria actually consume grease and oil from drain lines, grease traps, and interceptors. These bacteria are non-toxic, non-pathogenic, live, and vegetative. And best of all, they turn grease and oil waste into harmless carbon dioxide and water.

In addition to helping the environment, our bacteria will also save valuable time and money that might have otherwise been spent dealing with the drudgery of unexpected drain clogs, backups, and the foul odors normally associated with these problems.

The Service

GES bi-monthly service assures you of hassle-free drain line maintenance. Our service system includes the following:

- ❖ Facility visits by our Service Technician.
- ❖ Drain line dye inspection.
- ❖ Treatment of equipment drains with bacteria.
- ❖ Treatment of all floor drains with bacteria.
- ❖ Injection system inspection and maintenance.
- ❖ Temperature and pH testing on effluent.
- ❖ Visual evaluation of grease trap/interceptor.
- ❖ Consultation with Facility Manager concerning the inspection.
- ❖ Inspection Report provided on each visit.
- ❖ Troubleshooting service if necessary.

The Benefits

- ❖ Automatic and maintenance free.
- ❖ No more plumbers for grease related drain clogs.
- ❖ No more odour of rancid grease.
- ❖ Significant reduction in pumping grease traps/interceptors.
- ❖ No more costly cleaning of drain line backups.
- ❖ No more lost profit from kitchen disruptions.
- ❖ No more employee dissatisfaction with foul odours, health hazards, or cleaning messy backups.
- ❖ No more "unusual" or "nasty" odours escaping into your dining area.

(b) (SES) Sugar/Starch Eradication System

The Problem

Almost every food or drink production involves the use of sugar or starch. Bakeries, restaurants, breweries, and soft drink manufacturers send excess sugar and starch down their drains and into their drainage systems. Once there, these wastes settle in the bottoms of drain lines, accumulating until they completely block the pipes, often causing backups. This photo shows how sugar buildup from soft drink syrup can rapidly clog a drain line. This pipe is 70% blocked after just four months.

Additionally, sugar and starch buildup contributes to high BOD (biological oxygen demand) and TSS (total suspended solids) levels, which can add up to large fines and expensive plumbing bills.

The Solution

Desperate facility managers often turn to harsh chemicals and solvents or mechanical routing in their efforts to clear the blockages formed by waste sugar and starch. However, these methods only damage the pipes further, ultimately resulting in costly repairs and lost revenue from production "downtime." Worst of all, the problem inevitably reappears a few months later.

Environmental Biotech's *Sugar/Starch Eradication System (SES®)* provides a permanent solution to sugar and starch buildup. SES® harnesses the power of live, vegetative, non-toxic bacteria in a process called bioremediation. When pumped into drainage systems, these bacteria actually eat the clogs, converting them into harmless carbon dioxide and water. So the bacteria not only save you time and money in costly repairs, they help the environment as well.

The Service

SES® bi-monthly service assures you of hassle-free drain line maintenance. Our service system includes the following:

- ❖ Facility visits by our Service Technician
- ❖ Drain line dye inspection
- ❖ Treatment of equipment drains with bacteria
- ❖ Treatment of all floor drains with bacteria
- ❖ Injection system inspection and maintenance
- ❖ Temperature and pH testing on effluent
- ❖ Consultation with facility manager concerning the inspection
- ❖ Inspection Report provided on each visit
- ❖ Troubleshooting service if necessary

The Benefits

- ❖ Automatic and maintenance free
- ❖ No more plumbers for sugar/starch related drain clogs
- ❖ No more costly cleaning of drain line backups
- ❖ No more lost profit from kitchen disruptions
- ❖ No more employee dissatisfaction with foul odours, health hazards, or cleaning messy backups
- ❖ No more "unusual" or "nasty" odours escaping into your dining area

(c) GEL-OUT® Gelatin Eradication System

The Problem

Gelatin is a common product used in manufacturing, x-ray and photographic film development, pharmaceutical production, and food processing. Waste gelatin, however, can turn into uncommonly difficult problems—especially when it results in plugged drainage systems and high strength (BOD) wastewater discharges.

Waste gelatin has a sponge-like capacity to absorb up to ten times its own weight. The gelatin accumulates on the bottom of the drain line (1&2), eventually clogging (3) and resulting in a backup.

The Cause

X-Ray Systems

Gelatin is applied and then washed off during the development process, sending it into the drains where it becomes a food source for algae. The algae grow, clog the drains, and restrict water flow, ultimately causing backups and other maintenance problems.

Photo Processing Facilites

Commercial photo labs also produce waste gelatin during the development process. Excess gelatin washed from the film collects in the drainage systems, resulting in flow restrictions.

Food Processing and Packaging

Meat packing and other food processing facilities employ gelatin as a binding agent. The waste gelatin that escapes into the drains promotes backups and increases manufacturing downtime.

Pharmaceuticals

The application of gelatin to medications makes them easier for the consumer to swallow. Once again, the excess gelatin flushes into the drains, thereby reducing drain flow.

The Solution

In the past, clearing gelatin meant pouring harmful chemicals or solvents into the drainage system. The use of muriatic acid, bromine, chlorine, or copper sulfate is dangerous for employees and damaging to drain lines. These temporary solutions often result in expensive plumbing repairs.

The GEL-OUT® System uses live, vegetative, naturally-occuring bacteria, which are non-toxic, non-caustic, non-corrosive, non-pathogenic (do not cause disease), and totally safe for humans.

A pump injects the GEL-OUT® bacteria into the drain lines twelve times per day. The introduction of the bacteria breaks down the gelatin structure. Next, the bacteria actually consume the remaining compound, converting it into harmless carbon dioxide and water. This is the natural solution to operational difficulties!

The Service

> ❖ Facility visit by our Service Technician
> ❖ Injection System inspection
> ❖ Replenish vegetative bacteria
> ❖ Consultation with Facility Manager and Staff
> ❖ Troubleshooting
> ❖ Written service reports

The Benefits

> ❖ Reduction or elimination of related drain clogs
> ❖ Non-hazardous
> ❖ Non-caustic
> ❖ Totally natural
> ❖ Reduction or elimination of processing or manufacturing "downtime
> ❖ No chemicals
> ❖ Non-corrosive
> ❖ Environmentally correct
> ❖ Non-toxic
> ❖ Automatic and maintenance free

(d) OES™ Odour Elimination System

The Problem

Noxious odours: they linger around the garbage cans and dumpsters, float up from floor or sink drains, and take up residence in the restrooms, Cigarette smoke smells plague bar areas and hotel rooms. Worse, these odours can rapidly permeate a kitchen or preparation area and drift out into dining or service areas.

Odours can be more than an inconvenience they can deal painful blows to your business:

> ❖ Loss of even loyal customers.
> ❖ Negative first impressions.
> ❖ Employee complaints and dissatisfaction.
> ❖ Money and time spent on ineffective masking agents and heavy perfumes.

The Solution

OES Odour Elimination Service

Environmental Biotech's OES Odour Elimination System is a low cost, permanent solution to these and other odour-related problems.

The OES includes:

- ❖ Find the source of the odour(s)
- ❖ Installation of an atomizer or drip-feed chemical injection system.
- ❖ Bi-weekly service visits from a trained EB technician.
- ❖ Environmentally-friendly chemicals that actually contract and neutralize unwanted odours.
- ❖ Maintenance and housekeeping recommendations to ensure a fresh-smelling facility.

In sum, Environmental Biotech, provides its services to businesses that have operating problems associated with grease, sugars, starches, gelatins and odours. These operating industries include food service, waste water treatment, hospitals, hotels, bakeries, bottling plants, pharmaceutical companies, copper wire manufacturers, and others.

Bibliography

A. Baeza, J. Guillen, S. Hernandez, A. Salas, M. Bernedo, J.L. Manjon, G. Moreno. *Influence of the nutritional mechanism of fungi (mycorrhize/saprotroph) on the uptake of radionuclides by mycelium.* Radiochimica acta, 2005. vol. 93, no4, pp. 233-238

A.J.M. Baker, J. Proctor, M.M.J. van Balgooy, R.D. Reeves. *Hyperaccumulation of nickel by the flora of the ultramafics of Palawan, Republic of the Philippines.* pp. 291-304 in Baker AJM, Proctor J, Reeves RD (eds) The Vegetation of Ultramafic (Serpentine) Soils. GB-Andover: Intercept (1992).

A.J.M. Baker, R.R. Brooks. *Terrestrial higher plants which hyperaccumulate metallic elements—A review of their distribution, ecology and phytochemistry.* Biorecovery (1989), 1:81-126

A. Weintraub, C. Romero, T. Bjørndal, and R. Epstein (Editors) (2007). *Handbook of Operations Research in Natural Resources.* Springer. ISBN 0-387-71814-9.

Aber, J.D., and J.M., Melillo (1982). "Nitrogen immobilization in decaying hardwood leaf litter as a function of initial nitrogen and lignin content". In: *Cannadian Journal of Botany* 60:2263-2269.

Aldor, I.; Fourest, E.; Volesky, B. *Can. J. Chem. Eng.* 1995, 73, 516-522.

Barford, C.C., *et al.* (2001). "Factors controlling long and short term sequestration of atmospheric CO2 in a mid-latitude forest". In: *Science* 294: 1688-1691

Belovsky, G.E. and J.B. Slade. (2000). "Insect herbivory accelerates nutrient cycling and increases plant production". In: *Proceedings of the national academy of sciences* (USA). 97:14412-14417.

Benedict, B.; Pigford, T.H.; Levi, H.W. *Nuclear Chemical Engineering*, McGraw-Hill: New York, NY 1981.

Biotechnology for a clean environment. (1994) OECD. ISBN 92-64-14257-6

Biotechnology for clean industrial products and processes: Towards industrial sustainability. (1998) OECD. ISBN 92-64-16102-3

Biotechnology for non-specialists, a handbook of information sources. (1997) EFB Taskgroup on Public perceptions of Biotechnology. ISBN 90-76110-01-8

Byerley, J.J.; Scharer, J.M.; Charles, A.M. *Chem. Eng. Journal* 1987, 36, B49-B59.

Carpenter, S.A. (1981). "Decay of heterogeneous detritus: a general model". In: *Journal of theoretical biology* 89:539-547.

Chapin F.S. III, Matson, P.A., and Mooney, H.A. (2003). *Principles of terrestrial ecosystem ecology.* Springer-Verlag, New York, N.Y.

Chapin, F.S. III, B.H., Walker, R.J., Hobbs, D.U.ooper, J.H., Lawton, O.E., Sala, and D., Tilman. (1997). "Biotic control over the functioning of ecosystems". in: *Science* 277:500-504.

Chapman, S.K., Hart, S.C., Cobb, N.S., Whitham, T.G., and Koch, G.W. (2003). "Insect herbivory increases litter quality and decomposition: an extension of the acceleration hypothesis". in: *Ecology* 84:2867-2876.

Chrispeels, M.J. and Sadava, D. (1977). *Plants, food, and people*. W.H. Freeman and Company, San Francisco.

Conz, B.W. 2004. Continuity and Contestation: Conservation Landscapes in Totonicapán, Guatemala. University of Massachussetss Masters of Science thesis.

Council Directive on the Contained Use of Genetically Modified Micro-organisms. 1990, 90/219/EEC

Council Directive on the Deliberate Release into the Environment of Genetically Modified Organisms. 1990, 90/220/EEC

Coupled Model Intercomparison Project Control Run.

Crist, D.R.; Crist, R.H.; Martin, J.R.; Watson, J. In *Metals-Microorganisms Relationships and Applications, FEMS Symposium Abstracts, Metz, France, May*; Bauda, P., ed.; Societe Francaise de Microbiologie: Paris, France, 1993, p. 13.

D. Mishra, M. Kar. *Nickel in plant growth and metabolism*. Bot Rev (1974), 40:395-452

Das, S., Dutta, T.K. and Samanta, T.B. *J. Steroid Biochem. Mol.Biol.* 82: 257 (2002).

Datta, J. and Samanta, T.B. *Mol. Cell. Biochem.* 118: 31 (1992).

Datta, J., Dutta, T.K. and Samanta, T.B. *Biochem. Biophys. Res. Commun.* 203: 1508 (1994).

Defries, R.S., J.A. Foley, and G.P. Asner. (2004). "Land-use choices: balancing human needs and ecosystem function". in: *Frontiers in ecology and environmental science.* 2:249-257.

Dutta, D. and Samanta, T.B. *Biochem. Biophys. Res. Commun.* 155: 493 (1988).

Dutta, D., Ghosh, D.K., Mishra, A.K. and Samanta, T.B. *Biochem.Biophys. Res. Commun.* 115: 692 (1983).

Dutta, T.K. and Samanta, T.B. *Bioorg. Med. Chem. Lett.* 7: 629 (1997).

Dutta, T.K. and Samanta, T.B. *Curr. Microbiol.* 39: 309 (1999).

Dutta, T.K., Datta, J. and Samanta, T.B. *Biochem. Biophys. Res. Commun.* 192: 119 (1993).

Early Earth atmosphere favorable to life: study, University of Waterloo, April 7, 2005. Retrieved on 2007-07-30.

Earth's Radiation Balance and Oceanic Heat Fluxes.

Edgington, D.N.; Gorden, S.A.; Thommes, M.M.; Almodovar, L.R. *Limnol. Ocean.* 1970, *15*, 945-955.

Ehrenfeld, J.G. and Toth, L.A. (1997). "Restoration ecology and the ecosystem perspective". in: *Restoration Ecology* 5:307-317.

Environment Canada. "Testing the World's Drinking Water." Science and Technology Bulletin.www.ec.gc.ca/science/sandemar99/article6_e.html December 2001.

Eurobarometer 46.1, *The Europeans and modern biotechnology*, CEC DG XII, 1997

European Centre for Nature Conservation. "ELISA: Environmental Indicators for Sustainable Agriculture Final Report." European Centre for Nature Conservation. www.ecnc.nl/doc/projects/elisa.html December 2001.

European Initiative for Biotechnology Education (EIBE). *"Biotechnology and the Environment."* EIBE. www.rdg.ac.uk/EIBE/ENGLISH/U16.HTM December 2001.

Faculty of Medicine, University of Ottawa. *"Centre For Research on Environmental Biotechnology (CREM) Brochure."* University of Ottawa. www.uottawa.ca/academic/med/microbio/bmi/bmicrem.html. December 2001.

Foster, D.R. (1992). "Land-use history (1730-1990) and vegetation dynamics in central New England, USA". In: *Journal of Ecology* 80: 753-772.

Fourest, E.; Volesky, B. *Environ. Sci. Technol.* 1996, *30,* 277-282.

Frank *et al.* 2005.

Frank et al. 2005.

G.J.D. Kirk and S. Staunton. *On predicting the fate of radioactive caesium in soil beneath grassland,* Journal of Soil Science, 1989. 40: 71-84

Gerben J Zylstraa and Jerome J Kukor, *What is environmental biotechnology?* Current Opinion in Biotechnology 16(3):243-245, 2005

Ghosh, D.K. and Samanta, T.B. *Indian J. Biochem. Biophys.* 18: 51 (1981).

Ghosh, D.K. and Samanta, T.B. *J. Steroid Biochem.* 14: 1063 (1981).

Ghosh, D.K., Dutta, D., Mishra, A.K. and Samanta, T.B. *Biochem. Biophys. Res. Commun.* 113: 497 (1983).

Goulden, M.L., J.W. Munger, S.-M. Fan, B.C. Daube, and S.C. Wofsy, (1996). "Effects of interannual climate variability on the carbon dioxide exchange of a temperate deciduous forest". In: *Science* 271:1576-1578

Guibal, E.; Roulph, C.; Le Cloirec, P. *Water Res.* 1992, *26,* 1139-45.

Hagen, J.B. (1992). *An Entangled Bank: The origins of ecosystem ecology.* Rutgers University Press, New Brunswick, N.J.

Hannink N, Rosser SJ, French CE, Basran A, Murray JA, Nicklin S, Bruce NC. *Phytodetoxification of TNT by transgenic plants expressing a bacterial nitroreductase.* 1: Nat Biotechnol. 2001 Dec. 19 (12): 1168-72.

Hättenschwiler S. and P.M. Vitousek (2000). "The role of polyphenols in terrestrial ecosystem nutrient cycling". In: *Trends in Ecology and Evolution* 15: 238-243

Hench, P. ., Kendall, E.C., Slocumb, C.H. and Polley, H.F. *A.M.A.Arch. Internal Med.* 85: 545 (1950).

Horikoshi, T.; Nakajima, A.; Sakaguchi, T. *Agric. Biol. Chem.* 1979, *332,* 617.

How can Biotechnology benefit the Environment (1997) Report of a Workshop organized by the EFB Taskgroup on Public Perceptions of Biotechnology and The Green Alliance on 13 January 1997 at the Science useum in London, ISBN 90 76110 02 6.

Hu, M.Z.-C.; Norman, J.M.; Faison, N.B.; Reeves, M. *Biotechnol. Bioeng.* 1996, *51,* 237-47.

Hunter, M.D. (2001). "Insect population dynamics meets ecosystem ecology: effects of herbivory on soil nutrient dynamics". In: *Agricultural and Forest Entomology* 3:77-84.

Hussen, Ahmed, *Principles of Environmental Economics, 2e.* New York, NY: Routledge, 2004.

Huxman TE, ea.(2004). "Convergence across biomes to a common rain-use efficiency". *Nature.* 429: 651-654.

J.A. Entry, L.S. Watrud and M. Reeves, *Accumulation of cesium-137 and strontium-90 from contaminated soil by three grass species inoculated with mycorrhizal fungi.* Environmental Pollution, 1999. 104: 449-457. Cited in Westhoff99.

J.B. Wilson, A.D.Q. Agnew. *Positive-feedback switches in plant communities.* Adv Ecol Res (1992), 23:263-336.

J.L. Harley, *The significance of mycorrhizae.* Mycological Research 1989. 92: 129-134.

J.L. Maas et D.E. Stuntz. *Mycoecology on serpentine soil.* Mycologia 61:1106-1116 (1969). Cited in Boyd 98.

Jacobson, K. Bruce. "Biosensors and Other Medical and Environmental Probes." Oak Ridge National Laboratory. www.ornl.gov/ORNLReview/rev29_3/text/biosens.htm December 2001.

K. Killham. *Ecology of polluted soils.* Soil Ecology, 1995. (pp. 175-181) Cambridge: Cambridge University Press.

Keeling, C.D. and T.P. Whorf. (2005). "Atmospheric CO2 records from sites in the SIO air sampling network". In: *Trends: A Compendium of Data on Global Change.* Carbon Dioxide Information Analysis Center, Oak Ridge National Laboratory, U.S. Department of Energy, Oak Ridge, Tenn., U.S.A.

Keislich. K. in "Biotransformations" (K.Keislich, ed.) Velag Chemie, Weinheim 8: 1 (1984).

King, Dennis M. (University of Maryland) and Marisa Mazzotta (University of Rhode Island), Ecosystem Valuation.

Kuyucak, N.; Volesky, B. *Biorecovery* 1989, *1*, 189-204.

Kuyucak, N.; Volesky, B. In *Biosorption of Heavy Metals*; Volesky, B., ed.; CRC Press: Boca Raton, FL, 1990, pp. 173-198.

Laul, J.C. *Radioanal. Nucl. Chem. Articles* 1992, *156*, 235.

Leusch, A.; Holan, Z.R.; Volesky, B. *J. Chem. Tech. Biotechnol.* 1995, *62*, 279-288.

Likens, G.E., F.H. Bormann, N.M. Johnson, D.W. Fisher and R.S. Pierce. (1970). "Effects of forest cutting and herbicide treatment on nutrient budgets in the Hubbard Brook watershed-ecosystem". in: *Ecological Monographs* 40:23-47.

Lutgens, Frederick K. and Edward J. Tarbuck (1995) *The Atmosphere*, Prentice Hall, 6th ed., pp. 14-17, ISBN 0-13-350612-6.

M.A. Davis, J.F. Murphy, and R.S. Boyd. *Nickel Increases Susceptibility of a Nickel Hyperaccumulator to Turnip mosaic virus.* J. Environ. Qual., Vol. 30, January-February 2001.

M.F. Allen. *The Ecology of Mycorrhizae.* New York: Cambridge University Press (1991). Cité dans Boyd 1998.

M.T. Brown et I.R. Hall. *Ecophysiology of metal uptake by tolerant plants.* pp. 95-104 in Shaw AJ (ed.) *Heavy Metal Tolerance in Plants: Evolutionary Aspects.* Boca Raton: FL: CRC Press (1990).

Macaskie, L.E.; Empson, R.M.; Cheetham, A.K.; Grey, C.P.; Skarnulis, A.J. *Science* 1992, *257*, 782-784.

Marshall and Perry 1987

Meagher, RB (2000). "Phytoremediation of toxic elemental and organic pollutants". *CURRENT OPINION IN PLANT BIOLOGY* 3 (2): 153-162. PMID 10712958.

Meentemeyer, V. 1978 "Macroclimate and lignin control of litter decomposition rates". in: *Ecology* 59:465-472.

Melillo, J.M., Aber, J.D., and Muratore, J.F. (1982). "Nitrogen and lignin control of hardwood leaf litter decomposition dynamics". In: *Ecology* 63:621-626.

Mendez MO, Maier RM (2008). "Phytostabilization of mine tailings in arid and semiarid environments—an emerging remediation technology". *Environ Health Perspect* 116 (3): 278-83. doi:10.1289/ehp.10608.

Motzkin, G., D.R. Foster, A. Allen, J. Harrod, and R.D. Boone. (1996). "Controlling site to evaluate history: vegetation patterns of a New England sand plain". In: *Ecological Monographs* 66: 345-365.

Mullen, M.D.; Wolf, D.C.; Beveridge, T.J.; Bailey, G.W. "Sorption of heavy metals by soil fungi *Aspergillus niger* and *Mucor Rouxii*," In *Soil Biol. Biochem.* 1992, 24, 129-135.

Munroe, N.D.H.; Bonner, J.D.; Williams, R.; Pattison, K.F.; Norman, J.M.; Faison, B.D. In *Abstracts, American Society for Microbiology Annual Meeting*, 1993.

Murali Subramanian, David J. Oliver, and Jacqueline V. Shanks. *TNT Phytotransformation Pathway Characteristics in Arabidopsis: Role of Aromatic Hydroxylamines*. Biotechnol. Prog., 22 (1), 208-216, 2006.

N.A. Hopkins. *Mycorrhizae in a Californian serpentine grassland community*. Can Bot 1987, 65:484-487.

Odum, E.P 1969. "The strategy of ecosystem development". in: *Science* 164:262-270.

Odum, H.T. (1971). *Environment, Power, and Society*. Wiley-Interscience New York, N.Y.

Olson, J.S. (1963). "Energy storage and the balance of producers and decomposers in ecological systems". In: *Ecology* 44:322-331.

P.L. Klerks. *Adaptation to metals in animals*. pp 313-321 in Shaw AJ (ed.) *Heavy Metal Tolerance in Plants: Evolutionary Aspects*. Boca Raton:FL: CRC Press (1990).

Pal, A. and Samanta, T.B. *Curr. Microbiol.* 39: 244 (1999).

Peterson, D.H. and Murray, H.C. *J. Am. Chem. Soc.* 74: 1871, 5933 (1952).

Quentin Grafton and Robert Hill (University of New South Wales and W. Adamowicz, Diane Dupont, S Renzetti and Harry Nelson (University of British Columbia (2004). *The Economics of the Environment and Natural Resources*. Blackwell Punlishing. ISBN 0631215646.

Quinones, M.A., N.E. Borlaug, C.R. Dowswell. (1997). "A fertilizer-based green revolution for Africa". In: *Replenishing soil fertility in Africa*. Soil Science Society of America special publication number 51. Soil Science Society of America, Madison, WI.

R. Bradley, A.J. Burt et D.J. Read, *The biology of mycorrhizal infection in the Ericaceae. VIII. The role of mycorrhizal infection in heavy metal tolerance*. New Phytol 1982, 91:197-209.

R. Gabrielli, C. Mattioni, O. Vergnano. *Accumulation mechanisms and heavy metal tolerance of a nickel hyperaccumulator.* Plant Nutr (1991). 14:1067-1080.

R.D. MacNicol, P.H.T. Beckett. *Critical tissue concentrations of potentially toxic elements.* Plant Soil, 1985. 85:107-129.

Reich, P.B., Grigal, D.F., Aber, J.D., Gower, S.T. (1997). "Nitrogen mineralization and productivity in 50 hardwood and conifer stands on diverse soils". In: *Ecology* 78:335-347.

Roy, N., Chattopadhyay, S. and Samanta, T.B. *Biochem. J.* 176:593 (1978).

S.P. Hopkin. *Ecophysiology of Metals in Terrestrial Invertebrates.* GB-London: Elsevier Applied Science (1989).

Samanta, T.B. and Ghosh, D.K. *J Steroid Biochem.* 28: 327 (1987).

Schneider, J.J. *J. Steriod Biochem.* 5: 9 (1974).

Sen, R. and Samanta, T.B. *J. Steroid Biochem.* 14: 307 (1981).

Simonson, Sara. "Lichens and Lichen-Feeding Moths as Bioindicators of Air Pollution in the Rocky Mountain Front Range." Colorado State University. www.colostate.edu/Depts/Ent.es/en570/papers_1996/simonson.html December 2001.

Source for figures: Carbon dioxide, NASA Earth Fact Sheet, (updated 2007.01). Methane, IPCC TAR table 6.1, (updated to 1998). The NASA total was 17 ppmv over 100%, and CO_2 was increased here by 15 ppmv. To normalize, N_2 should be reduced by about 25 ppmv and O_2 by about 7 ppmv.

Stadler, B., Solinger, St., and Michalzik, B. (2001). "Insect herbivores and the nutrient flow from the canopy to the soil in coniferous and deciduous forests". In: *Oecologia* 126:104-113.

Stevens, Dr. Melita, Dr. Nicholas Ashbolt, and DR. David Cunliffe. "Microbial Indicators of Water Quality—An NHMRC Discussion Paper." National Health and Medical Research Council. < P>

Swank, W.T., Waide, J.B., Crossley, D.A., and Todd R.L. (1981). "Insect defoliation enhances nitrate export from forest ecosystems". In: *Oecologia* 51:297-299.

T.L. Rost, M.G. Barbour, C.R. Stocking and T.M. Murphy, *The root system.* Plant Biology, 1998. (pp. 68-84). California: Wadsworth Publishing Company. Cited in Westhoff99.

The significance of metal hyperaccumulation for biotic interactions, by R.S. Boyd and S.N. Martens. Chemoecology 8 (1998) pp. 1-7

Vannote, R.I., Minshall, G.W., Cummings, K.W., Sedell, J.R., and Cushing, C.E. (1980). "The river continuim concept". in: *Cannadian Journal of Fisheries and Aquatic Science* 37:120-137.

Veal, Duncan ed. "Microbial Indicators of River Health—1997 Workshop." Australian Land and Water Resources Research and Development Corporation. January 1998.

Vercheval, J. The thermosphere: a part of the heterosphere. (offline, see Internet Archive copy).

Vidya Sagar. K, *National Conference on Environmental Biotechnology,* Bangalore 2005.

Vitousek, P.M. and Howarth, R.W. (1991). "Nitrogen limitation on land and in the sea: how can it occur?" In: *Biogeochemistry* 13:87-115.

Volesky, B.; Holan, Z.R. *Biotechnol. Prog.* 1995, *11,* 235-250.

Volesky, B.; Tsezos, M. U.S. Patent 4320093, 1981. Canadian Patent 1143007, 1983.

White, S.K. *J. Am. Water Works Assoc.* 1983, *75,* 374.

Index